PRINCIPLES OF AEROELASTICITY

PRINCIPLES OF AEROELASTICITY

RAYMOND L. BISPLINGHOFF

CHANCELLOR
UNIVERSITY OF MISSOURI
ROLLA

HOLT ASHLEY

PROFESSOR, AERONAUTICS
AND ASTRONAUTICS
STANFORD UNIVERSITY

DOVER PUBLICATIONS, INC.
NEW YORK

Published in Canada by General Publishing Company, Ltd., 30 Lesmill Road, Don Mills, Toronto,
Ontario.
Published in the United Kingdom by Constable
and Company, Ltd., 10 Orange Street, London WC 2.

This Dover edition, first published in 1975, is an
unabridged, corrected republication of the work
originally published by John Wiley and Sons, Inc.,
New York, in 1962.

International Standard Book Number: 0-486-61349-6
Library of Congress Catalog Card Number: 74-20442

Manufactured in the United States of America
Dover Publications, Inc.
180 Varick Street
New York, N.Y. 10014

TO MANFRED RAUSCHER

OUR FORMER COLLEAGUE

AND TEACHER

PREFACE

Principles of Aeroelasticity constitutes an attempt to bring orderly arrangement to a group of problems which have coalesced into a distinct and
mature subdivision of flight-vehicle engineering. Aeroelasticity is perhaps
best described as an interface between solid and fluid mechanics, with
dynamics serving as the adhesive. Regardless of our choice of simile,
however, engineers in the aircraft and missile fields are aware of its
existence and influence on the success of their designs.

A degree of courage, if not rashness, is required to write two books on
the same subject, and some words of explanation are in order. The
present book is the outcome of what was initially an attempt to prepare a
handbook of theoretical aeroelasticity for the John Wiley and Sons, Inc.,
series on Aircraft and Missile Structures, and as such it was commissioned
by Professor Nicholas Hoff, general editor of that series. But the manuscript missed the mark of a handbook by a wide margin, and the effort has
not only produced another book on aeroelasticity but has led the authors
to conclude that there are many modern engineering disciplines for which
the only handbooks are texts on fundamentals. No longer is it possible
to be inclusive by example. This work also has important roots in two
weeks of lectures delivered at the Massachusetts Institute of Technology
by the authors and several of their colleagues during the summer of 1958.
Portions have been adapted as class notes for teaching advanced aeroelasticity and structural dynamics.

The relationship to the earlier *Aeroelasticity*, written in collaboration
with Professor Robert L. Halfman, is not difficult to describe. Beyond the
obvious consequences of a seven-year interval, such as the treatment of
higher flight speeds and thermal effects, the objectives and scope of the
present work are more modest. Many recent additions to the literature
have removed any need for detailed development of the aerodynamic and

structural tools; these are relegated to little more than lists of useful results and references in Chapters 4 and 5. Experimental methods and computation procedures are not covered, and no detailed numerical examples are given of the sort that were integrated into *Aeroelasticity*. For this and other reasons, the new book is more suitable as a reference for the engineer in industry than as a college text, although the instructor who is willing to make a critical selection of material will find that the latter purpose may also be served.

In the authors' view, their most significant innovation lies in the formulation of a unifying philosophy of the field. Their new point of view and scheme of presentation are elaborated in the final section of the Introduction, where the assignment of topics to chapters is also described. In essence, the concept of a collection of vaguely related problems is replaced by a continuum based on the equations of forced motion of the elastic flight vehicle. A distinction is made between static and dynamic phenomena, and beyond this the primary classification is by the number of independent space variables required to define the physical system. Major assistance toward attaining this unification has been drawn from Fung's idea of aeroelastic operators, and from the variational principles of solid and fluid mechanics.

The assumption of linear systems with properties independent of time underlies almost everything that has been written on aeroelasticity, and this is largely true of Chapters 2 through 9 which follow. Since the frontier is now passing beyond this restriction, however, Chapter 10 is included to discuss means of dealing both with rapidly time-varying parameters and with the (more formidable) effects of nonlinearity. Few illustrations are offered here because they are almost completely lacking in the literature. The brains behind Chapter 10 are those of the authors' admired and beloved colleague, Mr. Garabed Zartarian, and of a former student of the authors, Professor Eugene Brunelle of Princeton University. The wording is mainly Mr. Zartarian's, and the authors are everlastingly grateful to him for his assistance in filling this essential gap. The signed authors prepared the remaining material in close collaboration. The principal responsibility for an initial writing fell on R. L. Bisplinghoff for Chapters 2, 3, 5, 8 and 9, and on H. Ashley for Chapters 1, 4, 6 and 7, with both authors working over all chapters of the final manuscript.

As is always the case, many people participated in bringing the book to completion. All of them should share the recognition for anything that is good, whereas the blame for errors or imperfections falls squarely on the signed authors. Several colleagues on the M.I.T. faculty of Aeronautics and Astronautics and the Aeroelastic and Structures Research Laboratory staff read portions of the manuscript and offered valuable criticisms.

These include Professors John Dugundji, R. L. Halfman, Marten Landahl, James Mar, Theodore Pian, Paul Sandorff; Mr. Garabed Zartarian; and Dr. P. T. Hsu. Helpful suggestions came from Professor Hoff, who reviewed the entire first draft. Figures and associated calculations were contributed by Messrs. Marc Kolpin, William Loden, and David Stickler. The typing, preparation, and reproduction of the manuscript were skillfully handled by Mrs. Frances K. Bragg, Miss Dorothy Dube, Mrs. Barbara Marks, and Miss Theodate Coughlin. Heartfelt appreciation is due to all of those named and to the many students who struggled through the early versions as class notes.

Cambridge, Mass.   R. L. BISPLINGHOFF
April 1962   H. ASHLEY

CONTENTS

1

INTRODUCTION

1-1 THE LITERATURE OF AEROELASTICITY

A verbal explosion is today engulfing conscientious readers in the fields of aeronautics and astronautics. For example, the National Advisory Committee for Aeronautics issued approximately as many Technical Notes during the nine-year period from 1950 to its date of absorption into NASA as were released during the previous thirty-five years of its existence. A significant proportion of these reports deals with dynamic, structural, and aerodynamic problems that have direct importance for aeroelasticity.

From one viewpoint, this profusion of books, monographs, reports, and professional journal articles handicaps anybody undertaking a new presentation of the fundamentals, for it leaves much less that is original or constructive to be said. On balance, however, we regard it as a clear advantage. Aeroelasticity is now a recognized and reasonably well-defined discipline within the broader scope of flight engineering. There is agreement on both what it is and what it is not, with the emphasis coming to rest on *phenomena which exhibit appreciable reciprocal interaction (static or dynamic) between aerodynamic forces and the deformations induced thereby in the structure of a flying vehicle, its control mechanisms, or its propulsion system.* Furthermore, we are left with no uncertainty regarding our first task in writing an introduction for this work: we must remind our readers of the extensive, valuable contributions of those whose footsteps we are attempting to follow.

There are currently available several complete books in English dealing with aeroelasticity and its underlying dynamic, elastic, and aerodynamic tools. Among these the most comprehensive are by Fung (Ref. 1-1),

1

Scanlan and Rosenbaum (Ref. 1–3), and the present authors in collaboration with R. L. Halfman (Ref. 1–2). A five-volume manual being prepared by the NATO Advisory Group for Aeronautical Research and Development, under the editorship of W. P. Jones (Ref. 1–4), promises to be a storehouse of up-to-date information. Fully as illuminating, if less extensive, are treatments of particular areas in the books by Duncan (Ref. 1–5), whose context is dynamic stability of airplanes; Templeton (Ref. 1–6), who concentrates on the explanation and avoidance of control surface flutter at subsonic speeds; and Abramson (Ref. 1–7). A fine introductory monograph, written from the practical engineering standpoint with a high degree of physical insight, is Broadbent's (Ref. 1–8). Four other books, with considerable historical interest but also containing discussions of particular topics which have current validity, are the following: Myklestad (Ref. 1–9), Freberg and Kemler (Ref. 1–10), von Kármán and Biot (Ref. 1–11), and the translation from Russian of Grossman's report (Ref. 1–12). Collar's review and delineation of the field (Ref. 1–13) still makes enlightening reading.

Among works in languages of the European continent, we mention the proceedings of the 1957 colloquium in Göttingen (Ref. 1–14), especially the contibution by Küssner, the book in Polish by Fiszdon (Ref. 1–15), and the one in Russian by Nekrasov (Ref. 1–16). A variety of papers on topics vital in 1958 appear in the proceedings of the first national specialists' meeting sponsored by the Institute of the Aeronautical Sciences on dynamics and aeroelasticity (Ref. 1–17).

So many surveys have been prepared on the subject of flutter, its influence on design and associated analytical techniques, that we are impelled to list significant recent examples: the 1957 Minta Martin Lecture by Garrick (Ref. 1–18), Williams (Ref. 1–19), Templeton (Ref. 1–20), Goland (Ref. 1–21), Laidlaw (Ref. 1–22), and Collar (Ref. 1–23). Aerothermoelasticity, comprising the effects of aerodynamic heating on static and dynamic elastic phenomena of high-speed flight, is defined and broadly examined in publications of Dryden and Duberg (Ref. 1–24), and of one of the present authors (Refs. 1–25, 1–26, the latter in collaboration with Dugundji). See also Refs. 1–34 and 1–35.

A wealth of pertinent information will be found in a recent USAF report on aeroelastic effects in stability and control (Ref. 1–27). Finally, with an eye to the future, we cite the comprehensive look at problems of space flight vehicles (Ref. 1–28), taken as their last official act by the retiring members of the NACA Subcommittee on Vibration and Flutter.

No such general summary as the foregoing can be inclusive or entirely free from the unintentional oversight of contributions which have not been mentioned. In particular, we draw attention to numerous books and papers dealing exclusively with dynamics, theory of elasticity, or steady and unsteady aerodynamic theory; these are referenced in the discussions of the appropriate aeroelastic operators, Chaps. 2, 5, and 4, respectively.

1-2 SCOPE AND IMPORTANCE OF THE FIELD

In view of the several summaries already published on the role of aeroelasticity in the early history of aviation (Ref. 1–18 and Chap. 1 of Ref. 1–2 contain examples), we feel that it would be superfluous to include one here. The emergence of such problems may be said to have coincided almost exactly with the first achievement of powered flight.

We can bring forward a large body of evidence, however, to show that during the decade and a half since World War II the interaction between aerodynamic loads and structural deformations gained considerably in importance relative to many other factors affecting the design of aircraft and missiles. Most designers agree that this is a very undesirable development and that the most constructive task of the aeroelastician is to find ways of minimizing these effects or, in a few cases, of putting them to valuable use. Their emergence was not unpredicted (cf. Collar, Ref. 1–13), and the reasons for it are familiar to most aeronautical engineers. We know of no better way of summarizing the situation than to revive Figs. 1–1 and 1–2, which have been adapted from Ref. 1–25. The first of these

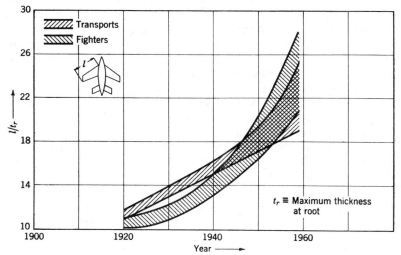

Fig. I–I. Slenderness ratio of fighter and transport aircraft as a function of year of first flight.

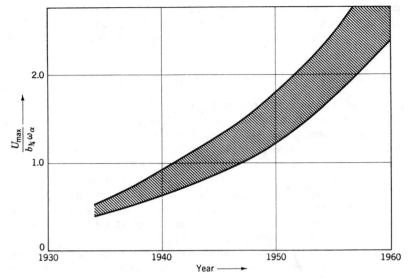

Fig. 1–2. Reduced velocity parameter for fighter aircraft aircraft as a function of year of first flight.

plots the increase with the passage of time of the slenderness ratio, or ratio of structural semispan (defined in the sketch) to maximum thickness at the root, of monoplane wings on fighter and transport-type aircraft. Manned bombers, at least those with maximum speeds in the subsonic or transonic ranges, follow closely the curve for transports. The implications of this trend with regard to bending flexibility are graphically illustrated by a composite photograph (Fig. 1–3) taken during static structural tests on the wing of an early version of the B-52 bomber.

Figure 1–2 plots, for fighters only, a reduced velocity parameter* $U_{max}/b_{3/4}\omega_\alpha$. This is defined as the maximum level flight speed at sea level, divided by the wing semichord at the three-quarter semispan station and by the fundamental frequency of torsional vibration. The startling growth of $U_{max}/b_{3/4}\omega_\alpha$ over the years has peculiar significance in connection with aeroelastic stability, because the avoidance of flutter or divergence (with certain other quantities fixed) involves not exceeding a prescribed value of just this parameter. For given levels of wing structural density, slenderness ratio, and aspect ratio, the square of $U_{max}/b_{3/4}\omega_\alpha$

---

* For a fuller explanation of aerodynamic and structural symbols, see Chaps. 4 and 5. We assume that the reader has some familiarity with the definitions of basic physical quantities used in aeronautics.

is proportional to another parameter ql^4/GJ_R, q being the flight dynamic pressure and GJ_R the torsional rigidity at some reference station. Laidlaw (Ref. 1–22) and others have found the size of this ratio a useful measure of the magnitude of aeroelastic effects.

Because of its dramatic and destructive consequences, the dynamic aeroelastic instability known as flutter has been primarily responsible for the recognition lately accorded the field in the aircraft industry. The pilot of the airplane shown in Fig. 1–4 succeeded in landing with roughly

Fig. 1–3. Composite photograph of the maximum upward and downward deflections, at limit load conditions, of the B-52 wing during static tests. (Courtesy of Boeing Airplane Company.)

Fig. 1–4. Rear view of empennage of jet fighter which was successfully landed after encountering flutter of the horizontal stabilizer in transonic flight. (Courtesy of North American Aviation, Inc.)

two-thirds of his horizontal tail surface out of action; some others have, unfortunately, not been so lucky. From being regarded as a minor nuisance in 1940, this phenomenon has grown in importance to a point where one industry representative recently made the following statement (Ref. 1–29): "The flutter problem is now generally accepted as a problem of primary concern in the design of current aircraft structures. Stiffness criteria based on flutter requirements are, in many instances, the critical design criteria."

In a study completed in late 1956, members of the NACA Subcommittee on Vibration and Flutter found that three times as many distinct flutter incidents occurred on United States military aircraft during the five years ending with 1956 as during the preceding five-year period. Concerning the future, the NACA report (Ref. 1–28) concludes: "There is no evidence that flutter will have less influence on the design of aerodynamically controlled booster vehicles and re-entry gliders than it has, for instance, on manned bombers." As a counterpoise to this gloomy prediction, we should cite the study of a hypothetical fighter in Sec. 9.4 of Ref. 1–25, which clearly shows ultimate strength to predominate over flutter in designing the wing structure throughout a large range of flight speeds where aerothermoelastic influences are important. From another stand-point, we suggest that winged vehicles which attain their maximum flight dynamic pressures at transonic Mach numbers are peculiarly sensitive to this sort of instability, so that performance increases may be bringing some aircraft types to a place where the *relative* danger is reduced.

On the other hand, flutter is not the only aeroelastic problem which can affect the success of a design. The frontispiece of Abramson's book (Ref. 1–7), for example, presents the catastrophic outcome of an en-counter with dynamic overstress during landing. Low divergence speeds have probably ruled sweptforward lifting surfaces out of practical considera-tion. It was revealed in an unclassified lecture (Ref. 1–30) that two Atlas ballistic missiles proved unsatisfactory because of an instability involving coupling between the automatic control system and a body-bending vibration. Figure 1–5 shows the missile leaving its launching pad, while Fig. 1–6 reproduces data recorded during one of the unsuccessful flights. To quote a letter from the manufacturer:

"The top trace is the rate gyro information. Roll, pitch and yaw rate outputs are shown in that order . . . for one second for each channel. The lower two traces are the engine position traces in the pitch and yaw channels. The last trace is the displacement gyro data which is presented in the same manner as the rate gyro informa-tion. The high frequency (17 c.p.s.) oscillation is a limit cycle result-ing from the autopilot coupling with the third lateral bending mode. The lower frequency (1 c.p.s.) is rigid body motion. Pre-flight simulations including the first three lateral bending modes did not uncover this problem because of the use of a linear third order hydraulic actuator simulation, which did not incorporate the effect of the missile vibration on the hydraulic servo system.

"As a result of this flight a highly non-linear hydraulic servo simulation was incorporated which permitted duplication of the flight test results. Changes were then made to the autopilot to attenuate the third lateral bending mode. All succeeding flights showed that the coupling of this mode with the autopilot was com-pletely eliminated."

When penetration beyond the confines of the lower atmosphere becomes a more routine operation, the aeroelastician may still expect to be fully occupied with the design of supporting vehicles. Space travel, as such, is excluded from consideration because of the required interaction with aerodynamic forces, but many similar dynamic problems will be encoun-tered there. Moreover, both the launching and entry phases present a

Fig. 1–5. Atlas missile rising from its launching pad. (Courtesy of Convair Astronautics Division of General Dynamics Corporation.)

host of unknowns. We reproduce on page 9 a table from Ref. 1–28 to emphasize this statement. The majority of subjects listed in the first and third columns are seen to fall within the domain of aeroelasticity.

For the purpose of defining the scope of the aeroelastic field, it has been customary in the past to name and describe a series of rather specific and distinct items: divergence, control effectiveness, control reversal, flutter, buffeting, dynamic response to various inputs such as gusts, aeroelastic effects on load distribution, and static and dynamic stability. This viewpoint is naturally related to Collar's famous triangle (Ref. 1–13), in which each item is connected to two or three of the vertices representing elastic, inertial, and aerodynamic forces.

We believe that today there are obsolescent features in such a categorization. As a practical criticism, it may have led to unnecessary duplication of methods of analysis and computation programs used by aeroelasticians, aerodynamicists, and elasticians working in the aircraft industry. The basic element for all cases is, after all, the flexible vehicle in flight. One

Dynamic and aeroelastic problems of space vehicles requiring increased research activity

Launching	Space flight	Entry and landing
Determination of free vibration characteristics	Determination of free vibration characteristics	Determination of free vibration characteristics
Aero-structural interaction (including aeroelastic stability and possible thermal effects)		Aero-thermal-structural interaction (including aeroelastic stability)
Dynamic modeling	Dynamic modeling	Dynamic modeling
Dynamics of stage separation, rocket firing, and burnout	Dynamics of stage separation, rendezvous, mass transfer, and construction in space	Recovery and impact dynamics
Dynamic problems of propulsion systems	Dynamic problems of propulsion systems (thrust-structure interaction)	
Flight-testing techniques	Flight-testing techniques	Flight-testing techniques
Nonlinear mechanics	Nonlinear mechanics and free-body mechanics	Nonlinear mechanics
Effects of noise and vibration	Effects of noise and vibration	Effects of noise and vibration
Influence of transient environment (time-dependent inputs, including atmospheric turbulence and blasts)	Influence of transient environment and dynamic response to fields (electromagnetic, gravitational, space detonations, X-ray, colliding particles)	Influence of transient environment
Pre-launch and launching dynamics, including crosswind effect		
Unsteady gasdynamics		Unsteady gasdynamics (hypersonic and viscous flow)
Structural-control-system interaction	Structural-control-system interaction (including stability)	Structural-control-system interaction
	Orbital perturbation dynamics	

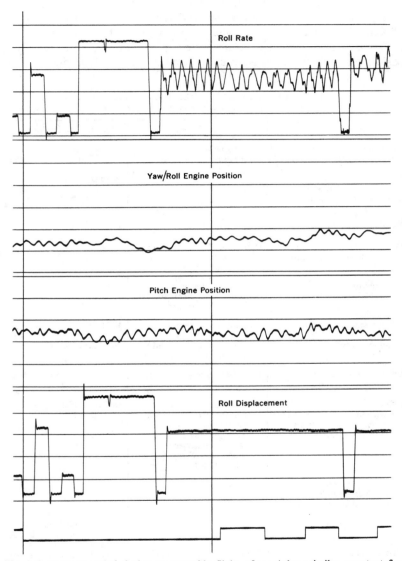

Fig. 1–6. Data recorded during an unstable flight of an Atlas missile: see text for details. (Courtesy of Convair Astronautics Division of General Dynamics Corporation.)

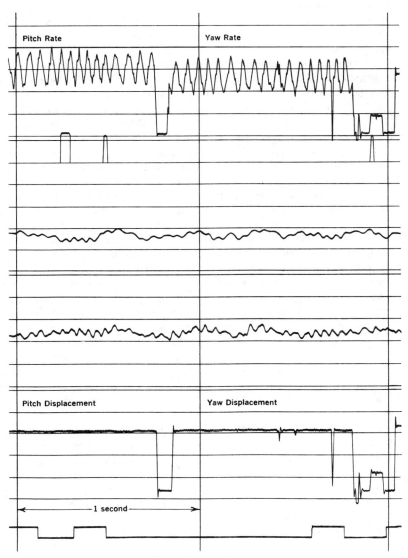

Fig. I-6. (Continued).

can distinguish with reasonable clarity between static problems, where time does not appear as an essential independent variable, and dynamic problems, where it does. But beyond this, excessive compartmentalization may be inefficient and artificial.

Considering time-dependent phenomena, for instance, there is one universal set of equations of motion appropriate to each vehicle. They may be subjected to a variety of inputs, such as impulsive or sinusoidal forcing of the controls, gust or blast loads, landing impacts, mechanical shaking, and so forth. Proceeding from one of these to the next involves changing only the "right-hand sides" of the equations, however; and the same system in homogeneous form possesses a set of eigenvalues, which describe the dynamic-stability and aeroelastic (or flutter) modes. In principle, over half of the problems of classical aeroelasticity can be analyzed from this single starting point. The authors are well aware of the simplifications that can be achieved in particular cases by resort to symmetry, by dropping certain degrees of freedom from the equations of motion, or by concentrating on individual structural components, such as wings or tail, for which the remainder of the vehicle is replaced by cantilever supports. Moreover, structural, aerodynamic, and dynamic nonlinearities may have to be accounted for in dynamic response calculations, whereas they might have less influence on stability. Nevertheless, the capacity and sophistication of computing equipment, both digital and analog, are hastening the day when it may be more efficient to work with a single, rather elaborate representation of the airplane or missile and its control system than with a variety of special-purpose programs.

The foregoing philosophy has guided us in writing much of the present book, although we have tried to avoid excessive complication in illustrative examples. Indeed, aeroelasticity seems to us to constitute one facet of the more general view of a flight vehicle suggested in Fig. 1–7. Here we see the airplane, missile, or space ship imagined as a set of interacting internal forces, simultaneously surrounded by external fields. The boundary of the system (it is usually an *open system* in the thermodynamic sense) is indicated by the inner large circle and coincides with the outer surface of the vehicle. A transfer of momentum, energy, and sometimes mass takes place continually across this boundary. Depending on the vehicle's location within or outside a planetary atmosphere, on its speed, and on many other factors, different types of environmental fields participate significantly in the various transfers. In some instances, the field is appreciably modified at the same time it is affecting the vehicle's motion and energy content; this is the case when surrounding atmospheric fluid interacts with a flexible structure, giving rise to aeroelastic phenomena. There are other examples, such as the force of gravity or the impingement

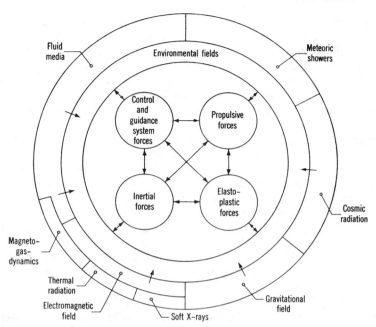

Fig. I–7. The flight vehicle conceived as a collection of interacting forces surrounded by environmental fields.

of solar radiation on the very cold skin and solar batteries of a space station, when the action is, for practical purposes, unidirectional.

1–3   A NEW SCHEME OF PRESENTATION

In this introductory chapter, we are purposely refraining from presenting a list of verbal or quantitative descriptions of typical problems in aeroelasticity. Chapter 6, in effect, amounts to such a breakdown, based on one very simple two-degree-of-freedom system. Pursuant to the point of view expressed in the previous section, we are laying out the book along the following lines. First, we give general methods of constructing the static and dynamic equations. The concept of aeroelastic operators, suggested by Fung (Ref. 1–1, Chap. 11), has proved very helpful here. Chapters 2 and 3 deal with the laws of mechanics for heated elastic solids and with associated ways of finding inertial operators for the equations of motion. Chapter 4 reviews the forms of aerodynamic operators which describe external loads on bodies and lifting surfaces in various ranges of flight speed. Chapter 5 discusses structural operators, including the influence of temperature variations throughout the solid. In these three presentations, we appeal as much as possible to the variational principles, which furnish compact statements of the fundamentals and sometimes also provide natural means of approximately solving practical problems. We admit that this approach is more successful in connection with elasticity than with fluid dynamics, but it does add an element of unity that deserves to be emphasized.

Particularly in the field of aerodynamic theory, we no longer feel there is a need for a full logical development of the tools in a book on aeroelasticity. An inordinate amount of space is consumed thereby, and such prodigality is better supplanted by judicious references to the excellent literature now available. Accordingly, parts of Chaps. 4 and 5 are no more than catalogues of operators for subsequent use. Chapter 3 completes the introductory material with further details on techniques of setting up and solving the more familiar equations, such as energy methods, normal coordinates, and other superposition schemes. There is no extensive review of the subject of mechanical vibrations. Although this subject certainly falls within the province of the aeroelastician, the profusion of excellent treatises already available relieves us of this task (see, for example: Den Hartog, Ref. 1–31; Timoshenko, Ref. 1–32; Rocard, Ref. 1–33; and also Refs. 1–9 and 1–11).

Chapters 6 through 9 form the heart of our survey of the current state of *linear* aeroelastic theory. The primary classification is by the number of independent space variables required to describe the physical system. It proceeds from simplified cases which have only a small, finite number of degrees of freedom, to one-dimensional systems (line structures), and finally to two-dimensional systems (plate- and shell-like structures). This would seem to cover all situations of major interest in aeroelasticity, so Chap. 9 combines some of the previous results by treating the unrestrained elastic vehicle in flight. Chapter 10 takes up the increasingly important but relatively unfamiliar (to theoretical aeroelasticians) subject of systems which must be represented by nonlinear equations or by equations with time-varying coefficients. For preparing the bulk of this final chapter, the authors are deeply indebted to two colleagues, Garabed Zartarian and Eugene J. Brunelle, Jr.

Within every category, it is possible only to describe a few typical structures and particular problems which characterize them. Each chapter starts with steady and quasi-steady phenomena and then goes on to dynamic phenomena. A practice is adopted of examining forced displacements and forced motions first, so that the homogeneous parts of the equations thus obtained will afterward apply directly to eigenvalue problems such as divergence and flutter. As long as the limitation to linear mathematical representations is preserved, this constitutes a consistent, efficient procedure. But we must always bear in mind the warning, so well expressed by Fung (Ref. 1–1, Introduction), that there is a "very important distinction between the response and stability problems, in regard to the justification of the linearization process." When examining the eigenvalues and eigenfunctions which determine stability, the amplitude is for the most part of little interest, and it is logical to consider infinitesimal departures from static equilibrium. In analyzing gust loads, control effectiveness and the like, however, the degree of finiteness of stresses, accelerations, and displacements becomes important. Linear theory has severe weaknesses in more such situations than are now recognized, and an urgent subject for future research is the clearer quantitative establishment of its limitations.

The contents of this book are essentially theoretical, not because of any prejudices on the part of the authors but because of the way in which the topic was assigned them. Where possible, the techniques outlined in what follows are chosen because favorable comparisons with measured data exist. Enormous advances have recently been made both in the experimental procedures of aeroelasticity and in the related instrumentation. Another book might be written entirely about dynamic modeling, and yet another on full-scale vibration and flight testing. The aeroelastician also has available to him several valuable additional tools, of which sled testing, rocket-model testing, and the direct measurement of unsteady airloads are illustrations.

Another important area which the authors have been able to touch on only in passing is that of digital and analog computation methods in aeroelasticity. The viewpoint of this book is somewhat colored by the authors' experience with high-speed digital machinery, as some readers will observe from the choice of examples. Nevertheless, the direct analogic representation of aeroelastic systems by electrical networks is perhaps the most promising technique that can be singled out for extensive exploitation in the future.

REFERENCES

1–1. Fung, Y. C., *An Introduction to the Theory of Aeroelasticity*, John Wiley and Sons, New York, 1955.

1–2. Bisplinghoff, R. L., H. Ashley, and R. L. Halfman, *Aeroelasticity*, Addison-Wesley Publishing Company, Cambridge, Mass., 1955.

1–3. Scanlan, R. H., and R. Rosenbaum, *Introduction to the Study of Aircraft Vibration and Flutter*, The Macmillan Company, New York, 1951.

1–4. Many Authors, *Manual on Aeroelasticity*, published in five parts by NATO Advisory Group for Aeronautical Research and Development, 1959.

1–5. Duncan, W. J., *Control and Stability of Aircraft*, Cambridge University Press, Cambridge, 1952.

1–6. Templeton, H., *Mass Balancing of Aircraft Control Surfaces*, Chapman and Hall Ltd., London, 1954.

1–7. Abramson, H. N., *An Introduction to the Dynamics of Airplanes*, The Ronald Press Company, New York, 1958.

1–8. Broadbent, E. G., *The Elementary Theory of Aeroelasticity*, "Aircraft Engineering" Monograph, Bunhill Publications Ltd., London, 1954.

1–9. Myklestad, N. O., *Vibration Analysis*, McGraw-Hill Book Company, New York, 1944.

1–10. Freberg, C. R., and E. N. Kemler, *Aircraft Vibration and Flutter*, John Wiley and Sons, New York, 1944. (Out of print.)

1–11. von Kármán, T., and M. A. Biot, *Mathematical Methods in Engineering*, McGraw-Hill Book Company, New York, 1940.

1–12. Grossman, E. P., *Flutter*, Joukowsky Mem. Central Aero-Hydrodynamic Institute Report 186, 1935, translated as Air Force Translation F-TS-1225-1A (GDAM-A9-T-44).

1–13. Collar, A. R., "The Expanding Domain of Aeroelasticity," *J. Royal Aero. Soc.*, Vol. L, August 1946, pp. 613–636.

1–14. Many Authors, *Aeroelastisches Kolloquium in Göttingen*, April 1957, Mitteilung Nr. 18, Max-Planck-Institut für Strömungsforschung, 1958.

1–15. Fiszdon, W., *Fundamentals of Aeroelasticity*, (P.W.N.) Polish Scientific Publications, Warsaw, 1951. (In Polish.)

1–16. Nekrasov, A. I., *Theory of Unsteady Flow Past a Wing*, Academy of Sciences of the USSR, Moscow, 1957. (In Russian.)

1–17. Many Authors, *Proceedings of the National Specialists Meeting on Dynamics and Aeroelasticity*, Ft. Worth, Texas, November 1958, published by the Institute of the Aeronautical Sciences, New York.

1–18. Garrick, I. E., *Some Concepts and Problem Areas in Aircraft Flutter*, the 1957 Minta Martin Aeronautical Lecture, Sherman Fairchild Fund Paper No. FF-15, Institute of the Aeronautical Sciences, March 1957.

1–19. Williams, J., *Aircraft Flutter*, British A.R.C. Reports and Memoranda No. 2492, 1951.

1–20. Templeton, H., *A Review of the Present Position on Flutter*, NATO Advisory Group for Aeronautical Research and Development Report 57, April 1956.

1–21. Goland, M., *An Appraisal of Aeroelasticity in Design, with Special Reference to Dynamic Aeroelastic Stability*, paper presented at Sixth Anglo-American Aeronautical Conference, London, September 1957.

1–22. Laidlaw, W. R., *The Aeroelastic Design of Lifting Surfaces*, Notes prepared for M.I.T. Summer Course on Aeroelasticity, June-July 1958, printed by North American Aviation, Inc.

1–23. Collar, A. R., *Aeroelasticity—Retrospect and Prospect*, The Second Lanchester Memorial Lecture, November 20, 1958.

1–24. Dryden, H. L., and J. E. Duberg, *Aeroelastic Effects of Aerodynamic Heating*, paper presented at the Fifth General Assembly of A.G.A.R.D., Ottawa, June 1955.

1–25. Bisplinghoff, R. L., "Some Structural and Aeroelastic Considerations of High Speed Flight," The Nineteenth Wright Brothers Lecture, *J. Aero. Sciences*, Vol. 23, No. 4, April 1956, pp. 289–321.

1-26. Bisplinghoff, R. L., and J. Dugundji, "Influence of Aerodynamic Heating on Aeroelastic Phenomena," Chapter 14 of Agardograph No. 28, *High Temperature Effects in Aircraft Structures*, N. J. Hoff, Ed., Pergamon Press, London, 1958.

1-27. J. B. Rea Company, Inc., *Aeroelasticity in Stability and Control*, USAF Wright Air Development Center Technical Report 55-173, March 1957.

1-28. NACA Subcommittee on Vibration and Flutter, *Dynamic and Aeroelastic Research for Space Flight Vehicles*, issued by National Aeronautics and Space Administration, Washington, D.C., September 1958.

1-29. Head, A. L., Jr., *A Philosophy of Design for Flutter*, Proceedings of the National Specialists Meeting on Dynamics and Aeroelasticity, Ft. Worth, Texas, November 1958, pp. 59–65, published by the Institute of the Aeronautical Sciences, New York.

1-30. Lecture by the late Dr. H. W. Friedrich, Convair Astronautics, to the student body of Course 16.919, "Aeroelasticity," Massachusetts Institute of Technology, June-July 1958.

1-31. Den Hartog, J. P., *Mechanical Vibrations*, 3rd ed., McGraw-Hill Book Company, New York, 1947.

1-32. Timoshenko, S., and D. H. Young, *Vibration Problems in Engineering*, 3rd ed., D. Van Nostrand Company, Princeton, N.J., 1955.

1-33. Rocard, Y., *Dynamique Générale des Vibrations*, Deuxième Éditions, Masson et Cie., Éditeurs, Paris, 1949.

1-34. Many Authors, *Proceedings of Symposium on Structural Dynamics of High Speed Flight*, Los Angeles, Calif., April 1961, sponsored by Aerospace Industries Assoc. and Office of Naval Research.

1-35. Many Authors, *Proceedings of Symposium on Aerothermoelasticity*, Dayton, Ohio, October–November 1961, sponsored by Aeronautical Systems Division, USAF.

2

MATHEMATICAL
FOUNDATIONS
OF AEROELASTICITY

2–1 INTRODUCTION

The quantitative treatment of any class of dynamic problems logically begins with a mathematical formulation of the fundamentals. The foremost matter that requires consideration in aeroelasticity is a study of the behavior of an unrestrained elastically deformable body under the simultaneous action of aerodynamic heating and pressures. Here we are forced to explore the borderlands among dynamics, elasticity, and thermodynamics. The general problem of analyzing the behavior of an elastic body under the combined influence of aerodynamic pressures and aerodynamic heating can be referred to as an aerothermoelastic problem. We assume that the character of the material and the magnitudes of the surface forces and heat inputs are such that the body returns to its original size and shape after they have ceased to act. The property of recovering size and shape, termed elasticity, is assumed for the most part throughout this book.

The general problem of aerothermoelasticity is one of computing the temperature, displacement, and stress distribution that result throughout the body from the boundary conditions and appropriate initial conditions. It is usually possible to separate this computation into two parts. The first is referred to as an aerothermal problem and the second as an aeroelastic problem. The aerothermal problem deals with the thermal equilibrium and heat transfer between the environment and the body and within the body. The end result of the aerothermal problem is the time history of the temperature distribution throughout the body. This is the starting point of the aeroelastic problem which deals with the equilibrium among aerodynamic, elastic, and inertial forces in the presence of the prescribed temperature distribution. In an aeroelastic problem, the strain distribution has a negligible influence on the temperature distribution. On the other hand, the temperature distribution may have a profound effect on the strain distribution. It is, of course, the aeroelastic problem which is of primary interest here, so we shall make only passing reference to the aerothermal problem.

2-2   EQUILIBRIUM AND COMPATIBILITY CONDITIONS FOR ELASTICALLY DEFORMABLE BODIES

The mathematical foundation of aeroelasticity rests upon the conditions of equilibrium and compatibility of a free elastic body together with appropriate force and displacement boundary conditions.

(a) Equilibrium equations

Referring to Fig. 2–1, we consider a three-dimensional elastic body which is unrestrained in space. The body is capable of assuming small displacements, with respect to an orthogonal x-y-z-axis system fixed to an arbitrary point.

A position vector $\mathbf{r}'$ is referred to an orthogonal axis system—x', y', z'— which is fixed in space. The body may assume large rigid body displacements with respect to the latter axis system.

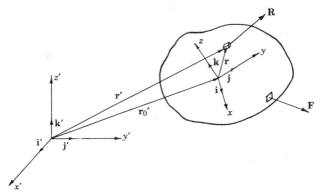

Fig. 2–1. Three-dimensional unrestrained elastic body.

If the body is acted on by surface tractions per unit of area designated by the vector $\mathbf{F}$ and by body forces per unit of volume designated by the vector $\mathbf{R}$, then the equations of force and moment equilibrium are given in vector form by Eqs. 2–1 and 2–2, respectively.

$$\frac{d}{dt} \iiint_V \rho \frac{d\mathbf{r}'}{dt} \, dV = \iiint_V \mathbf{R} \, dV + \iint_S \mathbf{F} \, dS \tag{2-1}$$

$$\frac{d}{dt} \iiint_V \left(\mathbf{r}' \times \rho \frac{d\mathbf{r}'}{dt} \right) dV = \iiint_V \mathbf{r}' \times \mathbf{R} \, dV + \iint_S \mathbf{r}' \times \mathbf{F} \, dS \tag{2-2}$$

where ρ is the mass per unit volume, assumed invariant with respect to time. V and S represent integrations throughout the volume and over the surface, respectively. When the origin of the x-y-z-coordinate system is taken at the center of gravity of the body, then we have the further condition that

$$M\mathbf{r}_0' = \iiint_V \rho \mathbf{r}' \, dV \tag{2-3}$$

where M is the total mass of the body and $\mathbf{r}_0'$ is the position vector to the center of gravity. If we introduce the vector equation

$$\mathbf{r}' = \mathbf{r}_0' + \mathbf{r} \tag{2-4}$$

where $\mathbf{r}$ is a position vector of a particle in the x-y-z-axis system, into Eq. 2–1 and make use of Eq. 2–3, we obtain the well known result that

$$\mathbf{P} = \frac{d\mathbf{G}}{dt} \tag{2-5}$$

where $\mathbf{P}$ is the resultant vector of the applied forces acting on the system

$$\mathbf{P} = \iiint_V \mathbf{R} \, dV + \iint_S \mathbf{F} \, dS$$

and $\mathbf{G}$ is the momentum vector

$$\mathbf{G} = M \frac{d\mathbf{r}_0'}{dt}$$

Similarly if we apply Eqs. 2–3 and 2–4 to Eq. 2–2, we find that

$$\mathbf{L} = \frac{d\mathbf{H}}{dt} \tag{2-6}$$

where $\mathbf{L}$ is the resultant vector of the applied moments about the center of gravity

$$\mathbf{L} = \iiint_V \mathbf{r} \times \mathbf{R} \, dV + \iint_S \mathbf{r} \times \mathbf{F} \, dS$$

and $\mathbf{H}$ is the moment of momentum about the center of gravity

$$\mathbf{H} = \iiint_V \mathbf{r} \times \rho \, \frac{d\mathbf{r}}{dt} \, dV$$

Equation 2–5 announces that the motion of the mass center, designated by the position vector $\mathbf{r}_0'$, follows the law of motion of a single mass particle equal to the total mass of the system and under the action of the resultant of all the forces. Equation 2–6 states a similar result that the rate of change of resultant moment of momentum about the mass center is equal to the resultant moment of the external forces about the mass center. Equations 2–5 and 2–6 provide a basis for computing the gross motion of an aeroelastic system, but they provide no information concerning the internal or elastic response. The latter must be obtained by appealing to the properties of stress and strain within the body.

Suppose that the surface vector $\mathbf{F}$ is represented in component form by

$$\mathbf{F} = \mathbf{i}F_x + \mathbf{j}F_y + \mathbf{k}F_z \tag{2–7}$$

where $\mathbf{i}$, $\mathbf{j}$, and $\mathbf{k}$ are unit vectors in the x-, y-, and z-directions, respectively, and F_x, F_y, and F_z are the components of $\mathbf{F}$. The latter quantities are related to the internal stresses at the surface by the boundary conditions

$$\begin{aligned}
F_x &= \sigma_x \mathbf{n} \cdot \mathbf{i} + \tau_{xy} \mathbf{n} \cdot \mathbf{j} + \tau_{xz} \mathbf{n} \cdot \mathbf{k} \\
F_y &= \tau_{yx} \mathbf{n} \cdot \mathbf{i} + \sigma_y \mathbf{n} \cdot \mathbf{j} + \tau_{yz} \mathbf{n} \cdot \mathbf{k} \\
F_z &= \tau_{zx} \mathbf{n} \cdot \mathbf{i} + \tau_{zy} \mathbf{n} \cdot \mathbf{j} + \sigma_z \mathbf{n} \cdot \mathbf{k}
\end{aligned} \tag{2–8}$$

where $\mathbf{n}$ is a unit vector normal to the surface in the outward direction and where

$$\mathbf{n} \cdot \mathbf{i} = \cos(x, n), \qquad \mathbf{n} \cdot \mathbf{j} = \cos(y, n), \qquad \mathbf{n} \cdot \mathbf{k} = \cos(z, n)$$

are direction cosines of this normal with respect to the x, y, and z axes, respectively. The stress quantities in Eq. 2–8 follow the notation of Timoshenko (Ref. 2–1).

Equations 2–7 and 2–8 can be represented by the single equation

$$\mathbf{F} = \mathbf{n} \cdot \Phi \tag{2–9}$$

where Φ is a second-order tensor of the stresses which can be written in the dyadic form

$$\Phi = \sigma_x \mathbf{ii} + \tau_{yx} \mathbf{ji} + \tau_{zx} \mathbf{ki}$$
$$+ \tau_{xy} \mathbf{ij} + \sigma_y \mathbf{jj} + \tau_{zy} \mathbf{kj} \qquad (2\text{--}10)$$
$$+ \tau_{xz} \mathbf{ik} + \tau_{yz} \mathbf{jk} + \sigma_z \mathbf{kk}$$

If we substitute Eq. 2–9 into the equations of equilibrium, we deduce from Eq. 2–1 that

$$\iiint_V \rho \frac{d^2 \mathbf{r}'}{dt^2} \, dV = \iiint_V \mathbf{R} \, dV + \iint_S \mathbf{n} \cdot \Phi \, dS \qquad (2\text{--}11)$$

and from Eq. 2–2 that

$$\iiint_V \left(\mathbf{r} \times \rho \frac{d^2 \mathbf{r}'}{dt^2} \right) dV = \iiint_V \mathbf{r} \times \mathbf{R} \, dV + \iint_S \mathbf{r} \times \mathbf{n} \cdot \Phi \, dS \qquad (2\text{--}12)$$

where we have made use of Eqs. 2–3 and 2–4.

Transforming the surface integrals into volume integrals by means of the divergence theorem (Ref. 2–2)

$$\iint_S \mathbf{n} \cdot \Phi \, dS = \iiint_V \nabla \cdot \Phi \, dV \qquad (2\text{--}13)$$

$$\iint_S \mathbf{r} \times \mathbf{n} \cdot \Phi \, dS = - \iint_S \mathbf{n} \cdot \Phi \times \mathbf{r} \, dS = - \iiint_V \nabla \cdot (\Phi \times \mathbf{r}) \, dV \qquad (2\text{--}14)$$

Eqs. 2–11 and 2–12 become

$$\iiint_V \left(\rho \frac{d^2 \mathbf{r}'}{dt^2} - \mathbf{R} - \nabla \cdot \Phi \right) dV = 0 \qquad (2\text{--}15)$$

$$\iiint_V \left[\mathbf{r} \times \rho \frac{d^2 \mathbf{r}'}{dt^2} - \mathbf{r} \times \mathbf{R} + \nabla \cdot (\Phi \times \mathbf{r}) \right] dV = 0 \qquad (2\text{--}16)$$

where ∇ is the divergence operator

$$\nabla = \mathbf{i} \frac{\partial}{\partial x} + \mathbf{j} \frac{\partial}{\partial y} + \mathbf{k} \frac{\partial}{\partial z}$$

Since the integrals of Eqs. 2–15 and 2–16 vanish for an arbitrary choice of V, their integrands must be identically zero; and the equations of equilibrium are, therefore,

$$\rho \frac{d^2 \mathbf{r}'}{dt^2} - \mathbf{R} - \nabla \cdot \Phi = 0 \qquad (2\text{--}17)$$

$$\mathbf{r} \times \rho \frac{d^2 \mathbf{r}'}{dt^2} - \mathbf{r} \times \mathbf{R} + \nabla \cdot (\Phi \times \mathbf{r}) = 0 \qquad (2\text{--}18)$$

Prior to reducing Eqs. 2–17 and 2–18 to component form, we assume that the acceleration vector has the form

$$\frac{d^2\mathbf{r}'}{dt^2} = a_x\mathbf{i} + a_y\mathbf{j} + a_z\mathbf{k} \tag{2–19}$$

and that the body force vector is

$$\mathbf{R} = X\mathbf{i} + Y\mathbf{j} + Z\mathbf{k} \tag{2–20}$$

where a_x, a_y, and a_z are accelerations and X, Y, and Z are body forces along the x, y, and z axes, respectively. Inserting Eqs. 2–19 and 2–20 into Eq. 2–17 and reducing to component form yields

$$\rho a_x = \frac{\partial \sigma_x}{\partial x} + \frac{\partial \tau_{yx}}{\partial y} + \frac{\partial \tau_{zx}}{\partial z} + X$$

$$\rho a_y = \frac{\partial \tau_{xy}}{\partial x} + \frac{\partial \sigma_y}{\partial y} + \frac{\partial \tau_{zy}}{\partial z} + Y \tag{2–21}$$

$$\rho a_z = \frac{\partial \tau_{xz}}{\partial x} + \frac{\partial \tau_{yz}}{\partial y} + \frac{\partial \sigma_z}{\partial z} + Z$$

Equation 2–18 yields the added result that the shear stresses on mutually perpendicular planes are equal, or that

$$\tau_{xy} = \tau_{yx}, \qquad \tau_{xz} = \tau_{zx}, \qquad \tau_{yz} = \tau_{zy} \tag{2–22}$$

Equations 2–21 and 2–22 are fully equivalent to 2–11 and 2–12, respectively, since we can pass from the former to the latter by integrating over the body.

(b) Compatibility equations and equations of state

In addition to the equations of equilibrium 2–21 and 2–22, we still require to know, in order to formulate unambiguously the laws of motion of a deformable body, the relationship between deformations and forces. Let us suppose that the elastic deformations of the body are represented by u, v, and w along the x, y, and z directions, respectively. Then the strains, assuming that they are so small that their squares and products can be neglected, are defined in terms of the displacements by

$$\epsilon_x = \frac{\partial u}{\partial x}, \qquad \epsilon_y = \frac{\partial v}{\partial y}, \qquad \epsilon_z = \frac{\partial w}{\partial z}$$

$$\gamma_{xy} = \frac{\partial v}{\partial x} + \frac{\partial u}{\partial y}, \qquad \gamma_{xz} = \frac{\partial w}{\partial x} + \frac{\partial u}{\partial z}, \qquad \gamma_{yz} = \frac{\partial w}{\partial y} + \frac{\partial v}{\partial z} \tag{2–23}$$

where ϵ_i and γ_{ij} are normal and shear strains according to the customary notation of theory of elasticity (Ref. 2–1).

The equations of state or stress-strain relations of a heated and strained homogeneous, isotropic, elastic body can be expressed in the following terms (Ref. 2–1):

$$\epsilon_x = \alpha\,\Delta T + \frac{1}{E}\left[\sigma_x - \nu(\sigma_y + \sigma_z)\right]$$

$$\epsilon_y = \alpha\,\Delta T + \frac{1}{E}\left[\sigma_y - \nu(\sigma_x + \sigma_z)\right] \qquad (2\text{–}24)$$

$$\epsilon_z = \alpha\,\Delta T + \frac{1}{E}\left[\sigma_z - \nu(\sigma_x + \sigma_y)\right]$$

$$\gamma_{xy} = \frac{1}{G}\tau_{xy}, \qquad \gamma_{xz} = \frac{1}{G}\tau_{xz}, \qquad \gamma_{yz} = \frac{1}{G}\tau_{yz}$$

where α is the coefficient of thermal expansion of the material, ΔT is the rise in temperature above a standard reference value, and E, G, and ν are Young's modulus, the shear modulus, and Poisson's ratio, respectively. The latter three quantities are related by

$$G = \frac{E}{2(1 + \nu)} \qquad (2\text{–}25)$$

In Eqs. 2–24, the normal strains ϵ_x, ϵ_y, and ϵ_z consist of two components. The first component $\alpha\,\Delta T$ is the strain resulting from the temperature change ΔT. This strain is uniform in all directions and is, therefore, a purely extensional strain. The second component of normal strain, represented by the term with brackets, comprises the strains which arise in order to preserve continuity of the body. The shear strains are seen to be unaffected by temperature.

In Eqs. 2–24 the strains are expressed in terms of the stresses. An inverted form, in which the stresses are expressed in terms of the strains, reads

$$\sigma_x = \lambda e + 2G\epsilon_x - \frac{\alpha E\,\Delta T}{1 - 2\nu}$$

$$\sigma_y = \lambda e + 2G\epsilon_y - \frac{\alpha E\,\Delta T}{1 - 2\nu} \qquad (2\text{–}26)$$

$$\sigma_z = \lambda e + 2G\epsilon_z - \frac{\alpha E\,\Delta T}{1 - 2\nu}$$

$$\tau_{xy} = G\gamma_{xy}, \qquad \tau_{xz} = G\gamma_{xz}, \qquad \tau_{yz} = G\gamma_{yz}$$

where e is the volume expansion

$$e = \epsilon_x + \epsilon_y + \epsilon_z$$

and λ is a Lamé elastic constant defined by

$$\lambda = \frac{\nu E}{(1 + \nu)(1 - 2\nu)}$$

(c) Equations and unknowns

When there is an *a priori* knowledge of the time history of the temperature distribution $\Delta T(x, y, z, t)$ obtained by a solution of the aerothermal problem, and when the force boundary conditions in Eq. 2–8 and appropriate initial conditions are specified, then the eighteen equations represented by 2–21, 2–22, 2–23, and 2–24 or 2–26 provide a sufficient basis for obtaining unique solutions of the time histories of stresses, strains, and displacements of the body. It may be said that the balance of the book is concerned largely with techniques of obtaining such solutions for certain classes of boundary and initial conditions which characterize the field of aeroelasticity.

2–3 THERMODYNAMIC BEHAVIOR OF LOADED AND HEATED ELASTICALLY DEFORMABLE BODIES

Besides the equations of equilibrium, compatibility, and state, it is profitable to consider the behavior of an elastic body in terms of thermodynamic variables. These considerations are useful in the application of variational principles to aeroelastic problems. When heat and external force are applied simultaneously to an unrestrained elastic body, there is a change of kinetic and internal energy. Let $\delta\tau$ be the total kinetic energy and δU_0 the internal energy per unit volume acquired during the time interval δt by the application of body and surface forces and heat. The first law of thermodynamics requires that for the entire body

$$\delta\tau + \iiint_V \delta U_0 \, dV = \delta W + \iiint_V \delta Q \, dV \qquad (2\text{–}27)$$

where δW is the work done by the body and surface forces in the interval of time δt, and δQ is the mechanical equivalent of the heat supplied per unit volume in the same interval of time. The kinetic energy is defined by the formula

$$\tau = \frac{1}{2} \iiint_V \rho \frac{d\mathbf{r}'}{dt} \cdot \frac{d\mathbf{r}'}{dt} \, dV \qquad (2\text{–}28)$$

We can focus our attention on a unit volume of the body and study its behavior by introducing the well known work and energy theorem (Ref. 2–3) which states that

$$\delta W = \delta \tau + \iiint_V (\sigma_x \delta \epsilon_x + \sigma_y \delta \epsilon_y + \sigma_z \delta \epsilon_z + \tau_{xy} \delta \gamma_{xy} + \tau_{xz} \delta \gamma_{xz} + \tau_{yz} \delta \gamma_{yz}) \, dV$$

$$(2\text{–}29)$$

Combining Eqs. 2–27 and 2–29 yields

$$\delta U_0 = \delta Q + \sigma_x \delta \epsilon_x + \sigma_y \delta \epsilon_y + \sigma_z \delta \epsilon_z + \tau_{xy} \delta \gamma_{xy} + \tau_{xz} \delta \gamma_{xz} + \tau_{yz} \delta \gamma_{yz}$$

$$(2\text{–}30)$$

The quantity δU_0 is a function of the quantities that determine the strained configuration of the unit volume and of the temperature. Its value, corresponding to any state, is a function of the internal energy in that state providing we take $U_0 = 0$ in some arbitrarily defined standard state.

For the change of entropy of a reversible system, also in the time interval δt, we have

$$\delta S = \frac{\delta Q}{T_1} \qquad (2\text{–}31)$$

where T_1 is the absolute temperature of the volume element.

Since U_0 is a function of the strains and the absolute temperature, δS can be expressed in the form

$$\delta S = \frac{1}{T_1}\left(\frac{\partial U_0}{\partial T_1}\right)_\epsilon \delta T_1 + \frac{1}{T_1}\left\{\left[\left(\frac{\partial U_0}{\partial \epsilon_x}\right)_T - \sigma_x\right]\delta \epsilon_x + \left[\left(\frac{\partial U_0}{\partial \epsilon_y}\right)_T - \sigma_y\right]\delta \epsilon_y\right.$$

$$+ \left[\left(\frac{\partial U_0}{\partial \epsilon_z}\right)_T - \sigma_z\right]\delta \epsilon_z + \left[\left(\frac{\partial U_0}{\partial \gamma_{xy}}\right)_T - \tau_{xy}\right]\delta \gamma_{xy}$$

$$\left. + \left[\left(\frac{\partial U_0}{\partial \gamma_{xz}}\right)_T - \tau_{xz}\right]\delta \gamma_{xz} + \left[\left(\frac{\partial U_0}{\partial \gamma_{yz}}\right)_T - \tau_{yz}\right]\delta \gamma_{yz}\right\} \qquad (2\text{–}32)$$

The subscripts ϵ and T indicate that the strain and temperature, respectively, are held constant during differentiation. The second law of thermodynamics requires that δS be a total differential in T_1 and the strains, and as a result,

$$\frac{1}{T_1}\left(\frac{\partial U_0}{\partial \epsilon_x}\right)_T = \frac{\sigma_x}{T_1} - \left(\frac{\partial \sigma_x}{\partial T_1}\right)_\epsilon, \qquad \frac{1}{T_1}\left(\frac{\partial U_0}{\partial \gamma_{xy}}\right)_T = \frac{\tau_{xy}}{T_1} - \left(\frac{\partial \tau_{xy}}{\partial T_1}\right)_\epsilon$$

$$\frac{1}{T_1}\left(\frac{\partial U_0}{\partial \epsilon_y}\right)_T = \frac{\sigma_y}{T_1} - \left(\frac{\partial \sigma_y}{\partial T_1}\right)_\epsilon, \qquad \frac{1}{T_1}\left(\frac{\partial U_0}{\partial \gamma_{yz}}\right)_T = \frac{\tau_{yz}}{T_1} - \left(\frac{\partial \tau_{yz}}{\partial T_1}\right)_\epsilon \qquad (2\text{–}33)$$

$$\frac{1}{T_1}\left(\frac{\partial U_0}{\partial \epsilon_z}\right)_T = \frac{\sigma_z}{T_1} - \left(\frac{\partial \sigma_z}{\partial T_1}\right)_\epsilon, \qquad \frac{1}{T_1}\left(\frac{\partial U_0}{\partial \gamma_{xz}}\right)_T = \frac{\tau_{xz}}{T_1} - \left(\frac{\partial \tau_{xz}}{\partial T_1}\right)_\epsilon$$

Introducing Eqs. 2–33 together with

$$\left(\frac{\partial U_0}{\partial T_1}\right)_\epsilon = c_\epsilon \qquad (2\text{–}34)$$

where c_ϵ is the heat capacity per unit volume under conditions of zero strain, into 2–32 yields for the change of entropy

$$\delta S = c_\epsilon \frac{\delta T_1}{T_1} - \left(\frac{\partial \sigma_x}{\partial T_1}\right)_\epsilon \delta\epsilon_x - \left(\frac{\partial \sigma_y}{\partial T_1}\right)_\epsilon \delta\epsilon_y - \left(\frac{\partial \sigma_z}{\partial T_1}\right)_\epsilon \delta\epsilon_z$$

$$- \left(\frac{\partial \tau_{xy}}{\partial T_1}\right)_\epsilon \delta\gamma_{xy} - \left(\frac{\partial \tau_{xz}}{\partial T_1}\right)_\epsilon \delta\gamma_{xz} - \left(\frac{\partial \tau_{yz}}{\partial T_1}\right)_\epsilon \delta\gamma_{yz} \quad (2\text{–}35)$$

Combining Eqs. 2–31 and 2–35, the heat supplied to the volume element in the interval δt can be represented by the following function of the increment of temperature and the increments of the strains:

$$\delta Q = c_\epsilon\, \delta T_1 - T_1\left\{\left(\frac{\partial \sigma_x}{\partial T_1}\right)_\epsilon \delta\epsilon_x + \left(\frac{\partial \sigma_y}{\partial T_1}\right)_\epsilon \delta\epsilon_y + \left(\frac{\partial \sigma_z}{\partial T_1}\right)_\epsilon \delta\epsilon_z\right.$$

$$\left. + \left(\frac{\partial \tau_{xy}}{\partial T_1}\right)_\epsilon \delta\gamma_{xy} + \left(\frac{\partial \tau_{xz}}{\partial T_1}\right)_\epsilon \delta\gamma_{xz} + \left(\frac{\partial \tau_{yz}}{\partial T_1}\right)_\epsilon \delta\gamma_{yz}\right\} \quad (2\text{–}36)$$

Let us consider next the results obtained when Eq. 2–36 is substituted into the energy balance represented by Eq. 2–30, and when adiabatic and isothermal changes of state take place. If the change of state takes place adiabatically—that is, if there is no heat gained or lost by the element and, consequently, $\delta Q = 0$—then we have simply the following equation:

$$\delta U_0 = \sigma_x\, \delta\epsilon_x + \sigma_y\, \delta\epsilon_y + \sigma_z\, \delta\epsilon_z + \tau_{xy}\, \delta\gamma_{xy}$$

$$+ \tau_{xz}\, \delta\gamma_{xz} + \tau_{yz}\, \delta\gamma_{yz} \text{ (Adiabatic)} \quad (2\text{–}37)$$

The expression on the right-hand side of Eq. 2–37 is, therefore, an exact differential; and we can postulate the existence of a function U_0 which has the properties

$$\frac{\partial U_0}{\partial \epsilon_x} = \sigma_x; \qquad \frac{\partial U_0}{\partial \epsilon_y} = \sigma_y; \qquad \frac{\partial U_0}{\partial \epsilon_z} = \sigma_z;$$

$$\frac{\partial U_0}{\partial \gamma_{xy}} = \tau_{xy}; \qquad \frac{\partial U_0}{\partial \gamma_{xz}} = \tau_{xz}; \qquad \frac{\partial U_0}{\partial \gamma_{yz}} = \tau_{yz}$$

(2–38)

The function U_0 represents the potential energy stored up per unit volume by the adiabatic strains, and its variation is the same as the variation of the internal energy. If the strains are very rapid, as in the case of vibrations of a structure, the process is nearly adiabatic.

In the previous paragraph, we have shown that a potential energy function exists in the case of adiabatic straining. Such a function may be shown also to exist in the case of isothermal straining. For isothermal straining we have from Eqs. 2–30 and 2–36

$$
\delta U_0 + T_1 \left\{ \left(\frac{\partial \sigma_x}{\partial T_1} \right)_\epsilon \delta\epsilon_x + \left(\frac{\partial \sigma_y}{\partial T_1} \right)_\epsilon \delta\epsilon_y + \left(\frac{\partial \sigma_z}{\partial T_1} \right)_\epsilon \delta\epsilon_z \right.
$$

$$
\left. + \left(\frac{\partial \tau_{xy}}{\partial T_1} \right)_\epsilon \delta\gamma_{xy} + \left(\frac{\partial \tau_{xz}}{\partial T_1} \right)_\epsilon \delta\gamma_{xz} + \left(\frac{\partial \tau_{yz}}{\partial T_1} \right)_\epsilon \delta\gamma_{yz} \right\}
$$

$$
= \sigma_x \, \delta\epsilon_x + \sigma_y \, \delta\epsilon_y + \sigma_z \, \delta\epsilon_z + \tau_{xy} \, \delta\gamma_{xy} + \tau_{xz} \, \delta\gamma_{xz} + \tau_{yz} \, \delta\gamma_{yz}
$$

$$
\text{(Isothermal)} \quad (2\text{–}39)
$$

In this case the work done by the stresses during small increments of strains of the volume element represented by the right-hand side of Eq. 2–39 is not equal to the change of internal energy, δU_0. The second term on the left-hand side of Eq. 2–39 represents the heat which must be added or extracted from the volume element in order to keep the temperature constant during the isothermal straining process. The magnitude of this term clearly depends upon such material properties as the thermal expansion of the material and the influence of temperature on modulus of elasticity. Equation 2–39 may also be written in the shorthand form

$$
\delta F_0 = \delta U_0 - T_1 \, \delta S = \sigma_x \, \delta\epsilon_x + \sigma_y \, \delta\epsilon_y + \sigma_z \, \delta\epsilon_z
$$

$$
+ \tau_{xy} \, \delta\gamma_{xy} + \tau_{xz} \, \delta\gamma_{xz} + \tau_{yz} \, \delta\gamma_{yz} \quad \text{(Isothermal)} \quad (2\text{–}40)
$$

where $F_0 = U_0 - T_1 S$ is a function of state called the free energy per unit volume. The work done isothermally by the stresses on the element is then equal to the change of this free energy. Since the right-hand side is an exact differential, we conclude that there exists an isothermal potential function F_0 with the properties

$$
\frac{\partial F_0}{\partial \epsilon_x} = \sigma_x; \qquad \frac{\partial F_0}{\partial \epsilon_y} = \sigma_y; \qquad \frac{\partial F_0}{\partial \epsilon_z} = \sigma_z;
$$

$$
\frac{\partial F_0}{\partial \gamma_{xy}} = \tau_{xy}; \qquad \frac{\partial F_0}{\partial \gamma_{xz}} = \tau_{xz}; \qquad \frac{\partial F_0}{\partial \gamma_{yz}} = \tau_{yz}
$$

$$
(2\text{–}41)
$$

When the strains are applied very slowly so that the temperature of the body is continuously adjusted to that of its surroundings, the changes of state are practically isothermal.

The foregoing development of adiabatic and isothermal potential functions, designated by U_0 and F_0, respectively, points to different coefficients for the terms in these two functions even though they are both expressed in the same functional form of the absolute temperature and the strains. Since the ordinary devices for testing materials apply the strains very slowly, the elastic constants obtained from such tests are, for all practical purposes, isothermal values. In fact the equations of state, expressed in terms of isothermal elastic constants, for a homogeneous isotropic element whose temperature has been raised ΔT above a standard level are those already given by Eqs. 2–24 and 2–26. Making use of Eq. 2–26, we observe that the requirements of Eq. 2–41 are met by the potential energy function

$$F_0 = \frac{1}{2} \left\{ \lambda(\epsilon_x + \epsilon_y + \epsilon_z)^2 + 2G(\epsilon_x^2 + \epsilon_y^2 + \epsilon_z^2) \right.$$

$$\left. + G(\gamma_{xy}^2 + \gamma_{yz}^2 + \gamma_{xz}^2) - \frac{2\alpha E \, \Delta T}{1 - 2\nu}(\epsilon_x + \epsilon_y + \epsilon_z) \right\} \quad (2\text{–}42)$$

2–4 HAMILTON'S PRINCIPLE

An introduction to the mathematical foundations of aeroelasticity would not be complete without a suitable variational statement for the equations of motion. A sentence from Morse and Feshbach (Ref. 2–4) is apt: "The use of superlatives enables one to express in concise form a general principle covering a wide range of phenomena." *Hamilton's principle* plays this role in dynamics. For a conservative system, which is one whose total *mechanical* energy is constant and whose forces can therefore be represented in terms of a scalar potential function, both the verbal and mathematical statements are charmingly concise: for the actual motion of the system, between specified, realizable initial and final conditions at times t_0 and t_1, the difference between the kinetic and potential energies will average out to be a minimum relative to any other dynamical path compatible with the physical constraints. In variational form,

$$\delta \int_{t_0}^{t_1} (\tau - U) \, dt = 0 \quad (2\text{–}43)$$

Here τ represents the instantaneous kinetic energy measured relative to any inertial set of coordinate axes as given, for example, by Eq. 2–28. The potential function represented by U in Eq. 2–43 is a volume integral of the state functions U_0 or F_0 derived in the previous section. In the case of adiabatic straining, we have seen from Eq. 2–28 that the appropriate definition would be

$$U = \iiint_V U_0 \, dV \tag{2-44}$$

and, in the case of isothermal straining,

$$U = \iiint_V F_0 \, dV \tag{2-45}$$

where F_0 is the free energy function defined by Eq. 2–42. Since the free energy function can be defined conveniently in terms of the isothermal elastic constants of the body and since it differs only slightly from the function representing adiabatic straining, it is usually employed in the formulation of aeroelastic problems.

The integrand of the single or multiple integral appearing in a variational principle, such as Eq. 2–43, is referred to as a Lagrange function. For Hamilton's principle, the Lagrange function $L = \tau - U$ bears the appropriate name *kinetic potential*. The reader will find many books which discuss those relatively elementary results from the calculus of variations that are needed here. Bliss (Ref. 2–5), Fox (Ref. 2–6), and Lanczos (Ref. 2–7) are good examples in addition to Chaps. 3 and 9 of Ref. 2–4.

Wholly conservative systems are rare in the field of aeroelasticity, so that a generalization of Eq. 2–43 to account for forces not derivable from a potential function is required. We express such a generalization in the following form:

$$\int_{t_0}^{t_1} \delta(\tau - U) \, dt + \int_{t_0}^{t_1} \delta W \, dt = 0 \tag{2-46}$$

where

$$\delta W = \iiint_V \mathbf{R} \cdot \delta \mathbf{r}' \, dV + \iint_S \mathbf{F} \cdot \delta \mathbf{r}' \, dS$$

is interpreted physically as the total virtual work done by the body and surface forces during the virtual displacement $\delta \mathbf{r}'$. The body and surface forces denoted by $\mathbf{R}$ and $\mathbf{F}$, respectively, are those forces in the system which are not derivable from a potential function; and they are regarded as known functions of the time, t, and of their position vector, $\mathbf{r}'$.

We may restate Hamilton's principle somewhat more completely in the following terms: Of all the admissible displacements, between specified initial and final conditions at times t_0 and t_1, which satisfy the prescribed displacement boundary conditions of the system, the displacements which also satisfy equilibrium and the stress boundary conditions are selected by the extremum principle (2–46). The Euler differential equations of (2–46) are the equations of equilibrium of the system. In addition, the stress boundary conditions are produced as a by-product of Eq. 2–46.

2–5 A SIMPLE APPLICATION OF HAMILTON'S PRINCIPLE

Let us take as an illustration of Hamilton's principle its application to the small motions of a restrained elastic body subjected to body forces, surface forces, and heating. We may adapt the general form of Eq. 2–46 to this case by putting

$$\tau = \frac{1}{2} \iiint_V \rho \, \frac{d\mathbf{r}}{dt} \cdot \frac{d\mathbf{r}}{dt} \, dV \tag{2-47}$$

$$U = \iiint_V F_0 \, dV \tag{2-48}$$

$$\delta W = \iiint_V \mathbf{R} \cdot \delta\mathbf{r} \, dV + \iint_S \mathbf{F} \cdot \delta\mathbf{r} \, dS \tag{2-49}$$

where F_0 is given by Eq. 2–42. If we carry out the variation of $\int_{t_0}^{t_1} \tau \, dt$, we have

$$\delta \int_{t_0}^{t_1} \tau \, dt = \int_{t_0}^{t_1} \left[\frac{1}{2} \iiint_V \rho \, \frac{d\mathbf{r}}{dt} \cdot \frac{d \, \delta\mathbf{r}}{dt} \, dV + \frac{1}{2} \iiint_V \rho \, \frac{d \, \delta\mathbf{r}}{dt} \cdot \frac{d\mathbf{r}}{dt} \, dV \right] dt$$

$$\delta \int_{t_0}^{t_1} \tau \, dt = \left[\iiint_V \rho \, \frac{d\mathbf{r}}{dt} \cdot \delta\mathbf{r} \, dV \right]_{t_0}^{t_1} - \int_{t_0}^{t_1} \iiint_V \rho \, \frac{d^2\mathbf{r}}{dt^2} \cdot \delta\mathbf{r} \, dV \, dt$$

Since t_0 and t_1 are the initial and final values of t, the first term may be omitted; and we have

$$\delta \int_{t_0}^{t_1} \tau \, dt = - \int_{t_0}^{t_1} \iiint_V \rho \, \frac{d^2\mathbf{r}}{dt^2} \cdot \delta\mathbf{r} \, dV \, dt \tag{2-50}$$

The variation of $\int_{t_0}^{t_1} U \, dt$ may be expressed in the following terms:

$$\delta \int_{t_0}^{t_1} U \, dt = \int_{t_0}^{t_1} \iiint_V \left[\frac{\partial F_0}{\partial \epsilon_x} \, \delta\epsilon_x + \frac{\partial F_0}{\partial \epsilon_y} \, \delta\epsilon_y + \frac{\partial F_0}{\partial \epsilon_z} \, \delta\epsilon_z + \frac{\partial F_0}{\partial \gamma_{xy}} \, \delta\gamma_{xy} \right.$$

$$\left. + \frac{\partial F_0}{\partial \gamma_{xz}} \, \delta\gamma_{xz} + \frac{\partial F_0}{\partial \gamma_{yz}} \, \delta\gamma_{yz} \right] dV \, dt \tag{2-51}$$

Substituting Eqs. 2–49, 2–50, and 2–51 into the general equation (2–46) yields the following *variational equation of motion*:

$$\iiint_V \rho \frac{d^2\mathbf{r}}{dt^2} \cdot \delta\mathbf{r} \, dV + \iiint_V \left[\frac{\partial F_0}{\partial \epsilon_x} \delta\epsilon_x + \frac{\partial F_0}{\partial \epsilon_y} \delta\epsilon_y + \frac{\partial F_0}{\partial \epsilon_z} \delta\epsilon_z + \frac{\partial F_0}{\partial \gamma_{xy}} \delta\gamma_{xy} \right.$$

$$\left. + \frac{\partial F_0}{\partial \gamma_{xz}} \delta\gamma_{xz} + \frac{\partial F_0}{\partial \gamma_{yz}} \delta\gamma_{yz} \right] dV - \iiint_V \mathbf{R} \cdot \delta\mathbf{r} \, dV - \iint_S \mathbf{F} \cdot \delta\mathbf{r} \, dS = 0$$

$$(2\text{–}52)$$

Since the body is restrained and the elastic displacements are small, we put

$$\delta\mathbf{r} = \delta u \mathbf{i} + \delta v \mathbf{j} + \delta w \mathbf{k} \tag{2–53}$$

and

$$\frac{d^2\mathbf{r}}{dt^2} = \frac{\partial^2 u}{\partial t^2} \mathbf{i} + \frac{\partial^2 v}{\partial t^2} \mathbf{j} + \frac{\partial^2 w}{\partial t^2} \mathbf{k} \tag{2–54}$$

Reducing Eq. 2–52 by partial integrations, we obtain finally,

$$\iiint_V \rho \left(\frac{\partial^2 u}{\partial t^2} \delta u + \frac{\partial^2 v}{\partial t^2} \delta v + \frac{\partial^2 w}{\partial t^2} \delta w \right) dV$$

$$+ \iint_S \left\{ \left[\frac{\partial F_0}{\partial \epsilon_x} \cos(x, n) + \frac{\partial F_0}{\partial \gamma_{yx}} \cos(y, n) + \frac{\partial F_0}{\partial \gamma_{zx}} \cos(z, n) \right] \delta u \right.$$

$$+ \left[\frac{\partial F_0}{\partial \gamma_{xy}} \cos(x, n) + \frac{\partial F_0}{\partial \epsilon_y} \cos(y, n) + \frac{\partial F_0}{\partial \gamma_{zy}} \cos(z, n) \right] \delta v$$

$$+ \left. \left[\frac{\partial F_0}{\partial \gamma_{xz}} \cos(x, n) + \frac{\partial F_0}{\partial \gamma_{yz}} \cos(y, n) + \frac{\partial F_0}{\partial \epsilon_z} \cos(z, n) \right] \delta w \right\} dS$$

$$- \iiint_V \left\{ \left[\frac{\partial}{\partial x} \frac{\partial F_0}{\partial \epsilon_x} + \frac{\partial}{\partial y} \frac{\partial F_0}{\partial \gamma_{yx}} + \frac{\partial}{\partial z} \frac{\partial F_0}{\partial \gamma_{zx}} \right] \delta u \right.$$

$$+ \left[\frac{\partial}{\partial x} \frac{\partial F_0}{\partial \gamma_{yx}} + \frac{\partial}{\partial y} \frac{\partial F_0}{\partial \epsilon_y} + \frac{\partial}{\partial z} \frac{\partial F_0}{\partial \gamma_{zy}} \right] \delta v$$

$$+ \left. \left[\frac{\partial}{\partial x} \frac{\partial F_0}{\partial \gamma_{xz}} + \frac{\partial}{\partial y} \frac{\partial F_0}{\partial \gamma_{yz}} + \frac{\partial}{\partial z} \frac{\partial F_0}{\partial \epsilon_z} \right] \delta w \right\} dV$$

$$- \iiint_V (X \, \delta u + Y \, \delta v + Z \, \delta w) \, dV$$

$$- \iint_S (F_x \, \delta u + F_y \, \delta v + F_z \, \delta w) \, dS = 0 \quad (2\text{–}55)$$

The coefficients of the variational quantities δu, δv, and δw under the volume and surface integrals must vanish separately. This allows us to derive three partial differential equations of motion which hold throughout the body and three boundary conditions which hold over the surface. The equations of motion are

$$\rho \frac{\partial^2 u}{\partial t^2} = \frac{\partial}{\partial x} \frac{\partial F_0}{\partial \epsilon_x} + \frac{\partial}{\partial y} \frac{\partial F_0}{\partial \gamma_{yx}} + \frac{\partial}{\partial z} \frac{\partial F_0}{\partial \gamma_{zx}} + X$$

$$\rho \frac{\partial^2 v}{\partial t^2} = \frac{\partial}{\partial x} \frac{\partial F_0}{\partial \gamma_{xy}} + \frac{\partial}{\partial y} \frac{\partial F_0}{\partial \epsilon_y} + \frac{\partial}{\partial z} \frac{\partial F_0}{\partial \gamma_{zy}} + Y \qquad (2\text{--}56)$$

$$\rho \frac{\partial^2 w}{\partial t^2} = \frac{\partial}{\partial x} \frac{\partial F_0}{\partial \gamma_{xz}} + \frac{\partial}{\partial y} \frac{\partial F_0}{\partial \gamma_{yz}} + \frac{\partial}{\partial z} \frac{\partial F_0}{\partial \epsilon_z} + Z$$

and the boundary conditions are

$$F_x = \frac{\partial F_0}{\partial \epsilon_x} \cos(x, n) + \frac{\partial F_0}{\partial \gamma_{yx}} \cos(y, n) + \frac{\partial F_0}{\partial \gamma_{zx}} \cos(z, n)$$

$$F_y = \frac{\partial F_0}{\partial \gamma_{xy}} \cos(x, n) + \frac{\partial F_0}{\partial \epsilon_y} \cos(y, n) + \frac{\partial F_0}{\partial \gamma_{zy}} \cos(z, n) \qquad (2\text{--}57)$$

$$F_z = \frac{\partial F_0}{\partial \gamma_{xz}} \cos(x, n) + \frac{\partial F_0}{\partial \gamma_{yz}} \cos(y, n) + \frac{\partial F_0}{\partial \epsilon_z} \cos(z, n)$$

When Eq. 2–41 is introduced, these results are seen to be identical with those already stated in Sec. 2–2a. Equations 2–56 correspond to Eqs. 2–21, and the boundary conditions expressed by Eqs. 2–57 correspond to those expressed by Eqs. 2–8.

It is evident from this example that Hamilton's principle depends upon the existence of a function such as F_0 having the special properties stated by Eqs. 2–41. It is also evident in this application of Hamilton's principle that the strain-displacement relations in Eq. 2–23 and the stress-strain relations in Eq. 2–24 or 2–26 are assumed to be known. They are required to construct the free energy function, F_0. The derived quantities are the differential equations of equilibrium (2–21) or (2–56) and the stress boundary conditions in Eq. 2–8 or 2–57.

2–6 LAGRANGE'S EQUATION

The systems envisioned in Secs. 2–4 and 2–5 may be said to have an infinite number of degrees of freedom, since the displacements and velocities are described as continuous functions of position and time. In

most aeroelastic applications the structure being analyzed has originally—
or may be forced by some approximate representation to have—only
a finite number n of independent degrees of freedom. This means that
there are n *generalized coordinates* $q_1(t), q_2(t), \cdots, q_n(t)$, such that the
structural displacements are completely known once the $q_i(t)$ are specified.
There exist transformation relations of the form

$$\mathbf{r}' = \mathbf{r}'(q_1, q_2, \cdots, q_n) \tag{2-58}$$

The term *generalized* is applied because the coordinates need not necessarily have dimensions of length; some may be angles or even quantities
with no direct observable physical significance at all.

The velocity of a point on a moving deformable body referred to a fixed
axis system (cf. Fig. 2–1) is

$$\frac{d\mathbf{r}'}{dt} = \sum_{i=1}^{n} \frac{\partial \mathbf{r}'}{\partial q_i} \dot{q}_i \tag{2-59}$$

This expression is a homogeneous linear function of the generalized
velocities, $\dot{q}_i$, with coefficients, $\partial \mathbf{r}'/\partial q_i$, which are functions of the displacements, q_i. When Eq. 2–59 is introduced into Eq. 2–28, we find that
the kinetic energy is a homogeneous quadratic function of the generalized
velocities,

$$\tau = \frac{1}{2} \sum_{i}^{n} \sum_{j}^{n} A_{ij} \dot{q}_i \dot{q}_j \tag{2-60}$$

where the coefficients, A_{ij}, may be functions of the generalized coordinates.

It is also apparent that the potential function, U, is a function of the
coordinates only and can be expressed in terms of the q_i alone. In fact the
functional form of this expression is

$$U = \frac{1}{2} \sum_{i}^{n} \sum_{j}^{n} B_{ij} q_i q_j \tag{2-61}$$

where the coefficients, B_{ij}, may also be functions of the generalized
coordinates.

In many important special cases, notably when the displacements are
in some sense small quantities, the A_{ij} and B_{ij} are constants. This is
necessary for the system to be linear.

The virtual work of the unconservative forces can be written also in
terms of the generalized coordinates by using relations such as

$$\delta \mathbf{r}' = \sum_{i}^{n} \frac{\partial \mathbf{r}'}{\partial q_i} \delta q_i \tag{2-62}$$

Substitution into the virtual work formula given in conjunction with Eq. 2–46 yields

$$\delta W = \sum_i^n \left[\iiint_V \mathbf{R} \cdot \frac{\partial \mathbf{r}'}{\partial q_i} \, dV + \iint_S \mathbf{F} \cdot \frac{\partial \mathbf{r}'}{\partial q_i} \, dS \right] \delta q_i$$

$$\delta W = \sum_i^n Q_i \, \delta q_i \tag{2-63}$$

In Eq. 2–63

$$Q_i = \iiint_V \mathbf{R} \cdot \frac{\partial \mathbf{r}'}{\partial q_i} \, dV + \iint_S \mathbf{F} \cdot \frac{\partial \mathbf{r}'}{\partial q_i} \, dS \tag{2-64}$$

is called the ith *generalized force* and equals the work that would be done by the force system $\mathbf{R}$ and $\mathbf{F}$ per unit displacement q_i, with all other generalized coordinates fixed. Separation of the contributions of the various coordinates is permissible because the variations are infinitesimal.

In view of the foregoing, Hamilton's principle reads

$$\int_{t_0}^{t_1} \left[\delta(\tau - U) + \sum_i^n Q_i \, \delta q_i \right] dt$$

$$= \int_{t_0}^{t_1} \sum_i^n \left\{ \frac{\partial \tau}{\partial \dot{q}_i} \delta \dot{q}_i + \left[\frac{\partial(\tau - U)}{\partial q_i} + Q_i \right] \delta q_i \right\} dt = 0 \tag{2-65}$$

An integration by parts with respect to time of the first term in braces, recognizing that $\delta q_i = 0$ at $t = t_0$ and t_1 because the state of the system is specified at these instants, leads to

$$\int_{t_0}^{t_1} \sum_i^n \left[-\frac{d}{dt}\left(\frac{\partial \tau}{\partial \dot{q}_i} \right) + \frac{\partial(\tau - U)}{\partial q_i} + Q_i \right] \delta q_i \, dt = 0 \tag{2-66}$$

Since the variations of the independent coordinates q_i may be chosen arbitrarily, Eq. 2–66 is equivalent to n differential equations of the form

$$\frac{d}{dt}\left(\frac{\partial \tau}{\partial \dot{q}_i} \right) - \frac{\partial \tau}{\partial q_i} + \frac{\partial U}{\partial q_i} = Q_i \qquad (i = 1, \cdots, n) \tag{2-67}$$

These are known as *Lagrange's equations of motion* for the mechanical system. In the terminology of the calculus of variations (Refs. 2–5 to 2–7), they are the Euler equations of the variational principle. Such equations exist whenever the state of the system can be represented in terms of a set of discrete coordinates.

Lagrange's equations are extraordinarily valuable when dealing with aeroelastic phenomena and possess far wider practical utility than the broader principle from which they spring. In an efficient, well-organized fashion, they generate both the exact equations for lumped parameter systems and approximate representations, to any desired degree of accuracy, for continuous or distributed parameter systems. They form the basis of the useful approximation technique, known as the Rayleigh-Ritz method. A typical Eq. 2–67 can be considered as a generalization of Newton's second law of motion by placing $\partial U/\partial q_i$ on the right-hand side. The force system then consists of two parts: the $-\partial U/\partial q_i$ terms describe any conservative generalized forces that have been included in the potential energy, with minus signs to account for U being the negative of the scalar potential function; and the Q_i terms express all remaining forces, particularly the unconservative ones. The acceleration terms are also divided into direct linear and angular "generalized inertia forces," contained in $d/dt(\partial \tau/\partial \dot{q}_i)$, and nonlinear inertial effects $-\partial \tau/\partial q_i$, involving products of velocities and going by such names as centrifugal, Coriolis, or gyroscopic forces.

It is not a difficult matter to broaden Hamilton's principle and Lagrange's equations to cover electrical systems, coupled electromechanical systems, or situations with heat conversion. All these matters possess at least subsidiary interest in the field of aeroelasticity.

There are certain restrictions on the comprehensive validity of what has just been presented, of which at least two deserve formal mention. The first concerns the physical constraints placed on the motion: it is required that the system be *holonomic*. Mathematically, the equations applying these constraints to the system coordinates must be algebraic. In a non-holonomic system, the same equations constitute relations among differentials which cannot be rendered exact and therefore integrated for arbitrary time intervals. This distinction is fully discussed by authors like Fox (Ref. 2-6) and Lanczos (Ref. 2–7).

Of greater practical interest is the requirement in Eqs. 2–43 and 2–46 that conditions be specified at both time t_0 and t_1. This is not characteristic of actual dynamic problems, where *initial conditions* are given at, say, $t_0 = 0$ and the motion is then to be calculated for times t thereafter. No difficulty is encountered when Lagrange's equations can be constructed, for these are differential equations which may, in principle, be integrated from instant to instant. But the question of how to handle the upper limit t_1 during direct application of Hamilton's principle is a more subtle one.

2–7   A SIMPLE APPLICATION OF LAGRANGE'S EQUATION

In a host of texts on dynamics and vibrations the single-degree-of-freedom mass on a linear spring (Fig. 2–2), or its mechanical or electrical analog, forms a first elementary illustration of how the laws of motion are

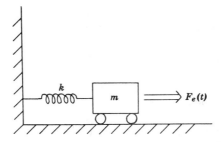

Fig. 2-2. Generalized single degree of freedom linear oscillator acted on by an external force.

applied to systems with and without external forces. The reader's experience can perhaps be counted on for supplying the details, and he may find it an interesting exercise to try to adapt Hamilton's principle and Lagrange's equations to this case.

A logical aeroelastic counterpart to the simple linear oscillator is shown in Fig. 2-3. It is the so-called *typical section* wing, a rigid airfoil section suspended in an airstream, and permitted degrees of freedom in bending

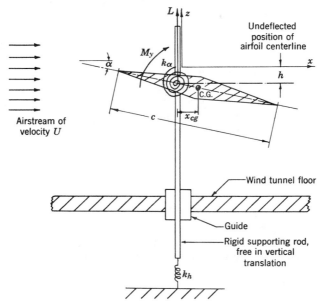

Fig. 2-3. Two-dimensional airfoil suspended in an airstream by bending and torsion springs.

and torsion by suitable suspension from two sets of springs. Both the motion and the airflow past the wing are arranged to be two-dimensional, that is, independent of the spanwise coordinate. Aerodynamically, this is done by mounting it wall-to-wall across a two-dimensional stream between plane, parallel sidewalls. With all the flexibility in "massless" springs, this is a true lumped parameter system. Only two generalized coordinates are needed to specify the instantaneous position fully; natural choices for these are the downward vertical displacement h of the line of attachment of the springs (*elastic axis*) and the leading-edge-up angular rotation α about this line. Both are measured from the unstrained positions of the corresponding springs, and gravity is neglected.

Let it be assumed that α and h/c are small, c being the wing chord, and also that the thickness ratio (ratio of maximum airfoil depth to c) is less than about 10%. x is a chordwise coordinate taken positive aft from the elastic axis, and z is normal to the chord in the plane of motion. The scalar equivalents of the transformation equation (2–58) then read

$$u = 0, \quad v = 0, \quad w = -h - x\alpha \qquad (2\text{–}68)$$

provided that the cubes of small quantities are negligible. If the bending and torsion springs have total stiffnesses k_h and k_α, respectively, the potential energy of strain is

$$U = \tfrac{1}{2}k_h h^2 + \tfrac{1}{2}k_\alpha \alpha^2 \qquad (2\text{–}69)$$

This is just the energy in Eqs. 2–67 associated with the conservative internal forces of the system. Equation 2–69 contains no cross-product term ($B_{12} = 0$ in Eq. 2–61), so the system is said to have no *static* and *elastic coupling* between the degrees of freedom.

In a similar way, the kinetic energy can be written as

$$\tau = \tfrac{1}{2}\int_{\text{chord}} \dot{w}^2 \, dm = \tfrac{1}{2}\int_{\text{chord}} (\dot{h} + x\dot{\alpha})^2 \, dm = \tfrac{1}{2}m\dot{h}^2 + S_\alpha \dot{h}\dot{\alpha} + \tfrac{1}{2}I_\alpha \dot{\alpha}^2$$

where $(2\text{–}70)$

$m =$ total mass,
$S_\alpha = mx_{cg} =$ static unbalance of wing about its elastic axis,
$I_\alpha =$ mass moment of inertia of wing about its elastic axis.

In the absence of air, the wing constitutes a conservative system. Choosing $q_1 = h$ and $q_2 = \alpha$, straightforward insertion of Eqs. 2–69 and 2–70 into Eqs. 2–67 yields

$$m\ddot{h} + k_h h + S_\alpha \ddot{\alpha} = 0$$
$$S_\alpha \ddot{h} + I_\alpha \ddot{\alpha} + k_\alpha \alpha = 0 \qquad (2\text{–}71)$$

These relations are useful for studying the undamped free vibrations, with special reference to the influence of the *inertia coupling*, S_α, upon them. The system, defined to be the rigid wing plus suspending springs, becomes unconservative in the presence of the airstream, which may be regarded as an inexhaustible source of energy capable of doing positive or negative work on the oscillating wing. Leaving small viscous shear stresses in the boundary layer out of consideration, the unconservative loads are distributed over the wing area and consist of pressures $p_U(x, t)$ and $p_L(x, t)$ on the upper and lower surfaces, respectively. If cubes of small quantities are again neglected, the net effect is a force per unit area

$$F_z = \Delta p = p_L - p_U \tag{2-72}$$

acting vertically upward in Fig. 2–3. Applying Eq. 2–63, in which we make use of only the surface integral term, and assuming two-dimensional flow characteristics, one obtains for the virtual work

$$\delta W = l \int_{\text{chord}} F_z \, \delta w \, dx$$

$$\delta W = -\delta h l \int_{\text{chord}} \Delta p \, dx - \delta \alpha l \int_{\text{chord}} \Delta p x \, dx \tag{2-73}$$

$$\delta W = Q_1 \, \delta h + Q_2 \, \delta \alpha$$

where l is the wing span.

Aeronautical engineers will recognize these generalized forces as the *lift* L (positive upward) and *pitching moment* M_y (positive leading-edge-up) about the elastic axis:

$$Q_1 = -L = -l \int_{\text{chord}} \Delta p \, dx \tag{2-74}$$

$$Q_2 = M_y = -l \int_{\text{chord}} \Delta p x \, dx \tag{2-75}$$

In Chap. 4 the relationships are discussed between aerodynamic loads, such as lift and moment, and the motions which give rise to them. Suffice it to say here that L and M_y are usually linear functions of α, h, and certain of their time derivatives. If there is some mean or initial *angle of attack* α_0 between the airstream and the zero-lift line of the airfoil in Fig. 2–3 when the torsion spring is in its unstrained position, L and M_y will have increments added to them which depend linearly on α_0. The existence of camber would also add another term to M_y. These loads are independent of time, however, and linearity permits one to separate out their (constant) contributions to h and α.

When Eqs. 2–69, 2–70, 2–74, and 2–75 are gathered together into Lagrange's equations (2–67), the motion of the unconservative system is found to be governed by

$$m\ddot{h} + k_h h + S_\alpha \ddot{\alpha} = -L \qquad (2\text{-}76a)$$

$$S_\alpha \ddot{h} + I_\alpha \ddot{\alpha} + k_\alpha \alpha = M_y \qquad (2\text{-}76b)$$

The introductory sections of Chap. 6 review the applications of Eqs. 2–76 to aeroelasticity. It is noted that much can be learned about the fundamental nature of the majority of aeroelastic phenomena without going far beyond this simple typical section wing.

REFERENCES

2–1. Timoshenko, S., and J. N. Goodier, *Theory of Elasticity*, McGraw-Hill Book Company, New York, 1951.
2–2. Brand, Louis, *Vector and Tensor Analysis*, John Wiley and Sons, New York, 1947.
2–3. Love, A. E. H., *The Mathematical Theory of Elasticity*, Dover Publications, New York, 1944.
2–4. Morse, P. M., and Herman Feshbach, *Methods of Theoretical Physics*, Vols. I and II, McGraw-Hill Book Company, New York, 1953.
2–5. Bliss, G. A., *Calculus of Variations*, Open Court, LaSalle, Ill., 1925.
2–6. Fox, C., *An Introduction to the Calculus of Variations*, Oxford University Press, London, 1950.
2–7. Lanczos, C., *The Variational Principles of Mechanics,* University of Toronto Press, Toronto, 1949.

3

AEROELASTIC
EQUATIONS AND
THEIR SOLUTIONS

3–1 INTRODUCTION

There are certain unifying features common to all aeroelastic problems
which provide a convenient framework for introducing and classifying
the entire subject. These features include the casting of the aeroelastic
equations in operator form and the generalized solution of such operator
equations. The present chapter is concerned with these features, and its
purpose is to present in concise form several methods of solution which
are commonly employed in aeroelastic analyses.

3–2 AEROELASTIC OPERATORS AND THEIR MANIPULATION

The concept of aeroelastic operators has proven useful, and an excellent
treatment can be found, among other places, in Chap. 11 of Fung's book
(Ref. 1–1). Such operators may form the basis of every analytical treatment
of aeroelasticity. Consider, for example, the spring appearing in Fig. 2–2.
As with any point on a linear elastic structure, there is a simple functional
relationship between the force F on the spring and the resulting displacement x from its unstrained length.

$$F = Kx \qquad (3\text{–}1)$$

The spring constant K may be regarded as a *structural operator*. Similarly, we have mentioned in Chap. 2 the functional relationship between the steady-state lift, L_0, on the two-dimensional wing in Fig. 2–3 and its angle of attack, α_0.

$$L_0 = (a_0 q S)\alpha_0 \tag{3-2}$$

Here $S = lc$ is the plan area, a_0 is the so-called two-dimensional *lift-curve slope*, and $q = \dfrac{\rho_\infty U^2}{2}$ is the flight *dynamic pressure*, ρ_∞ being the density in the undisturbed airstream. The quantity $(a_0 q S)$ serves as an *aerodynamic operator*, which transforms the angle of attack of the wing into the lift force which acts on it.

We can define a number of operators which may be employed as building blocks in the synthesis of aeroelastic problems. Whereas in Eqs. 3–1 and 3–2 we dealt with discrete forces, the majority of aeroelastic problems are concerned with continuous elastic bodies and with forces which are distributed over their surface or throughout their volume. The functional relation between the generalized displacement q, at a specified point and in a specified direction, and a distributed generalized force Q can be expressed symbolically by

$$Q = \mathscr{S}(q) \tag{3-3}$$

where $\mathscr{S}$ is a structural operator. We can conceive also of an inverse structural operator which we denote by $\mathscr{S}^{-1}$. Then the symbolic relationship,

$$q = \mathscr{S}^{-1}(Q) \tag{3-4}$$

represents one of the solutions, q, of Eq. 3–3. Aeroelastic operators may be algebraic, differential, or integral operators, or combinations thereof. For example, the structural operator $\mathscr{S}$ may be either a differential or integral operator or even a constant factor. If it is a differential operator, Eq. 3–4 represents an "indefinite" solution of Eq. 3–3 in much the same way that $\int f(x)\,dx$ is used to denote an indefinite function $y(x)$ for which $y'(x) = f(x)$. This indefiniteness arises because the inverse of a given differential operator can have several values, depending on the character of the boundary conditions which are introduced. This ambiguity is removed when the boundary conditions are specified, the simplest case being that of homogeneous boundary conditions.

The forces in an aeroelastic problem are of three principal types: aerodynamic, inertial, and external disturbance forces, represented symbolically by Q_A, Q_I, and Q_D, respectively. The total force in Eqs. 3–3 and 3–4 is, therefore, the sum of these three; and we may write symbolically

$$Q_A + Q_I + Q_D = \mathscr{S}(q), \qquad q = \mathscr{S}^{-1}(Q_A + Q_I + Q_D) \tag{3-5}$$

The disturbance force Q_D is usually assumed to be known explicitly, whereas the aerodynamic and inertial forces depend in some manner on the displacement and motion of the system.

When a three-dimensional body moves through the atmosphere, aerodynamic forces act over its surface. If the body is deformed, there is a change in the magnitude and distribution of these surface forces. We represent the functional relation between the deformation and the change in surface forces produced by the deformation with the symbolic forms

$$Q_A = \mathscr{A}(q), \qquad q = \mathscr{A}^{-1}(Q_A) \tag{3-6}$$

where $\mathscr{A}$ is an aerodynamic operator and $\mathscr{A}^{-1}$ its inverse.

In dynamic aeroelastic phenomena, there exist certain unique relations between the displacement of the system and the inertial forces which arise as a result of the time variation of this displacement. These relations can be expressed symbolically by

$$Q_I = \mathscr{I}(q), \qquad q = \mathscr{I}^{-1}(Q_I) \tag{3-7}$$

where $\mathscr{I}$ is the inertial operator and $\mathscr{I}^{-1}$ its inverse.

When Eqs. 3–6 and 3–7 are introduced into Eq. 3–5, we obtain the fundamental operator equations of aeroelasticity.

$$\mathscr{A}(q) + \mathscr{I}(q) + Q_D = \mathscr{S}(q) \tag{3-8}$$

$$q = \mathscr{S}^{-1}[\mathscr{A}(q) + \mathscr{I}(q) + Q_D] \tag{3-9}$$

Equations 3–8 and 3–9 may be employed alternatively as a basis for forming aeroelastic problems; the choice depends on the exact nature of the problem to be solved.

Aeroelastic operators may, in general, be nonlinear, and the remarks up to this point apply regardless of whether they are linear or nonlinear. However, in the great majority of problems, linearized operators are employed. In this case, we may write Eqs. 3–8 and 3–9 in the respective forms

$$(\mathscr{S} - \mathscr{A} - \mathscr{I})(q) = Q_D \tag{3-10}$$

$$q = \mathscr{S}^{-1}(\mathscr{A} + \mathscr{I})(q) + \mathscr{S}^{-1}(Q_D) \tag{3-11}$$

(a) A simple illustration of aeroelastic operators

The nature of aeroelastic operators may be simply illustrated by comparing Eq. 3–11 with the equation of aeroelastic equilibrium of a typical wing section. Suppose that the typical section of Fig. 2–3 is locked against vertical translation and permitted to deflect only in the pitching degree of freedom, α. If the initial angle of attack is α_0, the total angle of attack is

$$\alpha = \alpha_0 + \theta \tag{3-12}$$

where θ is the elastic angular deflection of the section using the same notation employed in Sec. 2–7. The aeroelastic equation of equilibrium which corresponds to the general form of Eq. 3–11 is

$$\theta = \frac{1}{K_\alpha}\left(a_0 q l c e - I_\alpha \frac{d^2}{dt^2}\right)\theta + \frac{1}{K_\alpha}(a_0 q l c e \alpha_0 + c_{\text{mAC}} q l c^2) \qquad (3\text{–}13)$$

where

a_0 = Lift curve slope of section
c_{mAC} = Pitching moment coefficient about the aerodynamic center
c = Section chord length
e = Distance from aerodynamic center to elastic axis.

It is evident, then, by comparing Eqs. 3–11 and 3–13 that, in this simple example, we have the following values for the aeroelastic operators and the disturbance force:

$$\mathscr{S} = K_\alpha, \qquad \mathscr{S}^{-1} = 1/K_\alpha$$
$$\mathscr{I} = -I_\alpha \frac{d^2}{dt^2}$$
$$\mathscr{A} = a_0 q l c e \qquad (3\text{–}14)$$
$$Q_D = a_0 q l c e \alpha_0 + c_{\text{mAC}} q l c^2$$

In the simple case of static aeroelasticity, Eq. 3–11 permits an immediate explicit solution for the displacement q in the operator form

$$q = [1 - \mathscr{S}^{-1}(\mathscr{A})]^{-1}\mathscr{S}^{-1}(Q_D) \qquad (3\text{–}15)$$

and Eq. 3–13 permits a corresponding solution for θ as follows:

$$\theta = \frac{\dfrac{1}{K_\alpha}(a_0 q l c e + c_{\text{mAC}} q l c^2)}{\left(1 - \dfrac{1}{K_\alpha}a_0 q l c e\right)} \qquad (3\text{–}16)$$

(b) Manipulation of linear aeroelastic operators

A few elementary rules govern the use of linear aeroelastic operators; and when these rules are followed, the operators may be subjected to algebraic manipulation.

Aeroelastic operators are, in general, not commutative with respect to multiplication; that is, we cannot change their order of multiplication. For example, we can indicate symbolically that the operators are not commutative by

$$\mathscr{S}^{-1}\mathscr{A} \neq \mathscr{A}\mathscr{S}^{-1} \qquad (3\text{–}17)$$

An aeroelastic operator and its inverse may, under certain conditions, be commutative. If we consider a differential structural operator $\mathscr{S}$ and its inverse $\mathscr{S}^{-1}$, we can always state that

$$\mathscr{S}\mathscr{S}^{-1} = 1 \qquad (3\text{--}18)$$

where 1 is the unity operator. However, the operators $\mathscr{S}$ and $\mathscr{S}^{-1}$ commute only under certain conditions. If we take $\mathscr{S}$ as an nth order differential operator, then the commutative law

$$\mathscr{S}\mathscr{S}^{-1} = \mathscr{S}^{-1}\mathscr{S} \qquad (3\text{--}19)$$

holds only when the function that is being operated on and its first $n - 1$ derivatives vanish at the boundary. The complete commutative properties of an operator and its inverse cannot be stated in general terms, but must be decided in each individual case. We can illustrate this latter point by referring to the structural operators $\mathscr{S}$ and $\mathscr{S}^{-1}$ for a uniform cantilever beam. It is evident that since the differential relation defining the lateral deflection w is*

$$EIw^{\text{IV}} = Z \qquad (3\text{--}20)$$

where EI is the bending stiffness and Z is the intensity of loading. The structural operator, $\mathscr{S}$, is given by

$$\mathscr{S} = EI\,\frac{d^4}{dy^4} \qquad (3\text{--}21)$$

The inverse structural operator $\mathscr{S}^{-1}$ can be derived by integration of Eq. 3–20 and application of the homogeneous boundary conditions

$$w(0) = w'(0) = w''(l) = w'''(l) = 0 \qquad (3\text{--}22)$$

This yields

$$w = \int_0^l C(y, \eta)Z(\eta)\,d\eta \qquad (3\text{--}23)$$

where $C(y, \eta)$ is an influence function defined by the differential equation

$$EIC''(y, \eta) = \eta - y \qquad (3\text{--}24)$$

in the interval $(y \leq \eta)$ and by the differential equation

$$EIC''(y, \eta) = 0 \qquad (3\text{--}25)$$

in the interval $(\eta \geq y)$. Integrating Eq. 3–24 and introducing the boundary conditions

$$C(0, \eta) = C'(0, \eta) = 0 \qquad (3\text{--}26)$$

* Superscript primes and Roman numerals denote differentiation with respect to y.

we obtain as the influence function for the range ($y \leq \eta$):

$$C(y, \eta) = \frac{y^2}{6EI}(3\eta - y), \qquad (y \leq \eta) \qquad (3\text{-}27)$$

Integrating Eq. 3–25 and evaluating the constants of integration by putting $y = \eta$ in Eq. 3–27 with the resulting requirements that

$$C(\eta, \eta) = \frac{\eta^3}{3EI}, \qquad C'(\eta, \eta) = \frac{\eta^2}{2EI} \qquad (3\text{-}28)$$

we obtain

$$C(y, \eta) = \frac{\eta^2}{6EI}(3y - \eta), \qquad (y \geq \eta) \qquad (3\text{-}29)$$

If y and η are interchanged in either Eqs. 3–27 or 3–29, the other equation is obtained. The latter fact verifies Maxwell's theorem of reciprocal deflections, which is expressed in this case by

$$C(y, \eta) = C(\eta, y) \qquad (3\text{-}30)$$

The inverse operator, $\mathscr{S}^{-1}$, may then be written as

$$\mathscr{S}^{-1}(Z) = \int_0^y \frac{\eta^2}{6EI}(3y - \eta)Z(\eta)\, d\eta + \int_y^l \frac{y^2}{6EI}(3\eta - y)Z(\eta)\, d\eta \quad (3\text{-}31)$$

Operating on Eq. 3–31 with the differential operator $\mathscr{S}$, we observe that

$$\mathscr{S}\mathscr{S}^{-1}(Z) = EI\frac{d^4}{dy^4}\int_0^l C(y, \eta)Z(\eta)\, d\eta = EI\frac{d^4w}{dy^4} = Z \qquad (3\text{-}32)$$

and therefore

$$\mathscr{S}\mathscr{S}^{-1} = 1 \qquad (3\text{-}33)$$

Similarly,

$$\mathscr{S}^{-1}\mathscr{S}(w) = \int_0^y \frac{\eta^2}{6EI}(3y - \eta)EI\frac{d^4w}{d\eta^4}\, d\eta + \int_y^l \frac{y^2}{6EI}(3\eta - y)EI\frac{d^4w}{d\eta^4}\, d\eta \quad (3\text{-}34)$$

By integration of Eq. 3–34, it becomes apparent that

$$\mathscr{S}^{-1}\mathscr{S}(w) = w \qquad (3\text{-}35)$$

and therefore

$$\mathscr{S}^{-1}\mathscr{S} = 1 \qquad (3\text{-}36)$$

Thus, in the case of a simple cantilever beam, we conclude that the structural operator and its inverse obey the commutative law

$$\mathscr{S}\mathscr{S}^{-1} = \mathscr{S}^{-1}\mathscr{S} \qquad (3\text{-}37)$$

Aeroelastic operators are associative; and we may state, for example, that

$$\mathscr{A}^{-1}(\mathscr{S}^{-1}\mathscr{A}) = (\mathscr{A}^{-1}\mathscr{S}^{-1})\mathscr{A} \qquad (3\text{–}38)$$

They are also commutative with respect to addition. For example, the first term of Eq. 3–10 remains unchanged by rearrangement of the order of addition of the operators, as follows:

$$(\mathscr{S} - \mathscr{I} - \mathscr{A})(q) = (-\mathscr{I} + \mathscr{S} - \mathscr{A})(q) = (\mathscr{S} - \mathscr{A} - \mathscr{I})(q) \quad (3\text{–}39)$$

Finally the operators satisfy the distributive law; and we may, for example, expand Eq. 3–11 to yield

$$q = \mathscr{S}^{-1}\mathscr{A}(q) + \mathscr{S}^{-1}\mathscr{I}(q) + \mathscr{S}^{-1}(Q_D) \qquad (3\text{–}40)$$

3–3 ADJOINT OPERATORS AND EQUATIONS

A question of some importance in the approximate solutions of aero-elastic problems relates to the adjoint properties of aeroelastic operators. Let us consider briefly in the present section some of the important features relating to these properties.

(a) Linear differential operators with a single independent variable

We can illustrate the adjoint properties of a linear differential operator with a single independent variable by considering the expression

$$\mathscr{P}(u) = P_0 u'' + P_1 u' + P_2 u \qquad (3\text{–}41)$$

where the primes represent differentiations with respect to the single independent variable y and where the P's are functions of y. There is a second differential expression

$$\widetilde{\mathscr{P}}(v) = Q_0 v'' + Q_1 v' + Q_2 v \qquad (3\text{–}42)$$

which is said to be adjoint to the first and related to it by means of the Lagrange identity

$$v\mathscr{P}(u) - u\widetilde{\mathscr{P}}(v) = \frac{d}{dy} R(u, v) \qquad (3\text{–}43)$$

where $R(u, v)$ is called the bilinear concomitant defined by

$$R(u, v) = P_0(u'v - uv') + (P_1 - P_0')uv$$

If we substitute Eq. 3–41 into Eq. 3–43 and integrate, we find that

$$\widetilde{\mathscr{P}}(v) = (P_0 v)'' - (P_1 v)' + P_2 v \qquad (3\text{–}44)$$

is the defining equation for the adjoint expression in terms of the elements of the original expression. In fact, we see from Eq. 3–44 that Eq. 3–42 can be constructed from Eq. 3–41 by applying the conditions that

$$Q_0 = P_0, \qquad Q_1 = 2P_0' - P_1, \qquad Q_2 = P_0'' - P_1' + P_2 \qquad (3\text{--}45)$$

The operator $\mathscr{P}$ ( ) is self-adjoint when $\mathscr{P}$ ( ) $= \widetilde{\mathscr{P}}($), that is, when the original operator is equal to its adjoint. The necessary and sufficient condition that $\mathscr{P}$ ( ) be self-adjoint is that

$$P_1 = P_0' \qquad (3\text{--}46)$$

If $\mathscr{P}$ ( ) is not self-adjoint and $P_0 \neq 0$, it is always possible to find a function $\lambda(y)$ such that $\lambda(y)\mathscr{P}$ ( ) is self-adjoint.

Equations 3–41, 3–42, and 3–45 provide the means for computing the operator which is adjoint to a given second-order differential operator. For an nth order differential operator given by

$$\mathscr{P}(u) = P_0 \frac{d^n}{dx^n}(u) \qquad (3\text{--}46a)$$

the corresponding adjoint operator is

$$\widetilde{\mathscr{P}}(v) = (-1)^n \frac{d^n}{dx^n}(P_0 v) \qquad (3\text{--}46b)$$

If we pass from differential expressions to differential equations, and if $\mathscr{P}(u)$ and $\widetilde{\mathscr{P}}(v)$ are adjoint expressions, then the linear differential equations

$$\mathscr{P}(u) = 0, \qquad \widetilde{\mathscr{P}}(v) = 0 \qquad (3\text{--}47)$$

are said to be adjoint to each other. A function $v(y)$ is said to be a multiplier of the equation $\mathscr{P}(u) = 0$ if it renders $v\mathscr{P}(u)$ exact, as follows:

$$v\mathscr{P}(u) = v(P_0 u'' + P_1 u' + P_2 u) = \frac{d}{dy}(Fu' + Gu) \qquad (3\text{--}48)$$

where F and G are functions of y. Equation 3–48 implies that

$$vP_0 = F, \qquad vP_1 = F' + G, \qquad vP_2 = G' \qquad (3\text{--}49)$$

and when we eliminate F and G from these equations we obtain

$$(P_0 v)'' - (P_1 v)' + P_2 v = \widetilde{\mathscr{P}}(v) = 0 \qquad (3\text{--}50)$$

Thus, the multipliers of $\mathscr{P}(u)$ are solutions of $\widetilde{\mathscr{P}}(v) = 0$. We see also that when v is a solution of $\widetilde{\mathscr{P}}(v) = 0$, the Lagrange identity (3–43) gives

$$v\mathscr{P}(u) = \frac{d}{dy}R(u, v) \qquad (3\text{--}51)$$

If the equation $\mathscr{P}(u) = 0$ is exact, it may be solved by quadratures. If $\mathscr{P}(u) = 0$ is not exact, $v\mathscr{P}(u) = 0$ will be exact if v is a solution of $\widetilde{\mathscr{P}}(v) = 0$. It is evident then that if any nonzero solution of $\widetilde{\mathscr{P}}(v) = 0$ is known, the general solution of $v\mathscr{P}(u) = 0$ may be obtained by quadratures. It is through these properties that the concepts of adjointness find application in the solution of aeroelastic equations. We shall have occasion to make use of these concepts in Sec. 3–5 where we illustrate the approximate solutions of nonself-adjoint aeroelastic systems.

(b) Linear differential operators with several independent variables

The same general principles just discussed apply when several independent variables are involved, although changes in notation are required. For example, the adjoint operator $\widetilde{\mathscr{L}}$ of a given partial differential operator $\mathscr{L}$ is defined by a form of Lagrange's identity analogous to Eq. 3–43 as follows (Ref. 2–4):

$$u\mathscr{L}(v) - v\widetilde{\mathscr{L}}(u) = \nabla \cdot \bar{\mathscr{P}}(u, v) \tag{3–52}$$

where $\bar{\mathscr{P}}$ is a generalized vector and ∇ is the gradient operator.

When several independent variables are involved, the most general differential operator would have the form (Ref. 2–4)

$$\mathscr{L} = D(x_1, x_2, \cdots, x_s) \frac{\partial^n}{\partial x_1{}^a \partial x_2{}^b \cdots \partial x_s{}^k}, \qquad (a + b + \cdots + k = n) \tag{3–53}$$

and its adjoint $\widetilde{\mathscr{L}}$ would be

$$\widetilde{\mathscr{L}} = (-1)^n \frac{\partial^n}{\partial x_1{}^a \partial x_2{}^b \cdots \partial x_s{}^k} [D(x_1, x_2, \cdots, x_s)] \tag{3–54}$$

(c) Adjoint integral operators

For integral operators, we define the adjoint operator by the following equation (Ref. 2–4):

$$\int_a^b v\mathscr{L}(u)\, dx - \int_a^b u\widetilde{\mathscr{L}}(v)\, dx = 0 \tag{3–55}$$

where we have assumed a one-dimensional form for simplicity. This definition follows from Eq. 3–43 when the boundary conditions on u and v are homogeneous. If we consider, for example, the integral operator

$$\mathscr{L}(u) = \int_a^b C(y, \eta)u(\eta)\, d\eta \tag{3–56}$$

and its adjoint

$$\widetilde{\mathscr{L}}(v) = \int_a^b \tilde{C}(y, \eta)v(\eta)\, d\eta \tag{3–57}$$

then the defining equation (3-55) becomes

$$\int_a^b dy \int_a^b d\eta [v(y)C(y,\eta)u(\eta) - u(y)\tilde{C}(y,\eta)v(\eta)] = 0 \qquad (3\text{-}58)$$

Interchanging variables of integration yields

$$\int_a^b dy \int_a^b d\eta \{v(\eta)u(y)[C(\eta,y) - \tilde{C}(y,\eta)]\} = 0 \qquad (3\text{-}59)$$

Since this is to hold for arbitrary u and v, we have

$$C(\eta, y) = \tilde{C}(y, \eta) \qquad (3\text{-}60)$$

Then we may write

$$\tilde{\mathscr{L}}(v) = \int_a^b C(\eta, y)v(\eta)\, d\eta \qquad (3\text{-}61)$$

Comparing Eqs. 3-56 and 3-61 we see that for $\mathscr{L} = \tilde{\mathscr{L}}$—that is, for the operator $\mathscr{L}$ to be self-adjoint—the kernel should have the symmetry property $C(\eta, y) = C(y, \eta)$. When an integral operator is derived from a self-adjoint differential operator, it is also self-adjoint and satisfies the symmetry property.

(d) Adjoint eigenvalue problems

Several useful properties can be stated which relate an eigenvalue problem to that of its adjoint. In order to illustrate these properties, suppose there is a nonself-adjoint eigenvalue problem having the form

$$(\mathscr{S} - \lambda \mathscr{A})(q) = 0 \qquad (3\text{-}62)$$

with real eigenvalues $\lambda_1, \lambda_2, \cdots, \lambda_n$. The eigenvalue problem adjoint to Eq. 3-62 is expressed

$$(\tilde{\mathscr{S}} - \lambda \tilde{\mathscr{A}})(\tilde{q}) = 0 \qquad (3\text{-}63)$$

where $\tilde{q}$ is chosen so that it satisfies the boundary conditions adjoint to those satisfied by q. Perhaps the most important mathematical connection between Eqs. 3-62 and 3-63 concerns their eigenvalues. It can be shown by the theory of adjoint equations (Ref. 2-4) that Eqs. 3-62 and 3-63 both have the same eigenvalues.

A second useful relation concerns the orthogonality between their eigenfunctions. Let us assume that two separate eigenvalues λ_i and λ_j correspond to eigenfunctions q_i and q_j, respectively, in Eq. 3-62 and to eigenfunctions $\tilde{q}_i$ and $\tilde{q}_j$, respectively, in Eq. 3-63. It can be shown (Ref. 2-4) that eigenfunctions q_i and q_j of the two equations corresponding to the two distinct eigenvalues λ_i and λ_j, respectively, enjoy the orthogonality property

$$\iiint_V q_i \tilde{q}_j \, dV = 0 \qquad (3\text{-}64)$$

whereas eigenfunctions of the same equation cannot be orthogonal; that is,

$$\iiint_V q_i q_j \, dV \neq 0, \qquad \iiint_V \tilde{q}_i \tilde{q}_j \, dV \neq 0, \quad (i \neq j) \qquad (3\text{-}65)$$

The variable of integration, V, in Eqs. 3–64 and 3–65 refers to an integration over the volume of a structural entity. The eigenfunctions q_i and their companion eigenfunctions $\tilde{q}_j$ are termed biorthogonal eigenfunctions. These two mathematical relations between adjoint eigenvalue problems are useful in applications of Galerkin's method and the Rayleigh-Ritz method to approximate solutions of aeroelastic equations.

As an illustration of the preceding discussion, let us consider the following eigenvalue differential equation which arises in the problems of sweptforward wing-bending divergence and chordwise divergence:

$$\mathscr{L}(w) = 0 \qquad (3\text{-}66)$$

where

$$\mathscr{L}(\quad) = \left\{ \frac{d^2}{d\bar{y}^2}\left[EI \frac{d}{d\bar{y}} \right] - \lambda c(\bar{y}) \right\}(\quad)$$

and where λ is a parameter. The boundary conditions are

$$\bar{y} = 0 \qquad w = 0$$

$$\bar{y} = l \qquad \frac{dw}{d\bar{y}} = \frac{d}{d\bar{y}}\left[EI \frac{dw}{d\bar{y}} \right] = 0$$

By using the definition of adjoint differential expression in the section including Eqs. 3–46a and 3–46b, the differential equation adjoint to Eq. 3–66 may be seen to be

$$\widetilde{\mathscr{L}}(\tilde{w}) = 0 \qquad (3\text{-}67)$$

with boundary conditions

$$\bar{y} = 0 \qquad \tilde{w} = \frac{d\tilde{w}}{d\bar{y}} = 0$$

$$\bar{y} = l \qquad EI \frac{d^2\tilde{w}}{d\bar{y}^2} = 0$$

where

$$\widetilde{\mathscr{L}}(\quad) = \left\{ \frac{d}{d\bar{y}}\left[EI \frac{d^2}{d\bar{y}^2} \right] + \lambda c(\bar{y}) \right\}(\quad)$$

The function $\tilde{w}$ which satisfies Eq. 3–67 is an integrating factor for the original differential equation (3–66); that is, $\tilde{w}\mathscr{L}(w)\,d\bar{y}$ is a perfect differential. Conversely, w is an integrating factor for $\widetilde{\mathscr{L}}(\tilde{w})$, and $w\widetilde{\mathscr{L}}(\tilde{w})$ is also a perfect differential. Forming the Lagrange identity (cf. Eq. 3–43), we have

$$\tilde{w}\mathscr{L}(w) - w\widetilde{\mathscr{L}}(\tilde{w}) = \frac{d}{d\bar{y}}\,R(w,\,\tilde{w}) \tag{3–68}$$

and integrating between the limits 0 and l yields

$$\int_0^l \tilde{w}\mathscr{L}(w)\,d\bar{y} - \int_0^l w\widetilde{\mathscr{L}}(\tilde{w})\,d\bar{y} = R(w,\,\tilde{w})\Big|_0^l \tag{3–69}$$

Using the definitions of $\mathscr{L}(w)$ and $\widetilde{\mathscr{L}}(\tilde{w})$ given above and performing integrations by parts, we find that

$$R(w,\,\tilde{w})\Big|_0^l = \left\{ \tilde{w}\frac{d}{d\bar{y}}\left(EI\frac{dw}{d\bar{y}}\right) - \frac{dw}{d\bar{y}}EI\frac{d\tilde{w}}{d\bar{y}} + EI\frac{d^2\tilde{w}}{d\bar{y}^2}w \right\}_0^l \tag{3–70}$$

It becomes apparent by examining the right-hand side of Eq. 3–70 that if the boundary conditions are those stated by Eqs. 3–66 and 3–67, then $R(w,\,\tilde{w})\big|_0^l = 0$. The construction of Eq. 3–70 and the requirement that $R(w,\,\tilde{w})\big|_0^l = 0$ in fact constitutes a method of deriving boundary conditions on $\tilde{w}$ which are adjoint to those on w.

Equations 3–66 and 3–67 are both homogeneous and have the trivial solutions $w = \tilde{w} = 0$ in addition to certain finite solutions for the eigenfunctions and eigenvalues. Physically, these eigenvalues and eigenfunctions correspond to divergence conditions of the wing. In those cases when $\mathscr{L}(\ ) = \widetilde{\mathscr{L}}(\ )$ and the system is therefore self-adjoint, such as the problem of torsional aeroelasticity of straight wings, it can be shown that there is an infinite denumerable number of pairs of real eigenvalues and eigenfunctions. However, in the case of nonself-adjoint systems, such as discussed above, the eigenvalues may be complex or may not exist at all. There is at present no general mathematical criterion for predicting the existence of real eigenvalues in nonself-adjoint problems, and each case must be evaluated individually.

The biorthogonal properties of the eigenfunctions can be demonstrated by supposing that ϕ_i is a solution of Eq. 3–66 for λ_i and $\tilde{\phi}_j$ is a solution of Eq. 3–67 for λ_j. Then we can write, based on Eqs. 3–66 and 3–67,

$$\frac{d^2}{d\bar{y}^2}\left[EI\frac{d\phi_i}{d\bar{y}}\right] - \lambda_i c(\bar{y})\phi_i = 0 \tag{3–71}$$

$$\frac{d}{d\bar{y}}\left[EI\frac{d^2\tilde{\phi}_j}{d\bar{y}^2}\right] + \lambda_j c(\bar{y})\tilde{\phi}_j = 0 \tag{3–72}$$

Multiplying Eq. 3–71 by $\tilde{\phi}_j$ and Eq. 3-72 by ϕ_i, integrating both from 0 to l, and subtracting the second from the first yields

$$\int_0^l \tilde{\phi}_j \frac{d^2}{d\bar{y}^2}\left[EI \frac{d\phi_i}{d\bar{y}}\right]d\bar{y} + \int_0^l \phi_i \frac{d}{d\bar{y}}\left[EI \frac{d^2\tilde{\phi}_j}{d\bar{y}^2}\right]d\bar{y}$$

$$= (\lambda_i - \lambda_j)\int_0^l \phi_i\tilde{\phi}_j c(\bar{y})\,d\bar{y} \quad (3\text{–}73)$$

The left-hand side of Eq. 3–73 is zero by virtue of an argument similar to that which was developed in connection with Eqs. 3–69 and 3–70. We conclude, therefore, that since $\lambda_i \neq \lambda_j$ we have the biorthogonality relation

$$\int_0^l \phi_i\tilde{\phi}_j c(\bar{y})\,d\bar{y} = 0 \qquad (i \neq j) \quad (3\text{–}74)$$

In the case of a self-adjoint system, this relation is seen to reduce to the familiar orthogonality condition between eigenfunctions or mode shapes.

3–4 CLASSIFICATION OF THE EQUATIONS OF AEROELASTICITY

In Sec. 3–2, the fundamental operator equations of aeroelasticity were introduced in the form of Eq. 3–8 and in the inverted form of Eq. 3–9. We may say that

$$\mathscr{S}(q) - \mathscr{A}(q) - \mathscr{I}(q) = Q_D \quad (3\text{–}75)$$

represents the most general form of aeroelasticity of dynamic systems. It includes within its scope the problems of the dynamic response of aircraft to natural and artificial gusts as well as to buffeting and forced vibrations in flight.

The homogeneous form of Eq. 3–75, obtained by putting $Q_D = 0$,

$$\mathscr{S}(q) - \mathscr{A}(q) - \mathscr{I}(q) = 0 \quad (3\text{–}76)$$

represents a more restricted class of problems in which the dynamic stability of an aeroelastic system is the foremost example. We include in this class the important problems of flutter and dynamic stability of flight vehicles.

Equation 3–75 reverts to static aeroelasticity by putting the inertia operator, $\mathscr{I}(q)$, equal to zero, which gives

$$\mathscr{S}(q) - \mathscr{A}(q) = Q_D \quad (3\text{–}77)$$

Equation 3–77 encompasses the phenomena of lifting surface static load distribution, control effectiveness, and control reversal.

Finally, the homogeneous form of Eq. 3-77,

$$\mathscr{S}(q) - \mathscr{A}(q) = 0 \qquad (3\text{-}78)$$

represents the static stability phenomenon of divergence.

3-5 SOLUTION OF THE EQUATIONS OF AEROELASTICITY

When we consider appropriate methods of solution of Eqs. 3-75 through 3-78, it must be remembered that flight vehicle structures are, more often than not, nonuniform in character. In fact, their geometrical and stiffness properties are usually known only in a numerical sense. Therefore, approximate numerical solutions will be required for the vast majority of applications. This does not imply that exact solutions are not obtainable or useful. Exact solutions of aeroelastic problems have been found in a few notable cases involving slender wings and two-dimensional aerodynamics. However, such solutions have proven more useful as guides in showing the relative influence of the various parameters than as a means of predicting the aeroelastic behavior of a given vehicle.

Most of the approximate methods of solution of the aeroelastic equations can be broken down into two steps. In the first, the space configuration of the deformed structure, which is actually an infinite-degree-of-freedom system, is approximated by an equivalent system with finite degrees of freedom. Once this initial step is taken, the equations of the continuous system are reduced to systems of simultaneous equations. These are algebraic equations in the problems of static aeroelasticity and differential equations with the independent variable time in dynamic aeroelasticity.

The second step in the finding of approximate solutions is one of solving the simultaneous equations. In the present section, we shall concentrate on some of the common methods of accomplishing the first step, that of reducing the continuous system to one of finite degrees of freedom.

(a) Direct collocation and matrix methods

One of the most useful approximate methods of solving aeroelastic equations is by direct collocation. A solution by collocation is one in which the equations are satisfied at a finite number of selected points on the structure. Let us consider, for example, the aeroelastic behavior of a fixed wing surface, as illustrated by Fig. 3-1. Equation 3-75 applied to this case takes the form

$$(\mathscr{S} - \mathscr{A} - \mathscr{I})(w) = Z_D \qquad (3\text{-}79)$$

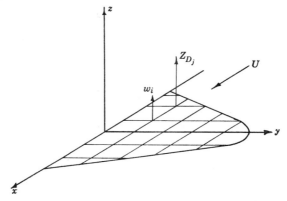

Fig. 3–1. Fixed wing surface.

where we have assumed linear operators and where w is the lateral displacement of the wing surface and Z_D is the net lateral disturbance pressure. When the wing surface is divided into a network of n points, as illustrated by Fig. 3–1, Eq. 3–79 can be satisfied at each point. Then its form changes to that of a matrix equation, as follows:

$$([\mathscr{S}] - [\mathscr{A}] - [\mathscr{I}]) \{w\} = \{Z_D\} \tag{3–80}$$

In Eq. 3–80 the square matrices $[\mathscr{S}]$, $[\mathscr{A}]$, and $[\mathscr{I}]$ are $n \times n$ matrices of structural, aerodynamic, and inertial influence coefficients, respectively. The column matrices $\{w\}$ and $\{Z_D\}$ are $n \times 1$ matrices of the deflection and the disturbance force, respectively. When the matrix form of Eq. 3–80 is formed, explicit solutions which depend on the exact nature of the problem can be obtained. By using the inverted form of the matrix of structural influence coefficients, we can rewrite Eq. 3–80 in a more suitable form for analysis:

$$\{w\} = [\mathscr{S}]^{-1}([\mathscr{A}] + [\mathscr{I}])\{w\} + [\mathscr{S}]^{-1}\{Z_D\} \tag{3–81}$$

where $[\mathscr{S}]^{-1}$ is a matrix reciprocal to $[\mathscr{S}]$.

In a dynamic aeroelastic problem, Eq. 3–81 represents a set of simultaneous ordinary differential equations with the independent variable time; but in a static problem, it represents a set of simultaneous algebraic equations.

Let us look in some detail at the formation of Eqs. 3–80 and 3–81 by applying them to the static torsional equilibrium of a slender straight wing as illustrated by Fig. 3–2. We make the fundamental assumption that the wing is a slender beam with rigid chordwise cross sections. The

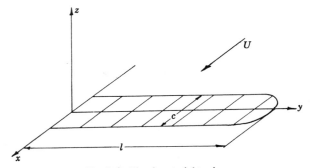

Fig. 3–2. Slender straight wing.

elastic twist angle, θ, can be expressed in terms of an integral equation of static torsional equilibrium as follows:

$$\theta(y) = \int_0^l C^{\theta\theta}(y, \eta)(ecc_l + c^2 c_{mAC})q \, d\eta \qquad (3\text{–}82)$$

We can regard the total angle of attack distribution, $\alpha(y)$, as a superposition of a rigid angle, $\alpha^r(y)$, and the elastic twist, $\theta(y)$:

$$\alpha(y) = \alpha^r(y) + \theta(y) \qquad (3\text{–}83)$$

and the corresponding local lift coefficient distribution, $c_l(y)$, a superposition of

$$c_l(y) = c_l^r(y) + c_l^e(y) \qquad (3\text{–}84)$$

where $c_l^r(y)$ is the local lift coefficient distribution resulting from rigid twist, $\alpha^r(y)$, and $c_l^e(y)$ is the local lift coefficient distribution resulting from elastic twist, $\theta(y)$. Substituting Eq. 3–84 into Eq. 3–82 yields

$$\theta(y) = q \int_0^l C^{\theta\theta}(y, \eta) ecc_l^e \, d\eta + f(y) \qquad (3\text{–}85)$$

where

$$f(y) = q \int_0^l C^{\theta\theta}(y, \eta)(ecc_l^r + c_{mAC}c^2) \, d\eta$$

In those cases where aerodynamic strip theory is admissible, we can put $a_0 \alpha(y) = c_l(y)$ and obtain the following simplified version of Eq. 3–85:

$$\theta(y) = q \int_0^l C^{\theta\theta}(y, \eta) a_0 ce\theta \, d\eta + f(y) \qquad (3\text{–}86)$$

where

$$f(y) = q \int_0^l C^{\theta\theta}(y, \eta)(a_0 ce\alpha^r + c_{mAC}c^2) \, d\eta$$

By means of Eq. 3–86, we can compute the angle of twist at a finite number of points, say n, as shown by Fig. 3–2. The elastic twist at the ith station is given by

$$\theta(y_i) = q \int_0^l C^{\theta\theta}(y_i, \eta) a_0 c e \theta \, d\eta + f(y_i), \qquad (i = 0, 1, 2, \cdots, n) \qquad (3\text{–}87)$$

An approximate numerical evaluation of the definite integral can be carried out which leads to a linear combination of the ordinates of the loading diagram, as follows:

$$\theta_i = q \sum_{j=1}^n C_{ij}^{\theta\theta} \overline{W}_j (a_0 c e \theta)_j + f_i, \qquad (i = 0, 1, 2, \cdots, n) \qquad (3\text{–}88)$$

where $C_{ij}^{\theta\theta}$ are the influence coefficients associated with the n points, and $\overline{W}_j$ are weighting numbers (Ref. 3–1) which depend on the method of numerical integration that is employed. Equation 3–88 becomes in matrix form,

$$\{\theta\} = q \left[C^{\theta\theta} \right] \left[\overline{W} \right] \left[a_0 c e \right] \{\theta\} + \{f\} \qquad (3\text{–}89)$$

where

$$\{\theta\} = \begin{bmatrix} \theta_0 \\ \theta_1 \\ \theta_2 \\ \cdot \\ \cdot \\ \cdot \\ \theta_n \end{bmatrix}, \qquad [C^{\theta\theta}] = \begin{bmatrix} 0 & 0 & 0 & \cdots & 0 \\ 0 & C_{11} & C_{12} & \cdots & C_{1n} \\ 0 & C_{21} & C_{22} & \cdots & C_{2n} \\ \cdot & \cdot & \cdot & & \cdot \\ \cdot & \cdot & \cdot & & \cdot \\ \cdot & \cdot & \cdot & & \cdot \\ 0 & C_{n1} & C_{n2} & \cdots & C_{nn} \end{bmatrix}$$

$$\left[\overline{W} \right] = \begin{bmatrix} \overline{W}_0 & 0 & 0 & 0 & \cdots & 0 \\ 0 & \overline{W}_1 & 0 & 0 & \cdots & 0 \\ 0 & 0 & \overline{W}_2 & 0 & \cdots & 0 \\ 0 & 0 & 0 & \overline{W}_3 & & \cdot \\ \cdot & \cdot & \cdot & & & \cdot \\ \cdot & \cdot & \cdot & & & \cdot \\ \cdot & \cdot & \cdot & & & \cdot \\ 0 & 0 & 0 & & \cdots & \overline{W}_n \end{bmatrix}$$

In order to examine more closely the nature of the weighting matrix, let us consider the evaluation of the definite integral

$$I = \int_a^b f(y) \, dy \qquad (3\text{–}90)$$

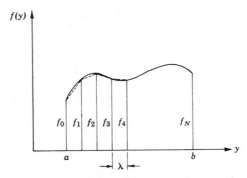

Fig. 3–3. Area represented by a sum of trapezoids.

If the ordinates to the $f(y)$ curve, $f_0, f_1, f_2, \cdots, f_n$, are known at $(n + 1)$ points within the interval (a, b), the definite integral may be expressed approximately, as we did with Eq. 3–88, by the summation

$$I = \sum_{i=0}^{n} f_i \overline{W}_i \qquad (3\text{–}91)$$

where the $\overline{W}_i$ are weighting numbers. A common and simple method of numerical integration is based upon the trapezoidal rule in which the assumption is made that the function between adjacent sets of ordinates can be replaced by straight lines as illustrated by Fig. 3–3. The definite integral can then be approximated by the sum of the areas of the trapezoids formed by the ordinates and the straight lines connecting the ordinates. If we choose $(n + 1)$ equally spaced ordinates, as illustrated by Fig. 3–3, the area, according to the trapezoidal rule, is

$$\int_a^b f(y)\, dy = \lambda(\tfrac{1}{2}f_0 + f_1 + f_2 + \cdots f_{n-1} + \tfrac{1}{2}f_n) \qquad (3\text{–}92)$$

where λ is the interval between ordinates. Thus, if the trapezoidal rule were employed in Eqs. 3–88 and 3–89 the weighting matrix would have the form of

$$\lceil \overline{W} \rfloor = \lambda \begin{bmatrix} \tfrac{1}{2} & 0 & 0 & \cdots & 0 \\ 0 & 1 & 0 & \cdots & 0 \\ 0 & 0 & 1 & & \\ \cdot & \cdot & & \cdot & \\ \cdot & \cdot & & \cdot & \\ \cdot & \cdot & & \cdot & \\ 0 & 0 & & & \tfrac{1}{2} \end{bmatrix} \qquad (3\text{–}93)$$

If the interval (a, b) is divided into an even number n of equal intervals λ, and if the curve $f(y)$ between the adjacent ordinates is approximated by parabolic arcs, the area is given by an approximate formula known as Simpson's rule, as follows:

$$\int_a^b f(y)\, dy = \frac{\lambda}{3}(f_0 + 4f_1 + 2f_2 + 4f_3 + 2f_4 + \cdots 2f_{n-2} + 4f_{n-1} + f_n)$$

$$(3\text{-}94)$$

Simpson's rule yields an exact result if the function which is being approximated is a polynomial of the third degree or less. The weighting matrix for Simpson's rule is

$$\left[W\right] = \frac{\lambda}{3}\begin{bmatrix} 1 & 0 & 0 & 0 & 0 & \cdots & 0 \\ 0 & 4 & 0 & 0 & 0 & \cdots & 0 \\ 0 & 0 & 2 & 0 & 0 & \cdots & 0 \\ 0 & 0 & 0 & 4 & 0 & \cdots & 0 \\ 0 & 0 & 0 & 0 & 2 & & \\ & \cdot & \cdot & \cdot & \cdot & & \cdot \\ & \cdot & \cdot & \cdot & \cdot & & \cdot \\ & \cdot & \cdot & \cdot & \cdot & & \cdot \\ 0 & 0 & 0 & 0 & & & 1 \end{bmatrix} \qquad (3\text{-}95)$$

Other forms of the weighting matrix are useful in aeroelasticity, but space does not permit us to delve further into this matter. The reader is referred, for example, to Ref. 3–2 for a discussion of Lagrangian interpolation functions and to Ref. 3–3 for Multhopp's quadrature formula.

(b) Collocation with generalized coordinates and assumed modes

In the collocation method described in the previous section we discussed the deformation of the structure in terms of discrete deflections at selected points on the structure. Referring to Fig. 3–1, let us now represent the deformed elastic surface in terms of a series of mode shapes

$$\gamma_1(x, y),\ \gamma_2(x, y),\ \cdots \gamma_n(x, y)$$

as follows:

$$w = \sum_{i=1}^{n} \gamma_i(x, y)q_i \qquad (3\text{-}96)$$

where q_i are generalized coordinates which are to be computed. Each mode shape is assumed to satisfy the boundary conditions of the elastic surface. In a static aeroelastic problem the q_i are constant coefficients, and in a dynamic aeroelastic problem they are functions of time. Inserting Eq. 3–96 into Eq. 3–79 yields

$$\sum_{i=1}^{n} (\mathscr{S} - \mathscr{A} - \mathscr{I})[\gamma_i(x, y)q_i] = Z_D \qquad (3\text{–}97)$$

The collocation method is applied by requiring that Eq. 3–97 be satisfied at n points over the surface. Thus we obtain the following n simultaneous equations in the variables, q_i,

$$\sum_{i=1}^{n} (\mathscr{S} - \mathscr{A} - \mathscr{I})[\gamma_i(x_j, y_j)q_i] = Z_{D_j} \qquad (j = 1, 2, \cdots, n) \quad (3\text{–}98)$$

the solution of which is substituted into Eq. 3–96 to determine the shape of the deformed surface.

There are other variations of the above process. Let us consider, for example, the *static* aeroelastic equilibrium of the wing surface of Fig. 3–1 as represented by the following simplified form of Eq. 3–79:

$$(1 - \mathscr{S}^{-1}\mathscr{A})(w) = \mathscr{S}^{-1}Z_D \qquad (3\text{–}99)$$

Suppose that for each assumed deformation shape, γ_i, we can obtain by means of an appropriate aerodynamic theory a corresponding pressure distribution $\Delta p_i(x, y)$. Then the total pressure distribution on the deformed surface can be represented by

$$\Delta p = \sum_{i=1}^{n} \Delta p_i(x, y)q_i \qquad (3\text{–}100)$$

If we assume that $\mathscr{A}(w) = \Delta p$, and if we make use of Eq. 3–96, we obtain from Eq. 3–99

$$\sum_{i=1}^{n} [\gamma_i(x, y) - \mathscr{S}^{-1}\Delta p_i(x, y)]q_i = \mathscr{S}^{-1}Z_D \qquad (3\text{–}101)$$

Applying the collocation method to Eq. 3–101 by satisfying it at n points on the surface, we obtain n algebraic equations in the variables q_i, as follows:

$$\sum_{i=1}^{n} [\gamma_i(x_j, y_j) - \mathscr{S}^{-1}\Delta p_i(x_j, y_j)]q_i = \mathscr{S}^{-1}Z_{D_j} \qquad (j = 1, 2, \cdots, n)$$
$$(3\text{–}102)$$

The final static deflection shape is obtained by solving Eq. 3–102 for the · q_i and substituting the results into Eq. 3–96.

(c) Galerkin's method

In 1915 Galerkin proposed a method of approximate solution of the boundary-value problems of mathematical physics that has wide application in aeroelasticity (Ref. 3–4). We can proceed to a solution of Eq.

3–79 by means of Galerkin's method by representing the deflection of the wing surface by a superposition of assumed modes according to Eq. 3–96 in the same manner as employed by the collocation method. The finite sum of Eq. 3–96 will, in general, not satisfy Eq. 3–79; and when we substitute the former into the latter we obtain

$$\sum_{i=1}^{n}(\mathscr{S} - \mathscr{A} - \mathscr{I})(\gamma_i q_i) - Z_D = \epsilon(x, y) \qquad (3\text{--}103)$$

where ϵ represents an error function which is small in size when Eq. 3–96 is a close approximation to the exact solution. Our task is one of selecting the q_i in such a way as to minimize ϵ. Galerkin proposed that this selection be made by imposing on the error function an orthogonality condition such that

$$\iint_S \epsilon \gamma_j \, dx \, dy = 0, \qquad (j = 1, 2, \cdots, n) \qquad (3\text{--}104)$$

This condition leads to a set of n simultaneous equations which read

$$\sum_{i=1}^{n}\left\{\iint_S (\mathscr{S} - \mathscr{A} - \mathscr{I})(\gamma_i q_i)\gamma_j \, dx \, dy\right\} = \iint_S Z_D \gamma_j \, dx \, dy \qquad (j = 1, 2, \cdots, n)$$
$$(3\text{--}105)$$

where the dependent variables are the generalized coordinates, q_i.

As a more specific illustration of Galerkin's method, let us consider a computation of the deformation under load of a slender sweptback wing free to bend without twisting, under the assumption of aerodynamic strip theory. The pertinent differential equation is

$$\frac{d^2}{d\bar{y}^2}\left[EI\frac{dw}{d\bar{y}}\right] + \lambda c(\bar{y})w = f(\bar{y}) \qquad (3\text{--}106)$$

where $f(\bar{y})$ is a loading function. Putting as a solution

$$w(\bar{y}) = \sum_{j=1}^{n} W_j(\bar{y})q_j \qquad (3\text{--}107)$$

where the $W_j(\bar{y})$ are arbitrarily assumed functions that satisfy the boundary conditions, we obtain the error function

$$\epsilon(\bar{y}) = \sum_{j=1}^{n}\left\{\frac{d^2}{d\bar{y}^2}\left[EI\frac{dW_j}{d\bar{y}}\right] + \lambda c W_j\right\}q_j - f(\bar{y}) \qquad (3\text{--}108)$$

Applying the Galerkin process of multiplying through by $W_i(\bar{y})$, integrating from 0 to l, and requiring that $\int_0^l \epsilon W_i \, d\bar{y} = 0$, we obtain the following set of simultaneous linear algebraic equations:

$$\sum_{j=1}^{n} A_{ij}q_j = B_i, \qquad (i = 1, 2, \cdots, n) \qquad (3\text{--}109)$$

where

$$A_{ij} = \int_0^l W_i \left\{ \frac{d^2}{d\bar{y}^2}\left[EI\frac{dW_j}{d\bar{y}} \right] + \overset{\bullet}{\lambda} c W_j \right\} d\bar{y}$$

$$B_i = \int_0^l f(\bar{y})W_i \, d\bar{y}$$

(3-110)

The bending deformation is obtained by solving Eqs. 3-109 simultaneously for $q_1, \cdots, q_n$ and substituting the results into Eq. 3-107.

The simultaneous Eqs. 3-109 can be reduced to diagonal form by expanding the solution in terms of biorthogonal eigenfunctions. Let us assume the existence of sets of biorthogonal eigenfunctions $\phi_1, \phi_2, \cdots, \phi_n$ and $\bar{\phi}_1, \bar{\phi}_2, \cdots, \bar{\phi}_n$ which satisfy Eqs. 3-71 and 3-72 and their associated boundary conditions, respectively. Physically, Eq. 3-71 represents the corresponding sweptforward wing. Then the bending deflection can be represented by

$$w(\bar{y}) = \sum_{j=1}^n \phi_j(\bar{y})q_j$$

(3-111)

Substituting Eq. (3-111) into Eq. (3-108) yields the error function

$$\epsilon(\bar{y}) = \sum_{j=1}^n \left\{ \frac{d^2}{d\bar{y}^2}\left[EI\frac{d\phi_j}{d\bar{y}} \right] + \lambda c \phi_j \right\} q_j - f(y)$$

(3-112)

Multiplying through by $\bar{\phi}_i(\bar{y})$, integrating from 0 to l, and requiring that $\int_0^l \epsilon \bar{\phi}_i \, d\bar{y} = 0$, yields

$$\sum_{j=1}^n \int_0^l \bar{\phi}_i \left[\frac{d^2}{d\bar{y}^2}\left(EI\frac{d\phi_j}{d\bar{y}} \right) + \lambda c \phi_j \right] d\bar{y} q_j - \int_0^l f(\bar{y})\bar{\phi}_i \, d\bar{y} = 0$$

(3-113)

Since ϕ_j satisfies Eq. 3-71, we can rewrite Eq. 3-113 as follows:

$$\sum_{j=1}^n (\lambda + \lambda_j) \int_0^l \bar{\phi}_i \phi_j c \, d\bar{y} q_j = \int_0^l f(\bar{y})\bar{\phi}_i \, d\bar{y}$$

(3-114)

Making use of the biorthogonality condition (3-74) and combining Eqs. 3-111 and 3-114, we conclude that the deflection shape is given by

$$w(\bar{y}) = \sum_{i=1}^n \phi_i(\bar{y}) \left\{ \frac{\displaystyle\int_0^l f(\bar{y})\bar{\phi}_i \, d\bar{y}}{(\lambda + \lambda_i)\displaystyle\int_0^l \phi_i \bar{\phi}_i c \, d\bar{y}} \right\}$$

(3-115)

(d) The Rayleigh-Ritz method

Among the most widely used methods of obtaining approximate solutions in aeroelasticity is the Rayleigh-Ritz method. This method was proposed originally by Rayleigh (Ref. 3–5), who applied it to eigenvalue problems. Later it was developed independently by Ritz (Ref. 3–5), who applied it to the bending of plates. It derives from a variational statement of the equations of aeroelasticity and involves the selection of a sequence of functions which are made to converge upon a solution.

Hamilton's principle forms a basis from which to apply the Rayleigh-Ritz method, and we shall illustrate how it can be applied in a specific case. Let us consider a three-dimensional restrained elastic body as discussed in Sec. 2–5. The variational principle which provides the equations of equilibrium is given by Eq. 2–46. This principle serves also as a starting point for applying the Rayleigh-Ritz method. Suppose that we choose sets of displacement functions $u_i(x, y, z)$, $v_i(x, y, z)$, and $w_i(x, y, z)$ such that they satisfy the prescribed displacement boundary conditions on the body. Then we can put as approximations

$$u = \sum_i^n u_i q_i, \qquad v = \sum_i^n v_i q_i, \qquad w = \sum_i^n w_i q_i \qquad (3\text{–}116a)$$

where the q_i are generalized coordinates. Substituting Eqs. 3–116a into Eq. 2–46 yields

$$\delta\left\{\int_{t_0}^{t_1}\left[\frac{1}{2}\sum_i^n\sum_j^n m_{ij}\dot{q}_i\dot{q}_j - U(q_1, q_2, \cdots, q_n)\right.\right.$$
$$\left.\left. + \sum_i^n q_i \iint_S (F_x u_i + F_y v_i + F_z w_i)\, dS\right] dt\right\} = 0 \qquad (3\text{–}116b)$$

where

$$\tau = \frac{1}{2}\sum_i^n\sum_j^n m_{ij}\dot{q}_i\dot{q}_j$$

is the kinetic energy and

$$m_{ij} = \iiint_V \rho(u_i u_j + v_i v_j + w_i w_j)\, dV$$

are generalized masses. It is evident that, within the framework of our assumptions, the quantities m_{ij} satisfy the reciprocal property, $m_{ij} = m_{ji}$. Carrying out the variational process leads to the equilibrium equations

$$\sum_j^n m_{ij}\ddot{q}_j + \frac{\partial U}{\partial q_i} = \iint_S (F_x u_i + F_y v_i + F_z w_i)\, dS$$
$$(i = 1, 2, \cdots, n) \qquad (3\text{–}116c)$$

where in general

$$\frac{\partial U}{\partial q_i} = \sum_j^n k_{ij} q_j$$

and the k_{ij} also enjoy the reciprocal property $k_{ij} = k_{ji}$. Simultaneous solution of the total differential equations (3–116c), subject to the prescribed initial conditions, yields explicit results for the generalized coordinates, q_i. The displacements u, v, and w are obtained by substituting the latter in Eqs. 3–116a.

If, in the previous derivation, the generalized coordinates, q_i, were selected as normal coordinates ξ_i, and the displacement functions u_i, v_i, and w_i were selected as natural mode shapes ϕ_{x_i}, ϕ_{y_i}, ϕ_{z_i}, respectively, then the form of Eq. 3–116c may be simplified considerably. Under these circumstances it is easily shown that since the natural mode shapes are orthogonal,* the kinetic energy has the form

$$\tau = \frac{1}{2} \sum_{i=1}^{n} M_i \dot{\xi}_i^2 \qquad (3\text{--}117a)$$

where

$$M_i = \iiint\limits_V (\phi_{x_i}^2 + \phi_{y_i}^2 + \phi_{z_i}^2) \rho \, dV$$

is the generalized mass. The potential energy reduces to the equally simple form

$$U = \frac{1}{2} \sum_{i=1}^{n} M_i \omega_i^2 \xi_i \qquad (3\text{--}117b)$$

where ω_i is the natural frequency of the system which corresponds to the natural mode shapes ϕ_{x_i}, ϕ_{y_i}, and ϕ_{z_i}.

When Eqs. 3–117a and 3–117b are introduced into Eq. 2–46 and the variations are performed, there is obtained

$$M_j \ddot{\xi}_j + M_j \omega_j^2 \xi_j = \Xi_j \qquad (j = 1, 2, \cdots, n) \qquad (3\text{--}118)$$

where Ξ_j is the generalized force given by

$$\Xi_j = \iint\limits_S (F_x \phi_{x_j} + F_y \phi_{y_j} + F_z \phi_{z_j}) \, dS$$

The simplicity of Eq. 3–118, which results from the elimination of inertial and elastic coupling of the coordinates, is evident by comparing it with Eq. 3–116c.

We have seen in the previous discussion that the Rayleigh-Ritz method rests upon a suitable variational statement of the equations of aeroelasticity.

---

* The orthogonality condition is expressed in the present notation as

$$\iiint\limits_V \rho(\phi_{x_i}\phi_{x_j} + \phi_{y_i}\phi_{y_j} + \phi_{z_i}\phi_{z_j}) \, dV = 0, \qquad (i \neq j)$$

Let us consider briefly the construction of such a statement from a given equilibrium equation, and in particular from an equilibrium equation which may be nonself-adjoint. Using again as a basis for discussion the two-dimensional wing surface of Fig. 3–1, we can state the variational condition from which the homogeneous form of Eq. 3–79 is derived (Ref. 3–6):

$$\delta \iint_S \tilde{w} \mathscr{L}(w)\, dS = 0 \tag{3–119}$$

where $\mathscr{L} = \mathscr{A} + \mathscr{I} - \mathscr{S}$ is an operator which is nonself-adjoint and which may be made up of a combination of differential or integral operators. $\tilde{w}$ is a solution of the homogeneous equation adjoint to $\mathscr{L}(w)$, namely

$$\tilde{\mathscr{L}}(\tilde{w}) = 0 \tag{3–120}$$

By multiplying $\tilde{w}$ times $\mathscr{L}(w)\, dS$, the quantity $\tilde{w} \mathscr{L}(w)\, dS$ is rendered exact, a necessary condition for the formation of a variational principle. Performing the variation in Eq. 3–119 yields

$$\iint_S [\mathscr{L}(w)\, \delta\tilde{w} + \tilde{w} \mathscr{L}(\delta w)]\, dS = 0 \tag{3–121}$$

By means of Lagrange's identity, Eq. 3–52, we can construct the integrated form

$$\iint_S [\tilde{w} \mathscr{L}(\delta w) - \delta w \tilde{\mathscr{L}}(\tilde{w})]\, dS = \iint_S \nabla \cdot \mathscr{P}(\tilde{w}, \delta w)\, dS \tag{3–122}$$

The right-hand side of Eq. 3–122 is zero for homogeneous boundary conditions on w and $\tilde{w}$, and we have the result that

$$\iint_S \tilde{w} \mathscr{L}(\delta w)\, dS = \iint_S \delta w \tilde{\mathscr{L}}(\tilde{w})\, dS \tag{3–123}$$

Inserting Eq. 3–123 into Eq. 3–121 yields

$$\iint_S [\mathscr{L}(w)\, \delta\tilde{w} + \tilde{\mathscr{L}}(\tilde{w})\, \delta w]\, dS = 0 \tag{3–124}$$

Since $\delta\tilde{w}$ and δw are arbitrary quantities, Eq. 3–124 can be satisfied only if

$$\mathscr{L}(w) = 0, \qquad \tilde{\mathscr{L}}(\tilde{w}) = 0 \tag{3–125}$$

We see, therefore, that the Euler equations of the variational condition (3–119) are $\mathscr{L}(w) = 0$ and its adjoint, $\tilde{\mathscr{L}}(\tilde{w}) = 0$.

In Eq. 3–119, we require that for equilibrium of the aeroelastic system a certain integral must have a stationary value. For nonself-adjoint systems, the integral is constructed as indicated by Eq. 3–119. In the case of self-adjoint systems, that is, when $\mathscr{L}(w) = \tilde{\mathscr{L}}(\tilde{w})$ and $w = \tilde{w}$, the variational condition reverts to

$$\delta \iint_S w\mathscr{L}(w)\, dS = 0 \qquad (3\text{–}126)$$

where, for a self-adjoint operator $\mathscr{L}(\quad)$, the quantity $w\mathscr{L}(w)\, dS$ is exact. In this case, the integral has a physical significance since it contains terms which describe the total energy of the system. However, the integral in Eq. 3–119 does not have this same physical significance.

In aeroelastic problems we find that the operator $\mathscr{L}(\quad)$ may frequently be nonself-adjoint. The problem of the divergence of a sweptforward wing is an example of such a system, but the torsional divergence of a straight wing is an example of a self-adjoint system. In many cases, the $\mathscr{I}(\quad)$ and $\mathscr{S}(\quad)$ portions of the operator $\mathscr{L}(\quad)$ are self-adjoint; whereas the $\mathscr{A}(\quad)$ portion is nonself-adjoint.

In applying the Rayleigh-Ritz method to self-adjoint systems—that is, to systems represented by Eq. 3–126—we select a series representation of the deformed elastic surface of the form

$$w(x, y) = \sum_{j=1}^{n} \gamma_j(x, y) q_j \qquad (3\text{–}127)$$

where $\gamma_j(x, y)$ are arbitrarily assumed functions that satisfy the boundary conditions of the surface. Substituting equation 3–127 into Eq. 3–126, carrying out the variational process, and making use of the fact that $\mathscr{L}(\quad)$ is a self-adjoint operator, we find that the generalized coordinates are specified by the homogeneous algebraic equations

$$\sum_{i=1}^{n} \left[\iint_S \gamma_j \mathscr{L}(\gamma_i)\, dS \right] q_i = 0 \; , \quad (j = 1, 2, \cdots, n) \qquad (3\text{–}128)$$

However, in the case of nonself-adjoint systems, a better approximation is obtained by assuming at the outset in addition to the series of Eq. 3–127 another series

$$\tilde{w}(x, y) = \sum_{i=1}^{n} \tilde{\gamma}_i(x, y) \tilde{q}_i \qquad (3\text{–}129)$$

where the $\tilde{\gamma}_i(x, y)$ are arbitrarily assumed functions which satisfy boundary conditions adjoint to those satisfied by $\gamma_i(x, y)$. When Eqs. 3–127 and 3–129 are substituted into Eq. 3–119 and the variational process is carried out, the following simultaneous algebraic equations are obtained which define the generalized coordinates q_i and $\tilde{q}_i$:

$$\sum_{i=1}^{n} \left[\iint_S \bar{\gamma}_j \mathscr{L}(\gamma_i)\, dS \right] q_i = 0, \qquad (j = 1, 2, \cdots, n) \qquad (3\text{--}130)$$

$$\sum_{i=1}^{n} \left[\iint_S \bar{\gamma}_i \mathscr{L}(\gamma_j)\, dS \right] \tilde{q}_i = 0, \qquad (j = 1, 2, \cdots, n) \qquad (3\text{--}131)$$

(e) Discussion of other methods and techniques of solving the aeroelastic equations

Among other ways of reducing aeroelastic problems to systems of simultaneous equations, there are the least-squares method of Boussinesq and the method of Trefftz (Ref. 3–7). Once the equations have been established, the techniques of analytical and numerical solution are far too numerous to be reviewed comprehensively. Three typical situations can be identified, however, depending on whether the problem is static, dynamic with simple harmonic time dependence, or dynamic with more general time dependence.

As pointed out previously, static problems, including those with an external forcing function, Z_D, reduce to systems of linear algebraic equations. The nominally infinite set is always replaced with a finite one by choosing a finite number of collocation points or mode shapes, and this is solved by any of the familiar techniques of matrix inversion. The homogeneous equations representing static eigenvalue problems, such as divergence, have a characteristic determinant, which can theoretically be expanded into a polynomial. More commonly the characteristic roots are extracted by any of several schemes that obtain them one by one, while successively reducing the degree of the equation. Matrix iteration is perhaps the best known of these.

When referring to iteration procedures, it is worth mentioning that they are generally very useful tools and are applicable both to self-adjoint and nonself-adjoint systems. The principal result of the work of Wielandt (Ref. 3–8) is the conclusion that the usual iteration procedures can be employed to find the eigenvalues and eigenfunctions of nonself-adjoint systems, providing these values actually exist. This discovery is significant in connection with certain static problems like divergence of swept wings and also in connection with the solution of flutter determinants. Not only can matrix iteration be applied to the algebraic equations, but in some cases iteration can also be used directly on the original integral or differential equations.

Dynamic aeroelastic problems are described by sets of simultaneous linear ordinary differential equations with time as the independent variable and constant coefficients in all but the aerodynamic terms, which may contain Duhamel superposition integrals. Two classes can be distinguished, depending on whether or not simple harmonic time-variation is involved. The former are reducible to algebraic systems, because it can be assumed that all variables are representable by $e^{i\omega t}$ multiplied with a complex constant, whose argument gives the phase lead. The factor $e^{i\omega t}$ is cancelled from all the equations, leaving a *complex* algebraic system. The aerodynamic coefficients are often very complicated functions of the circular frequency. In problems of forced motion, the right-hand sides are the complex amplitudes of the applied forces; the amplitudes of motion, responses, admittances, or system transfer functions are computed by familiar algebraic methods. Such results have many roles in aeroelasticity besides just describing the vehicle's behavior after direct application of sinusoidal forcing. They can be used to study stability by means of Nyquist plots, for analyzing the operation of automatic control systems, and for calculating response to random inputs through generalized harmonic analysis. The homogeneous equations of simple harmonic motion yield the flutter or aeroelastic stability determinant. There are at least as many methods of finding the complex eigenvalues of such determinants as there are major firms manufacturing aircraft.

When solving dynamic problems with transients, the techniques which were developed for the general treatment of response of linear systems come into play. It is usually a question of initial conditions which can be applied at the time origin $t = 0$. The method of Laplace transformation then yields the desired information. In this connection, the convolution theorem can be employed to produce algebraic equivalents for any aerodynamic terms expressed as superposition integrals. Unfortunately there are many systems where the unsteady aerodynamic theory has been worked out only for sinusoidal motion. Transients must then be determined numerically by some sort of Fourier series or Fourier integral approximation.

REFERENCES

3–1. Benscoter, S. U., and M. L. Gossard, *Matrix Methods for Calculating Cantilever-Beam Deflections*, NACA T.N. 1827, March, 1949.

3–2. *Tables of Lagrangian Interpolation Coefficients*, National Bureau of Standards, Columbia University Press, New York, 1944.

3–3. Multhopp, H., *Die Anwendung der Tragflügel Theorie auf Fragen der Flugmechanik*, Bericht S2 der Lilienthal-Gesellschaft für Luftfahrtforschung, Preisausschreiben, pp. 53–64, 1938–39.

3-4. Duncan, W. J., *Galerkin's Method in Mechanics and Differential Equations*, British A.R.C., R. and M. 1798, 1937.

3-5. Timoshenko, S. P., *Vibration Problems in Engineering*, D. Van Nostrand Co., New York, 1937.

3-6. Flax, A. H., "Aeroelastic Problems at Supersonic Speeds," *Proc. Second Intern. Aero. Conf.*, New York, 1949.

3-7. Bateman, H., *Partial Differential Equations of Mathematical Physics*, Dover Publications, New York, 1944.

3-8. Wielandt, H., *Das Iterationsverfahren Bei Nicht-Selbst-Adjungierten Linearen Eigenwert Aufgaben*, Aerodynamische Versuchsanstalt Goettingen, E.V., August 1943. (Translated by the Joint Intelligence Objectives Agency, Washington, D.C., June 1946.)

4

AERODYNAMIC
OPERATORS

4-1 INTRODUCTION: FUNDAMENTALS OF
AERODYNAMIC THEORY

Before launching into any extended exposition of aeroelastic problems, we need information on specific mathematical forms of the elastic and aerodynamic operators that appear in the governing equations. It is the objective of Chaps. 4 and 5 to set down all of those which we plan to use, along with key references to the literature. For the sake of completeness, we shall also give some additional operators whose utility will only be implied by analogy with the representative problems we are able to cover.

We cannot hope to offer here a full, connected account either of fluid mechanics or of structural theory. Fortunately, this is no longer necessary in a book on aeroelasticity. Regarding theoretical aerodynamics, which is the subject of the present chapter, an enormous amount of attention has been devoted in the last four decades to the general question of calculating steady and unsteady airloads on streamlined bodies and thin lifting surfaces in subsonic, transonic, or supersonic flight. Quite recently, theoretical results of unusual value to the aeroelastician were reviewed in the treatises by Garrick (Ref. 4–1) and Miles (Ref. 4–2), as well as in more comprehensive works on the field (Refs. 1–1, 1–2, and 1–3). Part 2 of the AGARD manual on aeroelasticity (Ref. 1–4) is wholly concerned with aerodynamic tools, and the chapters on unsteady flow in several other publications deserve special mention (Heaslet and Lomax, Ref. 4–3; Robinson and Laurmann, Ref. 4–4; Temple, Ref. 4–5; Frankl and Karpovich, Ref. 4–6; and Rodden and Revell, Ref. 4–166).

In some ways, the demands which the aeroelastician makes of the aerodynamicist are quite modest. He is satisfied with an inviscid, perfect fluid, because he is principally interested in the normal stress exerted across the surface of the elastic solid. On occasion, he can resort to the boundary-layer concept for data on heat-transfer rates and the like. He hopes for linear operators, which means that linearized small-perturbation theory yields most of the data which can really be put to practical use. On the other hand, the high vibration frequencies so often encountered mean that, of all customers, he is the one with the most regular, serious requirements in the area of *unsteady* external flows.

Bearing the foregoing prescription in mind, we now discuss briefly some gasdynamic fundamentals. Except in the special case of incompressible flow (density $\rho = $ const.), the medium is taken to be an inviscid, nonheat-conducting gas. Nearly all applications are made at altitudes where the molecular structure of the gas is not directly sensible, so that it behaves like a continuous medium. Its motion is then described by the space and time distributions of six scalar variables: density ρ, pressure p, temperature T, and three components of the velocity vector*

$$\mathbf{q} = u\mathbf{i} + v\mathbf{j} + w\mathbf{k} \qquad (4\text{-}1)$$

The former are related by an equation of state which, for atmospheric flight conditions, is the perfect gas law

$$p = R\rho T[1 + \alpha] \qquad (4\text{-}2)$$

Here $\alpha(p, T)$ is an effective mass fraction of diatomic molecules in the dissociated condition (cf. Liepmann and Roshko, Ref. 4-7, Chap. 1), and R is the "undissociated" gas constant. Dissociation can be neglected at sea-level pressures when ambient temperature T is less than about 4000°F, but dissociation gains greater importance with increasing altitude. Combinations of even lower p and higher T can lead also to ionization reactions of the gas atoms, but these are not expected to have any significant effect on the aeroelastic stability or loading of currently envisioned vehicles. Both dissociation and ionization must often be considered, however, in the viscous theory of heat transfer at high speeds, along with relaxation times of the various molecular changes which vitiate the usefulness of conventional equilibrium thermodynamics.

As readily proved, Eq. 4-2 with $\alpha = 0$ implies that the specific internal energy, e, and enthalpy, $h = e + (p/\rho)$, are functions of T only. From

---

* To conform with firmly established aerodynamic practice, we employ the same symbols here for velocity components as are used elsewhere in the book for elastic displacements. No confusion should arise, because the notational aberration is confined to Chap. 4.

this it follows that the specific heats, $c_p = (\partial h/\partial T)_p$ and $c_v = (\partial e/\partial T)_p$, are, at most, temperature-dependent, and that

$$c_p - c_v = R \qquad (4\text{--}3)$$

In most work, it has been customary to assume that c_p, c_v, and their ratio $\gamma \equiv c_p/c_v$ are all constants. This approximation is unsatisfactory in large-disturbance flows, such as those over the blunt nose of a wing or body at hypersonic speed. When viscosity is neglected, the thermal properties usually enter the theory only through the ratio γ. This is less sensitive to T changes and can be assigned some suitable average value in cases involving small or moderate perturbations from a given free-stream condition.

Another thermodynamic quantity which appears repeatedly is the ambient speed of sound, defined for small isentropic wave propagation by the equation

$$a = \sqrt{\left(\frac{\partial p}{\partial \rho}\right)_s} = \sqrt{\gamma R T} \qquad (4\text{--}4)$$

Many flows meet in whole or in part the condition of irrotationality

$$\text{curl } \mathbf{q} = 0 \qquad (4\text{--}5)$$

Equation 4–5 guarantees the existence of a velocity potential φ, such that

$$\mathbf{q} = \text{grad } \varphi \qquad (4\text{--}6)$$

By replacing three velocity components with a single scalar unknown, φ permits numerous mathematical simplifications. The majority of results of aerodynamic theory which have significant value in aeroelasticity depend on the fulfillment of Eq. 4–5, and its implications should be clearly understood. This question is discussed thoroughly in any modern book on fluid mechanics, so we need only summarize. It follows from a combination of geometrical and dynamical considerations (i.e., by simultaneous application of well-known theorems due to Stokes and Kelvin) that Eq. 4–5 holds throughout any continuous, initially irrotational flow field possessing a unique pressure-density relation. We write the relation

$$p = F(\rho) = \rho f'(\rho) - f(\rho) \qquad (4\text{--}7)$$

where the meaning of the function $f(\rho)$ will be brought out in connection with the variational principle that follows. Examples of these so-called *barotropic* or *piezotropic* fluid motions are the isothermal ($T = \text{const.}$), incompressible, and polytropic ($p \sim \rho^k$) flows. But by all odds the most important case is the one which, in thermodynamic terms, is both *adiabatic* and *reversible*. From the second law of thermodynamics, these two conditions assure that the specific entropy, s, is constant. For such *isentropic* flow of a perfect gas with fixed γ, it is easily shown that

$$p = e^{s/c_v}\rho^{\gamma} \qquad (4\text{–}8a)$$

provided the proper zero level is chosen for s. Equation 4–8a is usually recast in terms of some undisturbed reference state p_∞, ρ_∞ as

$$\frac{p}{p_\infty} = \left(\frac{\rho}{\rho_\infty}\right)^{\gamma} \qquad (4\text{–}8b)$$

Flows with appreciable viscous effects, strong curved shock waves, or heat addition are not isentropic and, therefore, cannot be analyzed by means of the velocity potential. The presence of weak shocks does not invalidate potential theory, however, since the entropy jump through such a discontinuity is proportional only to the cube of the velocity or pressure change. This observation has the fortunate consequence that steady or unsteady motions of streamlined wings and bodies can be successfully treated in terms of φ at all flight speeds short of hypersonic. The hypersonic range is defined as that where the product $M\delta$ is of the order of one or greater, $M \equiv U/a_\infty$ being the flight Mach number and δ the thickness ratio; within it, quite severely approximate aerodynamic theories must be used. Because of their increasing practical importance, some mention will be made below of aerodynamic operators for hypersonic airstreams.

Making the restriction to irrotational, barotropic flow, we can draw a parallel to the approach used for dynamics and elasticity in Chaps. 2 and 5 by giving a variational principle. Adapted by Bateman (Ref. 4–8 or 4–9) from an idea of Hargreaves (Ref. 4–10), this principle reads as follows:

$$\delta \int_{t_1}^{t_2} \iiint_V \left\{\rho\left[\frac{\partial \phi}{\partial t} + \tfrac{1}{2}\mathbf{q} \cdot \mathbf{q}\right] + f(\rho)\right\} dV \, dt = 0 \qquad (4\text{–}9)$$

Here V is the volume occupied by the fluid, t_1 and t_2 are time instants at which conditions are presumed to be known, and $f(\rho)$ is defined in Eq. 4–7. The quantities permitted independent variations are ρ and φ, $\mathbf{q}$ being related to φ through Eq. 4–6.

Without going into detail regarding the nature of the boundary conditions on the surface enclosing V, we can carry out the variations indicated in Eq. 4–9 by standard methods (Ref. 2–7) and obtain two Eulerian differential equations. The one associated with $\delta\varphi$ reads

$$\frac{\partial \rho}{\partial t} + \text{div}\,(\rho\mathbf{q}) = 0 \qquad (4\text{–}10)$$

which is recognized as the *continuity equation*, enforcing the law of conservation of mass. From the coefficient of $\delta\rho$, one finds

$$\frac{\partial\varphi}{\partial t} + \tfrac{1}{2}\mathbf{q}\cdot\mathbf{q} + f'(\rho) = 0 \qquad (4\text{-}11a)$$

Identification of Eq. 4–11a is accomplished by taking the derivative of Eq. 4–7,

$$\frac{dp}{d\rho} = \rho f''(\rho) \qquad (4\text{-}12)$$

and integrating the result between arbitrarily chosen points in the flow field at a particular instant of time to get

$$\int \frac{dp}{\rho} = f'(\rho) + F(t) \qquad (4\text{-}13)$$

When Eq. 4–13 is inserted, Eq. 4–11a becomes

$$\frac{\partial\varphi}{\partial t} + \tfrac{1}{2}\mathbf{q}\cdot\mathbf{q} + \int \frac{dp}{\rho} = F(t) \qquad (4\text{-}11b)$$

This is one of several forms of the so-called *Bernoulli equation* (the version displayed is sometimes called Kelvin's equation). It is the result of applying to the fluid Newton's second law of motion, in an inertial coordinate system and with negligible body forces acting. In fact, the vector gradient of Eq. 4–11b generates Euler's equations of conservation of momentum,

$$\frac{D\mathbf{q}}{Dt} \equiv \frac{\partial\mathbf{q}}{\partial t} + (\mathbf{q}\cdot\mathrm{grad})\mathbf{q} = -\frac{1}{\rho}\,\mathrm{grad}\,p \qquad (4\text{-}14)$$

Here D/Dt is the substantial derivative or rate of change of the quantity as a property of an individual fluid particle. The derivation of Eqs. 4–10, 4–11, and 4–14 from direct considerations of equilibrium forms an introductory part of any text on fluid mechanics (e.g., Ref. 4–7 or Lamb, Ref. 4–11).

The meaning of $F(t)$ in Eq. 4–11b is established by fixing the lower limit of the integral of dp/ρ at some reference point where $p = p_\infty$. Thus $F(t) = 0$ for flow from a reservoir, and $F(t) = U^2/2$ for uniform flight at speed U with coordinates fixed to the vehicle. For isentropic motion with constant or averaged specific heat and γ,

$$\int_{p\infty}^{p} \frac{dp}{\rho} = \frac{a^2}{\gamma-1} - \frac{a_\infty{}^2}{\gamma-1} \qquad (4\text{-}15)$$

so that Eq. 4–11*b* has the typical form

$$\frac{\partial \varphi}{\partial t} + \tfrac{1}{2}\mathbf{q} \cdot \mathbf{q} + \frac{a^2}{\gamma - 1} = \frac{U^2}{2} + \frac{a_\infty{}^2}{\gamma - 1} \tag{4-16}$$

When calculating pressures in a known isentropic flow field, under Eqs. 4–8, it is helpful to modify Eq. 4–11*b* into a *pressure coefficient* formula:

$$C_p = \frac{p - p_\infty}{\dfrac{\rho_\infty}{2} U^2}$$

$$= \frac{2}{\gamma M^2} \left\{ \left[1 + \frac{\gamma - 1}{2} M^2 \left(1 - \frac{\mathbf{q} \cdot \mathbf{q} + 2(\partial \varphi / \partial t)}{U^2} \right) \right]^{\gamma / \gamma - 1} - 1 \right\} \tag{4-17}$$

The factor in the denominator is called *dynamic pressure* $q \equiv \dfrac{\rho_\infty}{2} U^2$.

One fact having a certain physical interest is Hargreaves' interpretation (Ref. 4–10) of the Lagrange density, or integrand, of the variational principle (4–9). By means of Eqs. 4–7 and 4–11*a*, this function can be rewritten

$$\rho \left[\frac{\partial \varphi}{\partial t} + \tfrac{1}{2}\mathbf{q} \cdot \mathbf{q} \right] + f(\rho) = -\rho f'(\rho) + \rho f'(\rho) - p \equiv -p \tag{4-18}$$

Therefore, Eq. 4–9 says nothing more than that the volume integral of the pressure must have an extreme value. We have shown this result because one is at first inclined to believe that Eq. 4–9 represents the application of Hamilton's principle (see Sec. 2–4) to the field. This is not the case. Although $\dfrac{\rho}{2} \mathbf{q} \cdot \mathbf{q}$ equals the kinetic energy per unit volume, the remaining terms do not describe the potential energy of compression.

Hamilton's principle does furnish a variational equation for incompressible liquid, where the laws of thermodynamics do not affect the problem. Then the potential energy, U, is a constant or zero, unless there is an appreciable gravitational field (as with water); and Eq. 2–43 reduces to

$$\delta \int_{t_1}^{t_2} \tau \, dt = \delta \int_{t_1}^{t_2} \iiint_V \frac{\rho}{2} \mathbf{q} \cdot \mathbf{q} \, dV \, dt$$

$$= \frac{\rho}{2} \delta \int_{t_1}^{t_2} \iiint_V |\text{grad } \varphi|^2 \, dV \, dt = 0 \tag{4-19}$$

The Eulerian differential equation which corresponds to Eq. 4–19 is just Laplace's equation

$$\nabla^2 \varphi = 0 \tag{4-20}$$

This has long been known to govern both steady and unsteady irrotational motion of a liquid (Ref. 4–11).

Because of φ's role as principal dependent variable in the analysis of isentropic flows, we need a differential equation more comprehensive than 4–20. Its derivation is not difficult, if we proceed by writing Eq. 4–9 entirely in terms of φ. The integrand has already been recognized as the negative of the pressure, and p itself is expressible as follows upon substituting Eq. 4–8a into Eq. 4–11b:

$$-p = \text{const.} \left[\frac{a_\infty^2}{\gamma - 1} + \frac{U^2}{2} - \frac{\partial \varphi}{\partial t} - \tfrac{1}{2}\mathbf{q} \cdot \mathbf{q} \right]^{\gamma/\gamma - 1} \qquad (4\text{--}21)$$

Here the function $F(t)$ is specialized for a stream with velocity U and ambient speed of sound a_∞, but the restriction is nonessential. Furthermore, the constant factor in Eq. 4–21 may be dropped, since only extreme values of the integral are sought; hence, the variational principle reads

$$\delta \int_{t_1}^{t_2} \iiint_V \left[\frac{a_\infty^2}{\gamma - 1} + \frac{U^2}{2} - \frac{\partial \varphi}{\partial t} - \tfrac{1}{2} \operatorname{grad} \varphi \cdot \operatorname{grad} \varphi \right]^{\gamma/\gamma - 1} dV\, dt = 0 \quad (4\text{--}22)$$

Only φ is to be varied here. After some manipulation, one finds that the Euler-Lagrange equation is tantamount to

$$\left\{ a_\infty^2 + \frac{\gamma - 1}{2} U^2 - (\gamma - 1) \left[\frac{\partial \phi}{\partial t} + \tfrac{1}{2}\mathbf{q} \cdot \mathbf{q} \right] \right\} \nabla^2 \varphi$$

$$- \frac{\partial^2 \phi}{\partial t^2} - \frac{\partial}{\partial t}(\mathbf{q} \cdot \mathbf{q}) - \operatorname{grad} \varphi \cdot \operatorname{grad} \left(\frac{\mathbf{q} \cdot \mathbf{q}}{2} \right) = 0 \quad (4\text{--}23)$$

Equation 4–23 is the result we seek. Unfortunately, it is of the third degree in φ. Therefore, exact mathematical solutions to compressible flow problems can be found only in the most elementary situations, such as one-dimensional motion in channels. Equation 4–16 shows that the coefficient in braces of the first term is just the local speed of sound squared. Hence an equivalent form for Eq. 4–23 is

$$\nabla^2 \varphi - \frac{1}{a^2} \left[\frac{\partial^2 \varphi}{\partial t^2} + \frac{\partial}{\partial t}(\mathbf{q} \cdot \mathbf{q}) + \mathbf{q} \cdot \operatorname{grad} \left(\frac{\mathbf{q} \cdot \mathbf{q}}{2} \right) \right] = 0 \qquad (4\text{--}24)$$

In Ref. 4–1, Garrick has pointed out a physical interpretation by rewriting Eq. 4–24 as

$$\nabla^2 \varphi = \frac{1}{a^2} \frac{D_c^2 \varphi}{Dt^2} \qquad (4\text{--}25a)$$

Here the operator

$$\frac{D_c}{Dt} \equiv \frac{\partial}{\partial t} + \mathbf{q}_c \cdot \operatorname{grad} \qquad (4\text{--}25b)$$

where the subscript c means that $\mathbf{q}_c$ is to be treated as a constant with respect to the operations $\partial/\partial t$ and grad. Equation 4–25a states that the local field, observed in a coordinate system moving along with the fluid particle at velocity $\mathbf{q}$, obeys a wave equation with a propagation rate equal to local sonic speed, a.

In what follows, we shall generally assume that the fluid motion consists of a perturbation superimposed on a uniform stream U parallel to the x-axis of a rectangular, Cartesian system of coordinates. For this purpose, it is convenient to introduce the disturbance velocity potential $\varphi'(x, y, z, t)$ defined by

$$\mathbf{q} = U\mathbf{i} + \operatorname{grad} \varphi' \tag{4–26}$$

$(\mathbf{i}, \mathbf{j}, \mathbf{k})$ being the (x, y, z) unit vectors. The variational principle for φ' reads

$$\delta \int_{t_1}^{t_2} \iiint_V \left[\frac{a_\infty{}^2}{\gamma - 1} - \frac{\partial \varphi'}{\partial t} - U \frac{\partial \varphi'}{\partial x} - \tfrac{1}{2} \operatorname{grad} \varphi' \cdot \operatorname{grad} \varphi' \right]^{\gamma/\gamma - 1} dV \, dt = 0 \tag{4–27}$$

A useful form of the Eulerian differential equation for Eq. 4–27 is

$$\nabla^2 \varphi' - M^2 \left[\frac{\partial^2 \varphi'}{\partial x^2} + \frac{2}{U} \frac{\partial^2 \varphi'}{\partial x \, \partial t} + \frac{1}{U^2} \frac{\partial^2 \phi'}{\partial t^2} \right]$$

$$= M^2 \left\{ \left(\frac{\gamma + 1}{U} \right) \frac{\partial \varphi'}{\partial x} \frac{\partial^2 \varphi'}{\partial x^2} + \left(\frac{\gamma - 1}{U} \right) \frac{\partial \varphi'}{\partial x} \left[\frac{\partial^2 \varphi'}{\partial y^2} + \frac{\partial^2 \varphi'}{\partial z^2} \right] \right.$$

$$+ \frac{2}{U} \left[\frac{\partial \varphi'}{\partial y} \frac{\partial^2 \varphi'}{\partial x \, \partial y} + \frac{\partial \varphi'}{\partial z} \frac{\partial^2 \varphi'}{\partial x \, \partial z} \right] + \left(\frac{\gamma - 1}{U^2} \right) \frac{\partial \varphi'}{\partial t} \nabla^2 \varphi'$$

$$+ \frac{2}{U^2} \left[\frac{\partial \varphi'}{\partial x} \frac{\partial^2 \varphi'}{\partial x \, \partial t} + \frac{\partial \varphi'}{\partial y} \frac{\partial^2 \varphi'}{\partial y \, \partial t} + \frac{\partial \varphi'}{\partial z} \frac{\partial^2 \varphi'}{\partial z \, \partial t} \right]$$

$$\left. + \frac{1}{U^2} \operatorname{grad} \varphi' \cdot \operatorname{grad} (\tfrac{1}{2} \operatorname{grad} \varphi' \cdot \operatorname{grad} \varphi') \right\} \tag{4–28}$$

Here we have placed the terms linear in φ', which form the basis of small-perturbation subsonic and supersonic flow theory, on the left-hand side. All terms on the right are of the second degree except the last, which is of the third.

The linearized equation is evidently

$$\nabla^2 \varphi' - M^2 \left[\frac{\partial^2 \varphi'}{\partial x^2} + \frac{2}{U} \frac{\partial^2 \varphi'}{\partial x \, \partial t} + \frac{1}{U^2} \frac{\partial^2 \varphi'}{\partial t^2} \right] = 0 \tag{4–29}$$

In some applications involving accelerated or curved flight, it is more convenient to write Eq. 4–29 in coordinates fixed to the fluid at rest. The result is the wave equation of acoustics,

$$\nabla^2 \varphi' = \frac{1}{a_\infty^2} \frac{\partial^2 \varphi'}{\partial t^2} \qquad (4\text{–}30)$$

The complete Eq. 4–28 can be solved by an iteration procedure involving expansion in some small parameter, such as thickness ratio; and the first approximation is a solution of Eq. 4–29. All but the high Mach-number aerodynamic operators listed in Secs. 4–2 through 4–8 represent fully linearized solutions. It is significant for aeroelastic applications, however, that second-degree theory yields airload expressions which are still linear in the variables describing the motion, with coefficients determined by the wing thickness distribution, body shape, etc. This fact follows from considerations of symmetry. Second-degree theory has important practical implications for $M > 1$, and it is capable of accounting for the presence of weak shocks in the supersonic flow.

Compatible with the second-degree terms in Eq. 4–27 is the following approximate pressure formula, found by binomial expansion of Eq. 4–17:

$$C_p = -\left[\frac{2}{U}\frac{\partial \varphi'}{\partial x} + \frac{2}{U^2}\frac{\partial \varphi'}{\partial t} + \frac{\left(\dfrac{\partial \varphi'}{\partial x}\right)^2 + \left(\dfrac{\partial \varphi'}{\partial y}\right)^2 + \left(\dfrac{\partial \varphi'}{\partial z}\right)^2}{U^2}\right]$$
$$+ M^2\left[\frac{1}{U}\frac{\partial \varphi'}{\partial x} + \frac{1}{U^2}\frac{\partial \varphi'}{\partial t}\right]^2 \qquad (4\text{–}31)$$

The fully linearized version,

$$C_p = -\frac{2}{U}\frac{\partial \varphi'}{\partial x} - \frac{2}{U^2}\frac{\partial \varphi'}{\partial t} \qquad (4\text{–}32)$$

or

$$C_p = -\frac{2}{U^2}\frac{\partial \varphi'}{\partial t} \qquad (4\text{–}33)$$

for coordinates at rest in the air, is suitable for computing first-order airloads on nearly plane wings. When treating bodies of revolution and similar slender shapes, a consistent approximation requires that certain second-degree terms be retained (see Ward, Ref. 4–12, among other sources, for details).

In particular problems, boundary conditions must be specified on the body surface, at infinity, and sometimes on the wake or the remainder of a plane of symmetry. The infinity condition involves the uniform (or time-dependent) stream U, with disturbances decaying and propagating outward. Without viscosity, only the fluid velocity component $\partial\phi/\partial n$ normal to the instantaneous surface is fixed by the body's motion. If the body equation reads

$$F(x, y, z, t) = 0 \qquad (4\text{-}34)$$

the surface condition to be satisfied on $F = 0$ is

$$\frac{DF}{Dt} = \frac{\partial F}{\partial t} + (U\mathbf{i} + \text{grad } \varphi') \cdot \text{grad } F = 0 \qquad (4\text{-}35)$$

F assumes more explicit forms for particular shapes. The normal vector $\mathbf{n}$ to the body surface is often very nearly at right angles to the x-axis, permitting Eq. 4–34 to be written

$$n = n(x, y, t) \qquad \text{or} \qquad n = n(x, \theta, t) \qquad (4\text{-}36)$$

Here n is a direction in the y-z-plane and θ is the polar angle in that plane. When one of Eqs. 4–36 is substituted into Eq. 4–35, there results

$$\frac{\partial\varphi'}{\partial n} = \frac{\partial n}{\partial t} + \left(U + \frac{\partial\varphi'}{\partial x}\right)\frac{\partial n}{\partial x} + \left(\frac{\partial\varphi'}{\partial y} \text{ or } \frac{\partial\varphi'}{\partial\theta} \text{ term}\right) \qquad (4\text{-}37)$$

The linearized version of Eq. 4–37,

$$\frac{\partial\varphi'}{\partial n} = \frac{\partial n}{\partial t} + U\frac{\partial n}{\partial x} \qquad (4\text{-}38)$$

is adequate for many purposes.

A common example of Eq. 4–38 involves a thin lifting surface whose mean position is placed as close as possible to the x-y-plane (Fig. 4–1). Here $n \equiv z$, and the linearized boundary condition is split into parts for upper and lower surfaces:

$$\left.\begin{aligned}\frac{\partial\varphi'\,(x, y, 0+, t)}{\partial z} &= \frac{\partial z_U}{\partial t} + U\frac{\partial z_U}{\partial x} \\[2mm] \frac{\partial\varphi'\,(x, y, 0-, t)}{\partial z} &= \frac{\partial z_L}{\partial t} + U\frac{\partial z_L}{\partial x}\end{aligned}\right\} \ (x, y) \text{ in region } R_a \quad (4\text{-}39a, b)$$

Specification on the upper and lower sides of the x-y-plane at $z = \pm 0$ is permissible by virtue of a Taylor expansion in z and also by virtue of the fact that the slender wing is customarily represented by singularities distributed on this plane. Any linear problem can be separated into a flow symmetrical in z (thickness effect) and one antisymmetrical in z (camber or angle of attack). This is done by writing

$$\begin{aligned} z_U &= z_t(x, y) + z_a(x, y, t) \\ z_L &= -z_t(x, y) + z_a(x, y, t) \end{aligned} \qquad (4\text{-}40a, b)$$

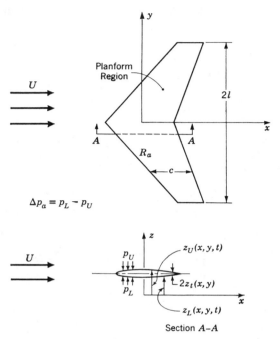

Fig. 4–1. Thin, almost plane wing performing small unsteady motions in a stream U, showing aerodynamic load Δp_a per unit area.

A physical interpretation of this separation is attempted in Fig. 4–2. The latter (load-distribution) problem is more important in aeroelasticity. Its boundary condition reads

$$\frac{\partial \varphi'\,(x, y, 0, t)}{\partial z} = \frac{\partial z_a}{\partial t} + U\,\frac{\partial z_a}{\partial x} \equiv w_a(x, y, t) \qquad (4\text{–}41)$$

Equivalent conditions are readily constructed for bodies of revolution, simple wing-body combinations, and more general slender shapes.

In connection with the calculation of aerodynamic operators for plane wings from the linearized Eqs. 4–29 and 4–41, Flax (Ref. 4–13) has given a variational principle that merits further study. His theorem involves solutions for the same planform in forward and reversed main streams. It is entirely different from principles for the complete flow field, such as Eqs. 4–9 and 4–22, and consists of integrals over the x-y-plane of Fig. 4–1.

During this introductory review of aerodynamic theory, we have emphasized the variational approach, not so much in an attempt to be different as to point out that its possibilities may have been neglected in aeroelastic applications. Very few examples of variational analyses of unsteady, compressible flow are known to the authors (see Fyfe and Klotter, Ref. 4–162, and Zartarian, Ref. 4–84). In a series of papers (Ref. 4–14 is typical), Wang and co-authors calculated some two-dimensional, *steady*, subsonic flows; and one can imagine that other investigators were discouraged by the algebraic complexity of their work. An important by-product of Wang's research was to clarify the manner of handling fields which extend outward to infinity. He thus identified a serious error in an earlier development of similar type and discovered how to avoid infinite values of integrals like that in Eq. 4–9.

In the theory of elasticity, one very fruitful procedure has been to approximate the Lagrange density of the variational principle by a series of functions that satisfy certain of the boundary conditions and to solve numerically for a set of undetermined coefficients. An analogous technique seem to hold promise for time-dependent fluid motions in finite domains, such as those produced by transient displacements of one- and two-dimensional pistons (Ref. 4–84). Wang's suggestion (Ref. 4–14) of using a conformal transformation to map the boundary into a simple geometrical shape would be helpful in the latter class of probems, and the small-perturbation concept is certain to yield great simplifications.

Even broader potential benefits are inherent in another, somewhat more visionary scheme: the exact or approximate treatment of aeroelastic situations based on a single or composite variational theorem, describing both the solid and gaseous (or liquid) phases. One interesting effort in this direction is Riparbelli's paper (Ref. 4–15). A second illustration of what can be done is found in Miles' study of fuel sloshing inside a flexible cylindrical tank (Ref. 4–16), wherein Hamilton's principle and Lagrange's equations are shown to furnish an efficient mathematical representation of a practically important type of hydroelastic phenomenon.

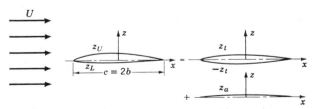

Fig. 4–2. Cross section of a thin wing or airfoil in a uniform flow of velocity U, showing the decomposition into a symmetrical shape at zero incidence and a cambered, inclined mean line which is permitted by linearization.

4-2 GENERAL FORMS OF THE OPERATORS

The archetypal aerodynamic operator used in setting up the equations of aeroelasticity is given by Eq. 3–6 as

$$Q_A = \mathscr{A}(q_i) \tag{4-42}$$

q_i being a generalized coordinate defining the motion or displacement of the elastic system, and Q_A being some sort of airload generated by q_i. For one- and two-dimensional systems whose deformations occur primarily in a single coordinate direction (i.e., normal to some reference line or surface in the structure), a convenient specification of Q_A is the pressure difference $\Delta p_a = (p_L - p_U)$ between the lower and upper surfaces. In Figs. 4–1 and 4–2, for example, $\Delta p_a(x, y, t)$ is the aerodynamic force in the positive z-direction per unit area of the x-y-plane, as it would also be if the wing were replaced by a body or wing-body combination. Since a constant z-displacement of the mean-plane of a flight vehicle causes no loading, but a change in its angle of attack does, the best way of choosing q_i is to equate it to the local streamwise incidence $\alpha(x, y)$ in steady flow and to the dimensionless upwash $w_a(x, y, t)/U$, defined by Eq. 4–41, for unsteady flow. (Note that w_a/U reduces to $(-\alpha)$ when the time dependence disappears from the boundary condition, which explains the minus sign in Eq. 4–43)*

$$q_i = \begin{cases} -\alpha(x, y) \equiv \dfrac{\partial z_a}{\partial x}(x, y), & \text{(steady flow)} \\[2ex] \dfrac{w_a(x, y, t)}{U} \equiv \left(\dfrac{\partial}{\partial x} + \dfrac{1}{U}\dfrac{\partial}{\partial t}\right) z_a(x, y, t), & \text{(unsteady flow)} \end{cases} \tag{4-43}$$

With these substitutions, we reformulate Eq. 4–42 as follows:

$$\frac{\Delta p_a}{q} = \mathscr{A}\left(\frac{w_a}{U}\right) \tag{4-44a}$$

$$\frac{w_a}{U} = \mathscr{A}^{-1}\left(\frac{\Delta p_a}{q}\right) \tag{4-44b}$$

Here

$$q \equiv \frac{\rho_\infty}{2} U^2 \tag{4-45}$$

is the dynamic pressure, introduced to make Δp_a dimensionless. $\mathscr{A}$ and its inverse are linear operators, according to either linearized or second-degree aerodynamic theory, with the thickness distribution $2z_t(x, y)$

* The structural symbol w is used in place of z_a in other chapters of this book.

(cf. Eq. 4–40) appearing as a known, time-independent modifier in the latter case. It is easy to harmonize Eqs. 4–44 with the definitions of operators used in Chaps. 2 and 5. Thus, if it is desired to signify the actual linear deflection w, one replaces $\mathscr{A}(\cdot\cdot\cdot)$ by

$$\mathscr{A}\left[\left(\frac{\partial}{\partial x}+\frac{1}{U}\frac{\partial}{\partial t}\right)(\cdot\cdot\cdot)\right];$$

the time derivative would be absent in steady-state problems and would be written $\dfrac{1}{U}\dfrac{\partial}{\partial t}(\cdot\cdot\cdot)=\dfrac{i\omega}{U}(\cdot\cdot\cdot)$ in cases of simple harmonic motion.

Because of their special significance in theories of the strip (two-dimensional flow) and lifting-line type, information is also furnished presently about aerodynamic operators that yield the lift per unit span (force normal to the flight direction, positive upward)

$$L=\int_{-b}^{+b}\Delta p_a\,dx \tag{4–46}$$

and pitching moment per unit span (positive in a nose-up sense about a spanwise axis at $x=ba$)

$$M_y=-\int_{-b}^{+b}\Delta p_a[x-ba]\,dx \tag{4–47}$$

The local chord of the lifting surface under consideration is $c\equiv 2b$, and the leading and trailing edges are located at $x=-b$ and $x=b$, respectively. L and M_y are related to two coordinates which completely describe the (small) motion of a chordwise-rigid airfoil, as follows:

$h(t)=$ displacement, positive downward, of the axis at $x=ba$ (4–48)

$\alpha(t)=$ rotation, position leading-edge upward, about the axis at x

$$=ba \tag{4–49}$$

These coordinates are illustrated in Fig. 6–5.

The lift and pitching moment per unit span are often presented in the form of dimensionless coefficients

$$c_l=\frac{L}{qc} \tag{4–50}$$

and

$$c_m=\frac{M_y}{qc^2} \tag{4–51}$$

On occasion, L and M_y are also used to denote total lift or pitching moment on a wing, tail, or an entire flight vehicle. Then their coefficients are distinguished by capital letters, as follows:

$$C_L = \frac{L}{qS} \qquad (4\text{-}52)$$

$$C_M = \frac{M_y}{qSc_R} \qquad (4\text{-}53)$$

S and c_R are the plan area (or other specified area) and some suitably chosen reference chordlength, respectively.

In Secs. 4–3 through 4–8, the various operators are listed without any effort to elaborate often long derivations which intervene between the theoretical fundamentals and their final forms. Selected references are cited, in which the reader can find all important details. The listing proceeds from cases of two- and three-dimensional steady flow (Secs. 4–3 and 4–4) to simple harmonic and transient motions of airfoils (Secs. 4–5 and 4–6), and finally to the complicated cases of three-dimensional unsteady flow (Secs. 4–7 and 4–8), where the current catalogue of operators is by no means complete. In each section the compilation goes from low subsonic to hypersonic speeds. $\mathscr{A}$ itself is displayed whenever available; otherwise, $\mathscr{A}^{-1}$ is given. Afterward, lift and moment operators are presented, along with information on loads due to a trailing edge flap in a few situations.

Most of the simple harmonic operators arise from integro-differential equations, containing a so-called *kernel function K*. These functions, which have been made dimensionless below, resemble structural influence functions because, in a loose sense, they represent the load per unit area at point (x, y) due to a unit impulse of sinusoidal upwash w_a concentrated at point (ξ, η). When only $\mathscr{A}^{-1}$ is available, the kernel function describes the upwash at (x, y) due to unit concentrated load at (ξ, η). Using the symbol ω for the circular frequency of sinusoidal motion, we can say that each K depends on four dimensionless quantities:

$$x_0 = \frac{x - \xi}{b}, \qquad y_0 = \frac{y - \eta}{b} \qquad (4\text{-}54a, b)$$

$$k = \frac{\omega b}{U} = \text{the } reduced\ frequency \qquad (4\text{-}55)$$

$$M = \frac{U}{a_\infty} = \text{the flight Mach number} \qquad (4\text{-}56)$$

Here b is an arbitrary reference length. It is taken to equal the semichord in cases of two-dimensional flow, when, of course, the dependence of K on y also disappears.

If the presence of the wing trailing edge must be accounted for during the derivation of $\mathscr{A}$, as is true for subsonic U, K may no longer be determined only by x_0, y_0 but also by the absolute locations of the points (x, y) and (ξ, η). It is then necessary to employ additional dimensionless variables of the form

$$\tilde{x}, \tilde{\xi} = \frac{x, \xi}{b} \tag{4-57a}$$

$$\tilde{y}, \tilde{\eta} = \frac{y, \eta}{b} \tag{4-57b}$$

Some remarks are in order about the question of experimental confirmation. The formulas given in Secs. 4–3 through 4–8 are purely theoretical, although adjustments to make them agree better with wind-tunnel data are often possible. Thus, steady-state measured lift coefficients, moment coefficients, and lift-curve slopes are available for an enormous variety of airfoils in a wide range of speeds. Aerodynamic operators for unsteady flow are sometimes corrected with an overall factor that assures good correlation in the limit of vanishing time dependence (e.g., $\omega = 0$), especially when strip methods are employed on a surface where properly they are not valid. Occasionally, even a phase shift is applied as a function of ω.

Regarding the accuracy of *unsteady* calculations, however, the authors are aware of few instances where reliable measurements show any *significant* disagreement, in the light of probable instrumental errors, with a theory which reasonably models the true experimental situation. One important exception involves the data of Greidanus, van de Vooren, and Bergh (Ref. 4–17) for oscillating airfoils in incompressible flow at the higher reduced frequencies. Fortunately, the parametric range where these investigators discovered large apparent differences is not often encountered in practice. A comprehensive review of unsteady airload measurements prior to 1956, with many references, including such topics as wind-tunnel wall effects at subsonic speeds, will be found in the AGARD report by Molyneux (Ref. 4–25).

Unadjusted theory usually proves unsatisfactory for controls with aerodynamic balance and for small trailing edge flaps or tabs, but here the influence of viscosity is obviously dominant. Another situation requiring special treatment arises when the mean angle of attack is so large that vibrations carry a surface into the stall range. The papers by Rainey (Ref. 4–26) and Halfman et al. (Ref. 4–27) contain extensive data and discussions of this subject and of the phenomenon of stall flutter, which is perhaps somewhat slighted in the present book. It is a continual danger on heavily loaded rotors, propellers, and rotating machinery.

With such qualifications as those just mentioned, however, the use of measured aerodynamic operators for predicting dynamic aeroelastic behavior of primary lifting surfaces cannot, within the current state of the art, be expected to yield results significantly more reliable than intelligently applied theory.

4–3   AIRFOILS IN TWO-DIMENSIONAL STEADY FLOW

(a) Subsonic flow ($M < 1$)

$$\mathscr{A} = \frac{4}{\pi\beta} \sqrt{\frac{1 - \tilde{x}}{1 + \tilde{x}}} \oint_{-1}^{1} \sqrt{\frac{1 + \tilde{\xi}}{1 - \tilde{\xi}}} \frac{1}{x_0} (\cdots) \, d\tilde{\xi} \qquad (4\text{–}58)$$

Here the Cauchy principal value of the integral must be taken because of the singularity in $1/x_0$ at $\tilde{\xi} = \tilde{x}$. Another quantity, appearing for the first time, that occurs commonly in compressible flow formulas is the *Prandtl-Glauert factor*

$$\beta = \sqrt{|1 - M^2|} \qquad (4\text{–}59)$$

Equation 4–58 and its supersonic counterpart are well-known results of thin-airfoil theory, which is discussed in such books as Ref. 4–7, Ref. 4–4, Glauert (Ref. 4–18), Pope (Ref. 4–19), Kuethe and Schetzer (Ref. 4–20), and Ferri (Ref. 4–21).

Forces and moments per unit span can be calculated by direct integration of $\Delta p_a(x)$, as in Eqs. 4–46 and 4–47. For example,

$$L = \frac{-4qb}{\beta} \int_{-1}^{1} \sqrt{\frac{1 + \tilde{\xi}}{1 - \tilde{\xi}}} \frac{w_a(\xi)}{U} \, d\xi \qquad (4\text{–}60)$$

$$M_y = \frac{-4qb^2}{\beta} \int_{-1}^{1} \left[\sqrt{1 - \tilde{\xi}^2} + a \sqrt{\frac{1 + \tilde{\xi}}{1 - \tilde{\xi}}} \right] \frac{w_a(\xi)}{U} \, d\xi \qquad (4\text{–}61)$$

Mathematical operations in Eqs. 4–58, 4–60, and 4–61 are facilitated by transforming to an angle variable θ, defined by $\tilde{\xi} = \cos\theta$.

The subsonic airfoil has one axis about which M_y is independent of changes in angle of attack, called the *aerodynamic center* (A.C.). This is the quarter-chord line, as can be seen from the vanishing of M_y when w_a is a constant and $a = -\frac{1}{2}$. The force system is generally decomposed into a moment about the A.C., M_{AC}, which for fixed q and M is a constant determined by the camber, and a lift acting at the A.C., which depends on the difference $\Delta\alpha$ between the angle of attack and its zero-lift value. From

Eqs. 4–50, 4–51, 4–60, and 4–61, the coefficients corresponding to these quantities are

$$c_{mAC} \equiv \frac{M_{AC}}{qc^2} = \frac{1}{\beta} \int_{-1}^{1} \left[-\sqrt{1 - \xi^2} + \frac{1}{2} \sqrt{\frac{1 + \xi}{1 - \xi}} \right] \frac{w_a(\xi)}{U} \, d\xi \quad (4\text{–}62)$$

$$c_l = \frac{\partial c_l}{\partial \alpha} \Delta \alpha = \frac{2\pi}{\beta} \Delta \alpha \quad (4\text{–}63a)$$

where

$$\Delta \alpha = -\frac{1}{\pi} \int_{-1}^{1} \sqrt{\frac{1 + \xi}{1 - \xi}} \frac{w_a(\xi)}{U} \, d\xi \quad (4\text{–}63b)$$

In static aeroelastic analyses, the lift-curve slope $\partial c_l / \partial \alpha$ is widely used as an aerodynamic operator, connecting the change in incidence with the incremental force produced thereby. It is often determined from tests. The subsonic value $2\pi/\beta$ usually comes out somewhat higher than what is measured, but the $1/\beta$ variation with Mach number agrees well for thin airfoils up to the critical M where sonic flow appears locally.

When the airfoil is fitted with a control surface, such as the trailing edge flap shown in Fig. 6–1, Eq. 4–58 is still useful for predicting the loadings. Without writing out all the forms explicitly, it can be stated that a flap rotation δ_f (positive trailing-edge downward) produces increments in running lift and M_{AC} that can be written

$$\Delta c_l = \frac{\partial c_l}{\partial \delta} \delta_f \quad (4\text{–}64)$$

$$\Delta c_{mAC} = \frac{\partial c_{mAC}}{\partial \delta} \delta_f \quad (4\text{–}65)$$

An aerodynamic moment H_f, positive in the same sense as δ_f, is generated about the flap hingeline due both to δ_f and the angle-of-attack change $\Delta \alpha$. Assuming linearity and omitting any contribution of airfoil camber, the hinge moment is expressed as follows:

$$H_f = qc_f^2 \left[\frac{\partial c_h}{\partial \delta} \delta_f + \frac{\partial c_h}{\partial \alpha} \Delta \alpha \right] \quad (4\text{–}66)$$

where c_f is the chordlength of the flap behind the hingeline.

To obtain loads on the full span of a rigid lifting surface whose aspect ratio is so large that three-dimensional effects are negligible, all the formulas (4–62) through (4–66) are directly applicable if the coefficients are written in capital letters. The reference area involved in their definitions is then S (S_f in the case of Eq. 4–66), replacing the area per unit span $c \equiv 2b$.

(b) Transonic flow $(M \cong 1)$

The theory of steady, two-dimensional transonic flow is essentially nonlinear, so that there exist no simple general forms for the aerodynamic operators. A recent book by Guderley (Ref. 4–22) contains a thorough treatment of this difficult analytical problem.

(c) Supersonic flow $(M > 1, \; M\delta \ll 1^*)$

$$\mathscr{A} = -\frac{4}{\beta}(\cdots) \qquad (4\text{--}67)$$

Equation 4–67 is the fully linearized result and exhibits a simple point-function relationship, under which the local loading on any area element is fixed entirely by the local incidence. It constitutes the basis of the well-known Ackeret formulas for supersonic airfoils, of which

$$\frac{\partial c_l}{\partial \alpha} = \frac{4}{\beta} \qquad (4\text{--}68)$$

$$c_{m_{AC}} = \frac{1}{\beta} \int_{-1}^{1} \tilde{x} \, \frac{w_a(\tilde{x})}{U} \, d\tilde{x} \qquad (4\text{--}69)$$

are typical. The A.C. is at midchord $(a = 0)$. Flap and hinge moment derivatives corresponding to those in Eqs. 4–64, 4–65, and 4–66 are easily worked out or can be found in books on supersonic wing theory, such as Ref. 4–21.

In contrast to the case of subsonic flow, the airfoil thickness distribution $2z_t(x)$ (see Fig. 4–2) often affects the A.C. location and chordwise load distribution to an important degree for such aeroelastic phenomena as divergence and flutter. Busemann's second-order supersonic theory or shock-expansion theory (Ref. 4–21, Chap. 7) provide the necessary corrections. They are also given in very simple form for M greater than roughly 2.5 (strictly, for $M^2 \gg 1$, $M\delta \ll 1$) by so-called piston theory (see Lighthill, Ref. 4–23, and Landahl, Ref. 4–24). The corresponding aerodynamic operator

$$\mathscr{A} = -\frac{4}{M}\left[1 + \left(\frac{\gamma + 1}{2}\right) M \frac{dz_t}{dx}\right](\cdots) \qquad (4\text{--}70)$$

is suggested by Lighthill to be accurate enough for practical purposes even up to $M\delta = 1$.

---

* In connection with subsection headings, some indications will be given of order-of-magnitude limitations on the important parameters in the range considered, although these are not intended to be mathematically complete. δ here denotes thickness ratio, but it may also be amplitude-to-chord ratio in cases of unsteady motion.

When applied to any airfoil with a closed trailing edge ($z_t(-b) = z_t(b) = 0$), neither Busemann's formula nor Eq. 4–70 causes any alteration in the lift-curve slope, Eq. 4–68. However, Eq. 4–70 shifts the A.C. forward to the point

$$a_{AC} = -\left(\frac{\gamma + 1}{8}\right) M \frac{A_w}{b^2} \qquad (4\text{--}71)$$

and gives

$$c_{m_{AC}} = \frac{1}{Mb}\left[z_a(+1) + z_a(-1) - \int_{-1}^{1} z_a(\tilde{x})\, d\tilde{x}\right]$$

$$+ \frac{\gamma + 1}{2}\left[\frac{A_w}{4b^2}\left(\frac{z_a(+1) - z_a(-1)}{b}\right) + \int_{-1}^{1} \tilde{x}\frac{dz_t}{dx}\frac{dz_a}{dx}\, d\tilde{x}\right] \qquad (4\text{--}72)$$

Here dz_a/dx has been substituted for w_a/U, and A_w is the cross-sectional area of the profile, given by the chordwise integral of $2z_t(x)$.

(d) Hypersonic flow ($M^2 \gg 1$, $M\delta = 0(1)$ or greater)

Numerous approximate procedures exist for calculating hypersonic airloads, most of them in nonlinear forms which do not easily permit the extraction of relations resembling Eqs. 4–44. An excellent summary, relevant to both two- and three-dimensional configurations, appears in a paper by Lees (Ref. 4–28). Perhaps most commonly employed in the United States is the shock-expansion technique (see, for example, Eggers, Syvertson, and Kraus, Ref. 4–29).

For our purposes, it serves as an illustration to set down a few formulas from Newtonian theory, which has been described as the limit when $M \to \infty$ and $\gamma \to 1$ of more rigorous hypersonic formulations. The basic physical idea (cf. Refs. 4–30 and 4–163) is that all fluid which impacts the front of a body remains within a layer of infinitesimal thickness adjacent to the surface as it flows by, so that the entire component of momentum normal to the surface is transmitted as a pressure force. Since the total momentum flux in a stream tube of unit cross section is $2q$, it follows that a surface element whose inward normal makes an angle $\bar{\theta}$ to the direction of U should experience a pressure

$$p - p_\infty = 2q \cos^2 \bar{\theta}, \qquad \left(\bar{\theta} \le \frac{\pi}{2}\right) \qquad (4\text{--}73)$$

Equation 4–73 has been refined in various ways. For instance, when the surface is curved, it ought to be corrected by a small centrifugal term (Ref. 4–30) proportional to the curvature and negative where the body is convex. Lees suggests (Ref. 4–28) finding the correct stagnation pressure p_0 at the nose, corresponding to a coefficient $C_{p_0} < 2$, and estimating p

locally by combining the known C_{p_0} with Newton's approximation,

$$\frac{C_p}{C_{p_0}} \cong \cos^2 \bar{\theta} \tag{4-74}$$

It is generally assumed that nearly vacuum conditions exist in the lee of hypersonic bodies $(\bar{\theta} > \pi/2)$, where the flow is probably separated.

Newtonian theory achieves its greatest accuracy in regions where $\bar{\theta}$ is small, and it therefore is especially useful for blunt noses and leading edges. When applied to the forward portion of the airfoil considered in this section, whether blunt or sharp, it yields the following nonlinear formula for steady-state load per unit x-y-area:

$$\frac{\Delta p_a}{q} = 2 \left[\frac{\left(\dfrac{dz_L}{dx}\right)^2}{1 + \left(\dfrac{dz_L}{dx}\right)^2} - \frac{\left(\dfrac{dz_U}{dx}\right)^2}{1 + \left(\dfrac{dz_U}{dx}\right)^2} \right] \tag{4-75}$$

Equation 4-75 can be linearized for small mean-line slope dz_a/dx, regardless of the magnitude of the thickness $2z_t$, and gives the aerodynamic operator

$$\mathscr{A} = \frac{-8 \dfrac{dz_t}{dx}(\cdots)}{\left[1 + \left(\dfrac{dz_t}{dx}\right)^2\right]^2} \tag{4-76}$$

The only terms resulting from the expansion of Eq. 4-75 which do not vanish when z_t is set equal to zero are $0\left(\left(\dfrac{dz_a}{dx}\right)^2\right)$, that is, $0(\alpha^2)$ and higher. This result implies the "sine-squared law" of lift that is sometimes associated with Newton's name.

More detail about hypersonic theory appears in Secs. 4-5 and 4-6, which treat unsteady motion of airfoils.

4-4 LIFTING SURFACES AND OTHER CONFIGURATIONS IN THREE-DIMENSIONAL STEADY FLOW

(a) Subsonic flow $(M < 1)$

For almost-plane surfaces of arbitrary planform and aspect ratio, the most compact statement of the operator is Multhopp's (Ref. 4-31),

$$\mathscr{A}^{-1} = \frac{1}{8\pi} \oint\!\!\!\oint_{R_a} K(x_0, y_0; M)(\cdots)\, d\xi\, d\tilde{\eta} \tag{4-77a}$$

where the kernel function reads

$$K(x_0, y_0; M) = \frac{1}{y_0^2}\left[1 + \frac{x_0}{\sqrt{x_0^2 + \beta^2 y_0^2}}\right] \tag{4-77b}$$

β is given by Eq. 4–59. As in all such problems, special consideration must be given to the singularity of the kernel function at $y_0 = 0$, as discussed in Appendix I of Ref. 4–31.

Although Eqs. 4–77 cannot be inverted in closed form,* the literature is replete with approximate methods for finding the load distribution corresponding to a given $\alpha(x, y)$, the majority of them biased to favor a particular shape of wing or aspect-ratio range. With high-speed digital computers now generally available, most such schemes would seem to be obsolete except at the extremes of aspect ratio. The needs of any organization which must carry out extensive airload calculations are best served by a general program which reduces Eq. 4–77a to an algebraic system,†

$$\{\alpha\} = [\mathscr{A}]^{-1}\left\{\frac{\Delta p_a}{q}\right\} \tag{4-78}$$

The two column matrices in Eq. 4–78 contain the local mean-line slopes and dimensionless pressure differences at a preassigned set of points over the area R_a. During the subsequent discussion of oscillatory loading (Sec. 4–7), methods and references are furnished on the algebraic reformulation of that integral equation, by direct numerical integration or series substitution; and steady-flow forms like Eq. 4–78 can be obtained therefrom by setting the reduced frequency $k = 0$. This solution procedure also includes the specification of Kutta's condition of smooth flow-off ($\Delta p_a \rightarrow 0$) from the trailing edge.

On nearly all wings and tails, α can be separated into symmetrical and antisymmetrical parts with respect to the vehicle's plane of symmetry (i.e., even and odd functions of y, respectively). For each of these, Eq. 4–78 is recast into a system involving points that cover just the right half $y \geq 0$ of R_a (Fig. 4–1). The notation $[\mathscr{A}^s]$ and $[\mathscr{A}^a]$ is used for the aerodynamic matrices reduced in this way. In any event, Eq. 4–78 or its equivalent is a real system of linear, simultaneous equations; and routine means now exist for inverting to obtain

$$\left\{\frac{\Delta p_a}{q}\right\} = [\mathscr{A}]\{\alpha\} \tag{4-79}$$

---

* The results which have been obtained by series solution of Laplace's equation for circular and elliptical planforms in incompressible flow might be regarded as exceptions to this statement. Reference 4–35 and several papers referred to therein contain the details.

† Note that in the reduction to matrix forms in Sec. 4–4, the known steady-state upwash distribution w_a/U is everywhere replaced by its negative, the angle $\alpha(x, y)$.

even when (reasonably well-conditioned) matrices of order 100 or more are encountered. In some cases, values of α may not be given at the same stations where the $\Delta p_a/q$ are required; it is then only necessary to introduce a standard two-way interpolation matrix, transferring the available α's to the pressure stations. When there are slope discontinuities, such as occur at the leading edge of a flap or control surface, they are handled either by refining the network in Eq. 4–78 or, more rigorously, by approximating $\Delta p_a/q$ with a series that includes the correct logarithmic singularity along the discontinuity line.

In the important special circumstance when the aspect ratio is large enough, Eqs. 4–77 can be simplified by choosing some elementary representation of the chordwise pressure variation and concentrating on more accurate computation of the spanwise load distribution. Sophisticated examples of this technique are those where various orders of chordwise moments of $\Delta p_a/q$ are made the unknowns, such as Reissner's L-and-M method for rectangular wings (Ref. 4–32) and Multhopp's own "lifting-surface" theory in Ref. 4–31. These yield algebraic equations, of which

$$\begin{Bmatrix} c_l \\ c_m \end{Bmatrix} = [\mathscr{A}] \begin{Bmatrix} \alpha_1 \\ \alpha_2 \end{Bmatrix} \tag{4-80}$$

is a typical example. Here α_1 and α_2 are the streamwise slopes at a pair of points along the chord of each spanwise station for which the lift and moment coefficients are being found. Alternatively, as in Ref. 4–32, α_1 and α_2 may be weighted chordwise integrals of w_a/U. Additional α_n and higher moments, such as hinge moments or integrals like

$$m_n(y) = \int_{\text{chord}} x^n \, \Delta p_a(x, y) \, dx \tag{4-81}$$

might also appear. Even with the two unknowns displayed in Eq. 4–80, very accurate results are obtained for aspect ratios in excess of $1\frac{1}{2}$ or 2.

In the past most airload estimates were based on *lifting-line* reductions of Eqs. 4–77, wherein $c_l(y)$ remains as the single quantity to be determined. Because of their special significance, we shall review two versions, first limiting ourselves to incompressible flow and then showing the simple extension to arbitrary subsonic M. Prandtl's classical equation for straight wings, which can be derived either by a weighted chordwise integration of the exact equation (Reissner, Ref. 4–32) or by the familiar physical reasoning (e.g., Ref. 4–18), reads

$$\alpha(y) = \left(\frac{c_R}{c\dfrac{\partial c_l}{\partial \alpha}}\right)\frac{cc_l(y)}{c_R} + \frac{c_R}{8\pi}\oint_{-l}^{l}\frac{d}{d\eta}\left(\frac{cc_l(\eta)}{c_R}\right)\frac{d\eta}{(y-\eta)} \tag{4-82}$$

Here l and c_R are the semispan and the chord at a reference station, respectively; $\partial c_l/\partial \alpha$ is the two-dimensional lift-curve slope, taken from experiment or assigned the theoretical value 2π; and $\alpha(y)$ is the sectional angle of attack, measured from zero-lift attitude (cf. Eq. 4–63b).

Equation 4–82 is most conveniently solved by a series substitution based on Gauss' quadrature formula (Multhopp, Ref. 4–33). Making the transformation

$$\eta = l\cos\theta, \qquad y = l\cos\phi \qquad (4\text{–}83)$$

and choosing the Gaussian stations

$$\varphi_n = \frac{n\pi}{m+1}, \qquad (n = 1, 2, \cdots, m; \quad m \text{ odd}) \qquad (4\text{–}84)$$

as collocation points, Multhopp approximates Eq. 4–82 by

$$\{\alpha\} = [\mathscr{A}]^{-1}\left\{\frac{cc_l}{c_R}\right\} \qquad (4\text{–}85)$$

Here the square aerodynamic matrix reads

$$[\mathscr{A}]^{-1} = \frac{c_R}{4l}\begin{bmatrix} \left(b_{11} + \dfrac{4l}{c_1\,\partial c_l/\partial\alpha}\right) & -b_{12} & \cdots & -b_{1m} \\[2ex] -b_{21} & \left(b_{22} + \dfrac{4l}{c_2\,\partial c_l/\partial\alpha}\right) & \cdots & -b_{2m} \\ \cdot & & & \cdot \\ \cdot & & & \cdot \\ \cdot & & & \\ -b_{m1} & & \cdots & \left(b_{mm} + \dfrac{4l}{c_m\,\partial c_l/\partial\alpha}\right) \end{bmatrix} \qquad (4\text{–}86)$$

where c_n is the chord at station n and b_{ij} are simple numerical coefficients tabulated in Ref. 4–33 for several values of m. Inversion of Eq. 4–85 is facilitated by the vanishing of b_{ij} when $|i - j|$ is an even nonzero integer. Reference 4–33 describes the reduction of Eq. 4–86 to $[\mathscr{A}^s]^{-1}$, of order $(m + 1)/2$, when the wing is symmetrically loaded, and to $[\mathscr{A}^a]^{-1}$, of order $(m - 1)/2$, when the wing is antisymmetrically loaded.

For a lifting surface whose line of aerodynamic centers has an appreciable sweep angle Λ (positive for sweepback), Weissinger has derived a useful integral equation of lifting-line type in Ref. 4–34:

$$\alpha(y) = \frac{c_R}{4\pi}\oint_{-l}^{l} \frac{d}{d\eta}\left(\frac{cc_l(\eta)}{c_R}\right)\frac{d\eta}{(y-\eta)}$$
$$+ \frac{c_R}{4\pi c(y)}\int_{-l}^{l} \frac{d}{d\eta}\left(\frac{cc_l(\eta)}{c_R}\right)L\left(\frac{y}{l}, \frac{\eta}{l} ; \Lambda, \frac{l}{c(y)}\right)d\eta \qquad (4\text{–}87)$$

where $\alpha(y)$ is taken in a cross section parallel to the flight direction. The dimensionless, nonsingular kernel L is a function of the four quantities indicated. Formulas and tables are given in Ref. 4–34 or in NACA publications dealing with this theory (e.g., Refs. 4–36 and 4–37). When the Multhopp-Gauss substitution is made in Eq. 4–87, another algebraic system resembling Eq. 4–85 emerges,

$$\{\alpha\} = [\overline{\mathscr{A}}]^{-1}\left\{\begin{matrix}cc_l\\c_R\end{matrix}\right\} \tag{4-88}$$

Here the bar signifies generalization to arbitrary sweep angle, and the square matrix, in Weissinger's notation, reads

$$[\overline{\mathscr{A}}]^{-1} = \frac{c_R}{2l}\begin{bmatrix} \left(b_{11}+\dfrac{l}{c_1}g_{11}\right) & \left(\dfrac{l}{c_1}g_{12}-b_{12}\right) & \cdots & \left(\dfrac{l}{c_1}g_{1m}-b_{1m}\right)\\[2mm] \left(\dfrac{l}{c_2}g_{21}-b_{21}\right) & \left(b_{22}+\dfrac{l}{c_2}g_{22}\right) & & \cdot\\ \cdot & \cdot & & \cdot\\ \cdot & \cdot & & \cdot\\ \cdot & \cdot & & \cdot\\ \left(\dfrac{l}{c_m}g_{m1}-b_{m1}\right) & \cdots\cdots\cdots\cdots\cdots & & \left(b_{mm}+\dfrac{l}{c_m}g_{mm}\right) \end{bmatrix} \tag{4-89}$$

The b_{ij} are the same coefficients appearing in Eq. 4–86, whereas the g_{ij} are algebraic sums containing values of the L-function.

For symmetrical loading, De Young and Harper (Ref. 4–36) rewrite Eqs. 4–88

$$\{\alpha\} = [a_{vn}]\left\{\frac{cc_l}{4l}\right\} \tag{4-90}$$

giving complete listings of the elements a_{vn} of the $(m+1)/2$-order matrix. Reference 4–37 deals similarly with the antisymmetrical case.

To summarize the foregoing, the lifting-line concept envisions an aerodynamic operator,

$$c_l(y) = \mathscr{A}\{\alpha(y)\} \tag{4-91}$$

relating dimensionless lift per unit y-distance to the angle from zero-lift incidence, which may include the effects of camber, deflected flaps,* and the like. No information is provided about section A.C. location,

---

* The flap contribution to α is usually written $\dfrac{\partial\alpha}{\partial\delta}\delta$, where δ is the actual flap rotation and $\dfrac{\partial\alpha}{\partial\delta} = \dfrac{\partial c_l}{\partial\delta}\Big/\dfrac{\partial c_l}{\partial\alpha}$ measures, on a two-dimensional basis, the rotation of the zero-lift direction per unit change in δ.

c_{mAC}, or zero-lift angle; so these must be presumed equal to their two-dimensional values. c_{mAC} should be corrected for sweep angle, as shown in Sec. 4–4(e); otherwise Eqs. 4–62 and 4–63b apply at subsonic speeds.

Both the lifting-line and lifting-surface theories just presented are valid for arbitrary subsonic M, in virtue of the Prandtl-Glauert transformation for planar systems. A very clear presentation will be found in Sears' article (Ref. 4–38). The key idea is that a change of variable such as

$$x', y', z' = \frac{x}{\beta}, y, z \qquad (4\text{–}92)$$

will transform the steady counterpart of the linearized differential equation (4–29) into Laplace's equation, so that every problem of compressible flow has an incompressible equivalent. For example, when lifting-line methods are used on a given wing of chord $c(y)$ and sweep Λ at Mach number M, exactly the same lift per unit span $L(y)$ will be found on a counterpart in a stream of incompressible fluid with the same density ρ_∞ and speed U. The new wing has the chord dimensions of its planform increased by $1/\beta$, so that

$$c'(y') = \frac{c(y)}{\beta} \qquad (4\text{–}93)$$

and

$$\tan \Lambda' = \frac{1}{\beta} \tan \Lambda \qquad (4\text{–}94)$$

but the angle of attack and slope distribution of the mean surface z_a remain unaltered. References 4–31, 4–36, and 4–37 illustrate the practical application of this useful result.

At the low end of the aspect-ratio range lie the very slender, pointed lifting surfaces (Fig. 4–3), to which R. T. Jones' adaptation (Ref. 4–39) of Munk's airship theory applies. As shown, for example, on pp. 244–248 of Ref. 1–2, the aerodynamic operator may be written in the direct form

$$\mathscr{A} = \frac{4}{\pi}\frac{\partial}{\partial x}\int_{-s(x)}^{y}\frac{1}{\sqrt{s^2(x)-y'^2}}\oint_{-s(x)}^{s(x)}\frac{\sqrt{s^2(x)-\eta^2}}{(y'-\eta)}(\cdots)\,d\eta\,dy'$$

$$= \frac{4}{\pi}\frac{\partial}{\partial x}\int_{-s(x)}^{s(x)}\ln\left[\frac{s^2(x)-y\eta-\sqrt{s^2(x)-y^2}\sqrt{s^2(x)-\eta^2}}{s(x)|y-\eta|}\right](\cdots)\,d\eta \qquad (4\text{–}95)$$

Here $s(x)$ is the local semispan, as pictured in the figure.

Equation 4–95 can be integrated analytically in several interesting practical cases, of which the most important are those when spanwise

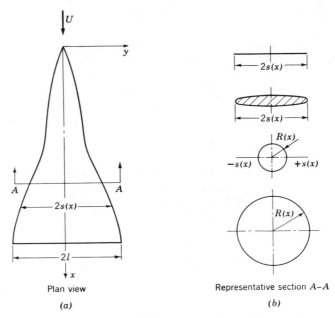

Plan view

(a)

Representative section $A-A$

(b)

Fig. 4–3. Various configurations which can be treated by slender-body theory. Typical shapes of spanwise cross sections are illustrated in (b) from top to bottom: wing of zero thickness, elliptical thickness distribution, midwing-body combination, body of revolution.

deformations are negligible and w_a/U (or α) is a function of x only. Then the operator reduces to

$$\mathscr{A} = -4 \frac{\partial}{\partial x} \left[\sqrt{s^2(x) - y^2}(\cdots) \right] \tag{4–96}$$

Another integration of Eq. 4–96 with respect to y, across the span at station x, yields the following relation between $\alpha(x)$ and the lift per unit of chord distance $L(x)$:

$$L(x) = 2q \frac{d}{dx} \left[S(x)\alpha(x) \right] \tag{4–97}$$

The quantity $\rho_\infty S(x) = \rho_\infty \pi s^2(x)$ is called the *virtual mass* of the plane spanwise cross section and is seen to equal the mass of fluid contained in a circular cylinder based on the local span as diameter. Other conclusions that can be drawn from Eq. 4–96 are that the total lift depends only on the angle of attack at the trailing edge and that the lift-curve slope of any slender planform with its maximum span at the trailing edge, as in Fig. 4-3, is

$$\frac{\partial C_L}{\partial \alpha} = \frac{\pi}{2} \mathcal{R} \qquad (4\text{-}98)$$

$\mathcal{R}$ denotes aspect ratio by the standard definition—wingspan-squared divided by plan area. All of Eqs. 4-95–4-98 are essentially independent of M.

It is of interest to complete this discussion of lifting surfaces by setting down Diederich's semi-empirical formula (Ref. 4-40) for the lift-curve slope of any straight or swept wing at any subsonic Mach number,

$$\frac{\partial C_L}{\partial \alpha} = \frac{\left(\dfrac{\partial c_l}{\partial \alpha}\right)_0 \cos \Lambda'}{\beta} \left\{ \frac{\beta \mathcal{R}}{\dfrac{\left(\dfrac{\partial c_l}{\partial \alpha}\right)_0 \cos \Lambda'}{\pi} + \beta \mathcal{R}\sqrt{1 + \left[\dfrac{\left(\dfrac{\partial c_l}{\partial \alpha}\right)_0 \cos \Lambda'}{\pi \beta \mathcal{R}}\right]^2}} \right\} \quad (4\text{-}99)$$

Here Λ' is the sweep angle of the equivalent incompressible planform, Eq. 4-94. Equation 4-99 reduces to Eq. 4-98 for slender wings, and measured data in Ref. 4-40 indicate excellent accuracy for aspect ratios above 1.5 and M up to 0.7.

Bodies and wing-body combinations having cross-sectional shapes, of which the last two in Fig, 4-3(b) are typical, constitute a second major category where aerodynamic operators are often needed for aeroelastic analyses. Nearly all aircraft and missile configurations can be built up by associating them together with almost-plane lifting surfaces. Except for interference effects, whose treatment unfortunately falls beyond the limited scope of the present chapter, wing and slender-body load data therefore fill most of the important requirements.

Regarding the body itself, the basic operator is one relating lift per unit length $L(x)$ to the local inclination $\alpha(x) \equiv -dz_a/dx$ between the mean line and the stream direction. Provided the fineness ratio is large enough and the nose is not too blunt, the aforementioned Munk-Jones theory yields this operator in a compact form identical with Eq. 4-97, except that the virtual mass $\rho_\infty S(x)$ assumes different mathematical expressions, depending on the cross-sectional shape. For instance, considering the lower three sections illustrated in Fig. 4-3(b), we have

$$S(x) = \pi s^2(x), \quad \text{(ellipse)} \qquad (4\text{-}100)$$

$$S(x) = \pi \left[s^2(x) - R^2(x) + \frac{R^4(x)}{s^2(x)} \right], \quad \text{(midwing-body combination)} \quad (4\text{-}101)$$

and

$$S(x) = \pi R^2(x), \quad \text{(body of revolution)} \tag{4-102}$$

Lateral loads due to sideslip angle are found in a similar fashion. The physical interpretation of the Munk-Jones formulas in terms of unsteady, two-dimensional, incompressible fluid motion in planes normal to the flight direction is well-known and will not be repeated here. Bryson's paper (Ref. 4–41) contains an excellent presentation of its application to static and dynamic problems of various configurations, along with a list of other key references. Calculation of the subsonic loading on bodies, including the extension to higher-order effects of fineness ratio, is also discussed in Refs. 4–42 and 4–43. The only shapes for which *exact* three-dimensional solutions exist are rigid ellipsoids in incompressible flow (Ref. 4–11), and for them the only aerodynamic operators in convenient linear forms are those giving total lifts and moments. For example, the pitching moment exerted on a prolate ellipsoid of revolution at a small, steady incidence α turns out to be a pure couple,

$$M_y = 2qV[k_2 - k_1]\alpha \tag{4-103}$$

Here k_1 and k_2 are *inertia coefficients*, expressed in Ref. 4–11 as elliptical integrals with the fineness ratio as argument. Formulas like Eq. 4–103 are useful for analyzing deformations of wings with external stores. When such bodies are relatively slender, the results are quite insensitive to Mach number changes throughout the subsonic range, but often interference cannot be neglected.

(b) Transonic flow ($M \cong 1$)

No explicit equations need be written for transonic aerodynamic operators, because whatever linear ones are valid constitute direct extrapolations of their subsonic counterparts. In the case of three-dimensional lifting surfaces, we conclude from the Prandtl-Glauert transformation as $M \to 1$ that the equivalent incompressible flow occurs about a wing of vanishing aspect ratio, to which R. T. Jones' theory (i.e., Eqs. 4–95, 4–96, etc., when the trailing edge is cut off normal to x) presumably applies. Equation 4–99, for instance, yields a limiting lift-curve slope of $\pi\mathcal{R}/2$, independent of sweep angle; and this result is quite well confirmed by wind-tunnel tests near $M = 1$.

Heaslet, Lomax, and Spreiter (Ref. 4–44) give one of the better discussions of transonic wing theory. An analytical approach to the problem (Ref. 4–45) suggests as one order-of-magnitude test for the validity of steady-state linearized theory the following inequality:

$$\mathcal{R}^3 \, \delta[\ln \mathcal{R} \, \delta^{1/3}]^2 \ll 1 \tag{4-104}$$

δ being the maximum thickness ratio of the surface.

This is perhaps too restrictively written here, since, properly speaking, the aspect ratio $Æ$ in (4–104) should be replaced by a more direct (and smaller) measure of the span-to-chord proportions, such as $2l/c_{\text{root}}$. When $Æ$ and δ are so large that linearized results are invalidated near $M = 1$, an essentially nonlinear mathematical situation arises, and the comments made in Sec. 4–3(b) apply.

For slender bodies and wing-body combinations flying transonically, results such as Eqs. 4–97, 4–100, 4–101, and 4–102 are probably more accurate than they are either above or below sonic speed. An illuminating physical explanation of this fact appears in a 1956 paper by R. T. Jones (Ref. 4–46).

(c) **Supersonic flow** ($M > 1$, $M\,\delta \ll 1$)

$$\mathscr{A} = \frac{4}{\pi} \frac{\partial}{\partial \tilde{x}} \iint\limits_{R_a'} K(x_0, y_0; M)(\cdots)\, d\xi\, d\tilde{\eta}, \qquad (4\text{–}105a)$$

where

$$K(x_0, y_0; M) = -\frac{1}{\sqrt{x_0{}^2 - \beta^2 y_0{}^2}} \qquad (4\text{–}105b)$$

β takes on its supersonic interpretation of $\sqrt{M^2 - 1}$. Equations 4–105 apply, without qualification, only to lifting surfaces with *simple planforms*, that is, those having all supersonic leading and trailing edges. The region of integration R_a' is then the *upstream zone of influence* of point $(\tilde{x}, \tilde{y})$, which is contained between the leading edge and the forward-going *Mach lines* $\tilde{\eta} = \tilde{y} \pm (\tilde{x} - \tilde{\xi})/\beta$ through $(\tilde{x}, \tilde{y})$ (see Refs. 4–21 and 4–47 for examples).

For wings with subsonic edges, R_a' consists of a portion of the projected planform and a portion of the disturbed *diaphragm region* adjacent to the leading, wingtip, and possibly trailing edges. Over the diaphragm, w_a/U is unknown but can be determined, during the process of integrating Eq. 4–105a, from the auxiliary condition that $\Delta p_a = 0$ there. Many techniques, such as doublet superposition, conical, and generalized conical flow theories, have been developed for calculating the loadings on particular supersonic planforms with particular slope distributions. Most of these have limited utility for the aeroelastician, who must deal with quite general deformation shapes. A full account, with many references, will be found in the article by Heaslet and Lomax (Ref. 4–47). These authors also furnish the derivation of Eqs. 4–105.

By an approximation procedure, whereby $\alpha(x, y)$ is assumed to be constant or to have some simple known variation in each of a large number of area elements distributed over the disturbed x-y-plane, Eq. 4–105a (for a fixed set of points at the centers of these areas) can be reduced to a system of algebraic equations,

$$\left\{\frac{\Delta p_a}{q}\right\} = [\mathscr{A}]\{\alpha\} \tag{4–79}$$

This result is analogous to the method of aerodynamic influence co-efficients for oscillatory motion, which is described in more detail in Sec. 4–7(c). Accordingly, we do not dwell on it here, except to point out that the matrices $\{\Delta p_a/q\}$ and $\{\alpha\}$ contain elements from both the planform and diaphragm regions. These can be ordered in such a way that, taking advantage of the many zeros in $[\mathscr{A}]$, the unknown $\alpha \equiv -w_a/U$ for each diaphragm element is computed in succession. The necessity of inverting any matrices is avoided.

The remarks of Sec. 4–4(a) on the use of symmetry to break down $[\mathscr{A}]$ to smaller square matrices $[\mathscr{A}^s]$ and $[\mathscr{A}^a]$, covering only the right half of the wing, are equally applicable at supersonic speeds. Flaps and controls present no special difficulties, since a discontinuity in α across a line which is less swept than the Mach lines causes no pressure singularity.

Running lift and moments are found by appropriate chord integrations of Eqs. 4–105a. For example,

$$c_l(\tilde{y}) = -\frac{b}{c} \iint\limits_{R'_{aTE}} K_{TE}(x_0, y_0; M)\alpha(\xi, \tilde{\eta})\, d\xi\, d\tilde{\eta} \tag{4–106}$$

where subscript TE means that the $\tilde{x}$ coordinate of the trailing edge at station $\tilde{y}$ is to be substituted into the function so labeled. If either α is independent of ξ or $K_{TE}\alpha$ can be suitably averaged across the chord, the ξ integration is eliminated from Eq. 4–106; and a relation is obtained which is reminiscent of the subsonic lifting-line operator, Eq. 4–91. A suitable approximation in the spanwise integral finally yields a form like Eq. 4–85, except that no inversion need be carried out.

These observations have formed the basis of supersonic lifting-line theories, of which Ref. 4–48 is typical. The familiar concept of spanwise induction is not so meaningful at $M > 1$, however, because the incidence of a given section of a supersonic lifting surface can influence only the loading of immediately adjacent sections and not the whole planform unless $\mathcal{R}$ is small or M is very close above unity. Indeed, the effective three-dimensionality of the flow over any wing or tail is measured, for a given sweep angle, by the parameter $\beta\mathcal{R}$. When Λ is small and $\beta\mathcal{R}$ exceeds 4 or 5, strip theory yields much better accuracy than might be expected in a similar case at subsonic speeds, especially if the loading is rounded off to zero in the small regions affected by the tips. Furthermore, strip theory is exact for determining certain properties of some simple planforms (cf. 4–49).

Away from the aforementioned tip regions, the second-order influence of thickness on the load distribution can be found exactly when $M^2 \gg 1$ by means of Landahl's "inverse Rayleigh-Jansen method" (Ref. 4–24). If Landahl's terms of orders δ/M, δ/M^3, and δ^2 are retained, the steady-state aerodynamic operator is as follows:

$$
\mathscr{A} = -\frac{4}{M}\left[1 + \left(\frac{\gamma+1}{2}\right)M\frac{dz_t}{dx}\right]\left(\frac{w_a}{U}\right)
$$
$$
-\frac{2}{M^3}\left(\frac{\partial^2}{\partial \tilde{x}^2} + \frac{\partial^2}{\partial \tilde{y}^2}\right)\int_{\tilde{x}_{LE}}^{\tilde{x}}[\tilde{x} - \xi]\frac{w_a(\xi, \tilde{y})}{U}\,d\xi \qquad (4\text{--}107)
$$

Here $\tilde{x}_{LE}(\tilde{y})$ is the local dimensionless coordinate of the leading edge, and w_a/U has been written explicitly in order to show its functional dependence under the integral sign.

When $\delta \gg 1/M^3$, the last term on the right of Eq. 4–107 can be neglected, and the piston theory operator, Eq. 4–70, is all that remains. This fact has led to the proposal that three-dimensional linearized theory might be roughly corrected for thickness effect by simply applying the over-all factor

$$
\left[1 + \left(\frac{\gamma+1}{2}\right)M\frac{dz_t}{dx}(\tilde{x}, \tilde{y})\right]
$$

to Eq. 4–105a. Although promising, this scheme has as yet no real theoretical or experimental justification.

As long as $M\delta \ll 1$, slender, pointed bodies and wing-body combinations in supersonic flow can be treated by Munk-Jones theory (see, for example, Heaslet and Lomax, Ref. 4–47). This statement applies only to the normal loads, which are of principal interest in aeroelasticity; but we can state that Eqs. 4–97, 4–100, 4–101, and 4–102 supply the necessary operators. A not-so-slender body theory, giving higher-order effects of δ and $Æ$, has been presented by Adams and Sears (Ref. 4–50).

(d) Hypersonic flow $(M^2 \gg 1,\ M\delta = 0(1)$ or greater)

The comments made in Sec. 4–3(d) about hypersonic aerodynamic operators need not be repeated. Three-dimensional configurations may be separated into those which are wing-like, having $Æ \gg \delta$ and all leading and trailing edges swept well ahead of the Mach lines, and those which are essentially slender bodies. For the former category, the flow is nearly two-dimensional in y-z-planes. Equations 4–75 and 4–76 constitute high

Mach-number limits on the operators; whereas in the lower hypersonic range, $M \delta = 0(1)$, we may take Lighthill's suggestion (Ref. 4–23) and employ Eq. 4–70 as an approximation.

Slender shapes—especially some of the complicated arrangements of wings, dorsal and ventral fins, end-plates, and central bodies that currently pass for re-entry gliders—do not lend themselves to easy theoretical treatment. In the range $M \delta = 0(1)$ we can mention van Dyke's "two-dimensional piston theory," (Ref. 4–51) which shows the fluid motion to occur principally in y-z-planes but with variations of both density and entropy. This has been adapted so far only to cones, ogives of revolution, and other simple forms, but it has great promise of future generalization. For blunt noses and rounded leading edges at $M^2 \gg 1$, Newtonian theory can be used. The generalized shock-expansion theory of Eggers and Syvertson (Ref. 4–29) serves well for pointed configurations, although linearized aerodynamic operators must be worked out separately in each case.

(e) Strip theory for swept wings

This section is concluded by restating the well-known method whereby strip theory may be extended to an infinite swept wing (Fig. 4–4), all of whose stations are performing the same steady or unsteady motion. By resolving the stream U into components normal and parallel to the swept span, it can be reasoned (e.g., Sec. 7–3 of Ref. 1–2) that

$$L_\Lambda = L \cos \Lambda \tag{4-108}$$

$$(M_y)_\Lambda = M_y \cos \Lambda \tag{4-109}$$

Here L and M_y are the lift and pitching moment per unit of spanwise distance on a straight wing whose sections move identically to section AA. L_Λ and $(M_y)_\Lambda$ are swept-wing loads per unit length normal to U, and $(M_y)_\Lambda$ is taken also about an axis normal to U.

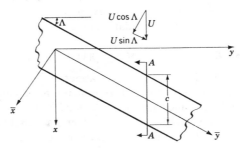

Fig. 4–4. Infinite swept wing in stream U, showing notation and section A–A.

In terms of coefficients, we may write

$$\Delta c_l = \frac{\partial c_l}{\partial \alpha} \cos \Lambda \, \Delta \alpha \qquad (4\text{–}110)$$

$$(c_{mAC})_\Lambda = c_{mAC} \cos \Lambda \qquad (4\text{–}111)$$

where $\partial c_l/\partial \alpha$ and c_{mAC} refer to the corresponding straight wing flying at a Mach number $M = U \cos \Lambda/a_\infty$. The incidence $\Delta \alpha$ is always measured in a plane containing U. Formulas referring to sections taken perpendicular to the swept y-axis can be worked out by a process of vector decomposition, since only small angles of attack are permissible within the framework of linearized theory. When the motion is oscillatory, the reduced frequency $k = \omega c/2U$ is the same on the swept and equivalent straight wings.

For strip theories applicable to large aspect-ratio swept wings in three dimensions, see Ref. 4–52, Sec. 7–3 of Ref. 1–2, or the more rigorous later developments of van de Vooren and Eckhaus (Refs. 4–53 and 4–54).

4–5 AIRFOILS IN TWO-DIMENSIONAL UNSTEADY FLOW; SIMPLE HARMONIC MOTION

Most of the aerodynamic operators take on less complicated forms and some can be written only when the unsteady motion consists of a simple harmonic oscillation, which has gone on at circular frequency ω long enough to assure the disappearance of transients. In view of the assumed linearity, we specify that the displacement of the mean line or mean surface of a wing or body is given by the real part of the complex function

$$z_a(x, y, t) = \bar{z}_a(x, y)e^{i\omega t} \qquad (4\text{–}112)$$

$\bar{z}_a$ is a function of x alone in two-dimensional problems. The thickness $2z_t$ is usually regarded as independent of time, since structural deformations equivalent to thickness variations are not often observed except on thin shells with widely spaced supports.

It follows, from the interchangeability of the operation of taking the real part with other linear mathematical operators, that all the dependent variables can be recast in complex notation similar to Eq. 4–112. Thus the linearized boundary condition, Eq. 4–41, becomes, after canceling the common factor $e^{i\omega t}$,

$$\frac{\partial \bar{\varphi}}{\partial z}(x, y, 0) = \left(U \frac{\partial}{\partial x} + i\omega\right)\bar{z}_a(x, y) \equiv \bar{w}_a(x, y),$$

$$\text{(for } (x, y) \text{ in region } R_a) \qquad (4\text{–}113)$$

On two-dimensional airfoils, of course, R_a is merely the strip $(-b \leqslant x \leqslant b)$. The loading per unit area of the x-y-plane may be written

$$\Delta p_a(x, y, t) = \Delta \bar{p}_a(x, y)e^{i\omega t}, \qquad (4\text{-}114)$$

and $e^{i\omega t}$ may be dropped from the definition of the aerodynamic operator, Eq. 4-44a, leaving

$$\frac{\Delta \bar{p}_a}{q} = \mathscr{A}\left(\frac{\bar{w}_a}{U}\right) \qquad (4\text{-}115)$$

As a general rule throughout Secs. 4-5, 4-7, and elsewhere in the book where sinusoidal motion is discussed, the bar over a dependent variable identifies it as a *complex amplitude*, from which the actual physical variable may be calculated by multiplying with $e^{i\omega t}$ and extracting the real part. For instance, the running lift, the pitching moment, and the two co-ordinates specifying the displacement of the wing in Fig. 6-5 would become

$$L = \bar{L}e^{i\omega t} \qquad (4\text{-}116)$$

$$M_y = \bar{M}_y e^{i\omega t} \qquad (4\text{-}117)$$

$$h(t) = \bar{h}e^{i\omega t} \qquad (4\text{-}118)$$

$$\alpha(t) = \bar{\alpha}e^{i\omega t} \qquad (4\text{-}119)$$

We now proceed to list and discuss the aerodynamic operators, using the same system of classification as for steady flow. One new parameter, the reduced frequency defined by Eq. 4-55, appears repeatedly through the sinusoidal cases.

(a) Incompressible flow ($M^2 \ll 1$)

$$\mathscr{A} = \frac{4}{\pi}\left\{[1 - C(k)]\sqrt{\frac{1 - \tilde{x}}{1 + \tilde{x}}}\int_{-1}^{1}\sqrt{\frac{1 + \tilde{\xi}}{1 - \tilde{\xi}}}(\cdots)\,d\tilde{\xi}\right.$$

$$+ \oint_{-1}^{1}\left[\sqrt{\frac{1 - \tilde{x}}{1 + \tilde{x}}}\sqrt{\frac{1 + \tilde{\xi}}{1 - \tilde{\xi}}}\frac{1}{x_0}\right.$$

$$\left.\left. - \frac{ik}{2}\ln\left(\frac{1 - \tilde{x}\tilde{\xi} + \sqrt{1 - \tilde{x}^2}\sqrt{1 - \tilde{\xi}^2}}{1 - \tilde{x}\tilde{\xi} - \sqrt{1 - \tilde{x}^2}\sqrt{1 - \tilde{\xi}^2}}\right)\right](\cdots)\,d\tilde{\xi} \quad (4\text{-}120)\right.$$

where

$$C(k) = \frac{H_1^{(2)}(k)}{H_1^{(2)}(k) + iH_0^{(2)}(k)} \qquad (4\text{-}121)$$

is called Theodorsen's function, $H_n^{(2)}$ being the Hankel function of the second kind and order n (Ref. 2-4, page 624). Equation 4-120 constitutes the inversion of a vortex-sheet integral equation, with the Kutta condition applied at the trailing edge during the process (Schwarz, Ref. 4-55).

Additional references and derivations from various points of view appear in Sec. 5–6 of Ref. 1–2 and in Ref. 4–1, the latter including Küssner's convenient Fourier series solution.

The incompressible operator may be thought of as the sum of a *noncirculatory* part,

$$\mathscr{A}_{NC} = \frac{4}{\pi} \oint_{-1}^{1} \left[\frac{\sqrt{1 - \xi^2}}{\sqrt{1 - \tilde{x}^2}} \frac{1}{x_0} \right.$$

$$\left. - \frac{ik}{2} \ln \left(\frac{1 - \tilde{x}\xi + \sqrt{1 - \tilde{x}^2}\sqrt{1 - \xi^2}}{1 - \tilde{x}\xi - \sqrt{1 - \tilde{x}^2}\sqrt{1 - \xi^2}} \right) \right] (\cdots) \, d\xi \quad (4\text{–}122)$$

which satisfies the instantaneous boundary conditions but gives zero total circulation around the airfoil, and a *circulatory* part ($\mathscr{A} - \mathscr{A}_{NC}$), which takes care of the Kutta condition and the influence of the vortex wake. $\mathscr{A}_{NC}$ can be generalized to arbitrary small motion by replacing ik with $\frac{b}{U} \frac{\partial}{\partial t}$ and including the time dependence in Δp_a and w_a. The noncirculatory loads involve derivatives of the displacements up to second order, with constant coefficients which are generally identified as *virtual masses* and *virtual inertias*. This concept of virtual mass is associated with the infinite speed of sound, however, and cannot be extended to compressible flow, where "memory" effects are present because of both the wake vortices and the finite rate of propagation of disturbances.

Extensive tables are available relating oscillatory lift, pitching moment, and hinge moment to rigid-body motion of the airfoil and flap rotation. The symbolism of Smilg and Wasserman (Ref. 4–56), whose table is commonly used in the United States, looks as follows:*

$$\bar{L} = -\pi \rho_\infty b^3 \omega^2 \left\{ L_h \frac{\bar{h}}{b} + [L_\alpha - (\tfrac{1}{2} + a)L_h]\bar{\alpha} + L_\beta \, \delta \right\} \quad (4\text{–}123)$$

$$\bar{M}_y = \pi \rho_\infty b^4 \omega^2 \left\{ [M_h - (\tfrac{1}{2} + a)L_h]\frac{\bar{h}}{b} + [M_\alpha - (\tfrac{1}{2} + a)(L_\alpha + M_h) \right.$$

$$\left. + (\tfrac{1}{2} + a)^2 L_h]\bar{\alpha} + [M_\beta - (\tfrac{1}{2} + a)L_\beta] \, \delta \right\} \quad (4\text{–}124)$$

$$\bar{H} = \pi \rho_\infty b^4 \omega^2 \left\{ T_h \frac{\bar{h}}{b} + [T_\alpha - (\tfrac{1}{2} + a)T_h]\bar{\alpha} + T_\beta \, \delta \right\} \quad (4\text{–}125)$$

The flap is here assumed to have zero aerodynamic balance, and the airfoil rotates about the moment axis $x = ba$. L_h, L_α, M_h, and M_α are dimensionless functions of k only, while the remaining coefficients L_β, etc., are

---

* To adapt Eqs. 4–123—4–125 to give total loads on a two-dimensional wing of plan area S, replace one of the factors b outside each set of braces by $S/2$.

tabulated versus k and the fractional-chord location of the flap leading edge. In Ref. 4–56, operators are also listed for an aerodynamically balanced flap and for a tab hinged at its nose. More complete tables for trailing edge controls and tabs appear in Ref. 4–57.

To give an indication of the simple mathematical forms in which the incompressible aerodynamic operators occur, we reproduce here the expression for the running lift $L(t)$ generated by the translational and pitching oscillations:

$$L = \pi \rho_\infty b^2 [\ddot{h} + U\dot{\alpha} - ba\ddot{\alpha}] + 2\pi \rho_\infty U b C(k)[\dot{h} + U\alpha + b(\tfrac{1}{2} - a)\dot{\alpha}]$$

$$(4\text{–}126)$$

The noncirculatory terms are those contained in the first bracket.

(b) Subsonic compressible flow $(M < 1)$

$$\mathscr{A}^{-1} = \frac{1}{8\pi} \oint_{-1}^{1} K(x_0; k, M)(\cdots)\, d\bar{\xi} \qquad (4\text{–}127a)$$

where

$$K(x_0; k, M) = \frac{\pi k e^{-ikx_0}}{\beta} \left\{ e^{i(kx_0/\beta^2)} \left[i\,\frac{M\,|x_0|}{x_0} H_1^{(2)}\left(\frac{Mk\,|x_0|}{\beta^2}\right) \right. \right.$$

$$\left. - H_0^{(2)}\left(\frac{Mk\,|x_0|}{\beta^2}\right) \right] + \frac{2i\beta}{\pi} \ln\left(\frac{1+\beta}{M}\right) + i[1 - M^2]$$

$$\times \int_0^{kx_0/\beta^2} e^{iu} H_0^{(2)}(M\,|u|)\, du \right\} \qquad (4\text{–}127b)$$

It is significant that this kernel function, given first by Possio (Ref. 4–58), can be divided by k and otherwise modified so as to produce a function of only two variables, M and kx_0. The earlier solutions of Eqs. 4–127 were accomplished by collocation, taking account of the Kutta condition and of singularities in both the kernel and $\Delta p_a/q$. Fettis' method (Ref. 4–59) is the most efficient; tables based on it were issued for the wing-flap combination at $M = 0.7$ (Ref. 4–60) in the notation of Eqs. 4–123–4–125. Other similar tables have been published by Luke (Ref. 4–61) and Turner and Rabinowitz (Ref. 4–62).

What amounts to an inversion of Eqs. 4–127 can be achieved by rewriting the two-dimensional partial differential equation in elliptic cylinder coordinates and solving it by infinite series of Mathieu functions. Among several versions of this approach, we mention the work of Timman, van de Vooren, and Greidanus (Ref. 4–63); Reissner (Ref. 4–64); and

Williams (Ref. 4–65). Tables based on Ref. 4–63 are given for the wing-flap with several flap-chord ratios at $M = 0$, 0.35, 0.6, 0.7, and 0.8 in two Dutch reports (Refs. 4–66 and 4–67). Finally, Jordan has used an interpolation procedure to compute derivatives for the airfoil without flap at $M = 0.8$, 0.9, and 0.95 (Ref. 4–68). Unfortunately but naturally, there is complete notational disagreement among the various tabulations, so we must refer the reader to the individual papers for information on this point.

(c) Transonic flow $(M \cong 1,\ k \gg |1 - M|)$

$$\mathscr{A} = 4\left(\frac{\partial}{\partial \tilde{x}} + ik\right)\int_{-1}^{\tilde{x}} K(x_0; k, M = 1)(\cdots)\, d\xi \qquad (4\text{–}128a)$$

where

$$K(x_0; k, M = 1) = -\frac{e^{-i(kx_0/2)}}{\sqrt{2\pi i k x_0}} \qquad (4\text{–}128b)$$

This form of the kernel function, which is directly related to the discontinuity $\Delta \bar{\phi}_a'$ of the velocity potential, follows from the work of Nelson and Berman (Ref. 4–69). The earliest research on the problem is reported, however, by Rott (e.g., Ref. 4–70).

Practical applications for particular upwash distributions require that Eq. 4–128a be integrated with various powers of ξ multiplying K, a process in which Fresnel integrals occur repeatedly. The full development is carried out in Ref. 4–69, where tables of lift, pitching moment, and hinge moment on the wing-flap combination without aerodynamic balance are presented. The notation is that originally used for supersonic coefficients (Garrick and Rubinow, Ref. 4–71):

$$\bar{L} = 4\rho_\infty b^3 \omega^2\left\{[L_1 + iL_2]\frac{h}{b} + [L_3 + iL_4]\bar{\alpha} + [L_5 + iL_6]\delta\right\} \qquad (4\text{–}129)$$

$$\bar{M}_y = -4\rho_\infty b^4 \omega^2\left\{[M_1 + iM_2]\frac{h}{b} + [M_3 + iM_4]\bar{\alpha} + [M_5 + iM_6]\delta\right\} \qquad (4\text{–}130)$$

$$\bar{H} = -4\rho_\infty b^4 \omega^2\left\{[N_1 + iN_2]\frac{h}{b} + [N_3 + iN_4]\bar{\alpha} + [N_5 + iN_6]\delta\right\} \qquad (4\text{–}131)$$

Here L_1 and L_2 are functions of k only; but the remaining coefficients depend on the axis location a, on the dimensionless coordinate of the hingeline, or on both. For purposes of tabulation, their dependence on a

is easily separated out. Thus, those associated with vertical translation and pitching are rewritten in Ref. 4–71 as follows:

$$[L_3 + iL_4] = [L'_3 + iL'_4] - [a + 1][L_1 + iL_2] \qquad (4\text{--}132)$$

$$[M_1 + iM_2] = [M'_1 + iM'_2] - [a + 1][L_1 + iL_2] \qquad (4\text{--}133)$$

$$[M_3 + iM_4] = [M'_3 + iM'_4] - [a + 1]$$
$$\times \{[M'_1 + L'_3 + i(M'_2 + iL'_4)] - [a + 1][L_1 + iL_2]\}$$
$$(4\text{--}134)$$

(Note that in Ref. 4–71 the symbol $x_0 = (a + 1)/2$ is used for the axis location.) The tabulation of forces and moments associated with the flap is simplified by certain relations which exist between them and those for the full airfoil.

In view of the essential nonlinearity of steady, two-dimensional transonic flow, the foregoing results become meaningless as $k \to 0$. By parallel physical and mathematical reasoning, Landahl establishes the condition $k \gg |1 - M|$ for validity of the linearized theory (Ref. 4–72). In practice, results of flutter calculations and the like at $M = 1$ appear consistent with subsonic and supersonic counterparts when k exceeds 0.1–0.15. Landahl suggests that, in the neighborhood of sonic speed, the correct linearized differential equation to use is Eq. 4–29 with the $(1 - M^2)\partial^2 \varphi'/\partial x^2$ term completely omitted. On this basis, it can be shown that the kernel function for M near but not exactly equal to unity reads

$$K(x_0; k, M \cong 1) = - \frac{e^{-i(kx_0/2)}}{M\sqrt{2\pi ikx_0}} \qquad (4\text{--}135)$$

Another interesting examination of the limiting behavior of the subsonic and supersonic solutions as $M \to 1$ has been given by Jordan (Ref. 4–68).

One further important question concerns the influence of shocks adjacent to the airfoil surfaces at very high subsonic speeds. These discontinuities are not accounted for in the conventional linearized solution, but they are known to dominate the control-surface instability called *aileron buzz*. Preliminary approximate analyses of oscillatory transonic airfoil motions with shocks were published in 1958 by Coupry and Piazzoli (Ref. 4–73) and in 1959 by Eckhaus (Ref. 4–74), the former paper including some suggestive correlations with measured unsteady flow properties.

(d) Supersonic flow $(M > 1, M\delta \ll 1)$

As in the transonic case, either the direct or inverse kernel is available at supersonic speeds, the former being more useful for computational purposes:

$$\mathscr{A} = 4\left(\frac{\partial}{\partial \bar{x}} + ik\right)\int_{-1}^{\bar{x}} K(x_0; k, M)(\cdots)\, d\xi \qquad (4\text{--}136a)$$

where

$$K(x_0; k, M) = -\frac{1}{\beta}\, e^{-i(kM^2 x_0/\beta^2)} J_0\!\left(\frac{kMx_0}{\beta^2}\right) \qquad (4\text{--}136b)$$

Again, this is a fully linearized form, and it is related to the velocity potential discontinuity across the airfoil. A straightforward derivation of Eqs. 4–136 is given by Garrick and Rubinow (Ref. 4–71). Applications require the evaluation of integrals containing K multiplied by various powers of ξ. These can, in turn, be expressed as combinations of functions of the following type:

$$f_\lambda = \int_0^1 u^\lambda e^{-i(2kM^2 u/\beta^2)} J_0\!\left(\frac{2kMu}{\beta^2}\right) du \qquad (4\text{--}137)$$

which are extensively tabulated by Huckel (Ref. 4–75) for $\lambda = 0, 1, 2, \cdots, 11$.

For the rigid airfoil with aerodynamically unbalanced flap, the tables of Refs. 4–71 and 4–76 provide coefficients in the notation of Eqs. 4–129–4–134 up to $M = 5$. Tables utilizing the British symbolism were published by Jordan (Ref. 4–68) and more recently, for airfoil and flap up to $M = 10/3$, by Minhinnick and Woodcock (Ref. 4–77). Reference 4–77 gives a useful comparison between British, French, American, and German notations. It should also be mentioned that Luke (Ref. 4–61) has tabulated data for the unflapped airfoil in the form of Eqs. 4–123 and 4–124.

In Sec. 4–3(c), we have already called attention to the effects of profile thickness in shifting the aerodynamic center forward. Thickness may also have important influences on the aerodynamic stiffness and damping of oscillating wings, and second-order approximate theories for supersonic speeds appeared as early as 1948 (W. P. Jones, Ref. 4–79). Van Dyke (Ref. 4–78) gives similar results for pitching and translation of a rigid airfoil, expanded in series up to the third power of reduced frequency. His theory has been systematized by Rodden and Revell (Ref. 4–80) for high-speed machine computation, with an approximate tip correction and sweep effect [in the manner of Sec. 4–4(e)] included. Even at low supersonic M, calculations for typical thickness ratios show large changes of damping-in-pitch from that predicted by linearized theory.

As the Mach number increases, Ref. 4–80 points out that van Dyke's aerodynamic operators approach the results of high-M expansions such as Landahl's (Ref. 4–24) and second-order piston theory. These latter forms are sufficiently general and elementary to be set down here. For instance, when terms of orders δ/M, δ/M^3, and δ^2 are retained in the two-dimensional, sinusoidal case from Ref. 4–24, we obtain

$$\mathscr{A} = -\frac{4}{M}\left[1 + \left(\frac{\gamma + 1}{2}\right)M\frac{dz_t}{dx}\right](\cdots)$$

$$-\frac{2}{M^3}\frac{\partial^2}{\partial\bar{x}^2}\int_{-1}^{\bar{x}}[\bar{x} - \xi]e^{-ik(\bar{x}-\bar{\xi})}(\cdots)\,d\xi \quad (4\text{–}138)$$

The first operator on the right is the point-function result of piston theory, which may be employed whenever $\delta \gg 1/M^3$ and $M^2 \gg 1$ but $M\,\delta \ll 1$. Landahl gives some computed values of aerodynamic lift and moment coefficients for the rigid airfoil in Ref. 4–24. Furthermore, a listing of all eighteen dimensionless piston-theory coefficients in the notation of Eqs. 4–129–4–134 will be found in Ref. 4–81. Those not involving the flap rotation are listed below:

$$L_1 = M_1 = 0 \quad (4\text{–}139)$$

$$L_2 = \frac{1}{Mk} \quad (4\text{–}140)$$

$$L_3' = \frac{1}{Mk^2} \quad (4\text{–}141)$$

$$L_4' = M_2' = \frac{1}{Mk} - \left(\frac{\gamma + 1}{4k}\right)\frac{A_w}{2b^2} \quad (4\text{–}142)$$

$$M_3' = \frac{1}{Mk^2} - \left(\frac{\gamma + 1}{4k^2}\right)\frac{A_w}{2b^2} \quad (4\text{–}143)$$

$$M_4' = \frac{4}{3Mk} - \left(\frac{\gamma + 1}{4k}\right)\frac{M_w}{b^3} \quad (4\text{–}144)$$

Equations 4–139–4–144 require both leading and trailing edges to be sharp-pointed, but the extension to a blunt trailing edge is trivial. The quantities A_w and M_w are the area of the cross section and the first moment of this area about the leading edge, respectively.

(e) Hypersonic flow $(M^2 \gg 1$, $M\,\delta = 0(1)$ or greater)

The comments made in the first paragraph of Sec. 4–3(d) regarding approximate hypersonic theories apply equally well to the unsteady motion of airfoils, except that considerably less work has been published to date on time-dependent problems. One attempt at a survey appears in the paper by Morgan, Runyan, and Huckel (Ref. 4–82).

We call attention here to three approximations which seem, among them, to cover the hypersonic range. For the lower values of $M\,\delta$, satisfactory results can be obtained from various orders of piston theory (Ref. 4–83). All such formulas are based on the assumption of isentropic flow, so that aerodynamic operators carried beyond the second order (the bracketed term on the right of Eq. 4–138) have no rigorous theoretical foundation. This order is accurate, however, only up to roughly $M\,\delta = \frac{1}{2}$. As suggested by Lighthill (Ref. 4–23) and elaborated in Ref. 4–83, useful operators, both linear and nonlinear, can be obtained by higher-degree binomial expansions of the expression

$$\frac{p}{p_\infty} = \left[1 + \frac{\gamma - 1}{2}\frac{w}{a_\infty}\right]^{2\gamma/\gamma-1} \tag{4-145}$$

Equation 4–145 gives the instantaneous isentropic pressure p on the face of a piston moving with velocity $w(t)$ into a perfect gas which is confined in a one-dimensional channel and has undisturbed properties identified by the subscript ∞.

As one typical illustration, consider the following operator:

$$\mathscr{A} = -\frac{4}{M}\left[1 + \left(\frac{\gamma + 1}{2}\right)M\frac{dz_t}{dx}\right]^{(\gamma + 1/\gamma - 1)}(\cdots) \tag{4-146}$$

Equation 4–146 is derived by assuming $\dfrac{dz_t}{dx} \gg \left|\dfrac{\bar{w}_a}{U}\right|$ and retaining linear terms in the binomial expansion of those parts of Eq. 4–145 that involve the motion but not the thickness distribution of the airfoil. This is the most exact linear form that can be extracted from Eq. 4–145; beyond it, effects of $(w_a/U)^3$ being to appear.

When $M\,\delta$ has intermediate values, say 1 or 2, some adaptation of the Eggers and Syvertson shock-expansion procedure (Ref. 4–29) is indicated. In their report, these authors suggest that unsteady airfoil motion can be handled by a stepwise process, in which individual fluid particles are traced along the wing surface and the time history of pressure computed for each. Since this method is unwieldy to apply and difficult to linearize, Zartarian has proposed (Ref. 4–84) a scheme for combining van Dyke's hypersonic similarity (Ref. 4–51) with the shock-expansion. Reference 4–51 demonstrates, when the velocity disturbances are not very large compared with U, that each y-z-fluid slab moves past a thin airfoil without appreciable distortion in the x-direction (i.e., $u \ll v,\ w$). On this basis, flow past the oscillating surface is equivalent to a series of steady flows past distorted surfaces, each one derived from the boundary conditions experienced by one particular slab. The steady flows are computed as in Ref. 4–29, taking advantage of the well-known fact that the oblique shock strengths at the leading edge (assumed sharp) are determined by the instantaneous effective turning angles on the upper and lower surfaces.

Zartarian's method involves considerable algebraic manipulation, and the aerodynamic operators differ from one configuration to another, so that no inclusive form can be set down here. To give some indication of the relative accuracies of two versions of piston theory and the shock expansion method, however, Table 4–1 shows the real and imaginary parts of the dimensionless moment due to pitching oscillation about the leading edge (cf. Eqs. 4–130 and 4–134) for a 12%-thick symmetrical double-wedge airfoil at $M = 10$.

TABLE 4–I
Dimensionless moment due to pitching about the leading edge of a 12%-thick symmetrical double-wedge airfoil at $M = 10$ ($M\delta = 1.24$)

Coefficient	Second-order piston theory	Third-order piston theory	Shock expansion
$k^2 M_3{}'$	0.0259	0.1174	0.0971
$k M_4{}'$	−0.0149	0.1072	0.076
C.P. location (% chord behind L.E.)	12.9	30.7	28.6

Negative $M_4{}'$ corresponds to positive work per cycle done by the airstream on the oscillation, but the pitching instability predicted by second-order piston theory for the airfoil about its leading edge is seen not to exist in the light of the more precise calculation. It is worth noting, in this connection, that on physical grounds the piston-theory model should never yield negative damping on a single-degree-of-freedom oscillation, since the normal velocity of any point on the airfoil surface produces an opposing change in pressure; this can be seen, for example, from the fact that the *linearized M_4* (Ref. 4–83) is positive irrespective of pitch-axis location.

For $M\delta \gg 1$, Newtonian theory (Ref. 4–30) supplies the limiting forms of the aerodynamic operators to be used on blunt noses and forward parts of airfoils, where the outward normal is tilted toward the oncoming airstream. Rearward portions are essentially in vacuum. Section 4–3(d) previously discussed the basic physical concept, and it can be adapted to unsteady flow simply by observing that there are additional relative velocities between the impinging fluid and the boundary produced by small time-dependent motions. Thus, when the upper and lower surfaces are described by $z_U(x, t)$ and $z_L(x, t)$, respectively, (Fig. 4–1) the pressure difference is given without linearization by

$$\frac{\Delta p_a}{q} = 2\left\{ \frac{\left[\dfrac{1}{U}\dfrac{Dz_L}{Dt}\right]^2}{1 + \left[\dfrac{1}{U}\dfrac{Dz_L}{Dt}\right]^2} - \frac{\left[\dfrac{1}{U}\dfrac{Dz_U}{Dt}\right]^2}{1 + \left[\dfrac{1}{U}\dfrac{Dz_U}{Dt}\right]^2} \right\} \qquad (4\text{--}147)$$

Here D/Dt is the substantial derivative, which has the following forms

$$\frac{Dz_L}{Dt} = \begin{cases} U\dfrac{\partial z_L}{\partial x} + \dfrac{\partial z_L}{\partial t}, & \text{(arbitrary motion)} \\[2ex] U\dfrac{\partial z_L}{\partial x} + i\omega z_L, & \text{(simple harmonic motion)} \end{cases} \qquad (4\text{--}148)$$

When Eq. 4–147 is linearized under the assumption of small w_a/U, although not necessarily small thickness slope dz_t/dx, one obtains the linear operator

$$\mathscr{A} = -\frac{8\dfrac{dz_t}{dx}(\cdots)}{\left[1 + \left(\dfrac{dz_t}{dx}\right)^2\right]^2} \qquad (4\text{--}149)$$

Equations 4–147 and 4–149 do not contain the centrifugal term associated with curved surfaces; it is often small but can be introduced following Refs. 4–30 and 4–82. Hayes and Probstein (Ref. 4–163, Sec. 3.7) have refined unsteady Newtonian theory to adjust in a more rational fashion for the pressure change that occurs through the thin shock layer. Their result is difficult to apply, however, since it calls for information on momentum flux in the layer which is not ordinarily available.

By comparing Eq. 4–149 with Eq. 4–76—or Eq. 4–146 with its steady-flow counterpart—we see that all the hypersonic aerodynamic operators for which we are able to supply simple, general expressions are *quasi-steady*. That is, the unsteadiness of the motion enters only through the time dependence of the boundary values w_a/U. Although this fact reflects, in part, our incomplete knowledge of the theory, there is a strong presumption that quasi-steady airloads can be used in the majority of aero-elastic analyses at these speeds. We may expect many calculations to be facilitated, as compared to previous experience with subsonic and supersonic flow. The need to distinguish simple harmonic from other types of motion disappears, and such operators as Eqs. 4–146 and 4–149 apply beyond the sinusoidal restriction made in the present section.

4–6   AIRFOILS IN TWO-DIMENSIONAL UNSTEADY FLOW; TRANSIENT MOTION, GUST LOADING, AND ACCELERATED FLIGHT

When a thin airfoil executes arbitrary small motions $z_a(x, t)$ normal to its chord-plane in a two-dimensional stream of constant velocity U, aerodynamic operators resembling Eq. 4–44 can be derived for some ranges of the flight Mach number.* Few similar results are available for three-dimensional wings, however, and in most cases the easiest way to calculate loads caused by maneuvers and gusts seems to be by superposition of sinusoidal or indicial quantities. Current methods of analyzing transient aeroelastic problems either start out from some elementary input, such as a step function, or rely on a statistical formulation which, in turn, requires information only on simple harmonic operators [cf. Secs. 6–4(c), 7–6(c), etc.]. Running or total lifts and pitching moments are usually needed, as are sometimes higher-order generalized forces; but the details of the pressure distribution have little intrinsic interest.

For these reasons, we follow the lead of the published literature and focus our attention here on resultant loads produced by certain step inputs. In the present section, the chordwise-rigid airfoil is considered; consequently the only loads to be determined are the lift $L(t)$ and pitching moment $M_y(t)$ per unit span. The manner in which these can be related to the displacements $h(t)$ and $\alpha(t)$ is set forth, among other places, in the general papers of Heaslet, Lomax, et al. (Refs. 4–129 and 4–130) or in Sec. 6–5 of Ref. 1–2. As a first step, we define two indicial motions commencing at $t = 0$: a vertical translation h_0 and an angular velocity q_0 about some fixed axis, such as the leading edge. These are pictured in Fig. 4–5, from which we see that q_0 involves maintaining the chordline tangent to the flight path at the chosen axis, so that the angle of attack remains zero there. In effect, this procedure means that we make a temporary transformation from the *wind-tunnel axes*, in which h and α are measured, to the *body axis* system used in dynamic stability. This is done in order to prevent the lift and moment generated by the angular motion from diverging toward infinity, as they would if we started with a

---

* Specifically, $\mathscr{A}$ has been given for sonic and supersonic (linearized theory), high supersonic (expansion in $1/M$, Ref. 4–24), and hypersonic (Newtonian theory) speeds. The incompressible-flow operator may be constructed by combining formulas due to Neumark (Ref. 4–127) and von Kármán and Sears (Ref. 4–128).

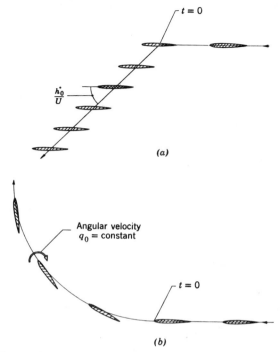

Fig. 4–5. Wings performing the two indicial motions referred to in the text. (*a*) Indicial vertical translation. (*b*) Indicial pitching about the leading edge.

step variation of α. The loads associated with h_0 and q_0 are written as follows

$$L_T(s) = 2\pi q(2b)\frac{h_0}{U}\,\phi(s) \qquad (4\text{--}150)$$

$$M_{yT}(s) = 2\pi q(2b)^2\,\frac{h_0}{U}\,\phi_M(s) \qquad (4\text{--}151)$$

$$L_q(s) = 4\pi q(2b)\left(\frac{q_0 b}{U}\right)\varphi_q(s) \qquad (4\text{--}152)$$

$$M_{yq}(s) = 4\pi q(2b)^2\left(\frac{q_0 b}{U}\right)\varphi_{Mq}(s) \qquad (4\text{--}153)$$

Here

$$s = \frac{Ut}{b} \qquad (4\text{--}154)$$

is the distance in semichords travelled after the start of the maneuver. To obtain total lift and pitching moment on a two-dimensional lifting surface of plan area S, one of the factors $(2b)$ should be replaced by S in each of Eqs. 4-150-4-153. (The same is true of many subsequent formulas.) The dimensionless indicial functions $\varphi(s)$, $\varphi_M(s)$, $\varphi_q(s)$, and $\varphi_{Mq}(s)$ are rather arbitrarily defined (there is no agreement in the literature), except that the 2π causes $\varphi(s)$ to approach an asymptote of unity in incompressible flow. One logical procedure, followed in several NACA publications, is to use indicial functions referred to their ultimate steady-state values. For example,

$$k_1(s) \equiv \frac{\varphi(s)}{\varphi(\infty)} \tag{4-155}$$

By means of the *Duhamel superposition integral*, it is a simple matter to show that the running loads resulting from arbitrary small $h(t)$ and $\alpha(t)$, commencing at $t = 0$, are

$$L(s) = 2\pi q(2b)\left\{\left[\alpha(0) + \frac{h(0)}{U}\right]\varphi(s)\right.$$

$$+ \int_0^s \frac{d}{d\sigma}\left[\alpha(\sigma) + \frac{h(\sigma)}{U}\right]\varphi(s - \sigma)\,d\sigma\right\}$$

$$+ 4\pi q(2b)\left\{\frac{b}{U}\dot{\alpha}(0)\varphi_q(s) + \int_0^s \frac{b}{U}\frac{d\dot{\alpha}(\sigma)}{d\sigma}\,\varphi_q(s - \sigma)\,d\sigma\right\} \tag{4-156}$$

$$M_y(s) = 2\pi q(2b)^2\left\{\left[\alpha(0) + \frac{h(0)}{U}\right]\varphi_M(s)\right.$$

$$+ \int_0^s \frac{d}{d\sigma}\left[\alpha(\sigma) + \frac{h(\sigma)}{U}\right]\varphi_M(s - \sigma)\,d\sigma\right\}$$

$$+ 4\pi q(2b)^2\left\{\frac{b}{U}\dot{\alpha}(0)\varphi_{Mq}(s) + \int_0^s \frac{b}{U}\frac{d\dot{\alpha}(\sigma)}{d\sigma}\,\varphi_{Mq}(s - \sigma)\,d\sigma\right\} \tag{4-157}$$

Here the superscript dot denotes the derivative with respect to physical time; replacing it in terms of an s- or σ-derivative requires another multiplication by U/b. The chordwise axis whose displacement is h and the moment axis are both supposed to coincide with the one used in defining angular velocity q_0. Transfer of the loading to another axis can be accomplished in the usual way.

A second fundamental transient aerodynamic problem arises when the airfoil meets atmospheric turbulence or a blast front, which causes a time-dependent change in angle of attack. One typical situation of this sort

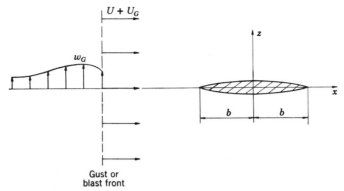

Fig. 4–6. Approaching gust or blast front observed in a coordinate system attached to the airfoil. The uniform airstream is moving with flight speed U. Generally, $w_G = f[x - (U + U_G)t]$.

is illustrated in Fig. 4–6. As usual, the airfoil itself has a flight speed U; but the front envelops the chord with an effective x-velocity $U + U_G$. It is convenient to set $t = 0$ as the instant when the front meets the leading edge $x = -b$. At any time thereafter, the normal z-component w_G of gust or blast velocity has exactly the same aerodynamic effect as a chordwise deformation which produces an upwash

$$w_a = -w_G \tag{4–158}$$

Moreover, since atmospheric motion occurs slowly relative to the envelopment rate, it is logical to assume that w_G is fixed to the front and, consequently, a function of the combination $[x - (U + U_G)t]$ only. Naturally, $w_G \ll U$; otherwise catastrophic accidents would befall aircraft much more commonly.

By analogy with h_0 and q_0, we can imagine a step gust of constant intensity w_0, such that

$$w_G(x, t) = w_0\, 1\!\left(t - \frac{x}{U + U_G}\right) \tag{4–159}$$

The running lift and moment associated with w_0 are written

$$L(s) = 2\pi q(2b)\frac{w_0}{U}\,\psi(s) \tag{4–160}$$

$$M_y(s) = 2\pi q(2b)^2 \frac{w_0}{U}\,\psi_M(s) \tag{4–161}$$

These indicial functions $\psi(s)$ and $\psi_M(s)$ are parametrically dependent on Mach number and the ratio U_G/U. In NACA literature, $\psi(s)$ is often replaced by

$$k(s) \equiv \frac{\psi(s)}{\psi(\infty)} \tag{4-162}$$

arranged to have a unit asymptote like $k_1(s)$; $k(s)$ is referred to as $k_2(s)$ for a *natural gust* when $U_G = 0$. Clearly it follows from Eqs. 4-160 and 4-161 that the loads due to an arbitrary w_G, for fixed values of U and U_G/U, are

$$L_G(s) = 2\pi q(2b)\left\{ \frac{w_G(0)}{U}\,\psi(s) + \int_0^s \frac{d}{d\sigma}\left[\frac{w_G(\sigma)}{U}\right]\psi(s-\sigma)\,d\sigma \right\} \tag{4-163}$$

$$M_{yG}(s) = 2\pi q(2b)^2\left\{ \frac{w_G(0)}{U}\,\psi_M(s) + \int_0^s \frac{d}{d\sigma}\left[\frac{w_G(\sigma)}{U}\right]\psi_M(s-\sigma)\,d\sigma \right\} \tag{4-164}$$

$w_G(\sigma)$ is defined here as *the normal velocity which strikes the leading edge* at time $t = b\sigma/U$.

Until recently, the gust problem centered around natural turbulence embedded in the atmosphere, for which the effective rate of convection relative to the flying vehicle is U, and $U_G = 0$. Therefore, most of the widely available literature concerns the calculation of $\psi(s)$ and $\psi_M(s)$ for this particular case. On the other hand, the design of many military aircraft today involves a study of the response to nuclear blasts. These may approach from any direction, with fronts which propagate at the speed of a shock wave. The contribution U_G to the chordwise envelopment rate, in two-dimensional cases, is equal to the propagation speed divided by the cosine of the angle between the x-direction and the direction toward which the front proceeds. For an actual airplane in flight, there may also be a spanwise component of frontal motion, so that the geometry of envelopment is a little more complicated. It is evident, nevertheless, that U_G can have values throughout most of the range between* $+\infty$ and $-\infty$ and that the blast may come from ahead ($U_G > -U$) or behind ($U_G < -U$). This situation is fully discussed by Drischler and Diederich (Ref. 4-131), who also point out that, according to linearized theory, only those components of blast disturbance which are parallel to the z-direction (i.e., w_G) produce appreciable loads. An additional source of blast loading is the overpressure behind the shock, which may produce high stresses locally in wing or fuselage skins; discussion of this effect exceeds the scope of what we are able to cover here (see Ludloff, Ref. 4-132).

---

* Note that, at both these limits, the indicial functions $\psi(s)$ and $\psi_M(s)$ must become identical with $\varphi(s)$ and $\varphi_M(s)$, respectively.

In connection with the subject of gusts, we mention one other formation that plays a major role in the statistical theory of turbulence and elsewhere. This is the so-called *sinusoidal gust*—a simple harmonic disturbance embedded in the atmosphere and described (in complex notation) by

$$w_G(x, t) = \bar{w}_G e^{i\omega[t - (x/U)]} \qquad (4\text{-}165)$$

After a wing or airfoil has proceeded through this gust for a sufficient length of time, the lift and moment settle down to the forms given by Eqs. 4–116–4–117 and may be calculated by the methods of Sec. 4–5. In anticipation of future needs, we shall reproduce below a few expressions for airloads associated with Eq. 4–165. The following notation is adopted:

$$L_G \equiv \bar{L}_G e^{i\omega t} = q(2b) \frac{\bar{w}_G}{U} e^{i\omega t} \bar{C}_{LG} \qquad (4\text{-}166)$$

$$M_G \equiv \bar{M}_G e^{i\omega t} = q(2b)^2 \frac{\bar{w}_G}{U} e^{i\omega t} \bar{C}_{MG} \qquad (4\text{-}167)$$

The remainder of this section is devoted largely to identifying sources where formulas and curves of the various indicial functions may be found. Certain equations are written out, either when they have unusual interest or fill special needs for applications in Chaps. 6 through 9. Section 4–6(e) treats briefly the problem of varying forward speed $U(t)$.

Beyond the references already cited, there are several general papers worthy of the attention of the reader who is attracted by the linearized theory of unsteady airfoil motion. Among others, we single out the following: for incompressible flow, Küssner (Ref. 4–133) and Fraeys de Veubeke (Ref. 4–134); for all flight speeds, Timman (Ref. 4–135) and Rott (Ref. 4–136); for applications of Rott's moving-source method, Ordway (Ref. 4–137).

(a) Incompressible flow ($M^2 \ll 1$)

Only when the fluid density is constant can a separation be made between circulatory and noncirculatory airloads, as pointed out in Sec. 4–5(a). Moreover, when the airfoil performs chordwise-rigid motions, the running circulatory lift depends only on the upwash at the $\frac{3}{4}$-chord station,

$$w_{3/4\,c} = -[\dot{h} + U\alpha + b(\tfrac{1}{2} - a)\dot{\alpha}] \qquad (4\text{-}168)$$

(cf. Eq. 4–126); and the lift associated with the function $C(k)$ always acts at the quarter-chord point. As a consequence, we need define only a single indicial function $\varphi(s)$, giving the *circulatory* force due to unit change $\alpha_0 1(t)$ in the angle of attack at the $\frac{3}{4}$-chord line:

$$L_{\text{circ}}(s) = 2\pi q(2b)\alpha_0 \varphi(s) \qquad (4\text{-}169)$$

In place of the four separate indicial functions needed for Eqs. 4–156 and 4–157, we discover the following simple formulas (for derivations, see Secs. 5–6 and 5–7 of Ref. 1–2 or the references given therein):

$$L(s) = \pi \rho_\infty b^2[\ddot{h} + U\dot{\alpha} - ba\ddot{\alpha}] \tag{4-170}$$

$$+ 2\pi \rho_\infty Ub \int_0^s \frac{d}{d\sigma}[h(\sigma) + U\alpha(\sigma) + b(\tfrac{1}{2} - a)\dot{\alpha}(\sigma)]\,\varphi(s - \sigma)\,d\sigma$$

$$M_y(s) = \pi \rho_\infty b^2[ba\ddot{h} - Ub(\tfrac{1}{2} - a)\dot{\alpha} \tag{4-171}$$

$$- b^2(\tfrac{1}{8} + a^2)\ddot{\alpha}] + 2\pi\rho_\infty Ub^2(\tfrac{1}{2} + a)\int_0^s \frac{d}{d\sigma}[h(\sigma)$$

$$+ U\alpha(\sigma) + b(\tfrac{1}{2} - a)\dot{\alpha}(\sigma)]\,\varphi(s - \sigma)\,d\sigma$$

Here, as previously, dots indicate differentiation with respect to physical time; and resultant loads can be found by replacing b with $S/2$.

The function $\varphi(s) \equiv k_1(s)$ was determined by Wagner (Ref. 4–138) in one of the earliest contributions to unsteady wing theory. Its calculation, using Fourier integral methods, is discussed by Garrick (Ref. 4–139) and, using Laplace transforms, by Sears (Ref. 4–140). Two convenient curve-fittings are

$$\varphi(s) \cong 1 - 0.165e^{-0.0455s} - 0.335e^{-0.3s} \tag{4-172a}$$

and

$$\varphi(s) \cong \frac{s + 2}{s + 4} \tag{4-172b}$$

the former more accurate in the intermediate range of s and the latter showing the correct asymptotic behavior.

For the natural gust ($U_G = 0$), the lift can also be proved to act at the quarter chord, so that only $\psi(s)$ is required. This is usually called Küssner's function, and numerical values appear in the paper by Sears (Ref. 4–140). Curve-fittings similar to Eqs. 4–172 are

$$\psi(s) \cong \begin{cases} 1 - 0.500e^{-0.130s} - 0.500e^{-s} \\ \dfrac{s^2 + s}{s^2 + 2.82s + 0.8} \end{cases} \tag{4-173a, b}$$

Both $\psi(s)$ and $\psi_M(s)$ for positive and negative U_G are extensively plotted in Ref. 4–131. Further information on the blast-wave problem will be found in a thesis by Hobbs, of which Ref. 4–141 is a précis, and in a paper by Miles (Ref. 4–142).

The sinusoidal gust loads for incompressible flow, in the notation of Eqs. 4–166 and 4–167, read as follows:

$$\bar{C}_{LG}(k, M = 0) = 2\pi\{C(k)[J_0(k) - iJ_1(k)] + J_1(k)\} \tag{4-174}$$

$$\bar{C}_{MG}(k, M = 0) = \left[\frac{a}{2} + \frac{1}{4}\right]\bar{C}_{LG}(k, M = 0) \tag{4-175}$$

These are also a development of Ref. 4–140.

(b) Subsonic flow $(M < 1)$

Some formulas and calculations on all six of the indicial functions for subsonic flight are given in Ref. 4–130. The most complete work, however, is that which has been done by Mazelsky and Drischler using Fourier-integral superposition of sinusoidal airloads. In Refs. 4–143, 4–144, and 4–145 appear numerical values and exponential curve-fittings at $M = 0.5$, 0.6, and 0.7 on $\varphi(s)$, $\varphi_M(s)$, $\varphi_q(s)$, $\varphi_{M_q}(s)$ (axis at the quarter-chord line), and $\psi(s)$ (for $U_G = 0$). At certain other values of U_G, Krasnoff (Ref. 4–145) reports $\psi(s)$ and $\varphi(s)$. Finally, Drischler has published $\bar{C}_{LG}(k, M)$ for a sinusoidal gust (Ref. 4–147).

(c) Transonic flow $(M \cong 1)$

Indicial functions for the sonic case $M = 1$ can easily be developed by taking the limit as $M \to 1$ of the supersonic formulas. In Sec. 6–8 of Ref. 1–2, for instance, φ, φ_M, φ_q, and φ_{M_q} are listed, along with ψ and ψ_M for the natural gust. As an illustration,

$$\varphi(s) = \begin{cases} \dfrac{2}{\pi}\dfrac{s}{}, \text{(for } 0 \le s \le 1) \\[2mm] \dfrac{2}{\pi^2}\left\{\cos^{-1}\left[1 - \dfrac{2}{s}\right] + 2\sqrt{s-1}\right\}, \text{(for } 1 \le s) \end{cases} \tag{4–176}$$

As would be expected from the essential nonlinearity of steady, two-dimensional transonic flow, all these functions blow up more or less rapidly as $s \to \infty$. This behavior does not seem to preclude their use in genuinely unsteady situations. At the lower end of the time scale, they constitute a reasonably continuous transition from their subsonic to supersonic counterparts.

(d) Supersonic flow $(M > 1, M\delta \ll 1)$

Fully linearized supersonic theory permits a relatively simple statement of the general aerodynamic operator, which we set down first:

$$\mathscr{A} = -\frac{4}{\pi\beta}\left(\frac{\partial}{\partial\tilde{x}} + \frac{b}{U}\frac{\partial}{\partial t}\right)\int_{-1}^{\tilde{x}}\int_{\tau_1}^{\tau_2}\frac{w_a(\xi, t - \tau)}{U}\frac{d\tau \, d\xi}{\sqrt{(\tau_2 - \tau)(\tau - \tau_1)}} \tag{4–177a}$$

where

$$\tau_1, \tau_2 = \frac{b(\tilde{x} - \xi)}{a_\infty(M + 1)}, \quad \frac{b(\tilde{x} - \xi)}{a_\infty(M - 1)} \tag{4–177b}$$

In view of Eqs. 4–177, all the indicial functions may be computed by straightforward but lengthy integrations. This has been done by Chang

(Ref. 4–148) and by Lomax et al. (Refs. 4–3, 4–129, and 4–130). Reference 4–130 presents φ, φ_M, φ_q, φ_{Mq} (for an axis at the leading edge), ψ, and ψ_M (for $U_G = 0$). Actually, the natural gust functions were first derived by Biot (Ref. 4–149). The formulas are too long to be reproduced in full here. Each of the indicial functions reaches its steady-state value (e.g., $\varphi = 2/\pi\beta$) after $s = 2M/(M-1)$, which represents the time required for all unsteady effects of starting to be swept off the trailing edge.

Regarding loads due to blast waves in supersonic flight, Ref. 4–131 plots ψ and ψ_M for $M = 2$ and many values of U_G/U. The sinusoidal gust functions, taken from Sec. 6–6 of Ref. 1–2, are the following:

$$\bar{C}_{LG}(k, M > 1) = \frac{4e^{ik}}{\beta} f_0 \tag{4-178}$$

$$\bar{C}_{MG}(k, M > 1) = \frac{e^{ik}}{\beta}\left[(2a + 1)f_0 - f_1\right] \tag{4-179}$$

where a is the dimensionless location of the moment axis and f_λ is the function defined by Eq. 4–137.

Because of their simplicity, the indicial functions according to the high supersonic theory of Landahl (Ref. 4–24) are noteworthy. The general aerodynamic operator, retaining as before the terms of order δ/M, δ/M^3, and δ^2, is*

$$\mathscr{A} = -\frac{4}{M}\left[1 + \left(\frac{\gamma + 1}{2}\right)M\frac{dz_t}{dx}\right]\frac{w_a}{U}$$

$$- \frac{2}{M^3}\frac{\partial^2}{\partial\tilde{x}^2}\int_{-1}^{\tilde{x}} [\tilde{x} - \xi]\frac{w_a}{U}\left(\tilde{\xi}, t - \frac{b(\tilde{x} - \tilde{\xi})}{U}\right)d\xi \tag{4-180}$$

Typical of the results which we can work out from Eq. 4–180 are the following:

$$\varphi(s) = \begin{cases} \dfrac{2}{\pi M}\left[1 + \dfrac{s}{4M^2}\right], & 0 \leq s \leq 2 \\[3mm] \dfrac{2}{\pi M}\left[1 + \dfrac{1}{2M^2}\right], & 2 < s \end{cases} \tag{4-181}$$

$$\psi(s) = \begin{cases} \dfrac{s}{\pi M}\left[1 + \dfrac{1}{2M^2}\right] + \left(\dfrac{\gamma + 1}{2\pi}\right)\dfrac{M}{b}z_t(x = b(s - 1)), & 0 \leq s \leq 2 \\[3mm] \dfrac{2}{\pi M}\left[1 + \dfrac{1}{2M^2}\right], & 2 < s \end{cases}$$

$$\tag{4-182}$$

---

* We remark that Eq. 4–180 may be adapted to wings of finite span, for regions uninfluenced by cut-off tips, by adding $\partial^2/\partial\tilde{y}^2$ to the $\partial^2/\partial\tilde{x}^2$ acting on the integral and including the dependence of w_a on $\tilde{y}$.

Here the second-order effects of thickness are accounted for in the usual way, but the airfoil is assumed to have pointed leading and trailing edges. For use in Chap. 6, we set down the very elementary reductions of Eqs. 4–181 and 4–182 which remain when both the thickness and the $1/M^3$-terms are negligible,

$$\varphi(s) = \frac{2}{\pi M} \quad \text{(a constant)} \tag{4–183}$$

$$\psi(s) = \begin{cases} \dfrac{s}{\pi M}, & 0 \le s \le 2 \\[2ex] \dfrac{2}{\pi M}, & 2 \le s \end{cases} \tag{4–184}$$

The problem of calculating indicial functions for hypersonic flight speeds has not been discussed in the literature. When $M\delta = 0(1)$, we may expect equations like (4–181) and (4–182) to furnish reasonable approximations, which can probably be improved by using an operator such as Eq. 4–146. The reader will have no difficulty in deriving corresponding forms from other quasi-steady hypersonic theories, such as the various versions of Newtonian theory described in Sec. 4–3(d).

(e) Unsteady effects of longitudinal acceleration

A subject of increasing interest is that of a lifting surface which undergoes rapid forward acceleration or deceleration while simultaneously performing lateral oscillations or some other unsteady motion. The general theoretical procedure for linearized treatment of the two-dimensional case was laid out by Lomax et al. (Ref. 4–130) and by Krasilshchikova (Ref. 4–151). A more recent discussion by Krasilshchikova (Ref. 4–152) deals with finite wings but is still somewhat remote from numerical applications. The acceleration problem is also discussed in one of the later chapters of Miles' monograph (Ref. 4–2), with emphasis on transonic drag calculation. When the flow can be regarded as incompressible, as during the launching of rockets from the ground or the catapulting of aircraft, the work of Ref. 4–150 may be useful.

We present here a brief review of the two-dimensional theory. Consider a coordinate system at rest with respect to the gas at infinity, the wing moving at speed $U(t)$ in the negative x-direction. The boundary condition of flow tangency must be satisfied along different segments of the x-axis at different instants:

$$\frac{\partial \varphi'(x, 0, t)}{\partial z} = w_a(x, t) \quad \text{(on } R_a) \tag{4–185}$$

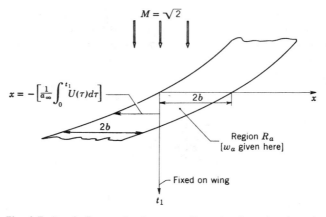

Fig. 4–7. Steady-flow analog for a two-dimensional accelerating wing.

Region R_a is shown in Fig. 4–7 for the case when the leading edge coincides with the origin at $t = 0$. A condition of zero pressure

$$\frac{\partial \varphi' \, (x, 0, t)}{\partial t} = 0 \qquad (4\text{–}186)$$

applies on the remainder of the x-axis (or x-t-plane).

Since small perturbations are assumed, the velocity potential is governed by the wave equation of acoustics, which reads as follows under the transformation $t_1 = a_\infty t$:

$$\frac{\partial^2 \varphi'}{\partial x^2} + \frac{\partial^2 \varphi'}{\partial z^2} = \frac{\partial^2 \varphi'}{\partial t_1^2} \qquad (4\text{–}187)$$

An exact mathematical analog for the problem defined by Eqs. 4–185 through 4–187 is a three-dimensional wing with the same upwash w_a but fixed on the x-t_1-plane in a stream moving at $M = \sqrt{2}$ in the positive t_1-direction (Fig. 4–7), t_1 being regarded as a third spatial coordinate. The flow is then steady.

In addition to Eq. 4–187, we have for boundary conditions

$$\frac{\partial \varphi'}{\partial z} = w_a(x, t_1)$$

$$\left(\text{for } \frac{-1}{a_\infty} \int_0^{t_1} U(\tau_1) \, d\tau_1 \le x \le 2b - \frac{1}{a_\infty} \int_0^t U(\tau_1) \, d\tau_1 \right) \qquad (4\text{–}188)$$

and

$$\frac{\partial \varphi'}{\partial t_1} = 0, \qquad \text{(elsewhere)} \qquad (4\text{–}189)$$

This analogous flow can be solved by aerodynamic influence coefficients, or by exact integration in simple cases, as in Ref. 4–130.

The effect of acceleration A, as a correction to the airloads computed on the basis of instantaneous values of $M(t)$ and $k(t) = \omega b/U(t)$, is governed by the parameter Ab/a_∞^2. The correction turns out to be very small if Ab/a_∞^2 is small, say less than 0.05 (corresponding to accelerations up to 200 g's on practical surfaces), and if $M \geq 1.5$. At lower supersonic speeds the effect is somewhat larger. These conclusions are based on some unpublished calculations.

A word of caution is in order here. The rapid accelerations of the order of 200 g's imply a rate of change of Mach number of order 6/sec. Thus for slowly harmonically oscillating airfoils, the airforces are far from harmonic, that is, w_a will change with time in a nonharmonic manner. This is generally the case in practice where during one cycle M can change quite appreciably. Thus the solution of the harmonically oscillating wing is by no means the only one of interest to the aeroelastician. However, the analog (Fig. 4–7) applies equally well to any arbitrary time-dependent motion. In such cases, the resultant equations of motion of the vehicle will be linear but will have time-dependent coefficients, a class of problems which we take up in Chap. 10. Moreover, the changes in ambient atmospheric properties during climbing or diving flight may have more significant effects than those of forward acceleration.

4–7 LIFTING SURFACES AND OTHER CONFIGURATIONS IN THREE-DIMENSIONAL UNSTEADY FLOW; SIMPLE HARMONIC MOTION

(a) Subsonic flow ($M < 1$)

$$\mathscr{A}^{-1} = \frac{1}{8\pi} \oint_{R_a} \oint K(x_0, y_0; k, M)(\cdots)\, d\xi\, d\bar{\eta} \qquad (4\text{–}190a)$$

where

$$K(x_0, y_0; k, M) = k^2 e^{-ikx_0} \left\{ \frac{1}{k|y_0|} K_1(k|y_0|) \right. \qquad (4\text{–}190b)$$

$$+ \frac{\pi i}{2k|y_0|}[I_1(k|y_0|) - L_1(k|y_0|)] - \frac{iMk|y_0| + \beta}{M\beta(ky_0)^2} e^{-i(Mk|y_0|/\beta)}$$

$$+ \int_0^{M/\beta} \sqrt{1+\tau^2}\, e^{-ik|y_0|\tau}\, d\tau - \frac{i}{M(ky_0)^2} \int_0^{kx_0} \exp\left(\frac{i[\lambda - M\sqrt{\lambda^2 + \beta^2(ky_0)^2}]}{\beta^2}\right) d\lambda$$

$$+ \frac{Mkx_0 + \sqrt{(kx_0)^2 + \beta^2(ky_0)^2}}{M(ky_0)^2\sqrt{(kx_0)^2 + \beta^2(ky_0)^2}} \exp\left(\frac{i[kx_0 - M\sqrt{(kx_0)^2 + \beta^2(ky_0)^2}]}{\beta^2}\right)\right\}$$

Here I_1 and K_1 are modified Bessel functions of the first order and the first and second kinds, respectively, while L_1 is a modified Struve function of the first order (see Watson, Ref. 4–85). Equation 4–190b relates to the lifting surface of Fig. 4–1 and is given in this form by Watkins, Runyan, and Woolston (Ref. 4–86). Note that K/k^2 is a function of only three distinct quantities, kx_0, $k|y_0|$, and M.

Methods of numerical solution of the subsonic load distribution problem, which involve no further approximations in the representation of the physical system and are particularly suited to high-speed machine computation, are developed in Refs. 4–87 through 4–90. Further information on recent NASA research and some comparisons with experiments appear in Refs. 4–91 and 4–92. We describe briefly here the method of Hsu (Ref. 4–89), which we consider to be typical and equally as efficient as, although not superior to, the other numerical procedures.

The solution proceeds by assuming an appropriate series of known functions with unknown coefficients to represent $\Delta \bar{p}_{a/q}$. These coefficients are related to the known upwash amplitudes at a number of points equal to the number of terms retained in the series, producing a set of linear algebraic equations. A suitable pressure formula, satisfying all edge conditions, is the following:

$$\frac{\Delta \bar{p}_a\,(\underline{\xi}, \eta)}{q} = \frac{8}{\tilde{b}(\eta)}\, s\sqrt{1 - \eta^2}\Big\{[a_{00} + a_{01}\eta + a_{02}\eta^2 + a_{03}\eta^3 + \cdots]\sqrt{\frac{1 - \underline{\xi}}{1 + \underline{\xi}}}$$

$$+ [a_{10} + a_{11}\eta + a_{12}\eta^2 + a_{13}\eta^3 + \cdots]\sqrt{1 - \underline{\xi}^2} \qquad (4\text{–}191)$$

$$+ [a_{20} + a_{21}\eta + a_{22}\eta^2 + a_{23}\eta^3 + \cdots]\underline{\xi}\sqrt{1 - \underline{\xi}^2} + \cdots\Big\}$$

Here $\underline{\xi}$ and η are coordinates chosen so as to transform the original wing (Fig. 4–8) into a square with sides at ξ, $\eta = \pm 1$. They are related to the dimensionless physical coordinates by

$$\underline{\xi} = \frac{\tilde{\xi} - \tfrac{1}{2}(\xi_t(\tilde{\eta}) + \xi_l(\tilde{\eta}))}{\tfrac{1}{2}(\xi_t(\tilde{\eta}) - \xi_l(\tilde{\eta}))}, \qquad \underline{x} = \frac{\tilde{x} - \tfrac{1}{2}(x_t(\tilde{\eta}) + x_l(\tilde{\eta}))}{\tfrac{1}{2}(x_t(\tilde{\eta}) - x_l(\tilde{\eta}))}$$

$$\eta = \frac{\tilde{\eta}}{s}, \qquad\qquad\qquad \underline{y} = \frac{\tilde{y}}{s} \qquad (4\text{–}192a, b, c, d)$$

where $\xi_l = \xi_l(\tilde{\eta})$ and $\xi_t = \xi_t(\tilde{\eta})$ define the locations of the leading and trailing edges.

Substituting the kernel function from Eq. 4–190b and the pressure series from Eq. 4–191 into equation 4–190a, we are faced with numerical integration problems that must be handled by a special technique because of the high-order singularities. Two distinct methods have been presented by Watkins et al. (Refs. 4–86, 4–87, and 4–89). In Ref. 4–86 the authors divide the wingspan into a number of strips (say 20). Within each strip the continuous chordwise pressure distribution is replaced in terms of a number of equivalent concentrated loads (say 4, at the $\frac{1}{8}$-, $\frac{3}{8}$-, $\frac{5}{8}$-, and $\frac{7}{8}$-chord points), following the "vortex-lattice" scheme.

Hsu's scheme is based on Chebysheff's numerical integration formula. Although derived in a different way, this procedure is very similar to Multhopp's (Ref. 4–31). The optimum chordwise and spanwise upwash collocation stations are arrived at by the condition that the pressure distribution given thereby will furnish the best possible approximation, in the sense that the total lift, the moment about the midchord, and the second and higher moments are exact for certain special cases. The accuracy achieved depends on the number of stations used and the complexity of the mode of wing vibration. After properly accounting for singularities, the numerical integrations both spanwise and chordwise are performed in terms of point values of the integrand. The locations of the stations for these point values follow from the condition that, for a given number of integration stations, the accuracy of the integration shall be a maximum under the circumstances. The stations turn out to be located, according to the Chebysheff formulas, in such a way that pressure and upwash points are interdigitated and never fall on the same spanwise lines. Hence no singularity is ever encountered when evaluating the kernel function. Values can readily be generated by a subroutine programmed for the high-speed computer, so that tables of the kernel function would appear to be unnecessary for most applications.

The transformation Eqs. 4–192 are first written out for the given planform. After choosing the numbers of upwash stations and integration

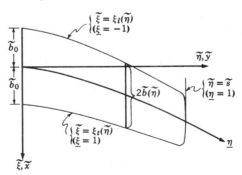

Fig. 4-8. Wing of arbitrary planform, showing dimensionless chordwise and spanwise variables.

stations, their locations follow from Chebysheff's formulas. Thus, if there are both m chordwise upwash collocation points and m chordwise integration stations, they are located, respectively, at

$$\underline{x}_i = -\cos\left(\frac{2i\pi}{2m+1}\right), \qquad i = 1, 2, \ldots, m \qquad (4\text{–}193a)$$

$$\underline{\xi}_j = -\cos\left(\frac{(2j-1)\pi}{2m+1}\right), \qquad j = 1, 2, \ldots, m \qquad (4\text{–}193b)$$

For R spanwise upwash collocation stations, there will be $(R+1)$ integration stations:

$$\underline{y}_r = -\cos\left(\frac{r\pi}{R+1}\right), \qquad r = 1, 2, \ldots, R \qquad (4\text{–}194a)$$

$$\underline{\eta}_s = -\cos\left(\frac{(2s-1)\pi}{2(R+1)}\right), \qquad s = 1, 2, \ldots, R+1 \qquad (4\text{–}194b)$$

The number of terms to be employed in the pressure expression (4–191) must equal the product mR; and it should be even or odd in the spanwise direction, depending on whether the motion is symmetrical or antisymmetrical. For the symmetrical case, the final algebraic system of equations is

$$4\pi\rho_\infty U \bar{w}_a(\tilde{x}_i, \tilde{y}_r) = 4\rho_\infty U^2 \Bigg\{ \frac{2\pi^2}{(R+1)(2m+1)} \sum_{s=1}^{R+1} \sum_{j=1}^{m} (1 - \underline{\eta}_s^2)$$

$$\cdot \Big([a_{00} + a_{02}\underline{\eta}_s^2 + a_{04}\underline{\eta}_s^4 + \cdots](1 - \underline{\xi}_j)$$

$$+ [a_{10} - a_{12}\underline{\eta}_s^2 + a_{14}\underline{\eta}_s^4 + \cdots](1 - \underline{\xi}_j^2)$$

$$+ [a_{20} + a_{22}\underline{\eta}_s^2 + a_{24}\underline{\eta}_s^4 + \cdots]\underline{\xi}_j(1 - \underline{\xi}_j^2) + \cdots \Big)$$

$$\cdot \left(\frac{s}{b_0}\right)^2 K(x_0, \underline{y}_0; k, M) \Bigg\}$$

$$+ 4\rho_\infty U^2 (R+1)\pi \Bigg\{ \sum_{j=1}^{m} \frac{2\pi}{2m+1} \Big([a_{00} + a_{02}\underline{y}_r^2 + a_{04}\underline{y}_r^4$$

$$+ \cdots](1 - \underline{\xi}_j)$$

$$+ [a_{10} + a_{12}\underline{y}_r^2 + a_{14}\underline{y}_r^4 + \cdots]$$

$$(1 - \underline{\xi}_j^2)$$

$$+ [a_{20} + a_{22}\underline{y}_r^2 + a_{24}\underline{y}_r^4 + \cdots]$$

$$\cdot \underline{\xi}_j (1 - \underline{\xi}_j^2) + \cdots \Big)$$

$$\cdot (e^{-ik(\underline{x}_i - \underline{\xi}_j)} - 1)\left[1 + \frac{\underline{x}_i - \underline{\xi}_j}{|\underline{x}_i - \underline{\xi}_j|}\right]\Bigg\}$$

$$+ 4\rho_\infty U^2(R + 1)2\pi\Bigg\{[a_{00} + a_{02}\underline{y}_r{}^2 + a_{04}\underline{y}_r{}^4 + \cdots]$$

$$\cdot \left[\sqrt{1 - \underline{x}_i{}^2} + \sin^{-1}\underline{x}_i + \frac{\pi}{2}\right]$$

$$+ [a_{10} + a_{12}\underline{y}_r{}^2 + a_{14}\underline{y}_r{}^4 + \cdots]$$

$$\cdot \left[\frac{\underline{x}_i}{2}\sqrt{1 - \underline{x}_i{}^2} + \tfrac{1}{2}\sin^{-1}\underline{x}_i + \frac{\pi}{4}\right]$$

$$+ [a_{20} + a_{22}\underline{y}_r{}^2 + a_{24}\underline{y}_r{}^4 + \cdots]$$

$$\cdot [-\tfrac{1}{3}(1 - \underline{x}_i{}^2)^{3/2}] + \cdots\Bigg\} \qquad (4\text{--}195)$$

In Eq. 4–195, $K(x_0, y_0; k, M)$ represents the kernel function evaluated for $x_0 = (\tilde{x}_i - \underline{\xi}_j)$ and $y_0 = (\tilde{y}_r - \tilde{\eta}_s)$.

Without further elaboration, it is evident that Eqs. 4–191 and 4–195 can be written at the mR upwash stations in the following matric forms:

$$\left\{\frac{\Delta \bar{p}_a}{q}\right\} = [K_1]\{a\} \qquad (4\text{--}196)$$

$$\left\{\frac{\bar{w}_a}{U}\right\} = [K_2]\{a\} \qquad (4\text{--}197)$$

There are mR elements in each column matrix. Square matrix $[K_1]$ depends on the planform geometry and the dimensionless coordinates $\tilde{x}_i, \tilde{y}_r$; whereas $[K_2]$ requires, additionally, the specification of k and M. Elimination of the pressure series between Eqs. 4–196 and 4–197 yields the algebraic representation of the aerodynamic operator,

$$\left\{\frac{\Delta \bar{p}_a}{q}\right\} = [\mathscr{A}]\left\{\frac{\bar{w}_a}{U}\right\} \qquad (4\text{--}198)$$

where

$$[\mathscr{A}] = [K_1][K_2]^{-1} \qquad (4\text{--}199)$$

Although complex algebra is involved in the inversion of $[K_2]^{-1}$, many successful, practical calculations of the type described above have been carried out. The presence of a control surface can be included either by refining the network of collocation points and artificially smoothing the upwash discontinuity at its leading edge or, more rigorously, by adding the proper logarithmic singularity along this line to the pressure series. Full details of the latter procedure have not yet been published.

Finally, it should be mentioned that many aeroelastic applications do not require a complete evaluation of the load distribution according to Eq. 4-191, but only certain weighted integrals (generalized forces) of it over the planform. Some details of the latter procedure are given by Hsu and Weatherill (Ref. 4-164).

Reference 4-90 shows that the amplitude of total lift on a wing of constant chord performing a vertical translation oscillation of amplitude $\bar{h}/b$ is

$$\bar{L} = 8\pi^2 q l^2 \left\{ \frac{1}{2}\left[a_{00} + \frac{a_{10}}{2} \right] + \frac{1}{8}\left[a_{02} + \frac{a_{12}}{2} \right] + \frac{1}{16}\left[a_{04} + \frac{a_{14}}{2} \right] \right\} \frac{\bar{h}}{b} \qquad (4\text{-}200)$$

Here l is the semispan, and three chordwise and three spanwise stations on the half-wing are used in the collocation process.

One can imagine many interesting special cases of the general method for subsonic oscillatory load determination just outlined; indeed, they effectively encompass the profusion of approximate theories in the published literature. Thus, when the fluid is incompressible, the kernel function reduces to

$$K(x_0, y_0; k, M = 0) = k^2 e^{-ikx_0} \left\{ \frac{1}{k|y_0|} K_1(k|y_0|) + \frac{\pi i}{2k|y_0|} [I_1(k|y_0|) - L_1(k|y_0|)] \right.$$

$$+ \frac{kx_0}{(ky_0)^2 \sqrt{(kx_0)^2 + (ky_0)^2}} - \frac{i\sqrt{(kx_0)^2 + (ky_0)^2}}{(ky_0)^2} e^{ikx_0}$$

$$\left. - \frac{1}{(ky_0)^2} \int_0^{kx_0} \sqrt{\lambda^2 + (ky_0)^2}\, e^{i\lambda}\, d\lambda \right\} \qquad (4\text{-}201)$$

There exist more than a score of methods designed to simplify the solution of the integral equation when $M = 0$ alone, covering the full range of aspect ratios from fractional to those where spanwise lift and moment distributions are the only quantities required.

An excellent summary of most of the incompressible-flow schemes appears in a report by Laidlaw (Ref. 4-93). A noteworthy example is that of W. P. Jones (Ref. 4-94), for which many applications have been published (e.g., Refs. 4-95, 4-96, 4-97), along with extensive tables of upwash coefficients (Ref. 4-98).

When the number m of chordwise collocation stations is set equal to unity, we obtain a theory of lifting-line type. Particularly numerous are the incompressible lifting-line approximations based on various modifications of the oscillating vortex sheet. The earliest of these is Cicala's

(Ref. 4–99), but perhaps most widely used in the United States is the one by Reissner and Stevens (Refs. 4–100). Their technique is well-systematized for computation and yields the fortunate result that the only change in the strip theory aerodynamic operator, Eq. 4–120, consists of an additive correction $\sigma(\bar{y})$ to $C(k)$. This function, which represents a purely circulatory effect, is found from the given planform shape and mode of motion by Fourier-series solution of a spanwise integral equation. Finite-span adjustments can be obtained in this way for the two-dimensional coefficients $L_h, L_\alpha, \cdots$ in Eqs. 4–123–4–125. On straight wings, such lifting-line theories are accurate down to aspect ratios of 3 or below when $M^2 \ll 1$.

The subsonic kernel function may be expanded in ascending powers of reduced frequency k (Eq. 54 of Ref. 4–86 is an illustration), and one is thus led to another sort of approximation. Reference 4–101 is concerned with low-frequency problems of dynamic stability and shows, for all Mach numbers, how important simplifications are achieved by retaining terms only up to $0(k)$. An extended Prandtl-Glauert compressibility correction has been worked out by Miles (Ref. 4–102). The subsonic theory of W. P. Jones (Ref. 4–103) also falls into this category, and typical calculations are given by Lehrian (Ref. 4–104).

Turning to the problem of simple harmonic motion of slender, pointed wings and bodies, we can rely in part on the discussion of Sec. 4–4(a), because the Munk-Jones hypothesis of two-dimensional, incompressible flow in planes normal to the flight direction retains considerable usefulness. In Ref. 4–2, Miles gives the following restrictions on its validity: for planar surfaces, $\beta R \ll 1$ and $kM^2 R^2 \ll 1$; for bodies,* $M\delta \ll 1$ and $k = 0(1)$. Assuming the reduced frequency small enough that these are met, we list several operators which are the unsteady counterparts of Eqs. 4–95–4–97. The notation for planform and cross-sectional geometry is defined in Fig. 4–3.

For almost-plane lifting surfaces of local span $2s(x)$,

$$\mathscr{A} = \frac{4}{\pi U} \frac{D}{Dt} \int_{-s(x)}^{y} \frac{1}{\sqrt{s^2(x) - y'^2}} \oint_{-s(x)}^{s(x)} \frac{\sqrt{s^2(x) - \eta^2}}{(y' - \eta)} (\cdots) \, d\eta \, dy'$$

$$= \frac{4}{\pi U} \frac{D}{Dt} \int_{-s(x)}^{s(x)} \ln \left(\frac{s^2(x) - \eta y - \sqrt{s^2(x) - y^2}\sqrt{s^2(x) - \eta^2}}{s(x)|y - \eta|} \right) (\cdots) \, d\eta$$

$$(4-202)$$

---

* In general, the amplitude ratio of time-dependent body motion must be small relative to thickness ratio δ; for further details see Miles' chapter, "Slender Non-planar Bodies."

where D/Dt is once more the substantial derivative (cf. Eq. 4–148), or rate of change following a fluid slab,

$$\frac{D}{Dt} \equiv \begin{cases} U\dfrac{\partial}{\partial x} + \dfrac{\partial}{\partial t}, & \text{(arbitrary motion)} \\[2ex] U\dfrac{\partial}{\partial x} + i\omega, & \text{(simple harmonic motion)} \end{cases}$$

A special case of Eq. 4–202 when w_a/U is independent of y reads

$$\mathscr{A} = -\frac{4}{U}\frac{D}{Dt}\left[\sqrt{s^2(x) - y^2}(\cdots)\right] \tag{4–203}$$

The lift amplitude per unit chordwise distance results from another integration across the span. Regardless of the cross-sectional shape,

$$\bar{L}(x) = -2q\left(\frac{\partial}{\partial x} + \frac{i\omega}{U}\right)\left[S(x)\frac{\bar{w}_a(x)}{U}\right] \tag{4–204}$$

Equation 4–204 has an obvious extension to more general unsteady motion. Expressions for the virtual mass $\rho_\infty S(x)$ appropriate to wings, bodies, and wing-body combinations are copied from Sec. 4–4(a) as follows:

$$S(x) = \pi s^2(x), \qquad \text{(ellipse or plane wing)} \tag{4–100}$$

$$S(x) = \pi\left[s^2(x) - R^2(x) + \frac{R^4(x)}{s^2(x)}\right], \qquad \text{(midwing-body combination)} \tag{4–101}$$

$$S(x) = \pi R^2(x), \qquad \text{(body of revolution)} \tag{4–102}$$

Further information on Munk-Jones theory and its applications will be found, among many other sources, in Ref. 4–2, Ref. 4–41, or Chaps. 5 and 7 of Ref. 1–2.

When $k \gg 1$ (according to Miles), slender wings can be analyzed by omitting x-derivatives but retaining the $\partial^2\varphi'/\partial t^2$ term in the potential differential equation

$$\frac{\partial^2\varphi'}{\partial y^2} + \frac{\partial^2\varphi'}{\partial z^2} = \frac{1}{a_\infty^2}\frac{\partial^2\varphi'}{\partial t^2} = -\frac{\omega^2}{a_\infty^2}\varphi' \tag{4–205}$$

An inverse aerodynamic operator resembling Possio's for two-dimensional, subsonic flow might be given in this case; but all the available solutions are actually developed by separating Eq. 4–103 in elliptic-cylinder coordinates. This process leads to infinite series of Mathieu functions. Examples of its application appear in the papers by Merbt and Landahl

(Ref. 4–105), Mazelsky (Ref. 4–106), and Milne (Ref. 4–107), to which the reader is directed for forms of the operators associated with particular planforms and modes of motion.

(b) Transonic flow $(M \cong 1)$

$$\mathscr{A}^{-1} = \frac{1}{8\pi} \oint \oint_{R_a'} K(x_0, y_0; k, M = 1)(\cdots) \, d\xi \, d\tilde\eta \qquad (4\text{-}206a)$$

where

$$K(x_0, y_0; k, M = 1)$$
$$= k^2 e^{-ikx_0} \left\{ \frac{1}{k|y_0|} K_1(k|y_0|) + \frac{\pi i}{2k|y_0|} \left[I_1(k|y_0|) - L_1(k|y_0|) - \frac{2}{\pi} \right] \right.$$
$$+ \frac{2}{k^2 y_0^2} \left\{ \exp \left[\frac{i}{2} \left(kx_0 - \frac{k^2 y_0^2}{kx_0} \right) \right] - \frac{1}{2} \right\}$$
$$\left. - \frac{i}{k^2 y_0^2} \int_{k|y_0|}^{kx_0} \exp \left[\frac{i}{2} \left(\lambda - \frac{k^2 y_0^2}{\lambda} \right) \right] d\lambda \right\} \qquad (4\text{-}206b)$$

Integration region R_a' consists of the disturbed upstream zone of influence of $(\tilde{x}, \tilde{y})$, which at sonic speed is limited to $\xi \leq \tilde{x}$ or $x_0 \geq 0$. There also exists a direct form of the aerodynamic operator in the three-dimensional sonic case, which may, after approximations reducing the upstream zone of influence to finite extent, form the basis of a numerical procedure like the one described below for supersonic flow. As can be worked out from Landahl's thesis (Ref. 4–165), the direct operator reads

$$\mathscr{A} = 4 \left(\frac{\partial}{\partial \tilde{x}} + ik \right) \iint_{R_a'} K(x_0, y_0; k, M = 1)(\cdots) \, d\xi \, d\tilde\eta \qquad (4\text{-}206c)$$

where

$$K(x_0, y_0; k, M = 1) = - \frac{\exp \left[-i\frac{k}{2} \left(x_0 + \frac{y_0^2}{x_0} \right) \right]}{2\pi x_0} \qquad (4\text{-}206d)$$

Under the restriction $k \gg |1 - M|$, Ref. 4–165 shows how Eq. 4–206d can be generalized to other transonic Mach numbers in the nieghborhood of unity.

Equation 4–206b first appeared in Ref. 4–86. A process of numerical solution can be worked out, paralleling that for the subsonic integral equation and leading to a matric relation like Eq. 4–198. The only fully documented technique now in the literature is that of Runyan and Woolston (Ref. 4–87).

The field of transonic flow past slender, pointed configurations has seen interesting recent developments of practical importance. To begin with, it can be stated that, for sufficiently small aspect ratio or fineness ratio, the Munk-Jones operators (Eqs. 4–202, 4–203, and 4–204) may be used right through sonic to supersonic flight speeds. The exact form of the limiting conditions on δ, $\mathcal{R}$, and k is quite complicated—as are the conditions permitting linearization itself—and they are thoroughly investigated by Landahl in Refs. 4–72 and 4–45. A general requirement for linearization in the case of a wing or wing-body having thickness ratio δ is stated as

$$k \gg \mathcal{R}\delta \left| \ln \mathcal{R}\delta^{\frac{1}{3}} \right| \qquad (4\text{--}207)$$

Equation 4–207 seems to imply that a wing without thickness ($\delta = 0$) is susceptible of linear treatment for arbitrarily small reduced frequency. This is true, however, only as regards in-phase or stiffness derivatives such as $\partial C_L/\partial \alpha$ and $\partial C_M/\partial \alpha$. Unless $\mathcal{R}$ is vanishingly small, damping derivatives become infinite as $|lnk|$. Nevertheless, linearized theory yields satisfactory results for damping at any k which might be encountered in practice.

Extension of the linearized formulation beyond Munk-Jones to solutions of the more general transonic differential equation

$$\frac{\partial^2 \varphi'}{\partial y^2} + \frac{\partial^2 \varphi'}{\partial z^2} - \frac{2U}{a_\infty^2} \frac{\partial^2 \varphi'}{\partial x \, \partial t} - \frac{1}{a_\infty^2} \frac{\partial^2 \varphi'}{\partial t^2} = 0 \qquad (4\text{--}208)$$

can be accomplished by a process of expansion in $\mathcal{R}$ and k resembling that of Adams and Sears (Ref. 4–50). Details of this development are almost entirely the work of Landahl, who has published data on total pitching moment and lift of pointed wings in Ref. 4–108; on midwing-body combinations and bodies of revolution in Ref. 4–109; on rectangular planforms and more general oscillatory motions in Ref. 4–110 and elsewhere. No inclusive statement of aerodynamic operators can be made here. It is significant, however, that Landahl finds damping derivatives which deviate appreciably from Munk-Jones theory even when $\mathcal{R}$ is quite small. In Refs. 4–108 and 4–109, for instance, he calls attention to a large influence of leading-edge curvature on fixed-axis damping in pitch; whereas calculations by Eq. 4–205 show none whatever.

(c) Supersonic flow ($M > 1$, $M\delta \ll 1$)

$$\mathcal{A} = 4\left(\frac{\partial}{\partial \tilde{x}} + ik\right) \iint\limits_{R_a'} K(x_0, y_0; k, M)(\cdots) \, d\xi \, d\tilde{\eta} \qquad (4\text{--}209a)$$

where

$$K(x_0, y_0; k, M) = - \frac{\exp\left(-i\,\dfrac{kM^2}{\beta^2}\,x_0\right)\cos\left(\dfrac{kM}{\beta^2}\sqrt{x_0{}^2 - \beta^2 y_0{}^2}\right)}{\pi\sqrt{x_0{}^2 - \beta^2 y_0{}^2}} \qquad (4\text{-}209b)$$

As discussed in Sec. 4–4(c), the region of integration $R_a{}'$ consists of that portion of the projected planform and disturbed diaphragm region contained between the Mach lines $\bar{\eta} = \bar{y} \pm (\tilde{x} - \tilde{\xi})/\beta$ going forward from $(\tilde{x}, \tilde{y})$. Equations 4–209 represent the fully linearized result for almost-plane surfaces, and K is actually the kernel function for the velocity potential discontinuity $\Delta\varphi_a{}'$. This form seems to have been given first by Garrick and Rubinow (Ref. 4–111).

The calculation of oscillatory supersonic airloads has been approached in two general ways: by analytic solution of the linearized differential equation for particular planforms and modes of motion; and by numerical integration of the operator $\mathscr{A}$, or its inverse, in more general cases. Miles' monograph (Ref. 4–2) furnishes the most complete account of solutions of the former class. He presents extended treatments of such cases as the wide delta with all supersonic edges, the quarter-infinite wing (rectangular wingtip without tip interaction), and tapered planforms with straight edges. Watkins et al. have analyzed rigid-body motions of rectangular surfaces in some detail by expanding the integral equation relating $\bar{\varphi}'$ and w_a in ascending powers of the reduced frequency (Refs. 4–112, 4–113).

The delta wing has also formed the subject of considerable work by NACA. Deltas with supersonic leading edges and related wings are treated, for example, in Refs. 4–114, 4–115, and 4–116. The same planform with subsonic edges poses greater practical difficulties, the series solution of Watkins and Berman (Ref. 4–117) being regarded as the most useful for applications. Theirs and subsequent airload computations are generally limited to elastic deformations which are, at most, quadratic in the coordinates $\tilde{x}$ and $\tilde{y}$.

A great virtue of such exact solutions is that they serve as standards for evaluating approximate methods. For the shapes to which they apply, a possibility also exists of building up tables from which generalized forces for actual flutter or dynamic response calculations can be constructed by curve-fitting the normal modes of the given structure.

When supersonic expansions in k are terminated at the first power, airloads for low-frequency motion can be obtained with relative ease by an extension of Evvard's theorem (Ref. 4–118). Two reports by Watson (Refs. 4–119 and 4–120) are more recent examples of this technique, which is also exploited in Ref. 4–101 and elsewhere.

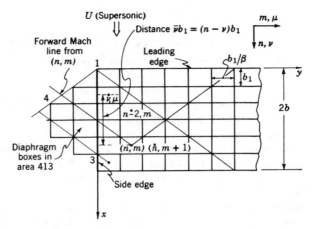

Fig. 4-9. Grid of boxes overlaid on a typical supersonic planform.

Considering the diverse planforms and aspect ratios of modern lifting surfaces, it is fortunate that numerical procedures like that of Pines and collaborators (Ref. 4–121) overcome nearly all limitations inherent in the analytical solutions. Except for some recent work on the inverse operator $\mathscr{A}^{-1}(\cdots)$ (Ref. 4–122), all such schemes start from the aerodynamic influence coefficient, which is defined as the velocity potential (or pressure) developed at a point on a wing due to constant amplitude of sinusoidal upwash $\bar{w}_a/U$ over an elementary area influencing that point. A grid of such areas or boxes is superimposed on the planform, as in Fig. 4–9.

For disturbed areas of the x-y-plane adjacent to a subsonic edge, the diaphragm concept is introduced by placing boxes there and requiring zero potential or pressure discontinuity. Details of box shapes and fineness of grid networks, which strongly affect the numerical accuracy, are discussed in two classified reports (Refs. 4–123 and 4–124). With an eye to high-speed computation, the whole airload determination is reduced to a large aggregate of elementary, repetitive operations.

Returning momentarily to the dimensional coordinates, the potential amplitude at point (x, y) on the upper wing surface is

$$\varphi_U'(x,y) = -\frac{1}{\pi} \iint_{R_a'} \frac{\bar{w}_a(\xi, \eta) \exp\left[-i(\omega M^2/U\beta^2)(x - \xi)\right]}{\sqrt{(x - \xi)^2 - \beta^2(y - \eta)^2}}$$

$$\times \cos\left(\frac{\omega M}{U\beta^2} \sqrt{(x - \xi)^2 - \beta^2(y - \eta)^2}\right) d\xi \, d\eta \quad (4\text{-}210)$$

As in Fig. 4–9, the disturbed area is split into rectangular boxes of chordwise dimension b_1, with diagonals parallel to the Mach lines. If over each box, $\bar{w}_a$ is made constant and equal to its value at the central point (x_ν, y_μ), then the potential discontinuity in the center of box (n, m) is

$$\overline{\varphi_L'}(x_n, y_m) - \overline{\varphi_U'}(x_n, y_m) = \overline{\Delta\varphi_a'}(x_n, y_m) = \frac{2}{\pi}\sum_\nu\sum_\mu \bar{w}_a(x_\nu, y_\mu)$$

$$\times \left\{ \iint_{A(\nu,\mu)} \frac{\exp\left[-i\dfrac{\omega M^2}{U\beta^2}(x_n - \xi)\right]\cos\left(\dfrac{\omega M}{U\beta^2}\sqrt{(x_n - \xi)^2 - \beta^2(y_m - \eta)^2}\right)}{\sqrt{(x_n - \xi)^2 - \beta^2(y_m - \eta)^2}} d\xi\, d\eta \right\}$$

$$\equiv -\frac{b_1}{\beta}\sum_\nu\sum_\mu \bar{w}_a(x_\nu, y_\mu)\{R_{\bar{\nu},\bar{\mu}} + iI_{\bar{\nu},\bar{\mu}}\} \qquad (4\text{–}211)$$

The quantity in braces is the influence coefficient. In dimensionless form, $C_{\bar{\nu},\bar{\mu}}$, is a function of M, $\bar{k}_1 = \omega b_1 M^2/U\beta^2$, and of the relative positions $\bar{\nu} = (n - \nu)$, $\bar{\mu} = (m - \mu)$ of the receiving and influencing boxes (Fig. 4–9). Although exact formulas are available (Ref. 4–123), it is usually adequate to approximate $C_{\bar{\nu},\bar{\mu}}$ by replacing (ξ, η) in the numerator with its central value. There then result

$$C_{\bar{\nu},\bar{\mu}} = R_{\bar{\nu},\bar{\mu}} + iI_{\bar{\nu},\bar{\mu}} = \begin{cases} \left\{-1 + \dfrac{2M^2 + 1}{48}\left(\dfrac{\bar{k}_1}{M}\right)^2 - \dfrac{8M^4 + 24M^2 + 3}{15,360}\left(\dfrac{\bar{k}_1}{M}\right)^4\right\} \\[2mm] + iM\left\{\dfrac{1}{4}\left(\dfrac{\bar{k}_1}{M}\right) - \dfrac{2M^2 + 3}{384}\left(\dfrac{\bar{k}_1}{M}\right)^3\right\} \quad (\text{for } \bar{\nu} = \bar{\mu} = 0), \\[3mm] -\dfrac{2}{\pi}(\cos \bar{\nu}\bar{k}_1 - i\sin \bar{\nu}\bar{k}_1)\cos\left(\dfrac{\bar{k}_1}{M}\sqrt{\bar{\nu}^2 - \bar{\mu}^2}\right)\cdot\beta_{\bar{\nu},\bar{\mu}} \\[2mm] \hspace{3cm} (\text{for } \bar{\nu} \geq \bar{\mu},\ \bar{\nu} \geq 2), \\[3mm] \dfrac{1}{5}\sum_{\bar{\nu}=3}^{7}(R_{\bar{\nu},0} + iI_{\bar{\nu},0})\Big|_{\bar{k}_1/5} + \dfrac{2}{5}\sum_{\bar{\nu}=3}^{7}\sum_{\bar{\mu}=1}^{2}(R_{\bar{\nu},\bar{\mu}} + iI_{\bar{\nu},\bar{\mu}})\Big|_{\bar{k}_1/5} \\[2mm] \hspace{3cm} (\text{for } \bar{\nu} = 1, \bar{\mu} = 0). \\[3mm] \dfrac{1}{5}\sum_{\bar{\nu}=3}^{7}\sum_{\bar{\mu}=3}^{\bar{\nu}}(R_{\bar{\nu},\bar{\mu}} + iI_{\bar{\nu},\bar{\mu}})\Big|_{\bar{k}_1/5} \quad (\text{for } \bar{\nu} = \bar{\mu} = 1), \end{cases}$$

and
$$C_{\bar{\nu},-\bar{\mu}} = C_{\bar{\nu},\bar{\mu}} \qquad (4\text{–}212a)$$

where

$$
\beta_{\bar{\nu},\bar{\mu}} = \begin{cases}
\cosh^{-1}(2\bar{\nu}+1) - \cosh^{-1}(2\bar{\nu}-1) + (\bar{\nu}+\tfrac{1}{2})\left[\pi - 2\cos^{-1}\dfrac{1}{2\bar{\nu}+1}\right] \\[2mm]
\quad - (\bar{\nu}-\tfrac{1}{2})\left[\pi - 2\cos^{-1}\dfrac{1}{2\bar{\nu}-1}\right] \quad \text{(for } \bar{\nu} \geq 2, \bar{\mu}=0), \\[3mm]
-(\bar{\nu}-\tfrac{1}{2})\cosh^{-1}\dfrac{2\bar{\nu}+1}{2\bar{\nu}-1} + (\bar{\nu}+\tfrac{1}{2})\cos^{-1}\dfrac{2\bar{\nu}-1}{2\bar{\nu}+1} \quad \text{(for } \bar{\nu}=\bar{\mu}\geq 2), \\[3mm]
(\bar{\mu}+\tfrac{1}{2})\left[\cosh^{-1}\dfrac{2\bar{\nu}+1}{2\bar{\mu}+1} - \cosh^{-1}\dfrac{2\bar{\nu}-1}{2\bar{\mu}+1}\right] \\[2mm]
\quad - (\bar{\mu}-\tfrac{1}{2})\left[\cosh^{-1}\dfrac{2\bar{\nu}+1}{2\bar{\mu}-1} - \cosh^{-1}\dfrac{2\bar{\nu}-1}{2\bar{\mu}-1}\right] \\[3mm]
+(\bar{\nu}+\tfrac{1}{2})\left[\cos^{-1}\dfrac{2\bar{\mu}-1}{2\bar{\nu}+1} - \cos^{-1}\dfrac{2\bar{\mu}+1}{2\bar{\nu}+1}\right] \\[2mm]
\quad -(\bar{\nu}-\tfrac{1}{2})\left[\cos^{-1}\dfrac{2\bar{\mu}-1}{2\bar{\nu}-1} - \cos^{-1}\dfrac{2\bar{\mu}+1}{2\bar{\nu}-1}\right] \quad \text{(for } \bar{\nu}\geq 2, \bar{\nu}>\bar{\mu}\geq 1)
\end{cases}
$$

$$(4\text{-}212b)$$

The computation represented by Eqs. 4–212 is normally relegated to a subprogram in high-speed computers. Equation 4–211 is complete for finding potential distributions on wings with all supersonic edges. When diaphragm regions are present, the systematic calculation of the un-known $\bar{w}_a$ on diaphragm boxes from the condition $\Delta\overline{\varphi_a'} = \overline{\varphi'} = 0$ is described in Ref. 4–123. Once $\bar{w}_a$ is everywhere known, the $\Delta\overline{\varphi_a'}$ deter-mination proceeds as before.

In nearly all dynamic analyses, the final aerodynamic step is to compute generalized forces or weighted integrals of the pressure distribution. A typical such force is described by the equation

$$
\bar{Q}_{ij} = \iint\limits_{R_a} \Delta\bar{p}_i(x,y)f_j(x,y)\,dx\,dy \tag{4-213}
$$

where f_j is the jth deflection and $\Delta\bar{p}_i$ is the pressure difference due to mode f_i. When velocity potential influence coefficients are adopted, a single integration by parts permits $\bar{Q}_{ij}$ to be expressed as a summation involving the predetermined $\Delta\overline{\varphi_a'}(x_n, y_m)$ at the centers of wing boxes and at points along the trailing edge. This summation is based on numerical evaluation, using stations at centers of the boxes, of the following:

$$\bar{Q}_{ij} = -\frac{\rho_\infty U}{b} \iint_{R_a} \overline{\Delta\varphi_i'}(x, y) \left[ikf_j(x, y) - b \frac{\partial f_j(x, y)}{\partial x} \right] dx \, dy$$

$$- \rho_\infty U \int_{\text{span}} \overline{\Delta\varphi_i'}(x_t, y) f_j(x_t, y) \, dy \quad (4\text{-}214)$$

since

$$\Delta\bar{p}_a = \frac{-\rho_\infty U}{b} \left[ik\overline{\Delta\varphi_a'} + b \frac{\partial(\overline{\Delta\varphi_a'})}{\partial x} \right] \quad (4\text{-}215)$$

and where the subscript t refers to points on the trailing edge.

Reduction of the foregoing operations to matrix notation in a manner similar to previous sections is an elementary process, which will be omitted here for brevity. As just described, the method of aerodynamic influence coefficients does not rigorously account for the presence of the singularity in $\Delta\bar{p}_a$ at a subsonic leading edge or that in $\bar{w}_a$ adjacent to a side edge or subsonic leading edge. Reference 4–123 discusses modifications to the basic theory for these purposes. When $\overline{\Delta\varphi_a'}$ is used in place of $\Delta\bar{p}_a$ as the principal unknown, however, only its slope is infinite at the leading edge; and an increase in the number of boxes covering the planform usually provides an accurate enough way of dealing with the singularities in practice. Flap-type control surfaces present no particular problem, so long as their leading and trailing edges are supersonic.

With regard to the second-order effect of wing thickness on oscillatory supersonic airloads, the comments made in Sec. 4–4(c) concerning three-dimensional steady flow retain their validity. No rigorous theory is available in the lower range of M. When $M^2 \gg 1$ and the area concerned is outside the influence of a cut-off wingtip, Landahl's results (Ref. 4–24) may be used. Including terms of order δ/M, δ/M^3, and δ^2, his aerodynamic operator is specialized as follows for sinusoidal motion:

$$\mathscr{A} = -\frac{4}{M} \left[1 + \left(\frac{\gamma + 1}{2} \right) M \frac{\partial z_t}{\partial x} \right] \left(\frac{\bar{w}_a}{U} \right)$$

$$- \frac{2}{M^3} \left(\frac{\partial^2}{\partial \tilde{x}^2} + \frac{\partial^2}{\partial \tilde{y}^2} \right) \int_{\tilde{x}_{LE}}^{\tilde{x}} [\tilde{x} - \xi] e^{-ik(\tilde{x} - \xi)} \frac{\bar{w}_a(\xi, \tilde{y})}{U} d\xi \quad (4\text{-}216)$$

Reference should also be made to the approximate adaptation of van Dyke's theory to wings of finite span by Rodden and Revell (Ref. 4–80).

As in steady flow, Eq. 4–216 suggests (provided $\delta \gg 1/M^3$) that linearized theory might be roughly corrected for thickness by applying the overall factor $\left[1 + \left(\dfrac{\gamma + 1}{2}\right) M \dfrac{\partial z_t}{\partial x}\right]$ to Eq. 4–209a.

When $M\delta$ is small enough and k is also limited, the Munk-Jones operators (4–202), (4–203), and (4–204) may be applied to slender, pointed wings, bodies, and wing-body combinations in supersonic flight. Linearized theory is still permissible for higher frequencies and aspect ratios, but then the additional terms beside $(\partial^2\varphi'/\partial y^2 + \partial^2\varphi'/\partial z^2)$ must be included in the governing differential equation. The various solutions of this type do not lend themselves to concise statements of aerodynamic operators, but we shall give a few references. The topic of low-aspect-ratio planar lifting surfaces is discussed at length in several chapters of Miles' monograph (Ref. 4–2). Especially useful for large k are the Mathieu-function series, of which Merbt and Landahl's is typical (Ref. 4–105).

Wing-body combinations and bodies of revolution have been analyzed by the method of expansion in k and $\mathcal{R}$ (Refs. 4–125). These papers deal mainly with total lift and moment due to rigid-body oscillations, although the second provides some information on chordwise deformations where $\bar{w}_a$ is represented by a polynomial in $\tilde{x}$. In the damping derivatives, Refs. 4–125 find large deviations from Munk-Jones theory even at relatively low values of $\beta\mathcal{R}$. In fact, this would appear to be a field deserving much further investigation. There are as yet unpublished studies, for example, which indicate that sometimes the nonlinear effects of thickness on supersonic wing-bodies may be as great as those due to higher-degree aspect-ratio terms in such expanded linearized solutions. Finally, it is essential that the transition be carried out to the nonlinear theory of van Dyke (Ref. 4–51) for slender bodies in hypersonic flow. A step in this direction is the technique of perturbing the steady-state characteristics solution (Holt, Ref. 4–126).

(d) Hypersonic flow $(M^2 \gg 1,\ M\delta = 0(1)$ or greater$)$

To avoid repetition, the reader is referred to Secs. 4–4(d) and 4–5(e) for discussions of this topic and literature citations. In particular, it appears that the comment regarding the quasi-steady nature of aerodynamic operators, made in Sec. 4–5(e), may be extended to three-dimensional flows. For hypersonic lifting surfaces which are neither too slender nor too highly sweptback, strip theory is valid; and results such as Eqs. 4–146, 4–147, and 4–149 are useful without modification.

4-8   OTHER   PROBLEMS   OF   UNSTEADY   MOTION
IN   THREE   DIMENSIONS

Any systematic presentation of time-dependent aerodynamic operators for finite wings and bodies is hampered by the rather incomplete development of the theory in this area. Indicial functions like those defined in the introduction of Sec. 4–6, along with certain generalized forms which are associated with elastic deflections, have been laboriously calculated on a few specific planforms in supersonic or incompressible flow; but there is little more in the literature. Except for setting down one or two generally interesting formulas, we therefore confine ourselves in this section to listing some of the important references.

From the standpoint of the aeroelastician faced with the routine analysis of transient phenomena where unsteady effects are significant, we can make some hopeful statements to ameliorate the apparently unpromising prospect of inadequate aerodynamic tools. In the past, strip theory, usually with a steady-state tip correction, has been employed for lifting surfaces in combination with slender-body theory for fuselages, external stores, missiles, and the like. There now exists a considerable body of experience on the determination and interpretation of dynamic loads and stresses in this way; by no means should this be hastily abandoned, except possibly on the moderate to low aspect-ratio surfaces, at subsonic, transonic, and low supersonic speeds, whose plate-like deformations render the strip concept wholly invalid. In high supersonic and hypersonic flow, two-dimensional aerodynamic methods have very broad applicability, as reviewed elsewhere in this chapter. But of even more direct concern is the validity of quasi-steady hypersonic theory, both two- and three-dimensional, for time-dependent motions which would certainly demand the full unsteady treatment at lower flight velocities. Intensified experimental and analytical research is called for to delineate more precisely the limits of such approximations.

There is a growing tendency to try to unify the study of transient behavior of flight vehicles on a foundation of simple harmonic inputs. This is a development to be encouraged. Sinusoidal aerodynamic operators, even for quite general three-dimensional configurations, are now very well systematized, the majority of them having already been programmed for high-speed digital computers. The use of sinusoidal responses in statistical treatment of the gust problem is an excellent example of this sort. Another is the evaluation of maneuvering performance by sinusoidally forcing various elements in the control system. Even where such inputs cannot be assumed directly (for instance, when complying with current military and civilian gust and maneuver specifications), Fourier's superposition principle can prove useful in an approximate way. Thus, on a stable aircraft, graded gusts of various lengths can be applied in a periodic fashion, with the fundamental period long enough to allow the motion to settle down. Then a Fourier series representation of the gust input permits the loads, stresses, or displacements to be synthesized from sinusoidal gust responses at a discrete set of wavelengths. In cases involving three-dimensional flow, when unsteady effects must be included, superposition may prove to be the only practicable approach here.

To review the literature on indicial functions for finite wings, we first consider incompressible linearized theory. The angle-of-attack and natural-gust lift functions $\varphi(s)$ and $\psi(s)$ [$k_1(s)$ and $k_2(s)$ in NACA notation] were first estimated by R. T. Jones (Ref. 4–153), using elliptical planforms with $\mathcal{R} = 3$ and 6 as examples. W. P. Jones presented an operational approach to the same problem in Ref. 4–154. His computations were on $\varphi(s)$ for rectangular wings of $\mathcal{R} = 4$ and 6 and for one tapered wing; additional results are given by Lehrian (Ref. 4–155). Further indicial lifts for triangular, rectangular, and elliptical planforms appear in Refs. 4–131 and 4–156, the latter including some data on the spanwise distribution of the load associated with $\varphi(s)$.

Apparently, no comparable quantities have been calculated for subsonic compressible or transonic flow. If required, this will almost certainly have to be done by a Fourier superposition of simple harmonic coefficients, analogous to the two-dimensional work of Mazelsky and Drischler (Refs. 4–143 to 4–145).

In connection with the starting values at $s = 0$ of any indicial function in any flight range, it is of interest that acoustical theory shows the initial overpressure to be $p - p_\infty = \rho_\infty a_\infty w$ on a plane surface element which proceeds into the gas with a normal velocity w. As shown by Heaslet and Lomax (Ref. 4–3, p. 404), this has the consequence that the local starting lift-curve slope of such an element is $4/M$. For instance, a flat wing of any planform, which suddenly assumes an incidence α_0, will at once experience a force*

$$L(0) \equiv qSC_L(0) = qS\,\frac{4\alpha_0}{M} \qquad (4\text{–}217)$$

S being the plan area. On a cambered wing, α_0 in Eq. 4–217 is replaced by the average chordwise slope $\overline{\alpha(x, y)}$ of the mean surface.

---

* We remark, in passing, that some so-called theories for calculating indicial functions do little more than fair a smooth curve between $L(0)$ and the final steady-state value $L(\infty)$.

The principal developments in supersonic linearized theory are those of Ref. 4–130, also lucidly summarized by two of the same authors in Ref. 4–3. The reader will find their concept of *acoustic planform* very helpful as an aid for clearer comprehension of unsteady compressible flow. Reference 4–130 gives a number of indicial functions for rigid wings of various planforms. Earlier work was also done by Miles on simple (Ref. 4–157) and rectangular (Ref. 4–158) planforms. Reference 4–2 furnishes considerable information on three-dimensional transients. Lomax et al. (Ref. 4–159) also present indicial functions for rectangular wings which deform according to power series shapes in the spanwise and chordwise coordinates. Recent research at Ames Research Center of NASA has related to aerodynamic influence coefficients for sudden downwash changes over area elements such as those discussed in Sec. 4–7(*d*), along with applications of these to transient aircraft responses (Ref. 4–160).

Under the important restriction that the time constants of the motion cannot be too short, Munk-Jones theory may be used to find airloads due to unsteady maneuvers of slender, pointed bodies and wing-body combinations at subsonic and supersonic speeds. The necessary modifications to Eqs. 4–202 through 4–204 are so elementary that they need not be set down here. It should be realized that step-function changes of incidence and sharp-edged gusts are ruled out by the aforementioned restriction. Fortunately, no such sudden situations occur in practice. For the graded gusts and continuous maneuvers which are actually encountered, there are usually no difficulties.

Miles has observed (Ref. 4–161) that a variable forward speed, $U(t)$, can be included in the Munk-Jones model. Thus it is deducible from his paper that a longitudinal acceleration or deceleration given by

$$U(t) = U_0[1 + f(t)] \qquad (4\text{--}218)$$

is accounted for by the following alteration in Eq. 4–204:

$$L(x, t) = -\rho_\infty U_0^2 [1 + f]^2 \left\{ \left[\frac{dS}{dx} \frac{\partial z_a}{\partial x} + S(x) \frac{\partial^2 z_a}{\partial x^2} \right] \right.$$

$$+ \frac{1}{U_0[1 + f]} \left[\frac{dS}{dx} \frac{\partial z_a}{\partial t} + 2S(x) \frac{\partial^2 z_a}{\partial x\, \partial t} \right] + \frac{S(x)}{U_0^2[1 + f]^2} \frac{\partial^2 z_a}{\partial t^2} \right\}$$

$$- \rho_\infty U_0 S(x) \frac{df}{dt} \frac{\partial z_a}{\partial x} \qquad (4\text{--}219)$$

Here $z_a(x,\ t)$ represents the time-dependent displacement of the centerline. The virtual mass $\rho_\infty S(x)$ in Eq. 4–219 can be found from Eq. 4–100,

4–101, 4–102, or the appropriate formula for the cross-sectional shape under consideration.

REFERENCES

4–1. Garrick, I. E., *Nonsteady Wing Characteristics*, Section F, Vol. VII, of *High Speed Aerodynamics and Jet Propulsion*, Princeton University Press, Princeton, N.J., 1957.

4–2. Miles, J. W., *The Potential Theory of Unsteady Supersonic Flow*, Cambridge Monographs on Mechanics and Applied Mathematics, Cambridge University, 1958. (Adapted from a monograph prepared for Air Research and Development Command, USAF, in 1955.)

4–3. Heaslet, M. A., and H. Lomax, *Supersonic and Transonic Small Perturbation Theory* (see Chap. 6), Section D, Vol. VI of *High Speed Aerodynamics and Jet Propulsion*, Princeton University Press, Princeton, N.J., 1954.

4–4. Robinson, A., and J. A. Laurmann, *Wing Theory*, Chap. 5, Cambridge University Press, Cambridge, 1956.

4–5. Temple, G., "Unsteady Motion," Chap. IX, Vol. I, pp. 325–374, of *Modern Developments in Fluid Dynamics—High-Speed Flow*, Clarendon Press, Oxford, 1953.

4–6. Frankl, F. L., and E. A. Karpovich, *Gas Dynamics of Slender Bodies*, Chap. IV, Moscow, 1948. (Translated from the Russian by Interscience Publishers.)

4–7. Liepmann, H. W., and A. Roshko, *Elements of Gasdynamics*, John Wiley and Sons, New York, 1957.

4–8. Bateman, H., "Notes on a Differential Equation Which Occurs in the Two-Dimensional Motion of a Compressible Fluid and the Associated Variational Problems," *Proc. Royal Soc.*, London, (A), Vol. 125, November 1929, pp. 598–618.

4–9. Bateman, H., *Partial Differential Equations of Mathematical Physics*, Dover Publications, New York, 1944.

4–10. Hargreaves, R., "A Pressure Integral as Kinetic Potential," *Philosophical Magazine*, Vol. XVI, September 1908, pp. 436–444.

4–11. Lamb, Sir Horace, *Hydrodynamics*, 6th ed., Dover Publications, New York, 1945.

4–12. Ward, G. N., *Linearized Theory of Steady High-Speed Flow*, Cambridge University Press, Cambridge, 1955.

4–13. Flax, A. H., "Reverse-Flow and Variational Theorems for Lifting Surfaces in Nonstationary Compressible Flow," *J. Aero. Sciences*, Vol. 20, No. 2, February 1953, pp. 120–126.

4–14. Wang, C. T., "Variational Method in the Theory of Compressible Fluid," *J. Aero. Sciences*, Vol. 15, No. 11, November 1948, pp. 675–685.

4–15. Riparbelli, C., "A Principle of Maximum Power for Real Fluids in Steady Motion," Readers' Forum, *J. Aero. Sciences*, Vol. 23, No. 10, October 1956, pp. 971–972.

4–16. Miles, J. W., "On the Sloshing of Liquid in a Flexible Tank," *J. Applied Mechanics*, Vol. 25, No. 2, June 1958, pp. 277–283.

4–17. Greidanus, J. H., A. I. van deVooren, and H. Bergh, *Experimental Determination of the Aerodynamic Coefficients of an Oscillating Wing in Incompressible, Two-Dimensional Flow*, Parts I–IV, Nationaal Luchtvaartlaboratorium, Amsterdam, Reports F-101, 102, 103, and 104, 1952.

4–18. Glauert, H., *The Elements of Aerofoil and Airscrew Theory*, 2nd ed., Cambridge University Press and The Macmillan Company, London, 1947.

4-19. Pope, A., *Basic Wing and Airfoil Theory*, McGraw-Hill Book Company, New York, 1951.
4-20. Kuethe, A. M., and J. D. Schetzer, *Foundations of Aerodynamics*, John Wiley and Sons, New York, 1950.
4-21. Ferri, A., *Elements of Aerodynamics of Supersonic Flows*, The Macmillan Company, New York, 1949.
4-22. Guderley, K. G., *Theorie schallnäher Strömungen*, Springer-Verlag, Berlin, 1957.
4-23. Lighthill, M. J., "Oscillating Airfoils at High Mach Number," *J. Aero. Sciences*, Vol. 20, No. 6, June 1953, pp. 402–406.
4-24. Landahl, M. T., "Unsteady Flow Around Thin Wings at High Mach Numbers," *J. Aero. Sciences*, Vol. 24, No. 1, January 1957, pp. 33–38.
4-25. Molyneux, W. G., *Measurement of the Aerodynamic Forces on Oscillating Aerofoils*, Advisory Group for Aeronautical Research and Development Report 35, April 1956.
4-26. Rainey, A. G., *Measurement of Aerodynamic Forces for Various Mean Angles of Attack on an Airfoil Oscillating in Pitch and on Two Finite-Span Wings Oscillating in Bending with Emphasis on Damping in the Stall*, NACA Technical Note 3643, May 1956.
4-27. Halfman, R. L., H. C. Johnson, and S. M. Haley, *Evaluation of High-Angle-of-Attack Aerodynamic-Derivative Data and Stall-Flutter Prediction Techniques*, NACA Technical Note 2533, 1951.
4-28. Lees, L., *Hypersonic Flow*, Institute of the Aeronautical Sciences Preprint No. 554, presented at Fifth International Aeronautical Conference, June 1955.
4-29. Eggers, A. J., Jr., C. A. Syvertson, and S. Kraus, *A Study of Inviscid Flow about Airfoils at High Supersonic Speeds*, NACA Report 1123, 1953.
4-30. Ivey, H. R., E. B. Klunker, and E. N. Bowen, *A Method for Determining the Aerodynamic Characteristics of Two- and Three-Dimensional Shapes at Hypersonic Speeds*, NACA Technical Note 1613, 1948.
4-31. Multhopp, H., *Methods for Calculating the Lift Distribution of Wings (Subsonic Lifting Surface Theory)*, British A.R.C. Reports and Memoranda 2884, 1950.
4-32. Reissner, E., "Note on the Theory of Lifting Surfaces," *Proc. Nat. Acad. of Sciences*, Vol. 35, No. 4, April 1949.
4-33. Multhopp, H., *The Calculation of the Lift Distribution of Aerofoils*, British Ministry of Aircraft Production, R.T.P. Translation 2392. (Originally Luftfahrtforschung, Bd. 15, Nr. 4, June 1938.)
4-34. Weissinger, J., *The Lift Distribution of Sweptback Wings*, NACA Technical Memorandum 1120, 1947.
4-35. Van Spiegel, E., and R. Timman, *Linearized Aerodynamic Theory for Wings of Circular Planform in Steady and Unsteady Incompressible Flow*, Nationaal Luchtvaartlaboratorium, Amsterdam, Report MP 134, 1956.
4-36. De Young, J., and C. W. Harper, *Theoretical Symmetrical Span Loading at Subsonic Speeds for Wings Having Arbitrary Planform*, NACA Report 921, 1948.
4-37. De Young, J., *Theoretical Antisymmetric Span Loading for Wings of Arbitrary Planform at Subsonic Speeds*, NACA Report 1056, 1951.
4-38. Sears, W. R., *Small Perturbation Theory*, Section C, Vol. VI of *High Speed Aerodynamics and Jet Propulsion*, Princeton University Press, Princeton, N.J., 1954.
4-39. Jones, R. T., *Properties of Low-Aspect-Ratio Pointed Wings at Speeds Below and Above the Speed of Sound*, NACA Report 835, 1946.
4-40. Diederich, F. W., *A Planform Parameter for Correlating Certain Aerodynamic Characteristics of Swept Wings*, NACA Technical Note 2335, 1951.

4-41, Bryson, A. E., "Stability Derivatives for a Slender Missile with Application to a Wing-Body-Vertical-Tail Configuration," *J. Aero. Sciences*, Vol. 20, No. 5, May 1953, pp. 297-308.

4-42. Brown, C. E., *Aerodynamics of Bodies at High Speeds*, Section B, Vol. VII of *High Speed Aerodynamics and Jet Propulsion*, Princeton University Press, Princeton, N.J., 1957.

4-43. Ferrari, C., *Interaction Problems* (see Chaps. C, 15 through C, 17), Section C, Vol. VII of *High Speed Aerodynamics and Jet Propulsion*, Princeton University Press, Princeton, N.J., 1957.

4-44. Heaslet, M. A., H. Lomax, and J. R. Spreiter, *Linearized Compressible-Flow Theory for Sonic Flight Speeds*, NACA Report 956, 1950.

4-45. Landahl, M. T., E. L. Mollö-Christensen, and H. Ashley, *Parametric Studies of Viscous and Nonviscous Unsteady Flows*, USAF Office of Scientific Research Technical Report No. 55-13, 1955.

4-46. Jones, R. T., "Some Recent Developments in the Aerodynamics of Wings for High Speeds," *Zeitschrift für Flugwissenschaften*, 4 Jahrgang, Heft 8, August 1956, pp. 257-262.

4-47. Heaslet, M. A., and H. Lomax, *Supersonic and Transonic Small Perturbation Theory*, Section D, Vol. VI of *High Speed Aerodynamics and Jet Propulsion*, Princeton University Press, Princeton, N.J., 1954.

4-48. Mirels, H., and R. C. Haefeli, "The Calculation of Supersonic Downwash Using Line Vortex Theory," *J. Aero. Sciences*, Vol. 17, No. 1, January 1950, pp. 13-21. (This paper emphasizes flow in the wake; see also references given herein.)

4-49. Walsh, J., G. Zartarian, and H. M. Voss, "Generalized Aerodynamic Forces on the Delta Wing with Supersonic Leading Edges," *J. Aero. Sciences*, Vol. 21, No. 11, November 1954, pp. 739-748.

4-50. Adams, M. C., and W. R. Sears, "Slender Body Theory—Review and Extension," *J. Aero. Sciences*, Vol. 20, No. 2, February 1953, pp. 85-98.

4-51. Van Dyke, M. D., *A Study of Hypersonic Small-Disturbance Theory*, NACA Report 1194, 1954.

4-52. Barmby, J. G., H. J. Cunningham, and I. E. Garrick, *Study of Effects of Sweep on the Flutter of Cantilever Wings*, NACA Report 1014, 1951.

4-53. Van de Vooren, A. I., and W. Eckhaus, *Strip Theory for Oscillating Swept Wings in Incompressible Flow*, Nationaal Luchtvaartlaboratorium, Amsterdam, Report F 146, 1954.

4-54. Eckhaus, W., *Strip Theory for Oscillating Swept Wings in Compressible Subsonic Flow*, Nationaal Luchtvaartlaboratorium, Amsterdam, Report F 159, 1955.

4-55. Schwarz, L., *Berechnung der Druckverteilung einer harmonisch sich verformenden Tragfläche in ebener Strömung*, Luftfahrtforschung, Bd. 17, Nr. 11 & 12, December 1940.

4-56. Smilg, B., and L. S. Wasserman, *Application of Three-Dimensional Flutter Theory to Aircraft Structures*, Air Force Technical Report 4798, 1942.

4-57. Wasserman, L. S., W. J. Mykytow, and I. N. Spielberg, *Tab Flutter Theory and Applications*, Air Force Technical Report 5153, 1944.

4-58. Possio, C., "L'Azione Aerodinamica sul Profilo Oscillante in un Fluido Compressibile a Velocitá Iposonora," *L'Aerotecnica*, t. XVIII, fazc. 4, April 1938.

4-59. Fettis, H. E., *An Approximate Method for the Calculation of Nonstationary Air Forces at Subsonic Speeds*, USAF Wright Air Development Center Technical Report 52-56, 1952.

4-60. Fettis, H. E., *Tables of Lift and Moment Coefficients for an Oscillating*

Wing-Aileron Combination in Two-Dimensional Subsonic Flow, Air Force Technical Report 6688 (with supplementary pages), 1951.

4–61. Luke, Y. L., *Tables of Coefficients for Compressible Flutter Calculations*, Air Force Technical Report 6200, 1950.

4–62. Turner, M. J., and S. Rabinowitz, *Aerodynamic Coefficients for an Oscillating Airfoil with Hinged Flap, with Tables for a Mach Number of 0·7*, NACA Technical Note 2213, 1950.

4–63. Timman, R., A. I. van deVooren, and J. H. Greidanus," Aerodynamic Coefficients of an Oscillating Airfoil in Two-Dimensional Subsonic Flow," *J. Aero. Sciences*, Vol. 18, No. 12, December 1951, pp. 797–802. (See other references given herein.)

4–64. Reissner, E., *On the Application of Mathieu Functions in the Theory of Subsonic Compressible Flow Past Oscillating Airfoils*, NACA Technical Note 2363, 1951.

4–65. Williams, D. E., *On the Integral Equations of Two-Dimensional Subsonic Flutter Derivative Theory*, British A.R.C. Reports and Memoranda 3057, 1955.

4–66. de Jager, E. M., *Tables of the Aerodynamic Aileron-Coefficients for an Oscillating Wing-Aileron System in a Subsonic, Compressible Flow*, Nationaal Luchtvaart-laboratorium, Amsterdam, Report F 155, 1954.

4–67. Anon., *Tables of Aerodynamic Coefficients for an Oscillating Wing-Flap System in a Subsonic Compressible Flow*, Nationaal Luchtvaartlaboratorium, Amsterdam, Report F 151, 1954.

4–68. Jordan, P. F., *Aerodynamic Flutter Coefficients for Subsonic, Sonic and Supersonic Flow (Linear Two-Dimensional Theory)*, British A.R.C. Reports and Memoranda 2932, 1953.

4–69. Nelson, H. C., and J. H. Berman, *Calculations on the Forces and Moments for an Oscillating Wing-Aileron Combination in Two-Dimensional Potential Flow at Sonic Speed*, NACA Report 1128, 1952.

4–70. Rott, N., "Oscillating Airfoils at Mach Number 1," *J. Aero. Sciences*, Vol. 16, No. 6, June 1949, pp. 380–381.

4–71. Garrick, I. E., and S. I. Rubinow, *Flutter and Oscillating Air-Force Calculations for an Airfoil in Two-Dimensional Supersonic Flow*, NACA Report 846, 1946.

4–72. Landahl, M. T., *Theoretical Studies of Unsteady Transonic Flow, Part I–Linearization of the Equations of Motion*, Flygtekniska Försöksanstalten, Stockholm, Report 77, May 1958.

4–73. Coupry, G., and G. Piazzoli, "Étude du flottement en régime transsonique," *La Recherche Aeronautique*, No. 63, March–April 1958, pp. 19-28.

4–74. Eckhaus, W., *Two-Dimensional Transonic Unsteady Flow with Shock Waves*, USAF Office of Scientific Research Technical Note 59–491, 1959.

4–75. Huckel, V., *Tabulation of the f-Functions Which Occur in the Aerodynamic Theory of Oscillating Wings in Supersonic Flow*, NACA Technical Note 3606, 1956.

4–76. Huckel, V., and B. J. Durling, *Tables of Wing-Aileron Coefficients of Oscillating Airforces for Two-Dimensional Supersonic Flow*, NACA Technical Note 2055, 1950.

4–77. Minhinnick, I. T., and D. L. Woodcock, *Tables of Aerodynamic Flutter Derivatives for Thin Wings and Control Surfaces in Two-Dimensional Supersonic Flow*, Royal Aircraft Establishment Report No. Structures 228, October 1957.

4–78. Van Dyke, M. D., *Supersonic Flow Past Oscillating Airfoils Including Nonlinear Thickness Effects*, NACA Report 1183, 1953.

4–79. Jones, W. P., *The Influence of Thickness/Chord Ratio on Supersonic Derivatives for Oscillating Airfoils*, British Aeronautical Research Council Reports and Memoranda 2679, 1948.

4–80. Rodden, W. P., and J. D. Revell, *Oscillatory Aerodynamic Coefficients for a Unified Supersonic-Hypersonic Strip Theory Including the Effects of Sweep, Thickness and Finite Span*, North American Aviation, Report Na-57-1549, December 1957.

4–81. Ashley, H., and G. Zartarian, *Supersonic Flutter Trends as Revealed by Piston Theory Calculations*, USAF Wright Air Development Center Technical Report 58-74, January 1958.

4–82. Morgan, H. G., H. L. Runyan, and V. Huckel, "Theoretical Considerations of Flutter at High Mach Numbers," *J. Aero. Sciences*, Vol. 25, No. 6, June 1958, pp. 371–381.

4–83. Ashley, H., and G. Zartarian, "Piston Theory—A New Aerodynamic Tool for the Aeroelastician," *J. Aero. Sciences*, Vol. 23, No. 12, December 1956, pp. 1109–1118.

4–84. Zartarian, G., *Unsteady Airloads on Pointed Airfoils and Slender Bodies at High Mach Numbers*, USAF Wright Air Development Center Technical Report 59-583, 1959.

4–85. Watson, G. N., *A Treatise on the Theory of Bessel Functions*, 2nd ed., The Macmillan Company, London, 1948.

4–86. Watkins, C. E., H. L. Runyan, and D. S. Woolston, *On the Kernel Function of the Integral Equation Relating the Lift and Downwash Distributions of Oscillating Finite Wings in Subsonic Flow*, NACA Report 1234, 1955. (Originally Technical Note 3131, 1954.)

4–87. Runyan, H. L., and D. S. Woolston, *Method for Calculating the Aerodynamic Loading on an Oscillating Finite Wing in Subsonic and Sonic Flow*, NACA Technical Note 3694, August 1956.

4–88. Richardson, J. R., *A Method for Calculating the Lifting Forces on Wings. (Unsteady Subsonic and Supersonic Lifting Surface Theory.)* Technical Office Report 165, The Fairey Aviation Company, Ltd., April 1955.

4–89. Hsu, P. T., *Flutter of Low-Aspect-Ratio Wings, Part I—Calculation of Pressure Distributions for Oscillating Wings of Arbitrary Planform in Subsonic Flow by the Kernel-Function Method*, Technical Report 64-1, M.I.T. Aeroelastic and Structures Research Laboratory, October 1957.

4–90. Hsu, P. T., "Some Recent Developments in the Flutter Analysis of Low-Aspect-Ratio Wings," *Proc.* National Specialists Meeting on Dynamics and Aeroelasticity, Fort Worth, November 1958, pp. 7–26.

4–91. Cunningham, H. J., and D. S. Woolston, "Developments in the Flutter Analysis of General Plan Form Wings Using Unsteady Air Forces from the Kernel Function Procedure," *Proc.* National Specialists Meeting on Dynamics and Aeroelasticity, Fort Worth, November 1958, pp. 27–36.

4–92. Woolston, D. S., and J. L. Sewall, *Use of the Kernel Function in a Three-Dimensional Flutter Analysis with Application to a Flutter-Tested Delta-Wing Model*, NACA Technical Note 4395, September 1958.

4–93. Laidlaw, W. R., *Theoretical and Experimental Pressure Distributions on Low-Aspect-Ratio Wings Oscillating in an Incompressible Flow*, Technical Report 51-2, M.I.T. Aeroelastic and Structures Research Laboratory, 1954.

4–94. Jones, W. P., *The Calculation of Aerodynamic Derivative Coefficients for Wings of Any Planform in Non-uniform Motion*, British A.R.C. Reports and Memoranda 2470, 1946.

4-95. Lehrian, D. E., *Aerodynamic Coefficients for an Oscillating Delta Wing*, British A.R.C. Reports and Memoranda 2841, 1951.

4-96. Woodcock, D. L., *Aerodynamic Derivatives for Two Cropped Delta Wings and One Arrowhead Wing Oscillating in Distortion Modes*, British R.A.E. Report No. Structures 201, April 1956.

4-97. Woodcock, D. L., *Calculated Aerodynamic Forces on a Sweptback Untapered Wing Oscillating in Incompressible Flow*, British R.A.E. Report No. Structures 217, December 1956.

4-98. Anonymous, *Downwash Tables for the Calculation of Aerodynamic Forces on Oscillating Wings*, British A.R.C. Reports and Memoranda 2956, 1952.

4-99. Cicala, P., *Comparison of Theory with Experiment in the Phenomenon of Wing Flutter*, NACA Technical Memorandum 887, 1939.

4-100. Reissner, E., and J. E. Stevens, *Effect of Finite Span on the Airload Distributions for Oscillating Wings, I and II*, NACA Technical Notes 1194 and 1195, 1947.

4-101. J. B. Rea Company, Inc., *Aeroelasticity in Stability and Control*, Chapter 5, USAF Wright Air Development Center Technical Report 55-173, March 1957.

4-102. Miles, J. W., "On the Compressibility Correction for Subsonic Unsteady Flow," *J. Aero. Sciences*, Readers' Forum, Vol. 17, No. 3, March 1950, pp. 181–182.

4-103. Jones, W. P., *Oscillating Wings in Compressible Subsonic Flow*, British A.R.C. Reports and Memoranda 2855, 1951.

4-104. Lehrian, D. E., *Calculation of Stability Derivatives for Oscillating Wings*, British A.R.C. Reports and Memoranda 2922, 1953.

4-105. Merbt, H., and M. T. Landahl, *Aerodynamic Forces on Oscillating Low-Aspect-Ratio Wings in Compressible Flow*, Swedish K.T.H. Aero Technical Note 30, 1953.

4-106. Mazelsky, B., "Theoretical Aerodynamic Properties of Vanishing Aspect Ratio Harmonically Oscillating Rigid Airfoils in a Compressible Medium," *J. Aero. Sciences*, Vol. 23, No. 7, July 1956, pp. 639–652.

4-107. Milne, R. D., *The Unsteady Aerodynamic Forces on Deforming, Low-Aspect-Ratio Wings and Slender Wing-Body Combinations*, Report No. 94, The College of Aeronautics, Cranfield, July 1955.

4-108. Landahl, M. T., *The Flow Around Oscillating Low-Aspect-Ratio Wings at Transonic Speed*, Swedish K.T.H. Aero Technical Note 40, 1954.

4-109. Landahl, M. T., *Forces and Moments on Oscillating Slender Wing-Body Combinations at Sonic Speed*, USAF Office of Scientific Research Technical Note 56-109, 1956.

4-110. Landahl, M. T., *The Flow Around Oscillating Low-Aspect-Ratio Wings and Wing-Body Combinations at Transonic Speeds*, Proc. Ninth International Congress of Applied Mechanics, Brussels, August 1956.

4-111. Garrick, I. E., and S. I. Rubinow, *Theoretical Study of Air Forces on an Oscil-, lating or Steady Thin Wing in a Supersonic Main Stream*, NACA Report 872, 1947.

4-112. Watkins, C. E., *Effect of Aspect Ratio on the Air Forces and Moments of Harmonically Oscillating Thin Rectangular Wings in Supersonic Potential Flow*, NACA Report 1028, 1951.

4-113. Nelson, H. C., R. A. Rainey, and C. E. Watkins, *Lift and Moment Coefficients Expanded to the Seventh Power of Frequency for Oscillating Rectangular Wings in Supersonic Flow and Applied to a Specific Flutter Problem*, NACA Technical Note 3076, April 1954.

4–114. Nelson, H. C., *Lift and Moment on Oscillating Triangular and Related Wings with Supersonic Edges*, NACA Technical Note 2494, 1951.

4–115. Cunningham, H. J., *Total Lift and Pitching Moment on Thin Arrowhead Wings Oscillating in Supersonic Potential Flow*, NACA Technical Note 3433, 1955.

4–116. Cunningham, H. J., *Lift and Moment on Thin Arrowhead Wings with Supersonic Edges Oscillating in Symmetric Flapping and Roll and Application to the Flutter of an All-Movable Control Surface*, NACA Technical Note 4189, January 1958.

4–117. Watkins, C. E., and J. H. Berman, *Velocity Potential and Air Forces Associated with a Triangular Wing in Supersonic Flow, with Subsonic Leading Edges, and Deforming Harmonically According to a General Quadratic Equation*, NACA Technical Note 3009, September 1953.

4–118. Evvard, J. C., *Use of Source Distributions for Evaluating Theoretical Aerodynamics of Thin Finite Wings at Supersonic Speeds*, NACA Report 951, 1950.

4–119. Watson, J., *Calculation of Derivatives for a Cropped Delta Wing with an Oscillating Constant-Chord Flap in a Supersonic Air Stream*, British A.R.C. Reports and Memoranda 3059, November 1955.

4–120. Watson, J., *Calculation of Derivatives for a Cropped Delta Wing with Subsonic Leading Edges Oscillating in a Supersonic Air Stream*, British A.R.C. Reports and Memoranda 3060, July 1956.

4–121. Pines, S., J. Dugundji, and J. Neuringer, "Aerodynamic Flutter Derivatives for a Flexible Wing with Supersonic and Subsonic Edges," *J. Aero. Sciences*, Vol. 22, No. 10, October 1955, pp. 693–700.

4–122. Garrick, I. E., *Some Concepts and Problem Areas in Aircraft Flutter*, The Minta Martin Lecture for 1957, Sherman M. Fairchild Publication Fund Paper No. FF 15, Inst. of Aeronautical Sciences, March 1957.

4–123. Zartarian, G., and P. T. Hsu, *Theoretical Studies on the Prediction of Unsteady Supersonic Airloads on Elastic Wings, Part 1, Investigations on the Use of Oscillatory Supersonic Aerodynamic Influence Coefficients*, WADC Technical Report 56–97, December 1955 (Confidential—Title Unclassified).

4–124. Zartarian, G., *Theoretical Studies on the Prediction of Unsteady Supersonic Airloads on Elastic Wings, Part 2, Rules for Application of Oscillatory Supersonic Aerodynamic Influence Coefficients*, WADC Technical Report 56–97, February 1956 (Confidential—Title Unclassified).

4–125. Zartarian, G., et al., *Forces and Moments on Oscillating Slender Wing-Body Combinations at Supersonic Speed, Part I—Basic Theory*, USAF Office of Scientific Research Technical Note No. 57–386, April 1957; *Part II—Applications and Comparison with Experiment*, USAF Office of Scientific Research Technical Note 58–114, December 1957.

4–126. Holt, M., "A Linear Perturbation Method for Stability and Flutter Calculations on Hypersonic Bodies," *J. Aero/Space Sciences*, Vol. 26, No. 12, December 1959, pp. 787–793.

4–127. Neumark, S., "Pressure Distribution on an Airfoil in Non-uniform Motion," *J. Aero. Sciences*, Readers' Forum, Vol. 19, No. 3, March 1952, pp. 214–215.

4–128. von Kármán, T., and W. R. Sears, "Airfoil Theory for Nonuniform Motion," *J. Aero. Sciences*, Vol. 5, No. 10, August 1938, pp. 379–390.

4–129. Heaslet, M. A., and H. Lomax, *Two-Dimensional Unsteady Lift Problems in Supersonic Flight*, NACA Report 945, 1949.

4–130. Lomax, H., M. A. Heaslet, F. B. Fuller, and L. Sluder, *Two- and Three-Dimensional Unsteady Lift Problems in High-Speed Flight*, NACA Report 1077, 1952.

4–131. Drischler, J. A., and F. W. Diederich, *Lift and Moment Responses to Penetration of Sharp-Edged Traveling Gusts, with Application to Penetration of Weak Blast Waves*, NACA Technical Note 3956, May 1957.

4–132. Ludloff, H. F., *On Aerodynamics of Blasts*, Vol. III, *Advances in Applied Mechanics*, Academic Press, New York, 1953, pp. 109–144.

4–133. Küssner, H. G., *Zusammenfassender Bericht über den instationären Auftrieb von Flügeln*, Luftfahrtforschung, Bd. 13, Nr. 12, December 1936, pp. 410–424.

4–134. Fraeys de Veubeke, *Aérodynamique instationnaire des profils minces déformables*, Bulletin du Service Technique de l'Aéronautique, Brussels, Belgium, No. 25, 1953.

4–135. Timman, R., *La théorie des profils minces en écoulement non stationnaire en fluide incompressible ou compressible*, Actes des Colloques du Centre National de la Recherche Scientifique, Colloque National No. 1, Journées de Mecanique des Fluides, Marseille, 1952.

4–136. Rott, N., "Flügelschwingungsformen in ebener kompressiblen Potentialströmung," *Zeitschrift für angewandte Mathematik und Physik*, Vol. 1, No. 6, 1950, pp. 380–410.

4–137. Ordway, D. E., *An Aerodynamic Theory of a Supersonic Propeller*, USAF Office of Scientific Research Technical Note No. 56–287, June 1956.

4–138. Wagner, H., "Über die Entstehung des dynamischen Auftriebes von Tragflügeln," *Zeitschrift für angewandte Mathematik und Mechanik*, Vol. 5, No. 1, February 1925.

4–139. Garrick, I. E., "On Some Fourier Transforms in the Theory of Nonstationary Flows," *Proceedings of the Fifth International Congress for Applied Mechanics*, John Wiley and Sons, New York, 1939, pp. 590–593. (Out of print.)

4–140. Sears, W. R., "Operational Methods in the Theory of Airfoils in Nonuniform Motion," *J. Franklin Institute*, Vol. 230, 1940, pp. 95–111.

4–141. Hobbs, N. P., "The Transient Downwash Resulting from the Encounter of an Airfoil with a Moving Gust Field," *J. Aero. Sciences*, Vol. 24, No. 10, October 1957, pp. 731–740, 754.

4–142. Miles, J. W., "The Aerodynamic Force on an Airfoil in a Moving Gust," *J. Aero. Sciences*, Vol. 23, No. 11, November 1956, pp. 1044–1050.

4–143. Mazelsky, B., *Numerical Determination of Indicial Lift of a Two-Dimensional Sinking Airfoil at Subsonic Mach Numbers from Oscillatory Lift Coefficients with Calculations for Mach Number 0.7*, NACA Technical Note 2562, 1951.

4–144. Mazelsky, B., *Determination of Indicial Lift and Moment of a Two-Dimensional Pitching Airfoil at Subsonic Mach Numbers from Oscillatory Coefficients with Numerical Calculations for a Mach Number of 0.7*, NACA Technical Note 2613, 1952.

4–145. Mazelsky, B., and J. A. Drischler, *Numerical Determination of Indicial Lift and Moment Functions for a Two-Dimensional Sinking and Pitching Airfoil at Mach Numbers 0.5 and 0.6*, NACA Technical Note 2739, 1952.

4–146. Krasnoff, E., "Subsonic Lift Response to Penetration of a Sharp-edged Gust Moving at Supersonic Speed," *J. Aero. Sciences*, Readers' Forum, Vol. 25, No. 3, March 1958, pp. 214–215.

4–147. Drischler, J. A., *Calculation and Compilation of the Unsteady Lift Function for a Rigid Wing Subjected to Sinusoidal Gusts and Sinusoidal Sinking Oscillations*, NACA Technical Note 3748, October 1956.

4–148. Chang, C. C., *Transient Aerodynamic Behavior of an Airfoil Due to Different Arbitrary Modes of Nonstationary Motions in a Supersonic Flow*, NACA Technical Note 2333, 1951.

4-149. Biot, M. A., *Loads on a Supersonic Wing Striking a Sharp-edged Gust*, Cornell Aeronautical Laboratory Report SA-247-S-7, 1948.

4-150. Ashley, H., J. Dugundji, and D. O. Neilson, "Two Methods for Predicting Air Loads on a Wing in Accelerated Motion," *J. Aero. Sciences*, Vol. 19, No. 8, August 1952, pp. 543–552.

4-151. Krasilshchikova, E. A., *Unsteady Motion of a Profile in a Compressible Fluid*, Doklady, A. N. USSR, Vol. 94, No. 3, 1954, pp. 397–400.

4-152. Krasilshchikova, E. A., *Unsteady Motion of a Finite-Span Wing in a Compressible Medium*, Doklady, A. N. USSR, Vol. 97, No. 5, September–October 1957.

4-153. Jones, R. T., *The Unsteady Lift of a Wing of Finite Aspect Ratio*, NACA Report 681, 1940.

4-154. Jones, W. P., *Aerodynamic Forces on Wings in Non-Uniform Motion*, British A.R.C. Reports and Memoranda 2117, 1945.

4-155. Lehrian, D. E., *Initial Lift of Finite Aspect-Ratio Wings Due to a Sudden Change of Incidence*, British A.R.C. Reports and Memoranda 3023, 1955.

4-156. Drischler, J. A., *Approximate Indicial Lift Functions for Several Wings of Finite Span in Incompressible Flow as Obtained from Oscillatory Lift Coefficients*, NACA Technical Note 3639, May 1956.

4-157. Miles, J. W., "On Simple Planforms in Supersonic Flow," *J. Aero. Sciences*, Readers' Forum, Vol. 17, No. 2, February 1950, Page 127.

4-158. Miles, J. W., "Transient Loading of Supersonic Rectangular Airfoils," *J. Aero. Sciences*, Vol. 17, No. 10, October 1950, pp. 647–652.

4-159. Lomax, H., F. B. Fuller, and L. Sluder, *Generalized Indicial Forces on Deforming Rectangular Wings in Supersonic Flight*, NACA Report 1230, 1954.

4-160. Warner, R. W., and B. B. Packard, *A Method for Calculating the Aerodynamic Forces due to Arbitrary, Time-Dependent Downwash for a Class of Thin, Flexible Wings at Supersonic Speeds*, NASA Technical Note D-142, February 1960.

4-161. Miles, J. W., *Unsteady Supersonic Flow Past Slender Pointed Bodies*, Navord Report 2031, U.S. Naval Ordnance Test Station, California, May 1953.

4-162. Fyfe, I. M., and K. Klotter, *Nonlinear Problems of One-Dimensional Wave Propagation in Gases*, USAF Wright Development Center Technical Report 58–293, August 1958.

4-163. Hayes, W. D., and R. F. Probstein, *Hypersonic Flow Theory*, Academic Press, New York, 1959.

4-164. Hsu, P. T., and W. H. Weatherill, *Pressure Distribution and Flutter Analysis of Low-Aspect-Ratio Wings in Subsonic Flow*, Technical Report 64-3, M.I.T. Aero-elastic and Structures Research Laboratory, June 1959.

4-165. Landahl, M. T., *Unsteady Transonic Flow*, Pergamon Press, New York, 1961.

4-166. Rodden, W. P., and J. D. Revell, *The Status of Unsteady Aerodynamic Influence Coefficients*, Fairchild Fund Paper No. FF–13, presented at 30th Annual Meeting of Institute of the Aerospace Sciences, New York, January 1962.

5

STRUCTURAL
OPERATORS

5-1 INTRODUCTION

Since the early 1930's, when aluminum alloys were first used, strong and rigid full-cantilever lifting surfaces have undergone substantial changes in their geometry and method of construction. Whereas a lifting surface invariably had the general appearance of a slender beam for many years, more recent trends in very high-speed military flight vehicles have been toward low-aspect-ratio plate-like surfaces. On the other hand, civilian jet transport aircraft retain exceptionally slender sweptback wing beams. We can observe the nature of some of these trends by studying the curves shown by Figs. 5–1 and 5–2.

Figure 5–1 illustrates the trend of wing thickness ratio in fighter and transport aircraft since 1920. The emphasis on drag reduction is evidenced by a continuous lowering of thickness ratio from about 0.2 in 1930 to something on the order of 0.05 in 1959. Accompanying the changes in thickness ratio, we find interesting changes in aspect ratio. In Fig. 5–2, the trend of wing structural aspect ratio is plotted, being defined as the ratio of the square of the structural span to the wing area. The term structural span denotes the total span of the wing measured along the beam axis, as illustrated by the inset of Fig. 5–2. In Fig. 5–2 we observe that lifting surfaces have had a continual upward trend in structural aspect ratio in the case of transport aircraft, with the swept wing jet transports exhibiting higher values than ever before. In the case of fighter aircraft, however, we find that lifting surfaces tended to a higher degree of slenderness in the chordwise bending direction until the advent of supersonic fighters, which

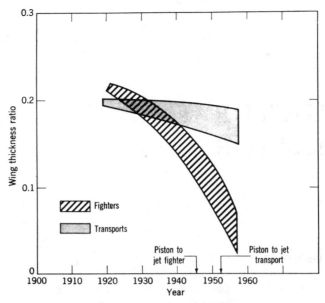

Fig. 5–1. Wing thickness ratio.

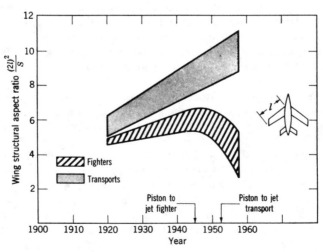

Fig. 5–2. Wing structural aspect ratio.

marked a reversal of the trend. This reversal evidently resulted from both aeroelastic and aerodynamic reasons.

It seems evident from these trends that aeroelasticians must deal not only with one-dimensional structures or slender beams, but also with two-dimensional structures or lifting surfaces. Whereas there has heretofore been comparatively little interest in the theory of bending and stretching of shells, this theory must now be employed as a basis for predicting the behavior of thin lifting surfaces.

In the case of missiles and rockets, lifting surfaces are frequently omitted entirely, stability and control being provided by gimbaled engines and obliquely mounted jet nozzles. In these cases, the aeroelastician may be required to consider the vehicle as a lifting body or as a three-dimensional structure. Thus modern applications of aeroelasticity can run the gamut from one- to two- to three-dimensional structures.

The effects of aerodynamic heating on structural behavior, heretofore neglected in aeroelastic analyses, now also assume importance. There are two principal effects. The first derives from the deterioration of the mechanical properties of materials at elevated temperature. This deterioration can be included in stiffness predictions by merely correcting the modulus of elasticity and Poisson's ratio for the influence of temperature. The second effect derives from the influence of thermal stresses which result from the fact that most engineering materials expand with increasing temperature. Temperature gradients within a solid body produce different degrees of expansion in different parts of the body. Thus, when the elements into which a solid body may be divided are interconnected and cannot expand freely, stresses will result. In the case of plates and shells it is possible for this self-equilibrating thermal stress system to have significant effects upon stiffness and buckling characteristics.

In the present chapter, it will not be possible to set forth a full account of the background of the structural theory, particularly as it applies to plates and shells. The reader is referred instead to a number of excellent comprehensive textbooks on these subjects, of which Refs. 5–1 through 5–3 are samples. This chapter will, however, be confined largely to a brief statement of the fundamentals and to a summary of the more important structural operators.

5–2 ONE-DIMENSIONAL STRUCTURES

We take up first, very briefly, the one-dimensional structure—that is, a structure whose state of deformation can be adequately described by a set of functions of a single space coordinate.

(a) Slender unswept wing

Let us consider a type of slender beam in which we can make the simplifying assumptions normally employed in engineering beam theory. First, the beam is permitted complete freedom to warp when torque loads are applied. This leads to the St. Venant solution of the torsion problem and permits the simplification that bending and twisting are separate uncoupled actions. Second, we assume that plane sections remain plane during bending. This allows application of the well-known engineering bending theory.

The beam may taper in the spanwise or y-direction and the properties of the cross section may vary with y. The reference axes are illustrated by Fig. 5–3. It is convenient in the present discussion to assume that the y-axis is placed so that it pierces each cross section at the centroid of the effective normal stress-carrying area and that the x- and z-axes are oriented so that the x-y- and y-z-planes pass through the principal bending axes.

Lateral beam deflections. The lateral deflection of a beam, that is, the deflection in the z-direction, according to the engineering theory, is made up of a bending deflection, w_B, and a shearing deflection, w_S. The former can be represented by

$$w_B = \int_0^y \frac{M_x(\lambda)(y - \lambda)}{EI}\, d\lambda \qquad (5-1)$$

where $M_x(\lambda)$ is the bending moment distribution about the x-axis and EI is the bending stiffness (Ref. 5–4). The positive directions of M_x and w are as illustrated by Fig. 5–3. Differentiation of Eq. 5–1 twice with respect to y produces the following well-known relation between beam bending moment and curvature:

$$\frac{d^2 w_B}{dy^2} = \frac{M_x}{EI} \qquad (5-2)$$

The shearing deflection, w_S, is given by

$$w_S = \int_0^y \frac{V(\lambda)\, d\lambda}{GK} \qquad (5-3)$$

where $V(\lambda)$ is the shear distribution along the beam positive upward, and GK is the shearing rigidity (Ref. 5–4). Differentiating Eq. 5–3 with respect to y gives a relation between the shear and the first derivative of the shearing deflection, as follows:

$$\frac{dw_S}{dy} = \frac{V}{GK} \qquad (5-4)$$

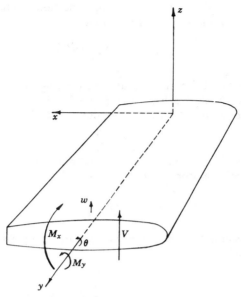

Fig. 5–3. Axis system for slender beam.

Making use of the fact that

$$Z = \frac{dV}{dy} = \frac{d^2M}{dy^2} \qquad (5\text{–}5)$$

where Z is the intensity of lateral loading, we can write the following relation for the lateral deflection of a slender beam:

$$Z = \mathscr{S}_B(w_B) + \mathscr{S}_S(w_S) \qquad (5\text{–}6)$$

where $\mathscr{S}_B$ is a structural operator for bending displacement defined by

$$\mathscr{S}_B = \frac{d^2}{dy^2}\left(EI\,\frac{d^2}{dy^2}\right)$$

and $\mathscr{S}_S$ is a structural operator for shearing displacement given by

$$\mathscr{S}_S = \frac{d}{dy}\left(GK\,\frac{d}{dy}\right)$$

The inverse structural operators for slender beams are obtained by integration of Eq. 5–6. We have already seen the form of the operator $\mathscr{S}_B^{-1}$ in connection with Eqs. 3–23 through 3–30.

Torsional beam deflections. The twisting deflection of a beam, according to St. Venant solution of the torsion problem, can be represented by

$$\theta(y) = \int_0^y \frac{M_y(\lambda)\, d\lambda}{GJ} \tag{5-7}$$

where $M_y(\lambda)$ is the distribution of applied twisting moment about the y axis, positive as illustrated by Fig. 5–3, and GJ is the torsional stiffness (Ref. 5–4). We can derive by differentiation of Eq. 5–7 the form of

$$m_y = \mathscr{S}(\theta) \tag{5-8}$$

where $m_y = dM_y/dy$ is the applied torsional moment per unit length and $\mathscr{S}$ is a structural operator defined by

$$\mathscr{S} = \frac{d}{dy}\left(GJ\,\frac{d}{dy}\right)$$

The inverse structural operator for twisting can be expressed as

$$\theta = \mathscr{S}^{-1}(m_y) \tag{5-9}$$

where

$$\mathscr{S}^{-1}(\quad) = \int_0^l C(y,\eta)(\quad)\, d\eta$$

and where

$$C(y,\eta) = \int_0^y \frac{d\lambda}{GJ}, \qquad (\eta \geq y)$$

$$C(y,\eta) = \int_0^\eta \frac{d\lambda}{GJ}, \qquad (y \geq \eta)$$

(b) Slender swept wings

In aeroelastic analyses of slender swept wings, it is frequently necessary to think in terms of linear and angular displacements of streamwise segments of the wing, that is, segments which are parallel to the airstream. We refer to Fig. 5–4, which illustrates a slender swept wing of large aspect ratio together with the axis system and other notation which will be employed. The deformation of a swept wing can be described in terms of linear and angular displacements of streamwise segments due to the running loads and torques on these segments. In the case of slender swept wings of large aspect ratio, an effective root can be assumed as shown by Fig. 5–4. Thus an elastic axis can be postulated; and the $\bar{y}$ reference axis can be located along the elastic axis. If $\bar{y}$ is a true elastic axis, we can employ slender beam theory to compute the wing deflections. We may,

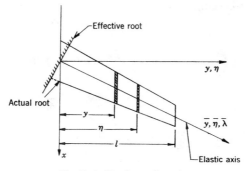

Fig. 5–4. Slender swept wing.

for example, visualize four distinct types of structural operators for a slender swept wing, which are expressible in the following inverted forms:

$$w = \mathscr{S}_{zz}^{-1}(Z) \tag{5-10}$$

$$\theta = \mathscr{S}_{\theta\theta}^{-1}(m_y) \tag{5-11}$$

$$\theta = \mathscr{S}_{\theta z}^{-1}(Z) \tag{5-12}$$

$$w = \mathscr{S}_{z\theta}^{-1}(m_y) \tag{5-13}$$

In Eq. 5–10 we have an inverse structural operator which transforms forces, Z, on streamwise segments into vertical displacements, w, of segments at the $\bar{y}$ axis.

$$\mathscr{S}_{zz}^{-1}(\quad) = \int_0^l C^{zz}(y, \eta)(\quad) \, d\eta \tag{5-14}$$

where

$$C^{zz}(y, \eta) = \int_0^{y/\cos\Lambda} \frac{\left(\dfrac{\eta}{\cos\Lambda} - \bar{\lambda}\right)\left(\dfrac{y}{\cos\Lambda} - \bar{\lambda}\right)}{EI} \, d\bar{\lambda}$$
$$+ \int_0^{y/\cos\Lambda} \frac{d\bar{\lambda}}{GK}, \quad (\eta \geq y)$$

$$C^{zz}(y, \eta) = \int_0^{\eta/\cos\Lambda} \frac{\left(\dfrac{\eta}{\cos\Lambda} - \bar{\lambda}\right)\left(\dfrac{y}{\cos\Lambda} - \bar{\lambda}\right)}{EI} \, d\bar{\lambda}$$
$$+ \int_0^{\eta/\cos\Lambda} \frac{d\bar{\lambda}}{GK}, \quad (y \geq \eta)$$

and where EI and GK are stiffness properties of the beam on sections normal to the elastic axis.

The inverse structural operator $\mathscr{S}_{\theta\theta}^{-1}(\ \ )$ transforms pitching moments, m_y, on streamwise segments into pitching displacements, θ, of the segments.

$$\mathscr{S}_{\theta\theta}^{-1}(\ \ ) = \int_0^l C^{\theta\theta}(y, \eta)(\ \ ) \, d\eta \qquad (5\text{--}15)$$

where

$$C^{\theta\theta}(y, \eta) = \int_0^{y/\cos\Lambda} \left(\frac{\cos^2 \Lambda}{GJ} + \frac{\sin^2 \Lambda}{EI} \right) d\bar{\lambda}, \qquad (\eta \geq y)$$

$$C^{\theta\theta}(y, \eta) = \int_0^{\eta/\cos\Lambda} \left(\frac{\cos^2 \Lambda}{GJ} + \frac{\sin^2 \Lambda}{EI} \right) d\bar{\lambda}, \qquad (y \geq \eta)$$

The inverse structural operator $\mathscr{S}_{\theta z}^{-1}(\ \ )$ transforms forces, Z, on streamwise segments into pitching displacements, θ, of the segments.

$$\mathscr{S}_{\theta z}^{-1}(\ \ ) = \int_0^l C^{\theta z}(y, \eta)(\ \ ) \, d\eta \qquad (5\text{--}16)$$

where

$$C^{\theta z}(y, \eta) = -\sin \Lambda \int_0^{y/\cos\Lambda} \frac{\left(\dfrac{\eta}{\cos \Lambda} - \bar{\lambda} \right)}{EI} \, d\bar{\lambda}, \qquad (\eta \geq y)$$

$$C^{\theta z}(y, \eta) = -\sin \Lambda \int_0^{\eta/\cos\Lambda} \frac{\left(\dfrac{\eta}{\cos \Lambda} - \bar{\lambda} \right)}{EI} \, d\bar{\lambda}, \qquad (y \geq \eta)$$

Finally, the inverse operator $\mathscr{S}_{z\theta}^{-1}(\ \ )$ transforms pitching moments on streamwise segments into vertical displacements, w, of segments at the $\bar{y}$-axis.

$$\mathscr{S}_{z\theta}^{-1}(\ \ ) = \int_0^l C^{z\theta}(y, \eta)(\ \ ) \, d\eta \qquad (5\text{--}17)$$

where we obtain the influence function $C^{z\theta}(y, \eta)$ from $C^{\theta z}(y, \eta)$, using the reciprocal relation, in the following way:

$$C^{z\theta}(y, \eta) = C^{\theta z}(\eta, y) \qquad (5\text{--}18)$$

Elastic axis. In Parts (a) and (b) above, we have outlined the structural operators for computing bending and twisting displacements of beams with straight elastic axes. In order to employ these operators in a given example it is necessary to know the location of the elastic axis. The elastic axis is located by drawing a spanwise line through the shear centers of the various cross sections of the beam. The shear center of each cross section is computed by establishing the point in the plane of the section at which a shear force can be applied to the section without producing a rate of twist at the section. Methods for computing the shear center location of multi-cell beams may be found, for example, in Ref. 5–4. Tests of full-scale wings and of models are frequently employed as means for determining elastic axis locations.

5–3 TWO-DIMENSIONAL STRUCTURES

As a basis for theoretical studies of lifting surfaces, we shall employ the theory of the bending and stretching of shell surfaces of variable thickness subjected to loads and temperature gradients. In some instances, the entire lifting surface may be regarded as a normally loaded shell surface. In other instances of built-up lifting surfaces or fuselage and missile bodies, the structure may be regarded as an assemblage of shell elements. In either case, a knowledge of shell theory is an important prerequisite to an understanding of structural behavior.

(a) Shell geometry

Let us confine our attention to thin shells in which the thickness h is small compared to the radius of curvature and other dimensions. We define a reference surface as the middle surface of the shell. Shell geometry is defined by specifying the form of the middle surface and the thickness at each point.

Referring to Fig. 5–5, we may state that any surface may be defined by a

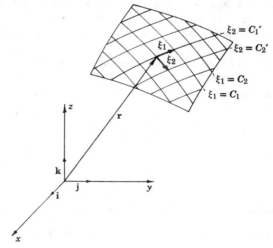

Fig. 5–5. Curvilinear coordinates on a surface.

position vector $\mathbf{r}$ relative to a fixed origin "O." The vector is a function of two independent parameters, ξ_1 and ξ_2, as follows:

$$\mathbf{r} = \mathbf{r}(\xi_1, \xi_2) \tag{5-19}$$

The Cartesian representation of the surface may be stated as

$$x = f_1(\xi_1, \xi_2), \qquad y = f_2(\xi_1, \xi_2), \qquad z = f_3(\xi_1, \xi_2) \tag{5-20}$$

and if we eliminate ξ_1 and ξ_2, we obtain the surface equation

$$F(x, y, z) = 0 \tag{5-21}$$

When a relation between ξ_1 and ξ_2 is derived, say, $g(\xi_1, \xi_2) = 0$, we obtain a curve on the surface known as a parametric curve. A surface can be defined by a doubly infinite set of such parametric curves. The parameters ξ_1 and ξ_2 constitute a system of curvilinear coordinates, as illustrated by Fig. 5–5; and the position of any point on the surface may be defined by the values of ξ_1 and ξ_2 at that point.

In the vast majority of practical applications of shell theory, it is assumed that the shell surface is described by a curvilinear coordinate system which lies along lines of curvature of the surface (Ref. 5–2). Lines of curvature of a surface are oriented along the directions of principal curvature. These are the directions along which the normal curvatures to the surface exhibit principal or maximum and minimum values. Directions of principal curvature may be shown to be orthogonal directions (Ref. 5–3).

Referring to Fig. 5–6, we form a right hand coordinate system on the undeformed middle surface such that the unit vectors corresponding to the curvilinear coordinates ξ_1 and ξ_2 and in the direction of the normal to the surface are represented by $\mathbf{t}_1$, $\mathbf{t}_2$, and $\mathbf{n}$, respectively. The radii of curvature in the directions of ξ_1 and ξ_2 are represented by R_1 and R_2, respectively; and these radii are taken positive when the centers of curvature lie in the positive direction of $\mathbf{n}$. By definition, the unit vectors $\mathbf{t}_i$

$$\mathbf{t}_i = \frac{1}{\alpha_i} \frac{\partial \mathbf{r}}{\partial \xi_i}, \qquad (i = 1, 2) \tag{5-22}$$

where

$$\alpha_i^2 = \frac{\partial \mathbf{r}}{\partial \xi_i} \cdot \frac{\partial \mathbf{r}}{\partial \xi_i}$$

are the so called "first fundamental magnitudes" of the surface.* There

---

* According to differential geometry, an element of arc ds on the shell surface is defined by

$$\overline{ds^2} = d\mathbf{r} \cdot d\mathbf{r} = \alpha_1^2 \, d\xi_1^2 + \alpha_2^2 \, d\xi_2^2$$

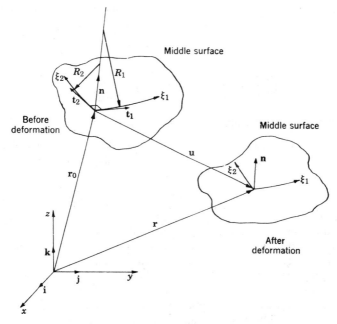

Fig. 5–6. Middle surface geometry.

are certain relations among the quantities defined above which are specified by the geometry of the middle surface. There are, first, the matrix relations

$$
\frac{\partial}{\partial \xi_1}
\begin{bmatrix} \mathbf{t}_1 \\ \mathbf{t}_2 \\ \mathbf{n} \end{bmatrix}
=
\begin{bmatrix}
0 & -\dfrac{1}{\alpha_2}\dfrac{\partial \alpha_1}{\partial \xi_2} & \dfrac{\alpha_1}{R_1} \\[2ex]
\dfrac{1}{\alpha_2}\dfrac{\partial \alpha_1}{\partial \xi_2} & 0 & 0 \\[2ex]
-\dfrac{\alpha_1}{R_1} & 0 & 0
\end{bmatrix}
\begin{bmatrix} \mathbf{t}_1 \\ \mathbf{t}_2 \\ \mathbf{n} \end{bmatrix}
\tag{5–23a}
$$

$$
\frac{\partial}{\partial \xi_2}
\begin{bmatrix} \mathbf{t}_1 \\ \mathbf{t}_2 \\ \mathbf{n} \end{bmatrix}
=
\begin{bmatrix}
0 & \dfrac{1}{\alpha_1}\dfrac{\partial \alpha_2}{\partial \xi_1} & 0 \\[2ex]
-\dfrac{1}{\alpha_1}\dfrac{\partial \alpha_2}{\partial \xi_1} & 0 & \dfrac{\alpha_2}{R_2} \\[2ex]
0 & -\dfrac{\alpha_2}{R_2} & 0
\end{bmatrix}
\begin{bmatrix} \mathbf{t}_1 \\ \mathbf{t}_2 \\ \mathbf{n} \end{bmatrix}
\tag{5–23b}
$$

and the conditions of Codazzi and Gauss represented, respectively, by

$$\frac{\partial}{\partial \xi_1}\left(\frac{\alpha_2}{R_2}\right) = \frac{1}{R_1}\frac{\partial \alpha_2}{\partial \xi_1}, \qquad \frac{\partial}{\partial \xi_2}\left(\frac{\alpha_1}{R_1}\right) = \frac{1}{R_2}\frac{\partial \alpha_1}{\partial \xi_2} \qquad (5\text{-}24a)$$

and

$$\frac{\partial}{\partial \xi_1}\left(\frac{1}{\alpha_1}\frac{\partial \alpha_2}{\partial \xi_1}\right) + \frac{\partial}{\partial \xi_2}\left(\frac{1}{\alpha_2}\frac{\partial \alpha_1}{\partial \xi_2}\right) + \frac{\alpha_1 \alpha_2}{R_1 R_2} = 0 \qquad (5\text{-}24b)$$

(b) Stress resultants and couples

Following the usual procedures of shell theory, we formulate the equations in terms of stress resultants and couples. The stress resultants and couples pertaining to shell theory may be defined with reference to Fig. 5–7, which illustrates an element of a shell of length $d\xi_1$, width $d\xi_2$, and thickness h. The orthogonal curvilinear coordinates ξ_1 and ξ_2 are on the midplane reference surface and lie along the directions of principal curvature. The coordinate axis ζ is perpendicular to the reference surface. The edges of the element are acted upon by the normal and shear stresses σ_{11}, σ_{22}, τ_{12}, τ_{21}, $\tau_{1\zeta}$, and $\tau_{2\zeta}$, as illustrated by Fig. 5–7.

Assuming that the thickness h is small compared to the principal radii of curvature and that the reference surface is a middle surface, we may state that a macroscopic description of the statics of the shell is provided by the three tangential stress resultants in the reference surface,*

$$N_{11} = \int_{-h/2}^{h/2} \sigma_{11}\, d\zeta \qquad (5\text{-}25)$$

$$N_{22} = \int_{-h/2}^{h/2} \sigma_{22}\, d\zeta \qquad (5\text{-}26)$$

$$N_{12} = \int_{-h/2}^{h/2} \tau_{12}\, d\zeta, \qquad N_{21} = \int_{-h/2}^{h/2} \tau_{21}\, d\zeta \qquad (5\text{-}27)$$

and the three stress couples

$$M_{11} = \int_{-h/2}^{+h/2} \sigma_{11}\zeta\, d\zeta \qquad (5\text{-}28)$$

$$M_{22} = \int_{-h/2}^{+h/2} \sigma_{22}\zeta\, d\zeta \qquad (5\text{-}29)$$

$$M_{12} = \int_{-h/2}^{h/2} \tau_{12}\zeta\, d\zeta, \qquad M_{21} = \int_{-h/2}^{h/2} \tau_{21}\zeta\, d\zeta \qquad (5\text{-}30)$$

Since $\tau_{21} = \tau_{12}$, it is evident that $N_{12} = N_{21}$ and $M_{12} = M_{21}$, and that Eqs. 5–25 through 5–30 represent, therefore, six independent quantities.

---

* In writing these equations we limit ourselves to Love's first approximation by setting $1 + \zeta/R_i \approx 1$.

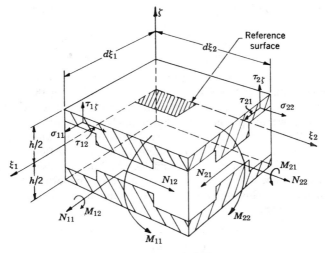

Fig. 5–7. Stiffened shell element.

The positive directions of the stresses are taken according to the usual convention of theory of elasticity (Ref. 5–5); and the positive directions of the stress resultants and couples are, therefore, as illustrated by Fig. 5–7.

(c) Strain-displacement relations

We assume at the outset that not only is the shell thin, that is, the thickness is small in comparison to the radii of curvature and the principal dimensions, but also that the transverse shear and normal strains are zero. Our concern, then, is with two normal strains and one shear strain in laminates parallel to the midplane reference surface. The normal strains in the ξ_1 and ξ_2 directions in laminates parallel to the reference surface are given by the vector form (Ref. 5–3)

$$\epsilon_n = \frac{1}{\alpha_n^2}\left(\frac{\partial \mathbf{r}}{\partial \xi_n} \cdot \frac{\partial \mathbf{u}}{\partial \xi_n} + \frac{1}{2}\frac{\partial \mathbf{u}}{\partial \xi_n} \cdot \frac{\partial \mathbf{u}}{\partial \xi_n}\right), \qquad (n = 1, 2) \qquad (5\text{–}31)$$

where $\mathbf{u}$ is the displacement vector of a point, defined by

$$\mathbf{u} = u_1\mathbf{t}_1 + u_2\mathbf{t}_2 + w\mathbf{n} \qquad (5\text{–}32)$$

and where u_1, u_2, and w are components of the displacement vector. The radius vector $\mathbf{r}$ in Eq. 5–31 is now a vector of position to a point in a laminate which is parallel to but not necessarily in the reference surface.

The shear strain in laminates parallel to the reference surface is given by (Ref. 5–3)

$$\gamma_{12} = \frac{1}{\alpha_1 \alpha_2} \left(\frac{\partial \mathbf{r}}{\partial \xi_1} \cdot \frac{\partial \mathbf{u}}{\partial \xi_2} + \frac{\partial \mathbf{u}}{\partial \xi_1} \cdot \frac{\partial \mathbf{r}}{\partial \xi_2} + \frac{\partial \mathbf{u}}{\partial \xi_1} \cdot \frac{\partial \mathbf{u}}{\partial \xi_2} \right) \tag{5-33}$$

Equations 5–31 and 5–33 apply in the case of large displacements and they may be specialized to small displacements by neglecting the final terms on the right-hand side of each equation. In reducing these equations to scalar form, we adopt the Euler-Bernoulli-Navier hypothesis that the normal to the undeformed reference surface is rotated without extension into the normal to the deformed reference surface. After substituting Eqs. 5–22 and 5–32 into Eqs. 5–31 and 5–33 and making use of Eqs. 5–23 and 5–24, together with the Euler-Bernoulli-Navier hypothesis, there may be obtained the following results for the strains in laminates parallel to the reference surface:

$$\epsilon_n = \bar{\epsilon}_n - \zeta \kappa_n, \quad (n = 1, 2) \tag{5-34}$$

$$\gamma_{12} = \bar{\gamma}_{12} - 2\zeta \kappa_{12} \tag{5-35}$$

where $\bar{\epsilon}_n$ and $\bar{\gamma}_{12}$ are the strains in, and κ_n and κ_{12} are the curvatures of, the reference surface. Adopting the notation of Washizu (Ref. 5–6), the reference surface strains and curvatures are related to the reference surface displacements u, v, and w by the general relations

$$\bar{\epsilon}_1 = l_{11} + \tfrac{1}{2}(l_{11}^2 + l_{21}^2 + l_{31}^2)$$

$$\bar{\epsilon}_2 = l_{22} + \tfrac{1}{2}(l_{12}^2 + l_{22}^2 + l_{32}^2) \tag{5-36}$$

$$\bar{\gamma}_{12} = (l_{12} + l_{21}) + (l_{11}l_{12} + l_{21}l_{22} + l_{31}l_{32})$$

$$\kappa_1 = \frac{1}{\alpha_1} \frac{\partial l_{31}}{\partial \xi_1} + \frac{1}{\alpha_1 \alpha_2} \frac{\partial \alpha_1}{\partial \xi_2} l_{32}$$

$$\kappa_2 = \frac{1}{\alpha_2} \frac{\partial l_{32}}{\partial \xi_2} + \frac{1}{\alpha_1 \alpha_2} \frac{\partial \alpha_2}{\partial \xi_1} l_{31} \tag{5-37}$$

$$2\kappa_{12} = \frac{1}{\alpha_1} \frac{\partial l_{32}}{\partial \xi_1} - \frac{1}{\alpha_1 \alpha_2} \frac{\partial \alpha_2}{\partial \xi_1} l_{32} + \frac{1}{\alpha_2} \frac{\partial l_{31}}{\partial \xi_2} - \frac{1}{\alpha_1 \alpha_2} \frac{\partial \alpha_1}{\partial \xi_2} l_{31}$$

where

$$l_{11} = \frac{1}{\alpha_1} \frac{\partial u}{\partial \xi_1} + \frac{v}{\alpha_1 \alpha_2} \frac{\partial \alpha_1}{\partial \xi_2} - \frac{w}{R_1} \qquad l_{12} = \frac{1}{\alpha_2} \frac{\partial u}{\partial \xi_2} - \frac{v}{\alpha_1 \alpha_2} \frac{\partial \alpha_2}{\partial \xi_1}$$

$$l_{21} = \frac{1}{\alpha_1} \frac{\partial v}{\partial \xi_1} - \frac{u}{\alpha_1 \alpha_2} \frac{\partial \alpha_1}{\partial \xi_2} \qquad l_{22} = \frac{1}{\alpha_2} \frac{\partial v}{\partial \xi_2} + \frac{u}{\alpha_1 \alpha_2} \frac{\partial \alpha_2}{\partial \xi_1} - \frac{w}{R_2}$$

$$l_{31} = \frac{u}{R_1} + \frac{1}{\alpha_1} \frac{\partial w}{\partial \xi_1} \qquad l_{32} = \frac{v}{R_2} + \frac{1}{\alpha_2} \frac{\partial w}{\partial \xi_2}$$

In Eqs. 5–36, we have retained the nonlinear terms in the reference surface strains; but we have retained only linear terms in the formulation of the curvatures in Eqs. 5–37. This combination of terms is especially useful in applications to aeroelastic problems of shells where small vibrations are executed about positions of equilibrium in which the static membrane strains may be large.

(d) Equations of state of an elastic shell element

Three sets of natural variables describe the state of an element of an elastic body: stress, strain, and temperature. Equations expressing a relation among these three are called equations of state. The equations of state of a shell element are those equations which relate the stress resultants and couples with the strains in the plane of the reference surface, the curvatures, and the temperature. Let us assume at the outset that the present discussion is concerned with orthogonally aeolotropic elastic shell elements. In this case, we may represent the state equations by the following matrix forms:

$$\{N_{11}N_{22}\} = [K]\{\bar{\epsilon}_1\bar{\epsilon}_2\} - \{\bar{T}_1\bar{T}_2\}$$

$$\{M_{11}M_{22}\} = -[D]\{\kappa_1\kappa_2\} - \{\bar{\bar{T}}_1\bar{\bar{T}}_2\} \tag{5–38}$$

$$\{N_{12}M_{12}\} = [E]\{\bar{\gamma}_{12}\kappa_{12}\}$$

where

$$[K] = \begin{bmatrix} K_{11} & K_{12} \\ K_{21} & K_{22} \end{bmatrix} \qquad [D] = \begin{bmatrix} D_{11} & D_{12} \\ D_{21} & D_{22} \end{bmatrix}$$

$$[E] = \begin{bmatrix} K_u & 0 \\ 0 & -D_u \end{bmatrix}$$

In these equations [K], [D], and [E] represent symmetrical arrays of influence coefficients for isothermal straining. Each coefficient may vary with its location on the surface of the shell. $\bar{T}$ and $\bar{\bar{T}}$ are functions of the temperature distribution, $T(\xi_1, \xi_2, \zeta)$. The $\bar{T}$ reflect the temperature distribution over the reference surface and the $\bar{\bar{T}}$ the temperature distribution throughout the thickness. $T(\xi_1, \xi_2, \zeta)$ is an increment of temperature above a reference absolute temperature T_0 for a state of zero stress and strain. The total absolute temperature, denoted by T_1, is the sum of T_0 and T.

The elastic coefficients in Eqs. 5–38 may take on a variety of forms if the shell element were orthogonally stiffened along the directions of the lines of·principal curvature. We shall not attempt to list these forms here, but will merely state explicit values for the coefficients when the shell is elastic, homogeneous, and isotropic. They are

$$K_{11} = K, \qquad K_{12} = K\nu, \qquad K_{22} = K$$

$$D_{11} = D, \qquad D_{12} = D\nu, \qquad D_{22} = D \qquad (5\text{--}39)$$

$$K_u = Gh, \qquad D_u = D(1 - \nu)$$

where $K = Eh/(1 - \nu^2)$ and $D = Eh^3/12(1 - \nu^2)$. The thermoelastic terms are

$$\bar{T}_1 = \bar{T}_2 = \alpha^t(1 + \nu)K \frac{1}{h} \int_{-h/2}^{h/2} T \, d\zeta$$

$$\bar{\bar{T}}_1 = \bar{\bar{T}}_2 = \alpha^t(1 + \nu)D \frac{12}{h^3} \int_{-h/2}^{h/2} T\zeta \, d\zeta \qquad (5\text{--}40)$$

In Eqs. 5–39 and 5–40, E, G, ν, and α^t are Young's modulus, the modulus of rigidity, Poisson's ratio, and the coefficient of thermal expansion, respectively.

(e) Equations of static equilibrium and boundary conditions of shell theory

In Secs. 5–2(c) and (d) above, we have recorded the strain-displacement relations and the stress-strain relations of shell theory. The mathematical statement of our problem is completed when we add the equilibrium equations and the boundary conditions. The equilibrium equations and boundary conditions may be formed by the direct summation of forces with reference to a sketch of an element of the shell or, alternatively, by means of the principle of minimum potential energy. We shall make use of the latter approach. The principle of minimum potential energy is a special case of Hamilton's principle applied to a statics problem. We assume that the shell is loaded statically by forces X_1, X_2, and X_ζ per unit of shell area and in the directions of the axes ξ_1, ξ_2, and ζ, respectively. It is subjected to a temperature distribution $T(\xi_1, \xi_2, \zeta)$, variable over the surface and throughout the thickness. Furthermore, the edge boundary conditions are divided into two categories. Over certain edges, designated by C_1, the edge forces and moments per unit length are prescribed; and over certain other edges, designated by C_2, the edge geometrical constraints are prescribed.

The variation of the potential energy of the shell is expressible as (cf. Eq. 2–46)

$$\delta(U - W)$$

$$= \iint_S (N_{11}\,\delta\bar{\epsilon}_1 + N_{22}\,\delta\bar{\epsilon}_2 + N_{12}\,\delta\bar{\gamma}_{12} - M_{11}\,\delta\kappa_{11} - M_{22}\,\delta\kappa_{22} - 2M_{12}\,\delta\kappa_{12})$$

$$\times \alpha_1\alpha_2 \, d\xi_1 \, d\xi_2$$

$$- \iint_S (X_1\,\delta u + X_2\,\delta v + X_\zeta\,\delta w)\alpha_1\alpha_2 \, d\xi_1 \, d\xi_2$$

$$- \int_{C_1} (\bar{N}_{1\nu}\,\delta u + \bar{N}_{2\nu}\,\delta v + \bar{V}_\nu\,\delta w - \bar{M}_{1\nu}\,\delta l_{31} - \bar{M}_{2\nu}\,\delta l_{32}) \, ds = 0 \quad (5\text{–}41)$$

In Eq. 5–41, the symbol ν in the last integral denotes an outward normal to the edge C_1 and s represents a line variable of integration along the edge C_1. $\bar{N}_1$ and $\bar{N}_2$ represent stress resultants applied to the edge C_1 with positive directions along ξ_1 and ξ_2. V_ν represents a stress resultant also applied to the edge C_1 with positive direction along the ζ axis. $\bar{M}_{1\nu}$ and $\bar{M}_{2\nu}$ represent stress couples applied to the edge C_1. The formulation of edge boundary conditions may be better understood by reference to Fig. 5–8. Let us suppose that $\bar{f}_1, \bar{f}_2$, and $\bar{f}_\zeta$ are unit stresses applied to the edge in the ξ_1, ξ_2, and ζ directions, respectively, as illustrated by the figure. It is

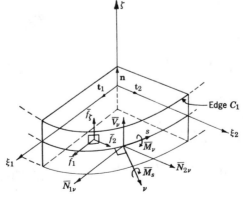

Fig. 5–8. Force boundary conditions on shell edge.

apparent, then, that for a thin shell—that is, a shell in which ζ/R_1 and ζ/R_2 may be neglected in comparison with one—we have

$$\bar{N}_{1v} = \int_{-h/2}^{h/2} f_1 \, d\zeta \qquad \bar{N}_{2v} = \int_{-h/2}^{h/2} f_2 \, d\zeta \qquad \bar{V}_v = \int_{-h/2}^{h/2} f_\zeta \, d\zeta$$

$$\bar{M}_{1v} = \int_{-h/2}^{h/2} f_1 \zeta \, d\zeta \qquad \bar{M}_{2v} = \int_{-h/2}^{h/2} f_2 \zeta \, d\zeta$$

$$\bar{M}_v = \bar{M}_{1v} l + \bar{M}_{2v} m \tag{5-42}$$

$$\bar{M}_s = \bar{M}_{2v} l - \bar{M}_{1v} m$$

where l and m are direction cosines of the normal, v, with reference to the local unit vectors $\mathbf{t}_1$ and $\mathbf{t}_2$.

The results obtained from Eq. 5–41 are dependent on the nature of the strain-displacement relations which are introduced. We shall adopt the strain-displacement relations of Eqs. 5–36 and 5–37. This leads to the following set of nonlinear thin shell equations (5–43) and boundary conditions (5–44) which we shall designate as Washizu's equations:

$$\frac{\partial}{\partial \xi_1} \{\alpha_2[N_{11}(1 + l_{11}) + N_{12}l_{12}]\} + \frac{\partial}{\partial \xi_2} \{\alpha_1[N_{22}l_{12} + N_{12}(1 + l_{11})]\}$$

$$+ [N_{11}l_{21} + N_{12}(1 + l_{22})] \frac{\partial \alpha_1}{\partial \xi_2} - [N_{22}(1 + l_{22}) + N_{12}l_{21}] \frac{\partial \alpha_2}{\partial \xi_1}$$

$$- (N_{11}l_{31} + N_{12}l_{32}) \frac{\alpha_1 \alpha_2}{R_1}$$

$$- \frac{1}{R_1} \left[\frac{\partial}{\partial \xi_1} (\alpha_2 M_{11}) - M_{22} \frac{\partial \alpha_2}{\partial \xi_1} + \frac{\partial}{\partial \xi_2} (\alpha_1 M_{12}) + \frac{\partial \alpha_1}{\partial \xi_2} M_{12} \right]$$

$$+ X_1 \alpha_1 \alpha_2 = 0$$

$$\frac{\partial}{\partial \xi_2} \{\alpha_1[N_{22}(1 + l_{22}) + N_{12}l_{21}]\} + \frac{\partial}{\partial \xi_1} \{\alpha_2[N_{11}l_{21} + N_{12}(1 + l_{22})]\}$$

$$+ [N_{22}l_{12} + N_{12}(1 + l_{11})] \frac{\partial \alpha_2}{\partial \xi_1} - [N_{11}(1 + l_{11}) + N_{12}l_{12}] \frac{\partial \alpha_1}{\partial \xi_2}$$

$$- [N_{22}l_{32} + N_{12}l_{31}] \frac{\alpha_1 \alpha_2}{R_2}$$

$$- \frac{1}{R_2} \left[\frac{\partial}{\partial \xi_2} (\alpha_1 M_{22}) - M_{11} \frac{\partial \alpha_1}{\partial \xi_2} + \frac{\partial}{\partial \xi_1} (\alpha_2 M_{12}) + \frac{\partial \alpha_2}{\partial \xi_1} M_{12} \right]$$

$$+ X_2 \alpha_1 \alpha_2 = 0 \tag{5-43a,b}$$

$$\frac{\partial}{\partial \xi_1} \{\alpha_2[N_{11}l_{31} + N_{12}l_{32}]\} + \frac{\partial}{\partial \xi_2} \{\alpha_1[N_{22}l_{32} + N_{12}l_{31}]\}$$

$$+ [N_{11}(1 + l_{11}) + N_{12}l_{12}]\frac{\alpha_1\alpha_2}{R_1} + [N_{22}(1 + l_{22}) + N_{12}l_{21}]\frac{\alpha_1\alpha_2}{R_2}$$

$$+ \frac{\partial}{\partial \xi_1} \left\{ \frac{1}{\alpha_1}\left[\frac{\partial}{\partial \xi_1}(\alpha_2 M_{11}) + \frac{\partial}{\partial \xi_2}(\alpha_1 M_{12}) + M_{12}\frac{\partial \alpha_1}{\partial \xi_2} - M_{22}\frac{\partial \alpha_2}{\partial \xi_1}\right]\right\}$$

$$+ \frac{\partial}{\partial \xi_2} \left\{ \frac{1}{\alpha_2}\left[\frac{\partial}{\partial \xi_2}(\alpha_1 M_{22}) + \frac{\partial}{\partial \xi_1}(\alpha_2 M_{12}) + M_{12}\frac{\partial \alpha_2}{\partial \xi_1} - M_{11}\frac{\partial \alpha_1}{\partial \xi_2}\right]\right\}$$

$$+ X_\zeta \alpha_1\alpha_2 = 0 \tag{5-43c}$$

$$[N_{11}(1 + l_{11}) + N_{12}l_{12}]l + [N_{12}(1 + l_{11}) + N_{22}l_{12}]m - \frac{1}{R_1}(M_{11}l + M_{12}\,m)$$

$$= \bar{N}_{1\nu} - \frac{\bar{M}_{1\nu}}{R_1}$$

$$[N_{11}l_{21} + N_{12}(1 + l_{22})]l + [N_{12}l_{21} + N_{22}(1 + l_{22})]m - \frac{1}{R_2}(M_{12}l + M_{22}\,m)$$

$$= \bar{N}_{2\nu} - \frac{\bar{M}_{2\nu}}{R_2}$$

$$[N_{11}l_{31} + N_{12}l_{32}]l + [N_{22}l_{32} + N_{12}l_{31}]m + V_\nu + \frac{\partial M_s}{\partial s} = \bar{V}_\nu + \frac{\partial \bar{M}_s}{\partial s}$$

$$M_\nu = \bar{M}_\nu \tag{5-44}$$

The equations of linear shell theory are obtainable from Eqs. 5-43 and 5-44 by putting $l_{11} = l_{12} = l_{21} = l_{22} = l_{31} = l_{32} = 0$.

(f) Formulation of the shell problem

The formulation of the shell problem, within the framework of the assumptions we have been making, is established by the equations stated above. In particular, we have six strain-displacement relations (5–36) and (5–37), six equations of state (5–38), and three equilibrium equations (5–43), making a total of fifteen equations. The unknown quantities which are defined by these equations are also fifteen in number, comprising six stress resultants and couples (N_{11}, N_{22}, N_{12}, M_{11}, M_{22}, and M_{12}), six reference surface strains ($\bar{\epsilon}_1$, $\bar{\epsilon}_2$, $\bar{\gamma}_{12}$, κ_1, κ_2, and κ_{12}), and three displacements (u, v, and w).

The problem of static shell theory is one of solving these equations, subject to the force boundary conditions on the edge C_1, given by Eqs.

5–44, and the displacement boundary conditions on the edge C_2. The latter boundary conditions may be stated simply by

$$u = \bar{u}, \qquad v = \bar{v}, \qquad w = \bar{w}, \qquad \frac{\partial w}{\partial v} = \frac{\partial \bar{w}}{\partial v} \qquad (5\text{–}45)$$

where the bar denotes a specified quantity, and differentiation with respect to v denotes differentiation in the direction of the normal drawn outward on the edge in the middle surface (cf. Fig. 5–8).

Subsequent paragraphs of the present section are devoted to the applications of shell theory to structural configurations of interest in aeroelasticity, and to the derivation of structural operators which derive from shell theory.

(g) Stiffened shallow shell theory

In the previous sections we have outlined a general theory applicable to shells of arbitrary curvature with curvilinear coordinates arranged in the directions of lines of curvature. Very frequently in aeroelasticity we are confronted with applications involving shells of small curvature. In these applications we may employ the shallow shell theory of Marguerre (Ref. 5–7), a special case of the theory described above with a simpler and more tractable mathematical form. We refer to Fig. 5–9, which illustrates a segment of a shallow shell oriented with respect to the rectangular coordinates (x, y, z). The equation for the reference surface of the shell is taken as

$$z = z(x, y) \qquad (5\text{–}46)$$

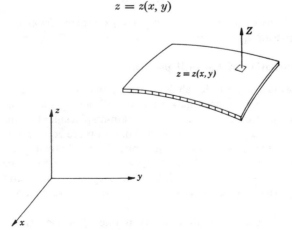

Fig. 5–9. Segment of a shallow shell.

A shallow shell is qualitatively one for which the slope of the reference surface is small at all points. A rule for the latitude of interpretation of the word "small" cannot be stated precisely; however, Reissner (Ref. 5–8) has suggested that shallow shell theory will be more than accurate enough as long as the slope is less than $\frac{1}{8}$ and often accurate enough for practical purposes up to $\frac{1}{2}$. In the case of shallow shell theory we may employ a rectangular coordinate system instead of the system of curvilinear coordinates of the more general theory. We put $\xi_1 = x$, $\xi_2 = y$, and $\alpha_2 = \alpha_1 = 1$ so that an element of arc on the reference surface is given by

$$(ds)^2 = (dx)^2 + (dy)^2 \tag{5-47}$$

and a radius vector to the reference surface is defined by

$$\mathbf{r} = x\mathbf{i} + y\mathbf{j} + z(x, y)\mathbf{k} \tag{5-48}$$

Quantitatively, we may say that a shallow shell is one for which

$$1 + \left(\frac{\partial z}{\partial x}\right)^2 \approx 1, \quad 1 + \left(\frac{\partial z}{\partial y}\right)^2 \approx 1, \quad 1 + \left(\frac{\partial z}{\partial x}\right)\left(\frac{\partial z}{\partial y}\right) \approx 1 \tag{5-49}$$

The unit tangent and normal vectors to the deformed surface are approximated by

$$\mathbf{t}_x = \mathbf{i} + \left(\frac{\partial z}{\partial x} + \frac{\partial w}{\partial x}\right)\mathbf{k}$$

$$\mathbf{t}_y = \mathbf{j} + \left(\frac{\partial z}{\partial y} + \frac{\partial w}{\partial y}\right)\mathbf{k} \tag{5-50}$$

$$\mathbf{n} = -\left(\frac{\partial z}{\partial x} + \frac{\partial w}{\partial x}\right)\mathbf{i} - \left(\frac{\partial z}{\partial y} + \frac{\partial w}{\partial y}\right)\mathbf{j} + \mathbf{k}$$

Shallow shell theory may be formulated in precisely the same way as the more general shell theory of the previous paragraphs by altering the nature of the assumed strain-displacement relations. In shallow shell theory these are assumed to be (Ref. 5–7)

$$\bar{\epsilon}_x = \frac{\partial u}{\partial x} + \frac{\partial z}{\partial x}\frac{\partial w}{\partial x} + \frac{1}{2}\left(\frac{\partial w}{\partial x}\right)^2$$

$$\bar{\epsilon}_y = \frac{\partial v}{\partial y} + \frac{\partial z}{\partial y}\frac{\partial w}{\partial y} + \frac{1}{2}\left(\frac{\partial w}{\partial y}\right)^2 \tag{5-51a}$$

$$\bar{\gamma}_{xy} = \frac{\partial u}{\partial y} + \frac{\partial v}{\partial x} + \frac{\partial z}{\partial x}\frac{\partial w}{\partial y} + \frac{\partial z}{\partial y}\frac{\partial w}{\partial x} + \frac{\partial w}{\partial x}\frac{\partial w}{\partial y}$$

$$\kappa_x = \frac{\partial^2 w}{\partial x^2}, \quad \kappa_y = \frac{\partial^2 w}{\partial y^2}, \quad \kappa_{xy} = \frac{\partial^2 w}{\partial x\,\partial y} \tag{5-51b}$$

An essential feature of the reduction of the strain-displacement relations to the relatively simple forms stated above is the assumption that bending displacements are significantly larger than stretching displacements (Ref. 5–7).

The stress-strain relations of an orthogonally stiffened shallow shell are identical to those of Eqs. 5–38. We shall formulate the equilibrium equations and boundary conditions by means of the principle of minimum potential energy as we did for the more general shell by Eq. 5–41. For a rectangular shallow shell element with dimensions $2a$ and $2b$, and loaded by a traction of intensity Z (cf. Fig. 5–9), we have

$$\delta(U - W)$$
$$= \int_{-a}^{a} \int_{-b}^{b} (N_{xx}\delta\bar{\epsilon}_x + N_{yy}\delta\bar{\epsilon}_y + N_{xy}\delta\bar{\epsilon}_{xy} - M_{xx}\delta\kappa_x$$
$$- M_{yy}\delta\kappa_y - 2M_{xy}\delta\kappa_{xy})\,dx\,dy - \int_{-a}^{a} \int_{-b}^{b} Z\,\delta w\,dx\,dy$$
$$- \int_{-a}^{a} \left| \bar{N}_{xy}\delta u + \bar{N}_{yy}\delta v + \bar{V}_y\,\delta w - \bar{M}_{yy}\delta\frac{\partial w}{\partial y} - \bar{M}_{yx}\delta\frac{\partial w}{\partial x} \right|_{-b}^{b} dx$$
$$- \int_{-b}^{b} \left| \bar{N}_{xx}\delta u + \bar{N}_{yx}\delta v + \bar{V}_x\,\delta w - \bar{M}_{xx}\delta\frac{\partial w}{\partial x} - \bar{M}_{xy}\delta\frac{\partial w}{\partial y} \right|_{-a}^{a} dy = 0$$

$$(5\text{–}52)$$

Employing the strain-displacement relations of Eqs. 5–51, we obtain from the variational condition (5–52) the differential equations of equilibrium

$$\frac{\partial N_{xx}}{\partial x} + \frac{\partial N_{xy}}{\partial y} = 0$$

$$\frac{\partial N_{xy}}{\partial x} + \frac{\partial N_{yy}}{\partial y} = 0$$

$$(5\text{–}53a)$$

$$\frac{\partial^2 M_{xx}}{\partial x^2} + 2\frac{\partial^2 M_{xy}}{\partial x\,\partial y} + \frac{\partial^2 M_{yy}}{\partial y^2} + \frac{\partial}{\partial x}\left[N_{xx}\frac{\partial}{\partial x}(z + w) + N_{xy}\frac{\partial}{\partial y}(z + w)\right]$$
$$+ \frac{\partial}{\partial y}\left[N_{yy}\frac{\partial}{\partial y}(z + w) + N_{xy}\frac{\partial}{\partial x}(z + w)\right] + Z = 0 \quad (5\text{–}53b)$$

and the following force boundary conditions:

(a) On edges $x = \pm a$:

$$\bar{N}_{xx} = N_{xx}, \qquad \bar{N}_{yx} = N_{yx}, \qquad \bar{M}_{xx} = M_{xx}$$

$$\bar{V}_x + \frac{\partial\bar{M}_{xy}}{\partial y} = N_{xx}\frac{\partial}{\partial x}(z + w) + N_{xy}\frac{\partial}{\partial y}(z + w) + \frac{\partial M_{xx}}{\partial x} + 2\frac{\partial M_{xy}}{\partial y}$$

(b) On edges $y = \pm b$:

$$\bar{N}_{xy} = N_{xy}, \qquad \bar{N}_{yy} = N_{yy}, \qquad \bar{M}_{yy} = M_{yy}$$

$$\bar{V}_y + \frac{\partial \bar{M}_{yx}}{\partial x} = N_{xy} \frac{\partial}{\partial x}(z + w) + N_{yy} \frac{\partial}{\partial y}(z + w) + \frac{\partial M_{yy}}{\partial y} + 2 \frac{\partial M_{xy}}{\partial x}$$

(c) At the corners

$$\left[[M_{xy} \, \delta w]_{-a}^{a} \right]_{-b}^{b} = 0$$

When a corner is free and unrestrained, so that δw is arbitrary, the quantities M_{xy} must be considered in formulating the boundary conditions.

Equations 5–53a and 5–53b may be reduced to a single equilibrium equation by introducing a stress function F defined as follows:

$$N_{xx} = \frac{\partial^2 F}{\partial y^2}, \qquad N_{yy} = \frac{\partial^2 F}{\partial x^2}, \qquad N_{xy} = -\frac{\partial^2 F}{\partial x \, \partial y} \qquad (5\text{–}54)$$

By means of Eqs. 5–54, we ensure the satisfaction of Eqs. 5–53a; and then Eq. 5–53b becomes

$$\frac{\partial^2 M_{xx}}{\partial x^2} + 2 \frac{\partial^2 M_{xy}}{\partial x \, \partial y} + \frac{\partial^2 M_{yy}}{\partial y^2} + Z = -\frac{\partial^2 F}{\partial x^2} \frac{\partial^2 w}{\partial y^2} - \frac{\partial^2 F}{\partial y^2} \cdot \frac{\partial^2 w}{\partial x^2}$$

$$+ 2 \frac{\partial^2 F}{\partial x \, \partial y} \frac{\partial^2 w}{\partial x \, \partial y} - \frac{\partial^2 F}{\partial x^2} \frac{\partial^2 z}{\partial y^2} - \frac{\partial^2 F}{\partial y^2} \cdot \frac{\partial^2 z}{\partial x^2} + 2 \frac{\partial^2 F}{\partial x \, \partial y} \cdot \frac{\partial^2 z}{\partial x \, \partial y} \qquad (5\text{–}55)$$

The three equations (5–51a), describing the relations between strains and displacements in the plane of the reference surface, can be combined into a single compatibility equation

$$\frac{\partial^2 \bar{\epsilon}_x}{\partial y^2} - \frac{\partial^2 \bar{\gamma}_{xy}}{\partial x \, \partial y} + \frac{\partial^2 \bar{\epsilon}_y}{\partial x^2} = \left(\frac{\partial^2 w}{\partial x \, \partial y} \right)^2 - \frac{\partial^2 w}{\partial x^2} \frac{\partial^2 w}{\partial y^2} - \frac{\partial^2 z}{\partial x^2} \cdot \frac{\partial^2 w}{\partial y^2}$$

$$- \frac{\partial^2 z}{\partial y^2} \frac{\partial^2 w}{\partial x^2} + 2 \frac{\partial^2 z}{\partial x \, \partial y} \frac{\partial^2 w}{\partial x \, \partial y} \qquad (5\text{–}56)$$

Equations 5–55 and 5–56 thus represent in two equations all of the equilibrium equations and strain-displacement relations of shallow shell theory except 5–51b. When Eqs. 5–55, 5–51b, and 5–56 are supplemented by the equations of state of an element of the shell, we obtain a complete set. For example, if we employ Eqs. 5–38, we obtain a mathematical statement of the problem of orthogonally aeolotropic elastic shallow shells with lateral loading and with arbitrary temperature variation. We proceed by introducing Eqs. 5–54, 5–51a, and 5–51b into Eqs. 5–38 and reducing them to the forms

$$\{\bar{\epsilon}_x \bar{\epsilon}_y\} = [K]^{-1}\left\{\frac{\partial^2 F}{\partial y^2} \frac{\partial^2 F}{\partial x^2}\right\} + [K]^{-1}\{T_x T_y\} \tag{5-57}$$

$$\{M_{xx} M_{yy}\} = -[D]\left\{\frac{\partial^2 w}{\partial x^2} \frac{\partial^2 w}{\partial y^2}\right\} - \{\bar{\bar{T}}_x \bar{\bar{T}}_y\} \tag{5-58}$$

$$\{\bar{\gamma}_{xy} M_{xy}\} = -\begin{bmatrix} 0 & \dfrac{1}{K_u} \\ D_u & 0 \end{bmatrix}\left\{\frac{\partial^2 w}{\partial x\, \partial y} \frac{\partial^2 F}{\partial x\, \partial y}\right\} \tag{5-59}$$

When Eqs. 5–57, 5–58, and 5–59 are substituted into Eqs. 5–55 and 5–56, we obtain coupled equilibrium and compatibility differential equations in terms of the dependent variables w and F. The equilibrium equation is

$$\frac{\partial^2}{\partial x^2}\left(D_{xx}\frac{\partial^2 w}{\partial x^2} + D_{xy}\frac{\partial^2 w}{\partial y^2}\right) + 2\frac{\partial^2}{\partial x\, \partial y}\left(D_u\frac{\partial^2 w}{\partial x\, \partial y}\right)$$

$$+ \frac{\partial^2}{\partial y^2}\left(D_{yx}\frac{\partial^2 w}{\partial x^2} + D_{yy}\frac{\partial^2 w}{\partial y^2}\right) = Z - \frac{\partial^2 \bar{\bar{T}}_x}{\partial x^2} - \frac{\partial^2 \bar{\bar{T}}_y}{\partial y^2} + \frac{\partial^2 F}{\partial x^2}\frac{\partial^2}{\partial y^2}(z + w)$$

$$-2\frac{\partial^2 F}{\partial x\, \partial y}\frac{\partial^2}{\partial x\, \partial y}(z + w) + \frac{\partial^2 F}{\partial y^2}\frac{\partial^2}{\partial x^2}(z + w) \tag{5-60}$$

and the compatibility equation is

$$\frac{\partial^2}{\partial x^2}\left(\bar{K}_{yy}\frac{\partial^2 F}{\partial x^2} + \bar{K}_{xy}\frac{\partial^2 F}{\partial y^2}\right) + \frac{\partial^2}{\partial x\, \partial y}\left(\frac{1}{K_u}\frac{\partial^2 F}{\partial x\, \partial y}\right)$$

$$+ \frac{\partial^2}{\partial y^2}\left(\bar{K}_{xy}\frac{\partial^2 F}{\partial x^2} + \bar{K}_{xx}\frac{\partial^2 F}{\partial y^2}\right) = -\frac{\partial^2}{\partial x^2}(\bar{K}_{yx}T_x + \bar{K}_{yy}T_y)$$

$$- \frac{\partial^2}{\partial y^2}(\bar{K}_{xx}T_x + \bar{K}_{xy}T_y) + \left(\frac{\partial^2 w}{\partial x\, \partial y}\right)^2 - \frac{\partial^2 w}{\partial x^2}\frac{\partial^2 w}{\partial y^2} - \frac{\partial^2 z}{\partial x^2}\frac{\partial^2 w}{\partial y^2}$$

$$- \frac{\partial^2 z}{\partial y^2}\cdot\frac{\partial^2 w}{\partial x^2} + 2\frac{\partial^2 z}{\partial x\, \partial y}\cdot\frac{\partial^2 w}{\partial x\, \partial y} \tag{5-61}$$

where $\bar{K}_{ij}$ are elements of the $[K]^{-1}$ matrix.

Equations 5–60 and 5–61 constitute two nonlinear partial differential equations in the unknown quantities w and F which apply to elastic orthogonally aeolotropic shallow shells with lateral loading and with temperature gradients over the surface and through the thickness. Because of their similarity to the von Kármán equations for the bending and stretching of flat plates, there are numerous known techniques of solution, and a comparatively simple approach is thus provided for the analysis of shells which can be regarded as shallow. A linear shallow shell theory is obtained by neglecting the squared and product terms involving w on the right-hand side of Eq. 5–61.

(h) Stiffened flat plate theory

Membrane and flat plate theories with and without temperature gradients can be derived as special cases of Eqs. 5–60 and 5–61. A theory for the bending and stretching of stiffened flat plates of variable thickness is of great interest in aeroelasticity since it is fundamental to the treatment of lifting surfaces. By placing the quantity z equal to a constant in Eqs. 5–60 and 5–61, we obtain the equilibrium equation

$$\frac{\partial^2}{\partial x^2}\left(D_{xx}\frac{\partial^2 w}{\partial x^2} + D_{xy}\frac{\partial^2 w}{\partial y^2}\right) + 2\frac{\partial^2}{\partial x \partial y}\left(D_u\frac{\partial^2 w}{\partial x \partial y}\right) + \frac{\partial^2}{\partial y^2}\left(D_{yx}\frac{\partial^2 w}{\partial x^2} + D_{yy}\frac{\partial^2 w}{\partial y^2}\right)$$

$$= Z - \frac{\partial^2 \overline{\overline{T}}_x}{\partial x^2} - \frac{\partial^2 \overline{\overline{T}}_y}{\partial y^2} + \frac{\partial^2 F}{\partial x^2}\frac{\partial^2 w}{\partial y^2} - 2\frac{\partial^2 F}{\partial x \partial y}\frac{\partial^2 w}{\partial x \partial y} + \frac{\partial^2 F}{\partial y^2}\frac{\partial^2 w}{\partial x^2} \quad (5\text{–}62)$$

and the compatibility equation

$$\frac{\partial^2}{\partial x^2}\left(K_{yy}\frac{\partial^2 F}{\partial x^2} + K_{yx}\frac{\partial^2 F}{\partial y^2}\right) + \frac{\partial^2}{\partial x \partial y}\left(\frac{1}{K_u}\frac{\partial^2 F}{\partial x \partial y}\right) + \frac{\partial^2}{\partial y^2}\left(K_{xy}\frac{\partial^2 F}{\partial x^2} + K_{xx}\frac{\partial^2 F}{\partial y^2}\right)$$

$$= -\frac{\partial^2}{\partial x^2}(K_{yx}T_x + K_{yy}T_y) - \frac{\partial^2}{\partial y^2}(K_{xx}T_x + K_{xy}T_y)$$

$$+ \left(\frac{\partial^2 w}{\partial x \partial y}\right)^2 - \frac{\partial^2 w}{\partial x^2}\frac{\partial^2 w}{\partial y^2} \quad (5\text{–}63)$$

The simultaneous Eqs. 5–62 and 5–63 can be simplified, if it is assumed that the lateral deflections are small. In this case, the bending and stretching actions are uncoupled. This uncoupling is accomplished mathematically by neglecting the last two terms in w on the right-hand side of Eq. 5–63. Thus, for a given temperature distribution, Eq. 5–63, together with the appropriate boundary conditions, can be solved for the stress function, F. By means of Eqs. 5–54, the stress function can be transformed into explicit initial stress resultants, $N_{xx}^{(0)}$, $N_{xy}^{(0)}$, and $N_{yy}^{(0)}$. Under these circumstances we may rewrite Eq. 5–62 in the form

$$\mathscr{S}(w) = Z - \frac{\partial^2 \overline{\overline{T}}_x}{\partial x^2} - \frac{\partial^2 \overline{\overline{T}}_y}{\partial y^2} \quad (5\text{–}64)$$

where $\mathscr{S}$ is a self-adjoint linear structural operator defined by

$$\mathscr{S} = \frac{\partial^2}{\partial x^2}\left(D_{xx}\frac{\partial^2}{\partial x^2} + D_{xy}\frac{\partial^2}{\partial y^2}\right) + 4\frac{\partial^2}{\partial x\,\partial y}\left(D_u\frac{\partial^2}{\partial x\,\partial y}\right)$$

$$+ \frac{\partial^2}{\partial y^2}\left(D_{yx}\frac{\partial^2}{\partial x^2} + D_{yy}\frac{\partial^2}{\partial y^2}\right) - N_{xx}^{(0)}\frac{\partial^2}{\partial x^2} - 2N_{xy}^{(0)}\frac{\partial^2}{\partial x\,\partial y} - N_{yy}^{(0)}\frac{\partial^2}{\partial y^2}$$

In Eq. 5–64, we have shown $\mathscr{S}$ as a differential structural operator. It is also possible to represent the inverted form of $\mathscr{S}$ in terms of an influence function $C(x, y; \xi, \eta)$. The latter represents the deflection of the plate at the point (x, y) due to a unit force applied at the point (ξ, η). When the boundary conditions on the plate are homogeneous, the inverted form of $\mathscr{S}$ is

$$\mathscr{S}^{-1}(\quad) = \iint_S C(x, y; \xi, \eta)(\quad)\,d\xi\,d\eta \qquad (5\text{--}65)$$

Except at the source point (ξ, η), the function $C(x, y; \xi, \eta)$ and its derivatives with respect to x and y up to the order of $\mathscr{S}$ are continuous. The function $C(x, y; \xi, \eta)$ satisfies the differential equation

$$\mathscr{S}(C) = \delta(x, y; \xi, \eta) \qquad (5\text{--}66)$$

and the prescribed homogeneous boundary conditions. The function $\delta(x, y; \xi, \eta)$ is the Dirac delta function. This is a function which has the unique property that it is zero everywhere except at $x = \xi$, $y = \eta$, where it is infinite. In addition, for any assumed function $g(x, y)$ the Dirac delta function has the integral properties

$$\iint_S \delta(x, y; \xi, \eta)\,dx\,dy = 1, \quad \iint_S g(x, y)\,\delta(x, y; \xi, \eta)\,dx\,dy = g(\xi, \eta) \quad (5\text{--}67)$$

if (ξ, η) is included in the region of integration. If not, the function has the properties that

$$\iint_S \delta(x, y; \xi, \eta)\,dx\,dy = 0, \quad \iint_S g(x, y)\,\delta(x, y; \xi, \eta)\,dx\,dy = 0 \quad (5\text{--}68)$$

(i) Solution of the plate equations

In the case of lifting surfaces of irregular shape or thickness and of plates with complex boundary conditions where precise results are desired, it is obviously necessary to solve Eqs. 5–62 and 5–63 or Eq. 5–64 by approximate numerical methods such as those illustrated in Chap. 3. Space does

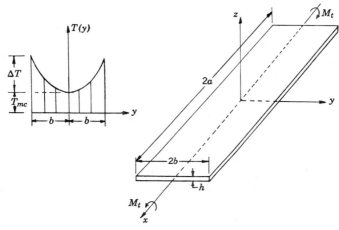

Fig. 5–10. Slender plate subjected to a chordwise temperature gradient and a pure twisting torque.

not permit us to include such solutions in detail here, so we shall confine our remarks in this section to a single relatively simple illustration of general interest.

The effective torsional stiffness of thin wings. Let us consider how the plate equations may be applied to compute a relationship between the applied torque, M_t, and the twist per unit length, θ, of a long flat plate with a chordwise temperature gradient. Figure 5–10 illustrates the axis system and some of the notation which will be employed.

We assume that a pure couple, M_t, is applied at the end and that the plate is subjected to a chordwise distribution of temperature, $T(y)$, as illustrated by Fig. 5–10. $T(y)$ is taken as an even function of y. For a solid uniform plate, the appropriate differential equations derived from Eqs. 5–62 and 5–63 are

$$D\nabla^4 w = \frac{\partial^2 F}{\partial y^2}\frac{\partial^2 w}{\partial x^2} - 2\frac{\partial^2 F}{\partial x\,\partial y}\frac{\partial^2 w}{\partial x\,\partial y} + \frac{\partial^2 F}{\partial x^2}\frac{\partial^2 w}{\partial y^2} \quad (5\text{–}69)$$

$$\frac{1}{Eh}\nabla^4 F = -\frac{\alpha^t}{h}\nabla^2\int_{-h/2}^{h/2} T\,dz + \left(\frac{\partial^2 w}{\partial x\,\partial y}\right)^2 - \left(\frac{\partial^2 w}{\partial x^2}\right)\left(\frac{\partial^2 w}{\partial y^2}\right) \quad (5\text{–}70)$$

The boundary conditions are the following:
For $y = \pm b$,

$$N_{yy} = N_{xy} = M_{yy} = V_y + \frac{\partial M_{yx}}{\partial x} = 0 \quad (5\text{–}71)$$

For $x = \pm a$,

$$(a) \int_{-b}^{b} N_{xx} \, dy = 0, \quad (b) \int_{-b}^{b} N_{xx} y \, dy = 0$$

$$(c) \; M_t = 4bD(1 - \nu)\theta + \theta \int_{-b}^{b} N_{xx} y^2 \, dy \tag{5-72}$$

Boundary conditions (5–71) indicate that the boundary conditions are to be satisfied exactly along the free edges, and (5–72) that they are satisfied in an average way along the loaded edges.

We assume that the chordwise temperature variation is described by $T(y) = T_{mc} + \Delta T g(y)$, where T_{mc} is the midchord temperature and $g(y)$ is a function of y as yet unspecified. Since the plate is long and narrow, we assume that the stresses are invariant in the x-direction; and we put $F = F(y)$. The form of the deflection shape is assumed to be the same as the St. Venant torsional solution $w = \theta xy$, where θ is the twist rate. When these functions are put into Eq. 5–69, we find that it is satisfied identically; and when they are put into Eq. 5–70, we obtain

$$\frac{\partial^4 F}{\partial y^4} = -E\alpha^t h \, \Delta T \frac{\partial^2 g}{\partial y^2} + Eh\theta^2 \tag{5-73}$$

Integrating Eq. 5–73 twice yields

$$\frac{\partial^2 F}{\partial y^2} = N_{xx} = -E\alpha^t h \, \Delta T g(y) + Eh\theta^2 \frac{y^2}{2} + A_0 + A_1 y \tag{5-74}$$

The constants of integration, A_0 and A_1, are evaluated by boundary conditions (5–72a and b). The following result for the self-equilibrating spanwise stress resultant is obtained:

$$N_{xx} = \frac{Eh\theta^2}{6}(3y^2 - b^2) + \frac{\alpha^t Eh \, \Delta T}{2b}\left[\int_{-b}^{b} g(y) \, dy - 2bg(y)\right] \tag{5-75}$$

When Eq. 5–75 is substituted into boundary condition (5–72c), we obtain the following relationship:

$$M_t = \mathscr{S}(\theta) \tag{5-76}$$

where $\mathscr{S}(\quad)$ is a nonlinear structural operator having the form of

$$\mathscr{S}(\quad) = 4bD(1 - \nu)\left[1 - \frac{\Delta T}{\Delta T_{cr}} + \frac{4}{15}(1 + \nu)\frac{b^4}{h^2}(\quad)^2\right](\quad)$$

and where

$$\Delta T_{cr} = \frac{bh^2}{3\alpha^t(1 + \nu)\left[\int_{-b}^{b} g(y)y^2 \, dy - \frac{b^2}{3}\int_{-b}^{b} g(y) \, dy\right]} \quad (5\text{--}77)$$

is the temperature difference required to buckle the plate in torsion. It is interesting to examine the physical significance of the various terms in Eqs. 5–75, 5–76, and 5–77. In Eq. 5–75 we observe that the first term, proportional to the square of the twist rate, gives the contribution of twist rate to the midplane stress resultant, N_{xx}. The second term, proportional to ΔT, gives the contribution of the temperature difference. In the structural operator of Eq. 5–76, the importance of the nonlinear term in θ is evident if we observe that when this term is neglected, the torsional rigidity vanishes for $\Delta T = \Delta T_{cr}$. The finiteness of the twist rate has, therefore, a profound influence on the torsional rigidity under these circumstances. Finally, in Eq. 5–77, we observe that, if $g(y)$ is a function such that it has higher values at the edges of the plate than at the center, then ΔT_{cr} is positive. For this type of temperature distribution, we see by Eq. 5–75 that the stress resultant N_{xx} is compressive in the vicinity of the plate edges and tensile in the center. It can be reasoned that, for this type of stress distribution, a mechanism exists for producing torsional instability. When the plate is twisted through a small angle, the effect of the compressive stresses near the free edges is to produce twisting moments which tend further to increase the twist angle. When these twisting moments exceed the elastic restoring moments in the plate, a torsional instability is produced. The temperature gradient required to produce such a torsional instability is given by Eq. 5–77.

(j) The circular cylindrical shell

Since fuselages and rocket and missile bodies are often of circular cross section, aeroelasticians are frequently concerned with studies of the deformation properties of circular cylindrical shells. The derivation of the appropriate equilibrium equations for circular cylindrical shells rests upon the three basic equilibrium equations (5–43). These equations are reduced to circular cylindrical equations by introducing the notation illustrated by Fig. 5–11 and by putting

$$\xi_1 = x, \quad \xi_2 = \theta, \quad \alpha_1 = 1, \quad \alpha_2 = R, \quad R_1 = \infty, \quad R_2 = R, \quad N_{11} = N_{xx}$$

$$N_{22} = N_{\theta\theta}, \quad N_{12} = N_{x\theta} = N_{21} = N_{\theta x}, \quad M_{11} = M_{xx}, \quad M_{22} = M_{\theta\theta}$$

$$M_{12} = M_{x\theta} = M_{21} = M_{\theta x}, \quad X_1 = X, \quad X_2 = Y, \quad X_\zeta = Z$$

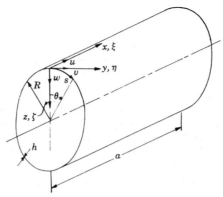

Fig. 5–11. Circular cylinder coordinate system.

This substitution yields the following nonlinear equations of equilibrium for a circular cylindrical shell.

$$\frac{\partial}{\partial x}\left[N_{xx}R\left(1+\frac{\partial u}{\partial x}\right)+N_{\theta x}\frac{\partial u}{\partial \theta}\right]$$

$$+\frac{\partial}{\partial \theta}\left[N_{\theta\theta}\frac{1}{R}\frac{\partial u}{\partial \theta}+N_{\theta x}\left(1+\frac{\partial u}{\partial x}\right)\right]+XR=0$$

$$\frac{\partial}{\partial x}\left[N_{xx}R\frac{\partial v}{\partial x}+N_{\theta x}R\left(1+\frac{1}{R}\frac{\partial v}{\partial \theta}-\frac{w}{R}\right)\right]$$

$$+\frac{\partial}{\partial \theta}\left[N_{\theta\theta}\left(1+\frac{1}{R}\frac{\partial v}{\partial \theta}-\frac{w}{R}\right)+N_{\theta x}\frac{\partial v}{\partial x}\right]-\left[N_{\theta\theta}\left(\frac{v}{R}+\frac{1}{R}\frac{\partial w}{\partial \theta}\right)\right.\quad(5\text{–}78)$$

$$\left.+N_{\theta x}\frac{\partial w}{\partial x}\right]-\frac{1}{R}\left(\frac{\partial M_{\theta\theta}}{\partial \theta}+2R\frac{\partial M_{x\theta}}{\partial x}\right)+YR=0$$

$$\frac{\partial}{\partial x}\left[N_{xx}R\frac{\partial w}{\partial x}+N_{\theta x}\left(v+\frac{\partial w}{\partial \theta}\right)\right]+\frac{\partial}{\partial \theta}\left[N_{\theta\theta}\left(\frac{v}{R}+\frac{1}{R}\frac{\partial w}{\partial \theta}\right)+N_{\theta x}\frac{\partial w}{\partial x}\right]$$

$$+\left[N_{\theta\theta}\left(1+\frac{1}{R}\frac{\partial v}{\partial \theta}-\frac{w}{R}\right)+N_{\theta x}\frac{\partial v}{\partial x}\right]$$

$$+\frac{\partial}{\partial x}\left[R\frac{\partial M_{xx}}{\partial x}+\frac{\partial M_{\theta x}}{\partial \theta}\right]+\frac{\partial}{\partial \theta}\left[\frac{1}{R}\frac{\partial M_{\theta\theta}}{\partial \theta}+\frac{\partial M_{\theta x}}{\partial x}\right]+ZR=0$$

These equations are especially useful when they are cast in terms of the dependent displacement variables u, v, and w and then linearized. This is accomplished for an isotropic elastic and homogeneous shell by introducing into Eqs. 5–78 the stress-strain relations obtained by combining

Eqs. 5–38 and 5–39, and linearizing the result. We obtain by this means the equilibrium equations

$$\frac{\partial^2 u}{\partial x^2} + \frac{1 - \nu}{2R^2} \frac{\partial^2 u}{\partial \theta^2} + \frac{1 + \nu}{2R} \frac{\partial^2 v}{\partial x \, \partial \theta} - \frac{\nu}{R} \frac{\partial w}{\partial x}$$

$$+ \frac{h^2}{12R^2} F_1(u, v, w) + \frac{1 - \nu^2}{Eh} X = 0$$

$$\frac{1 + \nu}{2R} \frac{\partial^2 u}{\partial x \, \partial \theta} + \frac{1}{R^2 \partial \theta^2} \frac{\partial^2 v}{} + \frac{1 - \nu}{2} \frac{\partial^2 v}{\partial x^2} - \frac{1}{R^2} \frac{\partial w}{\partial \theta}$$

$$+ \frac{h^2}{12R^2} F_2(u, v, w) + \frac{1 - \nu^2}{Eh} Y = 0 \quad (5\text{-}79)$$

$$\frac{\nu}{R} \frac{\partial u}{\partial x} + \frac{1}{R^2} \frac{\partial v}{\partial \theta} - \frac{w}{R^2} - \frac{h^2}{12R^4} \nabla^4 w$$

$$+ \frac{h^2}{12R^2} F_3(u, v, w) + \frac{1 - \nu^2}{Eh} Z = 0$$

where we define in this section

$$\nabla^4(\quad) = \left(R^4 \frac{\partial^4}{\partial x^4} + 2R^2 \frac{\partial^4}{\partial x^2 \, \partial \theta^2} + \frac{\partial^4}{\partial \theta^4} \right)(\quad)$$

and where

$$F_1(u, v, w) = 0$$

$$F_2(u, v, w) = 2(1 - \nu) \frac{\partial^2 v}{\partial x^2} + \frac{1}{R^2} \frac{\partial^2 v}{\partial \theta^2} + (2 - \nu) \frac{\partial^3 w}{\partial x^2 \, \partial \theta} + \frac{1}{R^2} \frac{\partial^3 w}{\partial \theta^3} \quad (5\text{-}80)$$

$$F_3(u, v, w) = -(2 - \nu) \frac{\partial^3 v}{\partial x^2 \, \partial \theta} - \frac{1}{R^2} \frac{\partial^3 v}{\partial \theta^3}$$

The temperature dependent terms in Eqs. 5–79 have been omitted for brevity. Numerous versions of the linearized circular cylinder equations (5–79) have been published in the literature by several investigators of shell theory (Refs. 5–7 through 5–15). All of these versions may be cast into the form given by Eqs. 5–79 in which the differences are reflected in the functions $F_1(u, v, w)$, $F_2(u, v, w)$, and $F_3(u, v, w)$, and all other terms are identical. The explicit values of $F_1(u, v, w)$, $F_2(u, v, w)$, and $F_3(u, v, w)$ given by Eq. 5–80 are those of Washizu (Ref. 5–6) and Goldenveizer (Ref. 5–13). In Table 5–1, the values of these functions for six of the more familiar theories of circular cylindrical shells are given. The simplest of these is Donnell's theory (Ref. 5–10), in which the functions $F_1(u, v, w)$, $F_2(u, v, w)$, and $F_3(u, v, w)$ are put equal to zero. Since these quantities

TABLE 5-I

Terms for various circular cylindrical shell theories

	$F_1(u, v, w)$	$F_2(u, v, w)$	$F_3(u, v, w)$
Donnell (Ref. 5-10)	0	0	0
Timoshenko (Ref. 5-1)	0	$\dfrac{\partial^3 w}{\partial x^2 \partial\theta} + \dfrac{1}{R^2}\dfrac{\partial^3 w}{\partial\theta^3} + (1-\nu)\dfrac{\partial^2 v}{\partial x^2} + \dfrac{1}{R^2}\dfrac{\partial^2 v}{\partial\theta^2}$	$-(2-\nu)\dfrac{\partial^3 v}{\partial x^2\partial\theta} - \dfrac{1}{R^2}\dfrac{\partial^3 v}{\partial\theta^3}$
Vlasov (Ref. 5-12)	$-\dfrac{1-\nu}{2R}\dfrac{\partial^3 w}{\partial x \partial\theta^2} + R\dfrac{\partial^3 w}{\partial x^3}$	$\dfrac{3-\nu}{2}\dfrac{\partial^3 w}{\partial x^2 \partial\theta}$	$-\dfrac{2}{R^2}\dfrac{\partial^2 w}{\partial\theta^2} - \dfrac{w}{R^2} - R\dfrac{\partial^3 u}{\partial x^3} + \dfrac{1-\nu}{2R}\dfrac{\partial^3 u}{\partial x \partial\theta^2} - \dfrac{3-\nu}{2}\dfrac{\partial^3 v}{\partial x^2\partial\theta}$
Flügge (Ref. 5-11)	$-\dfrac{1-\nu}{2R}\dfrac{\partial^3 w}{\partial x \partial\theta^2} + R\dfrac{\partial^3 w}{\partial x^3} + \dfrac{1-\nu}{2R^2}\dfrac{\partial^2 u}{\partial\theta^2}$	$\dfrac{3-\nu}{2}\dfrac{\partial^3 w}{\partial x^2 \partial\theta} + \dfrac{3}{2}(1-\nu)\dfrac{\partial^2 v}{\partial x^2}$	$-\dfrac{2}{R^2}\dfrac{\partial^2 w}{\partial\theta^2} - \dfrac{w}{R^2} - R\dfrac{\partial^3 u}{\partial x^3} + \dfrac{1-\nu}{2R}\dfrac{\partial^3 u}{\partial x \partial\theta^2} - \dfrac{3-\nu}{2}\dfrac{\partial^3 v}{\partial x^2\partial\theta}$
Washizu (Ref. 5-6) and Goldenveizer (Ref. 5-13)	0	$2(1-\nu)\dfrac{\partial^2 v}{\partial x^2} + \dfrac{1}{R^2}\dfrac{\partial^2 v}{\partial\theta^2} + (2-\nu)\dfrac{\partial^3 w}{\partial x^2\partial\theta} + \dfrac{1}{R^2}\dfrac{\partial^3 w}{\partial\theta^3}$	$-(2-\nu)\dfrac{\partial^3 v}{\partial x^2\partial\theta} - \dfrac{1}{R^2}\dfrac{\partial^3 v}{\partial\theta^3}$

are multiplied by $h^2/12R^2$ in Eqs. 5–79, it may be argued that their contribution is small and that they may therefore be neglected. It can be shown that Donnell's theory is equivalent to that which is obtained when the shallow shell theory is applied to a circular cylindrical panel (cf. Ref. 5–15).

When the tangential loading terms X and Y are put equal to zero, each of the theories of Eqs. 5–79 and Table 5–1 may be reduced to a single eighth-order differential equation in the displacement function ϕ (Ref. 5–12). It is then possible to express in the usual manner

$$\mathscr{S}(\phi) = Z \tag{5-81}$$

where the displacement function ϕ is connected with the displacements u, v, and w. The form of the structural operator $\mathscr{S}$ depends on the circular cylinder shell theory which is employed (Ref. 5–13). For example, for Donnell's theory, we have

$$\mathscr{S}(\quad) = -\frac{Eh}{R^2}\left(R^4\frac{\partial^4}{\partial x^4} + \frac{1}{1-\nu^2}\frac{h^2}{12R^2}\nabla^8\right)(\quad) \tag{5-82}$$

The connecting relations between the displacements u, v, and w and the displacement function ϕ are different in each case. In the case of Donnell's equation, they are simply

$$u = R\frac{\partial}{\partial x}\left(\frac{\partial^2\phi}{\partial\theta^2} - \nu R^2\frac{\partial^2\phi}{\partial x^2}\right)$$

$$v = -\frac{\partial}{\partial\theta}\left(\frac{\partial^2\phi}{\partial\theta^2} + (2+\nu)R^2\frac{\partial^2\phi}{\partial x^2}\right) \tag{5-83}$$

$$w = -\nabla^4\phi$$

In fact, by making use of Eqs. 5–83, it is evident that a further simplification in Donnell's equation is possible whereby

$$\frac{D}{R^8}\nabla^8 w + \frac{Eh}{R^2}\frac{\partial^4 w}{\partial x^4} = \frac{1}{R^4}\nabla^4 Z \tag{5-84}$$

Other relations connecting ϕ with u, v, and w may be found, for example, in Ref. 5–12.

5–4 HOMOGENEOUS ISOTROPIC ELASTIC SOLID

Let us consider the nature of the elastic operators which are appropriate for homogeneous isotropic elastic solids. In Eqs. 2–21, we have given the equations of equilibrium in component form. These equations may be altered by expressing them in terms of displacements rather than stresses. We may, for example, substitute for the normal stress components such expressions as (Ref. 5–5)

$$\sigma_x = \lambda\Delta + 2G\frac{\partial u}{\partial x} \tag{5-85}$$

and for the shear stress components such expressions as

$$\tau_{xy} = G\left(\frac{\partial w}{\partial y} + \frac{\partial v}{\partial z}\right) \tag{5-86}$$

where λ and G are defined by

$$\lambda = \frac{E\nu}{(1+\nu)(1-2\nu)}$$

$$G = \frac{E}{2(1+\nu)}$$

and Δ is the cubical dilation defined by

$$\Delta = \frac{\partial u}{\partial x} + \frac{\partial v}{\partial y} + \frac{\partial w}{\partial z} \tag{5-87}$$

When expressions such as (5–85) and (5–86) are substituted into Eqs. 2–21, we obtain the Navier equations

$$(\lambda + G)\left(\frac{\partial^2 u}{\partial x^2} + \frac{\partial^2 v}{\partial x\,\partial y} + \frac{\partial^2 w}{\partial x\,\partial z}\right) + G\nabla^2 u + X = \rho a_x$$

$$(\lambda + G)\left(\frac{\partial^2 u}{\partial x\,\partial y} + \frac{\partial^2 v}{\partial y^2} + \frac{\partial^2 w}{\partial y\,\partial z}\right) + G\nabla^2 v + Y = \rho a_y \tag{5-88}$$

$$(\lambda + G)\left(\frac{\partial^2 u}{\partial x\,\partial z} + \frac{\partial^2 v}{\partial y\,\partial z} + \frac{\partial^2 w}{\partial z^2}\right) + G\nabla^2 w + Z = \rho a_z$$

It is, of course, necessary to adjoin initial conditions and conditions at the boundary to Eqs. 5–88. When the surface displacements are given, the displacements u, v, and w are prescribed at the surface. When the surface tractions F_x, F_y, and F_z are prescribed, the boundary conditions are given by

$$F_x = \lambda\Delta\mathbf{n}\cdot\mathbf{i} + G\left[\frac{\partial u}{\partial n} + \frac{\partial u}{\partial x}\mathbf{n}\cdot\mathbf{i} + \frac{\partial v}{\partial x}\mathbf{n}\cdot\mathbf{j} + \frac{\partial w}{\partial x}\mathbf{n}\cdot\mathbf{k}\right]$$

$$F_y = \lambda\Delta\mathbf{n}\cdot\mathbf{j} + G\left[\frac{\partial v}{\partial n} + \frac{\partial u}{\partial y}\mathbf{n}\cdot\mathbf{i} + \frac{\partial v}{\partial y}\mathbf{n}\cdot\mathbf{j} + \frac{\partial w}{\partial y}\mathbf{n}\cdot\mathbf{k}\right] \tag{5-89}$$

$$F_z = \lambda\Delta\mathbf{n}\cdot\mathbf{k} + G\left[\frac{\partial w}{\partial n} + \frac{\partial u}{\partial z}\mathbf{n}\cdot\mathbf{i} + \frac{\partial v}{\partial z}\mathbf{n}\cdot\mathbf{j} + \frac{\partial w}{\partial z}\mathbf{n}\cdot\mathbf{k}\right]$$

where $\mathbf{n}$ is a unit vector normal to the surface in the outward direction.

Equations 5–88 may be expressed in the more compact vector form of

$$(\lambda + G)\nabla(\nabla \cdot \mathbf{q}) + G\nabla^2\mathbf{q} + \mathbf{R} = \rho\mathbf{a} \qquad (5\text{–}90)$$

where $\mathbf{q}$, $\mathbf{R}$, and $\mathbf{a}$ are the displacement, body force, and acceleration vectors, respectively; or they may be written in the following operator form:

$$\mathbf{R} - \rho\mathbf{a} = \widetilde{\mathscr{P}}(\mathbf{q}) \qquad (5\text{–}91)$$

where $\widetilde{\mathscr{P}}$ ( ) is a tensor differential operator with the form of

$$\widetilde{\mathscr{P}}(\quad) = -[(\lambda + G)\text{grad div} + G\nabla^2](\quad) \qquad (5\text{–}92)$$

in the case of a homogeneous, isotropic material.

5–5 THREE-DIMENSIONAL ELASTIC STRUCTURE

Finally, we take up the question of defining operators appropriate to the elastic deformation of three-dimensional structures in general. Since such structures vary widely in their geometry and general arrangement, we shall be content to state only the general form of these operators. The tensor operator inverse to the operator $\widetilde{\mathscr{P}} \cdot ($) of Eq. 5–91 relates the displacement vector and the body force vectors by

$$\mathbf{q} = \widetilde{\mathscr{P}}^{-1}(\mathbf{R} - \rho\mathbf{a}) \qquad (5\text{–}93)$$

where $\widetilde{\mathscr{P}}^{-1}$ ( ) is expressible in the integral form of

$$\widetilde{\mathscr{P}}^{-1}(\quad) = \int_V \Gamma \cdot (\quad) \, d\xi \, d\eta \, d\zeta \qquad (5\text{–}94)$$

The quantity Γ is a second order tensor of flexibility influence functions defined by

$$\Gamma = \begin{cases} C^{xx}(x, y, z; \xi, \eta, \zeta)\mathbf{ii} \\ + C^{xy}(x, y, z; \xi, \eta, \zeta)\mathbf{ij} + C^{xz}(x, y, z; \xi, \eta, \zeta)\mathbf{ik} \\ + C^{yx}(x, y, z; \xi, \eta, \zeta)\mathbf{ji} + C^{yy}(x, y, z; \xi, \eta, \zeta)\mathbf{jj} \\ + C^{yz}(x, y, z; \xi, \eta, \zeta)\mathbf{jk} + C^{zx}(x, y, z; \xi, \eta, \zeta)\mathbf{ki} \\ + C^{zy}(x, y, z; \xi, \eta, \zeta)\mathbf{kj} + C^{zz}(x, y, z; \xi, \eta, \zeta)\mathbf{kk} \end{cases} \qquad (5\text{–}95)$$

where, by Maxwell's law of reciprocal deflections,

$$C^{xy}(x, y, z; \xi, \eta, \zeta) = C^{yx}(\xi, \eta, \zeta; x, y, z)$$
$$C^{xz}(x, y, z; \xi, \eta, \zeta) = C^{zx}(\xi, \eta, \zeta; x, y, z)$$
$$C^{yz}(x, y, z; \xi, \eta, \zeta) = C^{zy}(\xi, \eta, \zeta; x, y, z)$$

We see, for example, that the quantity $C^{xy}(x, y, z; \xi, \eta, \zeta)$ represents the deflection of the structure in the x-direction at the point (x, y, z) due to a unit force in the y-direction at the point (ξ, η, ζ). Such influence functions may be computed, or in very complex cases, obtained by experiment. For example, in the special case of a homogeneous, isotropic, elastic solid, the influence functions C^{xx}, C^{yx}, and C^{zx} are computed by applying Eqs. 5–88 in the following way:

$$(\lambda + G)\left(\frac{\partial^2 C^{xx}}{\partial x^2} + \frac{\partial^2 C^{yx}}{\partial x\,\partial y} + \frac{\partial^2 C^{zx}}{\partial x\,\partial z}\right) + G\nabla^2 C^{xx} + \delta(x, y, z; \xi, \eta, \zeta) = 0$$

$$(\lambda + G)\left(\frac{\partial^2 C^{xx}}{\partial x\,\partial y} + \frac{\partial^2 C^{yx}}{\partial y^2} + \frac{\partial^2 C^{zx}}{\partial y\,\partial z}\right) + G\nabla^2 C^{yx} = 0 \qquad (5\text{–}96)$$

$$(\lambda + G)\left(\frac{\partial^2 C^{xx}}{\partial x\,\partial z} + \frac{\partial^2 C^{yx}}{\partial y\,\partial z} + \frac{\partial^2 C^{zx}}{\partial z^2}\right) + G\nabla^2 C^{zx} = 0$$

where $\delta(x, y, z; \xi, \eta, \zeta)$ is the Dirac delta function (cf. Eqs. 5–67 and 5–68).

REFERENCES

5–1. Timoshenko, S., and S. Woinowsky-Krieger, *Theory of Plates and Shells*, McGraw-Hill Book Company, New York, 1959.

5–2. Girkman, K., *Flachentragwerke*, Springer-Verlag, Vienna, 1956.

5–3. Wang, C. T., *Applied Elasticity*, McGraw-Hill Book Company, New York, 1953.

5–4. Peery, D. J., *Aircraft Structures*, McGraw-Hill Book Company, New York, 1950.

5–5. Timoshenko, S., and J. N. Goodier, *Theory of Elasticity*, McGraw-Hill Book Company, New York, 1951.

5–6. Washizu, K., *Some Considerations on Shell Theory*, Aeroelastic and Structures Research Laboratory Report 1001, M.I.T., October 1960.

5–7. Marguerre, K., "Zur Theorie der gekrummten Platte grosser Formanderung," *Proc. Fifth Internat. Congress of Applied Mechanics*. pp. 93–101, 1938.

5–8. Reissner, Eric, "On Some Aspects of the Theory of Thin Elastic Shells," *J. Boston Soc. Civil Engineers*, Vol. XLII, No. 2, pp. 100–133, April 1955.

5–9. Love, A. E. H., *A Treatise on the Mathematical Theory of Elasticity*, 4th ed., Dover Publications, New York, N.Y., 1944.

5–10. Donnell, L. H., *Stability of Thin-Walled Tubes Under Torsion*, NACA Report 479, 1933.

5–11. Flügge, W., *Statik und Dynamik der Schalen*, Julius Springer, Berlin, 1934.

5–12. Vlasov, V. S., *Basic Differential Equations in General Theory of Elastic Shells*, NACA TM 1241, 1955.

5–13. Goldenveizer, A. L., *Teoria uprugikh tonkikh obolochek* (*Theory of Thin Elastic Shells*), Gosizdat tekhnikoteoretich literatury, Moscow, 1953.

5–14. Novozhilov, V. V., *The Theory of Thin Shells*, edited by J. R. M. Radok, P. Noordhoff Ltd., The Netherlands, 1959.

5–15. Shulman, Yechiel, *Some Dynamic and Aeroelastic Problems of Plate and Shell Structures*, Sc.D. Thesis, M.I.T., 1959.

6

THE TYPICAL
SECTION

6–1 INTRODUCTION

The typical section (Fig. 2–3 and Sec. 2–7) was devised during the 1930's by such aeroelastic pioneers as Theodorsen and Garrick (Refs. 6–1 and 6–2), seeking a system suitable for elementary examination of the flutter problem. They suggested that the dynamics of an actual wing might be simulated by choosing the properties of the typical section to match those at a station 70–75% of the distance from the centerline to the tip. Subsequent experience has confirmed their judgment in situations where the aspect ratio is large, the sweep is small, and the sectional characteristics vary smoothly across the span.

Quantitative justification exists for this approach, because any two-degree-of-freedom lumped parameter approximation to a lifting surface which behaves elastically as a beam-rod will obey equations of motion in the same form as Eqs. 2–76a and b, and for unswept cases the correspondence is exact even in the relations between the airloads and the displacement coordinates. This is true, for example, of the *semi-rigid wing* used in early British investigations (e.g., Ref. 6–3), which is a structure artificially constrained so that its spanwise distributions of twist, flexure, etc., are independent of the manner of loading. There is no essential difference when either normal coordinates or arbitrarily selected modes of bending and torsion are adopted for setting up the analysis, as in Chap. 3, except that by the former method the elastic and inertia coupling terms are eliminated while the dependent variables lose part of their simple physical significance. In every case, the various coefficients of Eqs. 2–76a and b are replaced with weighted spanwise integrals of sectional properties and airloads; it can be said that there is always *some* system like the one in Fig. 2–3 which is aeroelastically equivalent to the true cantilever wing thus represented. The presence of an aerodynamic control surface just adds another degree of freedom or modifies the system inputs, as will be seen in subsequent sections. For further clarification of this equivalence, we can correlate the examples which follow against parallel treatments of the restrained beam-rod in Chap. 7.

Surely the best justification for devoting a full chapter to the typical section is the way in which it permits a swift, orderly, uncluttered presentation of the majority of classical aeroelastic problems. A few essential features are lost, such as the effects of three-dimensional flow and of rigid-body degrees of freedom. But the need for extensive approximation, some of it difficult to motivate or evaluate quantitatively, is wholly eliminated. The system will now be examined on its own merits, leaving to the reader much of the task of interpreting results as they illuminate what may take place on more complicated structures. We assume the linearity of all the aeroelastic operators.

6–2 STATIC PROBLEMS

Figure 6–1 reproduces the typical section, as adapted for the discussion of steady-state aeroelasticity with no time-dependent motion. A trailing edge flap has been added to the system of Fig. 2–3, although it is merely intended to be representative of any of the several devices currently in use for modifying lift on a wing. The C.G. position is not involved unless the weight has to be taken into account, but here the aerodynamic center—the axis about which pitching moment in steady flow is independent of angle of attack—has special interest; the A.C. is located a distance e ahead of the elastic axis. Bending spring K_h and its associated linear deflection, h, are not pictured in Fig. 6–1, since the equilibrium of *angular* displacements about the elastic axis primarily governs the static aeroelastic behavior. The bending degree of freedom may be described as uncoupled from α and δ, because a pure vertical translation causes no lift or moment. Thus the rotations are determined first; and, if desired, h can afterward be calculated from the steady-state version of Eq. 2–76a,

$$K_h h = -L(\alpha, \delta) \tag{6–1}$$

As will be seen in Chap. 7, the decoupling no longer occurs on a swept wing, where a bending slope gives rise to aerodynamic incidence.

In any aeroelastic system, the applied loads which play the role of inputs may be divided into (1) external loads of nonaerodynamic origin and (2) airloads due to the geometry of the structure of airflow. The former category embraces gravitational forces, certain inertial effects of maneuvers, forces due to various types of shakers, or even such things as the pull on a wire attached to the typical section. The aerodynamic loads arise generally from changes in angle of attack due to control deflections, maneuvers, gust or blast encounters, a turbulent wake acting on a tail, and the like. The rigid-body angle of attack and camber associated with trimmed flight constitute such inputs for the steady-state aeroelastic problem. Nonaerodynamic external loads will not be considered in this section, since they have little practical interest and their analysis is trivial.

(a) Treatment of the torsional degree of freedom and divergence

As a first illustration, we take the system in Fig. 6–1 with the flap locked out, $\delta = 0$. The angle of attack from zero lift includes a "rigid" part, α_0, and a twist, α_e, in the torsion spring:

$$\alpha = \alpha_0 + \alpha_e \qquad (6\text{--}2)$$

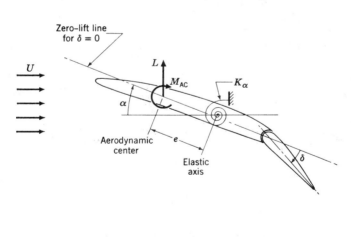

Fig. 6–1. Typical section airfoil, with trailing edge flap, elastically suspended in an airstream.

The structural operator relating the torque T_e to α_e is the spring constant,

$$T_e = K_\alpha \alpha_e \tag{6-3}$$

Equations 4–63 and 4–62 furnish the aerodynamic operators for lift and moment about the elastic axis,

$$L = C_L q S = q S \frac{\partial C_L}{\partial \alpha} (\alpha_0 + \alpha_e) \tag{6-4}$$

$$M_y = Le + M_{AC}$$

$$= q S \left[e \frac{\partial C_L}{\partial \alpha} (\alpha_0 + \alpha_e) + c C_{MAC} \right] \tag{6-5}$$

As before, q, c, and S denote the dynamic pressure, chord, and plan area of the typical section, respectively. Equating the elastic and aerodynamic torques from Eqs. 6–3 and 6–5, we find the equilibrium angle of twist to be

$$\alpha_e = \frac{\dfrac{qS}{K_\alpha}\left(e \dfrac{\partial C_L}{\partial \alpha}\alpha_0 + c C_{MAC}\right)}{1 - \dfrac{qSe}{K_\alpha}\dfrac{\partial C_L}{\partial \alpha}} \tag{6-6}$$

Both α_0 and C_{MAC} (which is directly proportional to the percentage camber of the thin airfoil) are seen to play roles as inputs to a linear system. With the other aerodynamic and structural parameters held constant, each one makes a contribution to α_e in proportion to its own magnitude.

Bearing in mind that $\partial C_L/\partial \alpha$, C_{MAC}, and e will usually be functions of the free-stream Mach number M, we can conclude a great deal from Eq. 6–6 about the influence of flight conditions on wing distortion. The second term in the denominator is often small compared with unity, suggesting an expansion of the form

$$\alpha_e = \frac{qS}{K_\alpha}\left[e \frac{\partial C_L}{\partial \alpha}\alpha_0 + c C_{MAC}\right]\left[1 + \frac{qSe}{K_\alpha}\frac{\partial C_L}{\partial \alpha} + \left(\frac{qSe}{K_\alpha}\frac{\partial C_L}{\partial \alpha}\right)^2 + \cdots\right]$$

$$= \alpha_e^{(1)} + \alpha_e^{(2)} + \cdots \tag{6-7}$$

Here

$$\alpha_e^{(1)} = \frac{qS}{K_\alpha}\left[e \frac{\partial C_L}{\partial \alpha}\alpha_0 + c C_{MAC}\right] \tag{6-8}$$

is the twist that would be produced by the pitching moment input, if no further aerodynamic torque resulted from the deformation itself. $\alpha_e^{(2)}$ is precisely the incremental twist caused by the moment from angle-of-attack change $\alpha_e^{(1)}$, and so on. The terms beyond $\alpha_e^{(1)}$ in Eq. 6–7 are a measure of the "aeroelastic effect," that is, of the interaction between airloads and deformations, which is the distinguishing feature of aeroelasticity. Evidently,

$$\frac{\alpha_e}{\alpha_e^{(1)}} = \frac{1}{1 - \dfrac{qSe}{K_\alpha}\dfrac{\partial C_L}{\partial \alpha}} \tag{6-9}$$

may be greater or less than unity, depending on whether the A.C. lies ahead or behind the elastic axis (e positive or negative).

There is an eigenvalue problem associated with the twisting wing. This can be observed from the blowing up of the solutions (6–6) and (6–9) at a certain value of the parameter $\dfrac{qSe}{K_\alpha}\dfrac{\partial C_L}{\partial \alpha}$; alternatively, it is inferred by causing the inputs α_0 and C_{MAC} to vanish in the moment-balance equation and noting that there is one condition where

$$\left[1 - \frac{qSe}{K_\alpha}\frac{\partial C_L}{\partial \alpha}\right]\alpha_e = 0 \tag{6-10}$$

may have a nonzero solution of arbitrary amplitude. This condition is known as torsional divergence. It occurs at the dynamic pressure

$$q_D = \frac{K_\alpha}{Se\dfrac{\partial C_L}{\partial \alpha}} \tag{6-11}$$

and comes about physically because the rate of change of pitching moment, M_y, with angle of twist increases with increasing q until it becomes just equal to the constant K_α and overpowers the torsional spring. The series (6–7) also diverges at this point.

Divergence-type instability is not ordinarily an important practical problem on straight wings, since flutter is likely to occur at a lower speed; but q_D provides a useful measure of the general stiffness level of the structure. As might be expected, Eq. 6–11 shows the eigenvalue to be independent of both initial incidence and camber. Except at the lower subsonic speeds, Eq. 6–11 is only an implicit expression for divergence dynamic pressure and Mach number, since both e and $\dfrac{\partial C_L}{\partial \alpha}$ depend on M_D. However, $q = \dfrac{\gamma}{2} p_\infty M^2$, so that an explicit solution can be developed either subsonically, where $\partial C_L/\partial \alpha \sim 1/\sqrt{1 - M^2}$ and e is fixed, or supersonically after similar substitutions. The latter case has little interest, because the A.C. is near midchord, e is usually negative, and M_D is, therefore, imaginary.

A result of some significance is derivable from Eq. 6–9 for any range of flight speed and altitude around M_D wherein variations of $e\, \partial C_L/\partial \alpha$ can be neglected:

$$\frac{\alpha_e}{\alpha_e^{(1)}} = \frac{1}{1 - (q/q_D)} = \frac{1}{1 - (M^2/M_D{}^2)} \tag{6–12}$$

For the third member of Eq. 6–12 to be valid, of course, flight must be at constant altitude; otherwise $M^2/M_D{}^2$ must be replaced by

$$p_\infty M^2/(p_\infty)_D M_D{}^2$$

(b) Effects of a control surface or spoiler

Let it first be assumed that the flap in Fig. 6–1, simulating any trailing edge control surface, is given a rigid rotation δ unaffected by any aerodynamic torques exerted on its own hinge. Since pitching moments about the elastic axis due to initial incidence and camber have effects which are additive linearly to those discussed here, they are omitted. According to Eqs. 4–64 and 4–65, the aerodynamic operators relating pitching moment to twist and flap angle are then

$$M_y = Le + M_{AC}$$
$$= qS\left[e\left(\frac{\partial C_L}{\partial \alpha}\alpha_e + \frac{\partial C_L}{\partial \delta}\delta\right) + c\,\frac{\partial C_{MAC}}{\partial \delta}\delta \right] \tag{6–13}$$

Equating moments between Eqs. 6–3 and 6–13, one calculates the ratio of elastic deformation to flap angle

$$\frac{\alpha_e}{\delta} = \frac{\dfrac{\partial C_L}{\partial \delta} + \dfrac{c}{e}\dfrac{\partial C_{MAC}}{\partial \delta}}{\dfrac{K_\alpha}{qSe} - \dfrac{\partial C_L}{\partial \alpha}} \tag{6–14}$$

and thence the total lift which is produced on the typical section by deflecting the flap,

$$L = \frac{qS\left[\dfrac{\partial C_L}{\partial \delta} + \dfrac{qSc}{K_\alpha}\dfrac{\partial C_L}{\partial \alpha}\dfrac{\partial C_{MAC}}{\partial \delta}\right]\delta}{1 - \dfrac{qSe}{K_\alpha}\dfrac{\partial C_L}{\partial \alpha}} \tag{6–15}$$

Equation 6–15 can be compared with the lift

$$L^r = qS\,\frac{\partial C_L}{\partial \delta}\,\delta \tag{6–16}$$

that would be developed by the same flap on an infinitely rigid wing. Their ratio reads

$$\frac{L}{L^r} = \frac{1 + \dfrac{qSc}{K_\alpha}\dfrac{\partial C_{MAC}}{\partial \delta}\Big/\dfrac{\partial \alpha}{\partial \delta}}{1 - \dfrac{qSe}{K_\alpha}\dfrac{\partial C_L}{\partial \alpha}} \tag{6-17a}$$

where

$$\frac{\partial \alpha}{\partial \delta} = \frac{\partial C_L}{\partial \delta}\Big/\frac{\partial C_L}{\partial \alpha} \tag{6-17b}$$

is the wing angle of attack giving the same lift as unit flap deflection on a rigid wing. L/L^r normally comes out less than unity, because $\partial C_{MAC}/\partial \delta$ is negative and the denominator of Eq. 6–17a falls off with increasing q much more slowly than the numerator; it is a measure of control surface effectiveness in the presence of losses due to torsional deformations of the main wing that tend to reduce the lift added by the control. The numerator of Eq. 6–17a passes through zero at a condition called *reversal*, which is characterized by the dynamic pressure

$$q_R = \frac{K_\alpha}{Sc}\frac{\dfrac{\partial \alpha}{\partial \delta}}{\left(-\dfrac{\partial C_{MAC}}{\partial \delta}\right)} \tag{6-18}$$

Like Eq. 6–11, this is an implicit formula for q_R (or M_R) whenever compressibility effects on $\partial \alpha/\partial \delta$ and $\partial C_{MAC}/\partial \delta$ must be accounted for.

The phenomenon of reversal is usually associated with ailerons or elevons mounted, as they often are, near the tips of wings having considerable torsional flexibility. Several aircraft have actually flown at q's where the antisymmetric lift distribution caused by moving the ailerons, and therefore the rolling moment, became opposite in sense to that called for by the control-column displacement (i.e., L/L^r became negative, at $q > q_R$). While not necessarily causing structural failure, this condition is obviously undesirable. It is aggravated by sweeping back the wings, in which case the designer often finds himself forced to adopt spoilers or some other lateral control device.

Horizontal and vertical tail surfaces do not experience complete reversal but only a partial loss of effectiveness. This is because fuselage vertical bending and fuselage torsion are more influential than local surface twisting in altering the control forces due to deflecting the elevator and rudder, respectively.

When the aerodynamic derivatives in Eq. 6–17a can be assumed constant, the following simple relation analogous to Eq. 6–12 is obtained:

$$\frac{L}{L^r} = \left(1 - \frac{q}{q_R}\right) \Big/ \left(1 - \frac{q}{q_D}\right) = \left(1 - \frac{M^2}{M_R{}^2}\right) \Big/ \left(1 - \frac{M^2}{M_D{}^2}\right) \quad (6\text{--}19)$$

(The third member again requires constant-altitude flight.) Equation 6–19 is plotted in Fig. 6–2 for several values of q_R/q_D. It reduces to the compact linear form

$$\frac{L}{L^r} \cong 1 - \frac{q}{q_R} \quad (6\text{--}20)$$

whenever both q_R and q are small compared to q_D. On the other hand, an interesting anomaly appears in the less common circumstance that $q_R = q_D$. The control then remains fully effective ($L/L^r = 1$) right up to the catastrophic onset of the divergence-reversal condition; this phenomenon has a simple physical explanation.

Reversal is not an eigenvalue problem in the same sense as divergence. Indeed, the characteristic dynamic pressure associated with the rigidly deflected flap is still q_D, as may be seen from the vanishing of the denominator in Eq. 6–14 or 6–17a.

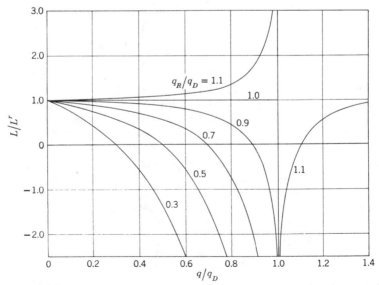

Fig. 6–2. Flap effectiveness L/L^r vs. dimensionless flight dynamic pressure q/q_D for several values of the ratio of reversal to divergence dynamic pressures. Aerodynamic properties are assumed independent of Mach number.

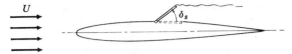

Fig. 6-3. Typical spoiler configuration, showing extended and retracted positions.

The foregoing analysis can readily be adapted to other control configurations on the typical section by making appropriate insertions for $\partial C_L / \partial \delta$ and $\partial C_{MAC} / \partial \delta$. For example, a leading-edge flap has both these aerodynamic derivatives positive if δ is defined to be the angle of nose-up rotation. Therefore, such a device would have its effectiveness enhanced by increasing dynamic pressure (Eq. 6-17a) and is not subject to reversal.

Most spoilers are highly nonlinear, since the incremental aerodynamic loads are not directly proportional to the control deflection. Thus the typical arrangement shown in Fig. 6-3 is recessed and has no effect at all except when δ_s is positive. A given value of δ_s produces (1) a negative contribution $\Delta C_L(\delta_s)$ to the lift coefficient and (2) a small moment $\Delta C_{MAC}(\delta_s)$, which can be negative or positive, depending on the configuration, although the latter sign is more common. Under these circumstances, Eq. 6-15 would be replaced by

$$L = \frac{qS\left[\Delta C_L(\delta_s) + \dfrac{qSc}{K_\alpha} \dfrac{\partial C_L}{\partial \alpha} \Delta C_{MAC}(\delta_s)\right]}{1 - \dfrac{qSe}{K_\alpha} \dfrac{\partial C_L}{\partial \alpha}} \qquad (6\text{-}21)$$

With positive $\Delta C_{MAC}(\delta_s)$, there is a remote possibility that the second term in the numerator might overpower the first and bring on reversal. In practice this eventuality turns out to be much less likely than with a trailing edge flap.

(c) The influence of flexibility in the controls

Moving a step closer to actual conditions on aircraft, let there be attached to the flap hinge a spring with torsional constant K_δ. It is so arranged that, in the absence of any aerodynamic moment, the flap would assume an angle δ_0. The spring torque tending to rotate the flap back toward neutral is then

$$H_e = K_\delta[\delta - \delta_0] \qquad (6\text{-}22)$$

δ being the equilibrium flap position. In the case of ailerons, for instance, δ_0 would be an angle proportional to the turning of the pilot's control wheel or the lateral displacement of the stick, and K_δ becomes a grossly simplified representation of cable flexibility and/or stiffness of an actuator and back-up structure.

According to Eq. 4–66, there is an aerodynamic operator giving the nose-up hinge moment due to the incidences of the wing and flap:

$$H_f = qS_f c_f \left[\frac{\partial C_H}{\partial \alpha} \alpha_e + \frac{\partial C_H}{\partial \delta} \delta \right] \qquad (6\text{–}23)$$

Here S_f and c_f are suitably defined flap area and chord, respectively. Since H_e and H_f must be numerically equal, Eqs. 6–22 and 6–23 can be rearranged to yield the following relation among the angles:*

$$-\frac{qS_f c_f}{K_\delta} \frac{\partial C_H}{\partial \alpha} \alpha_e + \left[1 - \frac{qS_f c_f}{K_\delta} \frac{\partial C_H}{\partial \delta} \right] \delta = \delta_0 \qquad (6\text{–}24)$$

A second equation is provided by the balance of torques about the E.A., Eq. 6–14, which is here rewritten

$$\left[\frac{\partial C_L}{\partial \alpha} - \frac{K_\alpha}{qSe} \right] \alpha_e + \left[\frac{\partial C_L}{\partial \delta} + \frac{c}{e} \frac{\partial C_{MAC}}{\partial \delta} \right] \delta = 0 \qquad (6\text{–}25)$$

The simultaneous Eqs. 6–24 and 6–25 furnish both the twist, α_e, and the resultant flap angle, δ, as functions of the control input angle, δ_0. Carrying out this solution, we can find the total lift on the elastically-mounted section,

$$L = qS \left[\frac{\partial C_L}{\partial \alpha} \alpha_e + \frac{\partial C_L}{\partial \delta} \delta \right]$$

$$= \frac{qS \left[\frac{\partial C_L}{\partial \delta} + \frac{qSc}{K_\alpha} \frac{\partial C_{MAC}}{\partial \delta} \frac{\partial C_L}{\partial \alpha} \right] \delta_0}{\left[1 - \frac{qSe}{K_\alpha} \frac{\partial C_L}{\partial \alpha} \right] \left[1 - \frac{qS_f c_f}{K_\delta} \frac{\partial C_H}{\partial \delta} \right]} \qquad (6\text{–}26)$$

$$- \left[\frac{\partial C_L}{\partial \delta} + \frac{c}{e} \frac{\partial C_{MAC}}{\partial \delta} \right] \left(\frac{qSe}{K_\alpha} \right) \left(\frac{qS_f c_f}{K_\delta} \right) \frac{\partial C_H}{\partial \alpha}$$

* Observe that, when the wing torsional stiffness is high ($\alpha_e \cong 0$), the negative sign of $\partial C_H / \partial \delta$ causes δ from Eq. 6–24 to be smaller than the desired deflection δ_0. This illustrates the phenomenon of *blowback*, which can interfere severely with the satisfactory operation of any aerodynamically unbalanced control surface on a high-speed aircraft.

Dividing by the "rigid" lift $L^r = qS \dfrac{\partial C_L}{\partial \delta} \delta_0$, we derive an expression for control effectiveness,

$$\frac{L}{L^r} = \frac{1 + \dfrac{qSc}{K_\alpha} \dfrac{\partial C_{MAC}}{\partial \delta} \bigg/ \dfrac{\partial \alpha}{\partial \delta}}{\left[1 - \dfrac{qSe}{K_\alpha} \dfrac{\partial C_L}{\partial \alpha}\right]\left[1 - \dfrac{qS_f c_f}{K_\delta} \dfrac{\partial C_H}{\partial \delta}\right] - \left[\dfrac{\partial C_L}{\partial \delta} + \dfrac{c}{e} \dfrac{\partial C_{MAC}}{\partial \delta}\right]\left(\dfrac{qSe}{K_\alpha}\right)\left(\dfrac{qS_f c_f}{K_\delta}\right)\dfrac{\partial C_H}{\partial \alpha}} \quad (6\text{-}27)$$

This more realistic system is seen also to display loss of effectiveness and reversal. In fact, the dynamic pressure q_R, obtained from the vanishing of the numerator in Eq. 6–27, is still given by Eq. 6–18 except in the unlikely event that the denominator goes to zero first. Equation 6–27 takes on a more meaningful form when the aerodynamic properties of the typical section are independent of flight speed and altitude. Let us assume this for the moment and define the following three constants:

(1) $$q_{D_0} = \frac{K_\alpha}{Se \dfrac{\partial C_L}{\partial \alpha}} \qquad (6\text{-}28a)$$

This is the divergence dynamic pressure for an infinitely stiff hingeline spring, $K_\delta \rightarrow \infty$.

(2) $$q_{Df} = \frac{K_\delta}{S_f c_f \left(-\dfrac{\partial C_H}{\partial \delta}\right)} \qquad (6\text{-}28b)$$

Since $\partial C_H / \partial \delta$ is negative in the large majority of cases, this is a positive constant proportional to K_δ. If the hinge moment derivative did happen to have an unstable, positive sense, $(-q_{Df})$ would be the dynamic pressure for torsional divergence of the uncoupled flap.

(3) $$A = \frac{\dfrac{\partial C_H}{\partial \alpha}}{\dfrac{\partial C_H}{\partial \delta}}\left[\dfrac{\partial \alpha}{\partial \delta} + \frac{c}{e} \dfrac{\dfrac{\partial C_{MAC}}{\partial \delta}}{\dfrac{\partial C_L}{\partial \alpha}}\right] \qquad (6\text{-}28c)$$

This is a purely aerodynamic constant, whose magnitude is normally less than unity. Its sign is opposite to that of the "floating tendency"

$$\frac{\partial C_H}{\partial \alpha} \bigg/ \frac{\partial C_H}{\partial \delta}$$

because the quantity in brackets is negative so long as $q_R < q_{D_0}$.

In terms of Eqs. 6–28, Eq. 6–27 can be rewritten

$$\frac{L}{L^r} = \frac{1 - \dfrac{q}{q_R}}{\left[1 - \dfrac{q}{q_{D_0}}\right]\left[1 + \dfrac{q}{q_{Df}}\right] + A\dfrac{q}{q_{D_0}}\dfrac{q}{q_{Df}}} \tag{6-29}$$

Several interesting conclusions about the behavior of the two-degree-of-freedom system can be drawn from a study of Eq. 6–29.

The eigenvalue of the typical section is modified by elastically mounting the flap. Divergence now coincides with the vanishing of the denominator in Eq. 6–26, 6–27, or 6–29. Although this is quadratic in q, it usually has only a single positive root. Retaining the restrictions* that underlie Eq. 6–29, we can easily show this root to be

$$(q_D)_{\text{wing-flap}} = q_{D_0}\frac{\left[1 - \dfrac{q_{Df}}{q_{D_0}}\right] + \sqrt{\left[1 - \dfrac{q_{Df}}{q_{D_0}}\right]^2 + 4[1 - A]\dfrac{q_{Df}}{q_{D_0}}}}{2[1 - A]} \tag{6-30}$$

The condition described by Eq. 6–30 is called *torsion-aileron divergence.* Generally it will be found to occur at a dynamic pressure below q_{D_0} when A is negative, because the flap then tends to float parallel to the wind and generates a nose-up moment, which assists in overpowering the spring K_α. Conversely, the dynamic pressure exceeds q_{D_0} when A is positive. Two special cases of Eq. 6–30 possess more than trivial interest:

(1) The unrestrained flap, $q_{Df} = 0$, for which

$$(q_D)_{\text{wing-flap}} = \frac{q_{D_0}}{1 - A} \tag{6-31}$$

(2) The case $A = 0$, which corresponds either to zero floating tendency or to $q_{D_0} = q_R$. This yields

$$(q_D)_{\text{wing-flap}} = q_{D_0} \tag{6-32}$$

for evident physical reasons. Figure 6–4 plots Eq. 6–30 for several values of the constant A.

---

* Note that Eq. 6–30 can be extended to cover the condition when the aerodynamic properties depend on the flight Mach number by resubstituting Eqs. 6–28. However, $(q_D)_{\text{wing-flap}}$ is then given only implicitly, as discussed above.

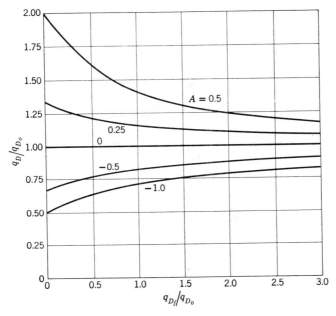

Fig. 6–4. Dimensionless dynamic pressure of torsion-aileron divergence plotted vs. the stiffness-ratio parameter q_{D_f}/q_{D_o} for five values of the constant A defined in Eq. 6–29c. Aerodynamic properties are assumed independent of Mach number.

6–3 THE SYSTEM UNDER PRESCRIBED TIME-DEPENDENT INPUTS

As an elementary introduction to the subject of forced motion, we examine the bending-torsion typical section driven by a prescribed external force, $F(t)$, in the absence of aerodynamic loading. Either the wind-tunnel test section may be assumed evacuated of air or the airspeed, U, may be set equal to zero; in the latter case, the inertial properties of the wing must include certain small *virtual inertias* of the air, whose presence does not alter the analysis significantly. Figure 6–5 shows the driving force to be acting at a distance d ahead of the E.A. In the notation of Sec. 2–2, the governing equations evidently read

$$m\ddot{h} + K_h h + S_\alpha \ddot{\alpha} = -F \qquad (6\text{–}33a)$$

$$S_\alpha \ddot{h} + I_\alpha \ddot{\alpha} + K_\alpha \alpha = Fd \qquad (6\text{–}33b)$$

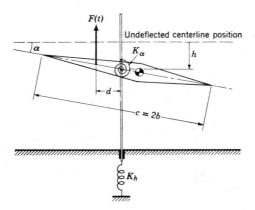

Fig. 6–5. The typical section driven by a prescribed external force $F(t)$, acting at a distance d ahead of the elastic axis.

The spring constants in Eqs. 6–33 are often replaced in terms of natural frequencies of *uncoupled* bending and torsional oscillation, defined by

$$\omega_h = \sqrt{K_h/m} \tag{6-34a}$$

$$\omega_\alpha = \sqrt{K_\alpha/I_\alpha} \tag{6-34b}$$

So modified, Eqs. 6–33 will be employed as follows in this section:

$$m[\ddot{h} + \omega_h^2 h] + S_\alpha \ddot{\alpha} = -F \tag{6-35a}$$

$$S_\alpha \ddot{h} + I_\alpha[\ddot{\alpha} + \omega_\alpha^2 \alpha] = Fd \tag{6-35b}$$

Three basic types of force input will now be considered, along with associated methods for finding the system response to more general inputs. The reader will recognize these techniques as the fundamental material of any modern book on vibrations in linear systems, but they must be introduced in preparation for the analysis of more particularly aeroelastic problems.

(a) Sinusoidal forcing; the mechanical admittance method

We assume first that a simple harmonic driving force of circular frequency ω has been acting for a sufficient time to ensure the disappearance of any transients associated with its application. (Damping is neglected in Eqs. 6–35 but must be present to some degree in any passive system.) Linearity permits the complex representation $e^{i\omega t}$ to be used throughout; so that a force which, in physical actuality, is the real part of

$$F(t) = \bar{F}e^{i\omega t} \tag{6-36}$$

is known to generate *permanent* or *steady-state* responses which are the real parts of

$$h = \bar{h}e^{i\omega t} \qquad (6\text{--}37)$$

and

$$\alpha = \bar{\alpha}e^{i\omega t} \qquad (6\text{--}38)$$

Generally $\bar{F}$, $\bar{h}$, and $\bar{\alpha}$ would be complex numbers, to allow for phase differences; they may be regarded as real here, since the sinusoidal response of an undamped system is always either in phase or 180° out of phase with the input.

Combination of Eqs. 6–35 through 6–38 yields the matrix equation

$$\begin{bmatrix} mb[\omega_h^2 + (i\omega)^2] & S_\alpha(i\omega)^2 \\ S_\alpha b(i\omega)^2 & I_\alpha[\omega_\alpha^2 + (i\omega)^2] \end{bmatrix} \begin{Bmatrix} \dfrac{\bar{h}}{b} \\ \bar{\alpha} \end{Bmatrix} = \begin{Bmatrix} -\bar{F} \\ \bar{F}d \end{Bmatrix} \qquad (6\text{--}39)$$

where the semichord $b = c/2$ has been introduced as a reference length for reasons that will become clearer in Secs. 6–4 and 6–5. The simultaneous system (6–39) is then solved for dimensionless ratios of outputs to input, one such pair being the following:

$$\frac{\bar{h}/b}{\bar{F}/K_h b} \equiv H_{hF}(i\omega)$$

$$= \frac{-\left[1 + \left(\dfrac{i\omega}{\omega_\alpha}\right)^2\right] - \dfrac{d}{b}\dfrac{x_\alpha}{r_\alpha^2}\left(\dfrac{i\omega}{\omega_\alpha}\right)^2}{\left[1 + \left(\dfrac{i\omega}{\omega_\alpha}\right)^2\right]\left[1 + \left(\dfrac{i\omega}{\omega_h}\right)^2\right] - \dfrac{x_\alpha^2}{r_\alpha^2}\left(\dfrac{i\omega}{\omega_\alpha}\right)^2\left(\dfrac{i\omega}{\omega_h}\right)^2} \qquad (6\text{--}40)$$

$$\frac{\bar{\alpha}}{\bar{F}/K_h b} \equiv H_{\alpha F}(i\omega)$$

$$= \frac{\left[1 + \left(\dfrac{i\omega}{\omega_h}\right)^2\right]\dfrac{d}{br_\alpha^2}\left(\dfrac{\omega_h}{\omega_\alpha}\right)^2 + \dfrac{x_\alpha}{r_\alpha^2}\left(\dfrac{i\omega}{\omega_\alpha}\right)^2}{\left[1 + \left(\dfrac{i\omega}{\omega_\alpha}\right)^2\right]\left[1 + \left(\dfrac{i\omega}{\omega_h}\right)^2\right] - \dfrac{x_\alpha^2}{r_\alpha^2}\left(\dfrac{i\omega}{\omega_\alpha}\right)^2\left(\dfrac{i\omega}{\omega_h}\right)^2} \qquad (6\text{--}41)$$

Equations 6–40 and 6–41 contain two convenient dimensionless parameters,

$$x_\alpha = \frac{S_\alpha}{mb} \qquad (6\text{--}42)$$

the distance in semichords by which the C.G. lies behind the E.A., and

$$r_\alpha = \sqrt{\frac{I_\alpha}{mb^2}} \qquad (6\text{--}43)$$

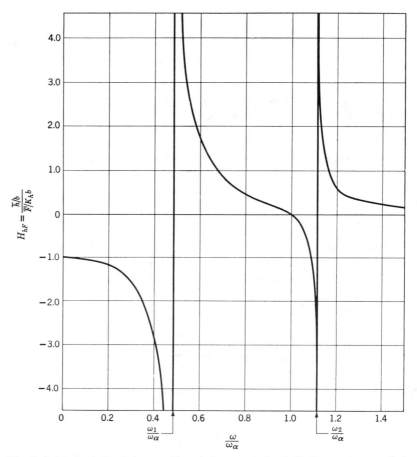

Fig. 6–6. Mechanical admittance H_{hF} relating translational displacement to applied force on the typical section in vacuum, plotted vs. frequency ratio ω/ω_α for the following set of parameters: $d/b = 0$, $x_\alpha = 0.2$, $r_\alpha = 0.5$, $\omega_h/\omega_\alpha = 0.5$.

the radius of gyration of the typical section about its E.A., also expressed in semichords.

The symbol $H(i\omega)$ is frequently termed the *mechanical admittance*—the complex amplitude of sinusoidal response per unit sinusoidal input—of the linear system, generalizing the concept of admittance used so commonly in electrical work. The *impedance* $Z(i\omega)$ is the inverse of $H(i\omega)$. In Eqs. 6–40 and 6–41 the admittances are given subscripts suggesting the output and input which they relate; such identification is even more important when treating complicated systems having many inputs and outputs, but a unique function of $i\omega$ exists for each pair.

For illustrative purposes, the admittance $H_{hF}(i\omega)$ given by Eq. 6-40 is plotted versus the frequency ratio ω/ω_α in Fig. 6-6 for a typical set of the four section parameters ω_h/ω_α, d/b, x_α, r_α which govern its behavior. Damping being absent, the response is seen to become infinite and change sign at each of the two abscissas ω_1/ω_α and ω_2/ω_α. These mark the eigenvalues of the typical section in vacuum, equal to the roots of the denominator polynomial in Eq. 6-40 or 6-41. They represent physically the natural frequencies of vibration of the coupled bending-torsion system and can be expressed in the attractively symmetrical forms

$$\frac{\omega_1{}^2}{\omega_h\omega_\alpha}, \frac{\omega_2{}^2}{\omega_h\omega_\alpha} = \frac{\left[\dfrac{\omega_h}{\omega_\alpha} + \dfrac{\omega_\alpha}{\omega_h}\right] \mp \sqrt{\left[\dfrac{\omega_h}{\omega_\alpha} + \dfrac{\omega_\alpha}{\omega_h}\right]^2 - 4\left[1 - \dfrac{x_\alpha{}^2}{r_\alpha{}^2}\right]}}{2\left[1 - \dfrac{x_\alpha{}^2}{r_\alpha{}^2}\right]} \qquad (6\text{-}44)$$

Equation 6-44 shows that the natural frequencies (but not, however, the associated normal mode shapes) depend only on the two parameters ω_h/ω_α and

$$\frac{x_\alpha{}^2}{r_\alpha{}^2} = \frac{S_\alpha{}^2}{mI_\alpha} = \frac{1}{1 + \dfrac{r_{CG}{}^2}{x_{CG}{}^2}} \qquad (6\text{-}45)$$

(Here r_{CG} and x_{CG} are the radius of gyration about the C.G. and the C.G.–E.A. distance, respectively.) Since $x_\alpha{}^2/r_\alpha{}^2$ according to Eq. 6-45 must have a value between 0 and $\frac{1}{2}$, Eq. 6-44 always yields real, positive frequencies.

The presence of any dissipative force out of phase with the displacements in Eqs. 6-35 will eliminate the singular resonance peaks from Fig. 6-6. But then, as will be elaborated in Sec. 6-4, the admittance becomes a complex function of $i\omega$ and must be illustrated by plotting both its real and imaginary parts versus frequency. More commonly, curves of amplitude and phase angle or complex polars with frequency as a parameter are used.

Any advanced text on linear systems (e.g., Gardner and Barnes, Ref. 6-4, or Vol. 2 of Draper, McKay, and Lees, Ref. 6-5) demonstrates how the admittance function H, with $i\omega$ replaced by a general complex argument, contains all the information about system characteristics needed to calculate the response to any arbitrary input, $F(t)$. In particular, the Fourier series and Fourier integral methods rely directly on a knowledge of $H(i\omega)$. The former method is limited to periodically repeating inputs, all of which are known to be representable by the series

$$F(t) = \sum_{n=-\infty}^{\infty} c_n e^{in\omega_0 t} \qquad (6\text{-}46a)$$

Here $\omega_0 = 2\pi/T_0$, T_0 being the fundamental period. The complex constant c_n is computed by techniques of harmonic analysis,

$$c_n = \frac{1}{T_0} \int_0^{T_0} F(t) e^{-in\omega_0 t} \, dt \qquad (6\text{-}46b)$$

Evidently c_{-n} is the conjugate of c_n, which permits Eq. 6–46a to be reduced to the more familiar, but less compact, real Fourier series containing $\cos n\omega_0 t$ and $\sin n\omega_0 t$.

The system's response to the nth harmonic of the input *spectrum* is $H(in\omega_0)c_n e^{in\omega_0 t}$, by the definition of mechanical admittance. However, one basic consequence of the property of linearity is that outputs can be superimposed to find the effect of a combined input. It follows that the bending and torsional responses to the force described by Eq. 6–46a are

$$\frac{h(t)}{b} = \frac{1}{K_h b} \sum_{n=-\infty}^{\infty} H_{hF}(in\omega_0)c_n e^{in\omega_0 t} \qquad (6\text{-}47)$$

$$\alpha(t) = \frac{1}{K_h b} \sum_{n=-\infty}^{\infty} H_{\alpha F}(in\omega_0)c_n e^{in\omega_0 t} \qquad (6\text{-}48)$$

Convergence of all these series may safely be assumed, since no difficulty occurs in practice with physically realizable inputs. There is no question of taking real parts in Eqs. 6–47 and 6–48; the vanishing of imaginaries is easily proved from the conjugate character of c_n and c_{-n}, and of $H(i\omega)$ and $H(-i\omega)$.

Conceptually it is a short step from the Fourier series or *line spectrum* of a periodic function to the Fourier integral or *continuous spectrum* of an aperiodic one. Indeed, many books furnish proofs that, whenever the integral of $|F(t)|$ over all time exists, we can find a representation of the form

$$F(t) = \int_{-\infty}^{\infty} G(i\omega) e^{i\omega t} \, d\omega \qquad (6\text{-}49a)$$

where $G(i\omega)$ is the *spectral density* of $F(t)$ and may be obtained from the formula

$$G(i\omega) = \frac{1}{2\pi} \int_{-\infty}^{\infty} F(t) e^{-i\omega t} \, dt \qquad (6\text{-}49b)$$

For real or complex ω, Eq. 6–49b yields what is called the *Fourier transform* of $F(t)$. The extended theory of such transforms appears in Sneddon's book (Ref. 6–6), while Campbell and Foster have published an excellent table (Ref. 6–7).

By obvious extension of the reasoning leading to Eqs. 6–47 and 6–48, the typical section responses to the force described by Eq. 6–49a are

$$\frac{h(t)}{b} = \frac{1}{K_h b} \int_{-\infty}^{\infty} H_{hF}(i\omega)G(i\omega)e^{i\omega t}\, d\omega \tag{6–50}$$

$$\alpha(t) = \frac{1}{K_h b} \int_{-\infty}^{\infty} H_{\alpha F}(i\omega)G(i\omega)e^{i\omega t}\, d\omega \tag{6–51}$$

Integrals such as those in Eqs. 6–50 and 6–51 will exist if the one in Eq. 6–49a does, for normally $H(i\omega) = 0(1/i\omega t)$ or higher at the limits. However, because of difficulties with both the convergence of Eq. 6–49 and evaluation of integrals, relations like 6–50 and 6–51 are rarely useful for direct solution of practical problems. One exception involves random inputs, described under (c) below. When the mechanical admittance is known, these relations are often employed to find the response to some elementary input, such as a step or impulse function. The latter then forms the basis for analyzing more complicated inputs, as will now be discussed.

(b) Transient forcing; solution by Laplace transformation

Returning to Eqs. 6–35, we consider a force $F(t)$ initially applied to the typical section at $t = 0$. At this starting instant the system is either quiescent or in a known state, characterized by a set of initial conditions on h, $\dot{h}$, α, and $\dot{\alpha}$. The concept of Laplace transformation, defined in the case of $F(t)$ by*

$$\mathscr{L}\{F(t)\} = \bar{F}(p) = \int_0^{\infty} e^{-pt}F(t)\, dt \tag{6–52}$$

proves valuable when computing responses. For the complete theory and for derivations of several formulas that will be exploited below, the reader is referred to Ref. 6–4, Churchill (Ref. 6–8), Doetsch (Ref. 6–9), or another of the excellent books on the subject.

In this section we concentrate on zero initial conditions, since motion

* Observe that, unless a dimensionless t is employed, this transformation adds the dimension of time to any quantity on which it operates. The new dimension is dropped in the process of *inversion* or return to the real time domain. p has dimensions of inverse time. Similar comments apply to Fourier transformation.

at $t = 0$ complicates the analysis without contributing anything of significance. Under these circumstances, the transformation of Eqs. 6–35
becomes

$$
\begin{bmatrix} m[p^2 + \omega_h{}^2] & S_\alpha p^2 \\ S_\alpha p^2 & I_\alpha[p^2 + \omega_\alpha{}^2] \end{bmatrix} \begin{Bmatrix} \bar{h}(p) \\ \bar{\alpha}(p) \end{Bmatrix} = \begin{Bmatrix} -\bar{F}(p) \\ \bar{F}(p)d \end{Bmatrix} \qquad (6\text{–}53)
$$

(The argument p of each dependent variable is always written explicitly,
to distinguish from the complex amplitudes of simple harmonic quantities,
which were also identified by the bar.) In terms of previously defined
dimensionless quantities, Eqs. 6–53 have the algebraic solutions

$$
\bar{h}(p)/b = - \left[\frac{\left[1 + \dfrac{p^2}{\omega_\alpha{}^2}\right] + \dfrac{d}{b}\dfrac{x_\alpha}{r_\alpha{}^2}\dfrac{p^2}{\omega_\alpha{}^2}}{\left[1 + \dfrac{p^2}{\omega_\alpha{}^2}\right]\left[1 + \dfrac{p^2}{\omega_h{}^2}\right] - \dfrac{x_\alpha{}^2}{r_\alpha{}^2}\dfrac{p^2}{\omega_\alpha{}^2}\dfrac{p^2}{\omega_h{}^2}} \right] \frac{\bar{F}(p)}{K_h b} \qquad (6\text{–}54)
$$

$$
\bar{\alpha}(p) = \left[\frac{\left[1 + \dfrac{p^2}{\omega_h{}^2}\right]\dfrac{d}{b r_\alpha{}^2}\left(\dfrac{\omega_h}{\omega_\alpha}\right)^2 + \dfrac{x_\alpha}{r_\alpha{}^2}\dfrac{p^2}{\omega_\alpha{}^2}}{\left[1 + \dfrac{p^2}{\omega_\alpha{}^2}\right]\left[1 + \dfrac{p^2}{\omega_h{}^2}\right] - \dfrac{x_\alpha{}^2}{r_\alpha{}^2}\dfrac{p^2}{\omega_\alpha{}^2}\dfrac{p^2}{\omega_h{}^2}} \right] \frac{\bar{F}(p)}{K_h b} \qquad (6\text{–}55)
$$

There is a manifest parallelism with Eqs. 6–40 and 6–41, illustrating the
general result that mechanical admittances can be found by replacing p
with $i\omega$ in the Laplace transform of the corresponding transient input-
output relationship (under zero initial conditions). Moreover, the roots
of the denominator polynomial in p, which are $\pm i\omega_1$ and $\pm i\omega_2$, according
to Eqs. 6–44, reveal by their pure imaginary character that the typical
section is undamped and neutrally stable.

The bracketed coefficients on the right-hand sides of Eqs. 6–54 and 6–55
are closely connected with the responses to two elementary types of input:

(1) *The step function*

$$
\frac{F(t)}{K_h b} = 1(t) = \begin{cases} 0, & t < 0 \\ 1, & t \geq 0 \end{cases} \qquad (6\text{–}56)
$$

(a dimensionless quantity). The Laplace transform of this function is

$$
\mathscr{L}\{1(t)\} = \frac{1}{p} \qquad (6\text{–}57)
$$

so that each of the two responses has as its transform the product of
$1/p$ by the corresponding bracket.

(2) *The unit impulse function or Dirac delta*

$$\frac{F(t)}{K_h b} = \delta(t) \qquad (6\text{–}58a)$$

$\delta(t)$ being defined to have an infinite peak, with unit area underneath, centered around $t = 0$:

$$\int_{(\text{through } t=0)} \delta(t)\, dt = 1 \qquad (6\text{–}58b)$$

$$\delta(t) = 0, \qquad t \neq 0 \qquad (6\text{–}58c)$$

(dimensions of inverse time). Since the Laplace transform is just unity,

$$\mathcal{L}\{\delta(t)\} = 1 \qquad (6\text{–}59)$$

the brackets in Eqs. 6–54 and 6–55 are themselves the transformed responses.

The name *indicial admittance* is customarily applied to the step-function response, its symbol being $A_{ij}(t)$, where i and j identify the output and input, respectively. By means of expansions in partial fractions, by direct application of transform tables, or by resort to the general inversion formula for the Laplace transformation (Ref. 6–4, etc.), we can find the indicial admittances of the typical section:

$$A_{hF}(t) = \mathcal{L}^{-1}\left\{ -\frac{1}{p} \frac{\left[1 + \dfrac{p^2}{\omega_\alpha^2}\right] + \dfrac{d}{b}\dfrac{x_\alpha}{r_\alpha^2}\dfrac{p^2}{\omega_\alpha^2}}{\left[1 + \dfrac{p^2}{\omega_\alpha^2}\right]\left[1 + \dfrac{p^2}{\omega_h^2}\right] - \dfrac{x_\alpha^2}{r_\alpha^2}\dfrac{p^2}{\omega_\alpha^2}\dfrac{p^2}{\omega_h^2}} \right\}$$

$$= -1 + \frac{r_\alpha^2 \omega_h^2}{(r_\alpha^2 - x_\alpha^2)(\omega_2^2 - \omega_1^2)}\left\{\left[1 + \frac{d}{b}\frac{x_\alpha}{r_\alpha^2} - \frac{\omega_\alpha^2}{\omega_2^2}\right]\cos \omega_2 t \right.$$

$$\left. - \left[1 + \frac{dx_\alpha}{br_\alpha^2} - \frac{\omega_\alpha^2}{\omega_1^2}\right]\cos \omega_1 t\right\} \qquad (6\text{–}60)$$

$$A_{\alpha F}(t) = \mathcal{L}^{-1}\left\{ \frac{1}{p} \frac{\left[1 + \dfrac{p^2}{\omega_h^2}\right]\dfrac{d}{br_\alpha^2}\left(\dfrac{\omega_h}{\omega_\alpha}\right)^2 + \dfrac{x_\alpha}{r_\alpha^2}\dfrac{p^2}{\omega_\alpha^2}}{\left[1 + \dfrac{p^2}{\omega_\alpha^2}\right]\left[1 + \dfrac{p^2}{\omega_h^2}\right] - \dfrac{x_\alpha^2}{r_\alpha^2}\dfrac{p^2}{\omega_\alpha^2}\dfrac{p^2}{\omega_h^2}} \right\}$$

$$= \frac{1}{r_\alpha^2}\frac{d}{b}\frac{\omega_h^2}{\omega_\alpha^2} - \frac{d\omega_h^2}{b(r_\alpha^2 - x_\alpha^2)(\omega_2^2 - \omega_1^2)}\left\{\left[1 + x_\alpha\frac{b}{d} - \frac{\omega_h^2}{\omega_2^2}\right]\cos \omega_2 t \right.$$

$$\left. - \left[1 + x_\alpha\frac{b}{d} - \frac{\omega_h^2}{\omega_1^2}\right]\cos \omega_1 t\right\} \qquad (6\text{–}61)$$

Each of these equations is arranged with the so-called *permanent* part (particular solution) first, followed by the *transient* or homogeneous solution. The *unit impulse responses* $h_{ij}(t)$ are determined either by inverting the foregoing p-functions without the $1/p$ factors, or by observing that since their transforms differ only by the derivative operator p from the $A_{ij}(t)$,

$$h_{ij}(t) = \frac{dA_{ij}(t)}{dt} \qquad (6\text{-}62)$$

The resulting forms are

$$h_{hF}(t) = -\frac{r_\alpha^2 \omega_h^2 \omega_2}{(r_\alpha^2 - x_\alpha^2)(\omega_2^2 - \omega_1^2)} \left\{ \left[1 + \frac{d}{b} \frac{x_\alpha}{r_\alpha^2} - \frac{\omega_\alpha^2}{\omega_2^2} \right] \sin \omega_2 t \right.$$

$$\left. - \frac{\omega_1}{\omega_2} \left[1 + \frac{d}{b} \frac{x_\alpha}{r_\alpha^2} - \frac{\omega_\alpha^2}{\omega_1^2} \right] \sin \omega_1 t \right\} \qquad (6\text{-}63)$$

$$h_{\alpha F}(t) = \frac{d}{b} \frac{\omega_h^2 \omega_2}{(r_\alpha^2 - x_\alpha^2)(\omega_2^2 - \omega_1^2)} \left\{ \left[1 + x_\alpha \frac{b}{d} - \frac{\omega_h^2}{\omega_2^2} \right] \sin \omega_2 t \right.$$

$$\left. - \frac{\omega_1}{\omega_2} \left[1 + x_\alpha \frac{b}{d} - \frac{\omega_h^2}{\omega_1^2} \right] \sin \omega_1 t \right\} \qquad (6\text{-}64)$$

The indicial and unit impulse responses of the bending degree of freedom are plotted in Fig. 6–7 for the same set of parameters as Fig. 6–6. In practical cases damping would, of course, bring about the gradual decay of these motions.

In view of an important transform known as the *convolution* or *faltung theorem*,

$$\bar{x}(p)\bar{y}(p) = \mathscr{L}\left\{ \int_0^t x(\tau) y(t - \tau) \, d\tau \right\} \qquad (6\text{-}65)$$

we can find three very useful expressions for the response, starting from rest at $t = 0$, to an arbitrary time-dependent input. Using the torsional degree of freedom as an example, these so-called *Duhamel integrals* read

$$\alpha(t) = \frac{1}{K_h b} \int_0^t F(\tau) h_{\alpha F}(t - \tau) \, d\tau$$

$$= \frac{1}{K_h b} \int_0^t F(\tau) A_{\alpha F}{}'(t - \tau) \, d\tau$$

$$= \frac{1}{K_h b} \left[F(0) A_{\alpha F}(t) + \int_0^t F'(\tau) A_{\alpha F}(t - \tau) \, d\tau \right] \qquad (6\text{-}66)$$

The elementary physical interpretation of these forms is discussed in Appendix C of Ref. 1–2, among other places. In situations where the

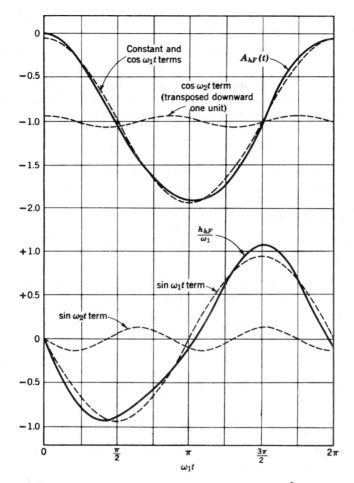

Fig. 6–7. Indicial admittance $A_{hF}(t)$ and unit impulse response $\dfrac{1}{\omega_1} h_{hF}(t)$ for bending degree of freedom of the typical section in vacuum (Fig. 6–5), showing superposition of the ω_1 and ω_2 contributions. Dimensionless parameters are $d/b = 0$, $x_\alpha = 0.2$, $r_\alpha = 0.5$, $\omega_h/\omega_\alpha = 0.5$.

driving force is not known in analytical form, numerical, graphical, or analogue evaluations of Duhamel's integrals offer convenient, systematic techniques for obtaining responses.

Finally, we point out a connection between the mechanical and indicial admittances, which is especially useful when the sinusoidal behavior of the system is readily available. As shown in books on Fourier integrals, the step function can be written

$$1(t) = \frac{1}{2\pi} \int_{-\infty}^{\infty} \frac{e^{i\omega t}}{i\omega} d\omega \qquad (6\text{–}67)$$

where convergence is ensured by requiring the integration path to make a small semicircular loop below the origin. Therefore, analogously with Eqs. 6–50 and 6–51, the step function response must be

$$A_{ij}(t) = \frac{1}{2\pi} \int_{-\infty}^{\infty} \frac{H_{ij}(i\omega)e^{i\omega t}}{i\omega} d\omega \qquad (6\text{–}68)$$

Again in Eq. 6–68 we must be careful of the pole at $\omega = 0$, but this inconvenience can be avoided by adding and subtracting in the integrand:

$$A_{ij}(t) = \frac{H_{ij}(0)}{2\pi} \int_{-\infty}^{\infty} \frac{e^{i\omega t}}{i\omega} d\omega + \frac{1}{2\pi} \int_{-\infty}^{\infty} \frac{[H_{ij}(i\omega) - H_{ij}(0)]e^{i\omega t}}{i\omega} d\omega$$

$$= H_{ij}(0)1(t) + \frac{1}{2\pi} \int_{-\infty}^{\infty} \frac{[H_{ij}(i\omega) - H_{ij}(0)]e^{i\omega t}}{i\omega} d\omega \qquad (6\text{–}69a)$$

By means of the substitution $H(i\omega) = R(\omega) + iI(\omega)$ and full exploitation of conjugate properties in Eq. 6–69a, an extensive simplification is possible, resulting in

$$A_{ij}(t) = \tfrac{1}{2}R_{ij}(0) + \frac{1}{\pi} \int_{0}^{\infty} \left[\frac{R_{ij}(\omega)}{\omega} \sin \omega t + \frac{I_{ij}(\omega)}{\omega} \cos \omega t \right] d\omega, \qquad (t \geq 0+)$$

$$(6\text{–}69b)$$

Equation 6–62 leads us to the even simpler result

$$h_{ij}(t) = \frac{1}{\pi} \int_{0}^{\infty} [R_{ij}(\omega) \cos \omega t - I_{ij}(\omega) \sin \omega t] \, d\omega, \qquad (t \geq 0+) \quad (6\text{–}70)$$

The last two formulas are helpful for both analytical evaluations and numerical approximations.

(c) Random forcing

We close this section with a short introduction to the effects of *random inputs*. $F(t)$ is said to be random when it varies irregularly and unpredictably, so that it can be defined only by certain probabilities or average values and manipulated only by the tools of statistical theory. A full exposition of the subject of probability (cf. Ref. 6–10, for instance) is beyond our scope and unnecessary for our purposes. As a matter of fact, certain semi-intuitive concepts prove wholly adequate for aeroelastic applications, along with a few formulas relating them, whose rigorous proofs, under the actual conditions where they are used, exceed the authors' mathematical comprehension.

$F(t)$ can be fully defined by giving its *probability density*, the fraction of observations which result in an arbitrarily chosen sample exceeding any given numerical value; this is often assumed to have a simple exponential form known as the *Gaussian* or *normal distribution* (Refs. 6–10, 6–11). There are, however, several other properties that are easier to measure or estimate and which are usually adequate for structural design purposes, etc. One of these is the *n*th moment or *n*th mean value

$$\overline{F^n} = \lim_{T \to \infty} \frac{1}{2T} \int_{-T}^{T} F^n(t)\, dt \qquad (6\text{--}71)$$

$2T$ is an interval of observation, which in practice should be chosen long enough for the averaged F^n to settle down to a clearly defined value. A *stationary random process* is one having $\overline{F^n}$ independent of where the time origin is chosen; all random variables treated in this book are assumed stationary. Where possible, the origin of F is selected so as to make the first moment or arithmetic mean $\overline{F} = 0$. The second moment or mean-square value

$$\overline{F^2} = \lim_{T \to \infty} \frac{1}{2T} \int_{-T}^{T} F^2(t)\, dt \qquad (6\text{--}72)$$

provides a convenient measure of intensity and plays a central role in applications.

The degree to which the magnitude of F at any instant depends on its preceding history is estimated by the autocorrelation function φ, whose double subscript is supposed to identify the integrand,

$$\varphi_{FF}(\tau) = \lim_{T \to \infty} \frac{1}{2T} \int_{-T}^{T} F(t)F(t + \tau)\, dt \qquad (6\text{--}73)$$

$\varphi_{FF}(\tau)$ is an even function, because the stationary random statistical properties should be independent of direction in time; it falls to zero as $\tau \to \infty$ because widely-separated observations are unrelated; and $\varphi_{FF}(0)$ is obviously equal to the mean-square value.

The Fourier transform of twice the autocorrelation function proves to have valuable mathematical properties and a reasonably clear physical interpretation:

$$\Phi_{FF}(\omega) = \frac{1}{\pi} \int_{-\infty}^{\infty} \varphi_{FF}(\tau)e^{-i\omega\tau}\, d\tau = \frac{2}{\pi} \int_{0}^{\infty} \varphi_{FF}(\tau) \cos \omega\tau\, d\tau \quad (6\text{--}74)$$

As shown in Appendix C of Ref. 1–2, among other places, we are led to $\Phi_{FF}(\omega)$ by discovering that $F(t)$ itself has a divergent transform, whereas the transform of the product of F by itself (in the sense of $\varphi_{FF}(\tau)$) is well-defined. $\Phi_{FF}(\omega)$ is proportional to the square of the magnitude of the spectral component of F having frequency ω and is therefore called *power spectral density*. This interpretation is clarified by inverting Eq. 6–74,

$$\varphi_{FF}(\tau) = \frac{1}{2} \int_{-\infty}^{\infty} e^{i\omega\tau} \Phi_{FF}(\omega) \, d\omega \qquad (6\text{–}75)$$

and setting $\tau = 0$ to obtain

$$\overline{F^2} = \frac{1}{2} \int_{-\infty}^{\infty} \Phi_{FF}(\omega) \, d\omega = \int_{0}^{\infty} \Phi_{FF}(\omega) \, d\omega \qquad (6\text{–}76)$$

Thus the total power or intensity is, in a sense, the integrated sum of all its spectral components. The *wave analyzer* is a standard item of electronic equipment designed to make a direct determination of $\Phi_{FF}(\omega)$ for any suitably random electrical signal.

To illustrate the utility of the power spectral concept, let us consider a stationary random force applied to the typical section. The outputs h and α will also be random and will possess autocorrelation functions $\varphi_{hh}, \varphi_{\alpha\alpha}$ and power spectral densities $\Phi_{hh}, \Phi_{\alpha\alpha}$. Concentrating on the twist α, it can be calculated from the known F by means of the first member of Eq. 6–66

$$\alpha(t) = \frac{1}{K_h b} \int_{-\infty}^{t} F(\tau) h_{\alpha F}(t - \tau) \, d\tau \qquad (6\text{–}77)$$

where the lower limit is now $-\infty$ because the process must have been going on a long time. The unit impulse response must vanish when its argument is negative, so the upper limit in Eq. 6–77 may be replaced by ∞. Making the change of variable $t' = t - \tau$,

$$\alpha(t) = \frac{1}{K_h b} \int_{-\infty}^{\infty} F(t - t') h_{\alpha F}(t') \, dt' \qquad (6\text{–}78)$$

Now let us compute the autocorrelation function of the output.

$$\varphi_{\alpha\alpha}(\tau) = \lim_{T \to \infty} \frac{1}{2T} \int_{-T}^{T} \alpha(t)\alpha(t + \tau) \, dt$$

$$= \frac{1}{(K_h b)^2} \lim_{T \to \infty} \frac{1}{2T} \int_{-T}^{T} \int_{-\infty}^{\infty} \int_{-\infty}^{\infty} F(t - t')F(t + \tau - t'')$$

$$\times h_{\alpha F}(t') h_{\alpha F}(t'') \, dt' \, dt'' \, dt$$

$$= \frac{1}{(K_h b)^2} \int_{-\infty}^{\infty} \int_{-\infty}^{\infty} h_{\alpha F}(t') h_{\alpha F}(t'') \left\{ \lim_{T \to \infty} \frac{1}{2T} \int_{-T}^{T} F(t - t') \right.$$

$$\left. \times F(t + \tau - t'') \, dt \right\} dt' \, dt''$$

$$= \frac{1}{(K_h b)^2} \int_{-\infty}^{\infty} \int_{-\infty}^{\infty} h_{\alpha F}(t') h_{\alpha F}(t'') \varphi_{FF}(\tau + t' - t'') \, dt' \, dt'' \qquad (6\text{–}79)$$

The substitution in the last line is justified by making $(t - t')$ into a new integration variable and noting that the value of φ_{FF} is unaffected by a shift of the range of integration wherein it is calculated. Equation 6–79 has some direct interest, but a much simpler relation is uncovered when the autocorrelation functions are replaced by their Fourier transforms (cf. Eq. 6–75).

$$\frac{1}{2} \int_{-\infty}^{\infty} \Phi_{\alpha\alpha}(\omega) e^{i\omega\tau} \, d\omega$$

$$= \frac{1}{2} \int_{-\infty}^{\infty} \int_{-\infty}^{\infty} \int_{-\infty}^{\infty} \frac{\Phi_{FF}(\omega)}{(K_h b)^2} e^{i\omega\tau} h_{\alpha F}(t') e^{i\omega t'} h_{\alpha F}(t'') e^{-i\omega t''} \, dt'' \, dt' \, d\omega$$

$$= \frac{1}{2} \int_{-\infty}^{\infty} \frac{\Phi_{FF}(\omega)}{(K_h b)^2} e^{i\omega\tau} \left\{ \left[\int_{-\infty}^{\infty} h_{\alpha F}(t') e^{i\omega t'} \, dt' \right] \left[\int_{-\infty}^{\infty} h_{\alpha F}(t'') e^{-i\omega t''} \, dt'' \right] \right\} \, d\omega$$

$$(6\text{–}80)$$

Each of the two integrals in brackets expresses the torsional response to a sinusoidal force of unit amplitude which has acted for a very long time, during which period the transient would die out in the real physical system. Hence they are related to the mechanical admittance,

$$\int_{-\infty}^{\infty} h_{\alpha F}(t'') e^{-i\omega t''} \, dt'' = \int_{0}^{\infty} e^{-i\omega t''} h_{\alpha F}(t'') \, dt''$$

$$= e^{-i\omega t} \int_{-\infty}^{t} e^{i\omega\tau} h_{\alpha F}(t - \tau) \, d\tau = H_{\alpha F}(i\omega) \qquad (6\text{–}81a)$$

Equation 6–81a can also be derived by constructing the inverse of Eq. 6–68 and integrating by parts with respect to time. Similarly,

$$\int_{-\infty}^{\infty} h_{\alpha F}(t') e^{i\omega t'} \, dt' = H_{\alpha F}(-i\omega) \qquad (6\text{–}81b)$$

We substitute Eqs. 6–81 into Eq. 6–80,

$$\int_{-\infty}^{\infty} \Phi_{\alpha\alpha}(\omega) e^{i\omega\tau} \, d\omega = \int_{-\infty}^{\infty} |H_{\alpha F}(i\omega)|^2 \frac{\Phi_{FF}(\omega)}{(K_h b)^2} e^{i\omega\tau} \, d\omega \qquad (6\text{–}82)$$

and observe that the equality of the two sides for all values of τ is equivalent to

$$\Phi_{\alpha\alpha}(\omega) = |H_{\alpha F}(i\omega)|^2 \frac{\Phi_{FF}(\omega)}{(K_h b)^2} = \frac{\Phi_{FF}(\omega)/(K_h b)^2}{|Z_{\alpha F}(i\omega)|^2} \qquad (6\text{–}83)$$

Stated verbally, Eq. 6–83 asserts that the power spectral densities differ only by a factor which is the square of the magnitude of the mechanical

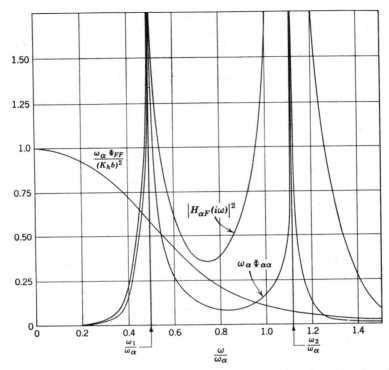

Fig. 6–8. Dimensionless input and output power spectral densities for torsional motion produced by applying a random force to the typical section, as related by Eq. 6–83. System parameters are $d/b = 0$, $x_\alpha = 0.2$, $r_\alpha = 0.5$, $\omega_h/\omega_\alpha = 0.5$.

admittance. A similar relation can be written for any input-output pair. In Fig. 6–8, we illustrate Eq. 6–83 by presenting a hypothetical $\Phi_{FF}(\omega)/(K_h b)^2$ and showing how the spectrum is modified by passage through the two-degree-of-freedom system. Incidentally, Eq. 6–83 also applies when a simple harmonic rather than random input is involved. Then the spectral densities must be replaced by the squares of the amplitudes of α and F, as can be seen by multiplying Eq. 6–41 into its complex conjugate equation.

When studying the linear system under random forcing, it is also possible to determine *cross-correlation functions* such as

$$\varphi_{\alpha F}(\tau) = \lim_{T \to \infty} \frac{1}{2T} \int_{-T}^{T} \alpha(t + \tau) F(t) \, dt \qquad (6\text{–}84)$$

The Fourier transform of $\varphi_{\alpha F}$ would be the *cross-spectral density* (a complex quantity)

$$\Phi_{\alpha F}(i\omega) = \frac{1}{\pi} \int_{-\infty}^{\infty} \varphi_{\alpha F}(\tau) e^{-i\omega\tau} \, d\tau \tag{6–85}$$

Several useful relations containing these functions can be derived, such as

$$\frac{\Phi_{\alpha F}(i\omega)}{K_h b} = H_{\alpha F}(i\omega) \frac{\Phi_{FF}(\omega)}{(K_h b)^2} \tag{6–86a}$$

and

$$\Phi_{\alpha\alpha}(\omega) = H_{\alpha F}(-i\omega) \frac{\Phi_{\alpha F}(i\omega)}{K_h b} \tag{6–86b}$$

An excellent treatment of engineering applications of all the foregoing material and much more on random processes will be found in the book by Laning and Battin (Ref. 6–11).

6–4 FORCED MOTION IN THE PRESENCE OF EXTERNAL LOADS DEPENDING ON THE MOTION

Now we turn on the air in the wind tunnel which contains the typical section and expose the system to another assortment of time-varying inputs. The wing may be thought of as moving about a certain constant mean position, determined as in Sec. 6–2 by the equilibrium of static loads due to initial incidence, camber, and weight. Following standard practice in dynamic studies, the displacements $h(t)$ and $\alpha(t)$ dealt with here are superimposed upon their static values; time-independent terms have already been subtracted out of the equations of motion treated below.

Three types of external load may act on the typical section: (1) the force $F(t)$ pictured in Fig. 6–5; (2) aerodynamic lifts and moments due to gust-like disturbances in the airstream, identified by the superscript D; and (3) aerodynamic effects of the motions $h(t)$ and $\alpha(t)$ themselves, identified by the superscript M. In keeping with the general pattern of linearity, the last are presumed to be linearly related to the displacements and/or their time derivatives of any order. With the foregoing conventions we may write the equations of motion

$$m[\ddot{h} + \omega_h{}^2 h] + S_\alpha \ddot{\alpha} = -F(t) - L^D(t) - L^M(t) \tag{6–87a}$$

$$S_\alpha \ddot{h} + I_\alpha[\ddot{\alpha} + \omega_\alpha{}^2 \alpha] = Fd + M_y{}^D(t) + M_y{}^M(t) \tag{6–87b}$$

For the purpose of organizing a series of illustrative problems, we adopt the same order as in Sec. 6–3. The various concepts and techniques introduced there—in connection then with situations of negligible practical importance—will prove their worth as we analyze some more realistic aeroelastic phenomena.

(a) Sinusoidal forcing

Two examples of steady-state, simple harmonic response will be investigated. This type of motion has special significance because, as is almost too well-known by aeroelasticians, the required aerodynamic theory has been developed and tabulated most thoroughly.

First, we re-examine the oscillating concentrated force of amplitude F, described by Eq. 6–36. Once again Eqs. 6–37 and 6–38 define the outputs, but now h and $\bar{\alpha}$ are unquestionably complex constants. L^D and $M_y{}^D$ are zero. The amplitudes of L^M and $M_y{}^M$ receive linear contributions from h and $\bar{\alpha}$, as set forth in Chap. 4, but the coefficients in these relations are complex, transcendental functions of the Mach number M and reduced frequency $k = \omega b/U$. The following, conventional symbolism is used in this chapter, consolidating Eqs. 4–123, 4–129, 4–124, and 4–130:

$$
\overline{L^M} = \begin{cases} -\dfrac{\pi}{2}\rho_\infty b^2 S \omega^2 \left\{ L_h \dfrac{h}{b} + [L_\alpha - (\tfrac{1}{2} + a)L_h]\bar{\alpha} \right\}, & 0 \le M < 1 \\[3mm] 2\rho_\infty b^2 S \omega^2 \left\{ [L_1 + iL_2] \dfrac{h}{b} + [L_3 + iL_4]\bar{\alpha} \right\}, & 1 \le M \end{cases} \tag{6–88}
$$

$$
\overline{M_y{}^M} = \begin{cases} \dfrac{\pi}{2}\rho_\infty b^3 S \omega^2 \Big\{ [M_h - (\tfrac{1}{2} + a)L_h] \dfrac{h}{b} \\[2mm] \quad + [M_\alpha - (\tfrac{1}{2} + a)(L_\alpha + M_h) + (\tfrac{1}{2} + a)^2 L_h]\bar{\alpha} \Big\}, & 0 \le M < 1 \\[3mm] -2\rho_\infty b^3 S \omega^2 \Big\{ [M_1 + iM_2] \dfrac{h}{b} + [M_3 + iM_4]\bar{\alpha} \Big\}, & 1 \le M \end{cases}
$$
$$
\tag{6–89}
$$

Here a is the distance in semichords by which the E.A. lies *aft* of the midchord line. L_h, L_α, M_h, and M_α are complex numbers independent of a. The supersonic coefficients $(L_1 + iL_2)$, $(L_3 + iL_4)$, etc., have been separated into real and imaginary parts, but all of them except L_1 and L_2 do vary with the E.A. location (in a simple linear or quadratic fashion). This general choice of notation is neither hallowed nor especially desirable, but it has the advantage of broad usage throughout the United States and will continue to have such for some time to come. Explicit mathematical expressions for the aerodynamic coefficients will be reproduced only occasionally in this and subsequent chapters. The two-dimensional forms are presented in Chap. 4 and are fully tabulated.

Equations 6–88 and 6–89 are substituted, along with the other simple harmonic quantities, into Eqs. 6–87, and the factor $e^{i\omega t}$ is canceled. Rearranging the homogeneous portions to the left-hand sides and resorting to matrix algebra, we obtain

$$
\begin{bmatrix}
\{mb[\omega_h^2 - \omega^2] + 2\rho_\infty b^2 S\omega^2[L_1 + iL_2]\} & \{-S_\alpha\omega^2 + 2\rho_\infty b^2 S\omega^2[L_3 + iL_4]\} \\
\{-S_\alpha b\omega^2 + 2\rho_\infty b^3 S\omega^2[M_1 + iM_2]\} & \{I_\alpha[\omega_\alpha^2 - \omega^2] + 2\rho_\infty b^3 S\omega^2[M_3 + iM_4]\}
\end{bmatrix}
\begin{Bmatrix} \dfrac{h}{b} \\ \bar{\alpha} \end{Bmatrix}
= \begin{Bmatrix} -F \\ Fd \end{Bmatrix}
$$

$$(6\text{--}90)$$

Supersonic notation is used in Eqs. 6–90, but the reader will have no difficulty in writing out the subsonic counterpart of the square matrix. To identify the individual aeroelastic operators on the left side, one need only separate the matrix into the sum of structural, inertial, and aerodynamic parts. For example, the matric structural operator is

$$[\mathscr{S}] = \begin{bmatrix} m\omega_h{}^2 & 0 \\ 0 & I_\alpha\omega_\alpha{}^2 \end{bmatrix} = \begin{bmatrix} K_h & 0 \\ 0 & K_\alpha \end{bmatrix} \tag{6–91}$$

Equation 6–91 holds for any motion of the typical section, but the other two operators are uniquely sinusoidal.

In terms of dimensionless quantities which parallel those appearing in Eqs. 6–40 and 6–41, Eqs. 6–90 can be solved for the mechanical admittances modified by the presence of the airstream:

$$\frac{\bar{h}/b}{\bar{F}/K_h b} \equiv H_{hF}(i\omega, M)$$

$$= \frac{-\left(\dfrac{m}{2\rho_\infty bS}\right)\left(\dfrac{\omega_h}{i\omega}\right)^2 \left\{ \left(\dfrac{m}{2\rho_\infty bS}\right) r_\alpha{}^2 \left[1 + \left(\dfrac{\omega_\alpha}{i\omega}\right)^2\right] - [M_3 + iM_4] \right\}}{\Delta}$$

$$+ \frac{\dfrac{d}{b}\left(\dfrac{m}{2\rho_\infty bS}\right)\left(\dfrac{\omega_h}{i\omega}\right)^2 \left\{ -\left(\dfrac{m}{2\rho_\infty bS}\right) x_\alpha + [L_3 + iL_4] \right\}}{\Delta} \tag{6–92}$$

$$\frac{\bar{\alpha}}{\bar{F}/K_h b} \equiv H_{\alpha F}(i\omega, M)$$

$$= \frac{-\left(\dfrac{m}{2\rho_\infty bS}\right)\left(\dfrac{\omega_h}{i\omega}\right)^2 \dfrac{d}{b} \left\{ -\left(\dfrac{m}{2\rho_\infty bS}\right)\left[1 + \left(\dfrac{\omega_h}{i\omega}\right)^2\right] + [L_1 + iL_2] \right\}}{\Delta}$$

$$+ \frac{\left(\dfrac{m}{2\rho_\infty bS}\right)\left(\dfrac{\omega_h}{i\omega}\right)^2 \left\{ -\left(\dfrac{m}{2\rho_\infty bS}\right) x_\alpha + [M_1 + iM_2] \right\}}{\Delta} \tag{6–93}$$

where

$$\Delta = \left\{ \left(\dfrac{m}{2\rho_\infty bS}\right)\left[1 + \left(\dfrac{\omega_h}{i\omega}\right)^2\right] - [L_1 + iL_2] \right\}$$

$$\times \left\{ \left(\dfrac{m}{2\rho_\infty bS}\right) r_\alpha{}^2 \left[1 + \left(\dfrac{\omega_\alpha}{i\omega}\right)^2\right] - [M_3 + iM_4] \right\}$$

$$- \left\{ -\left(\dfrac{m}{2\rho_\infty bS}\right) x_\alpha + [L_3 + iL_4] \right\}\left\{ -\left(\dfrac{m}{2\rho_\infty bS}\right) x_\alpha + [M_1 + iM_2] \right\}$$

$$\tag{6–93a}$$

The reduction to Eqs. 6–40 and 6–41 is readily carried out after setting all aerodynamic terms equal to zero. A new parameter appears explicitly in Eqs. 6–92 and 6–93: the mass ratio $m/2\rho_\infty bS$, which is a measure of the relative inertias of the wing and a representative volume of air surrounding it; the symbol μ is often used for this ratio, or for the same combination with 2 replaced by $\pi/2$. Altogether, *four* additional dimensionless quantities enter the picture when the wind is turned on. These are $m/2\rho_\infty bS$ and three geometrical aerodynamic parameters M, k, and the E.A. location, a. In a given experimental situation (e.g., the measurement of airloads by driving a two-dimensional wing in a tunnel), the ambient state of gas in the test section furnishes ρ_∞ and the speed of sound $\sqrt{\gamma RT_\infty}$. Knowledge of airspeed U then yields M; the driving frequency ω determines k; and the inertial and geometrical properties of the model fix $m/2\rho_\infty bS$ and the remaining parameters.

In Fig. 6–9, the bending response is plotted versus ω/ω_α at subsonic, sonic, and supersonic Mach numbers for a representative system. Here it is necessary to show both the amplitude ratio and the phase angle φ_h by which the (downward) displacement leads the (upward) driving force, that is,

$$\frac{h/b}{F/K_h b} = \left| \frac{h/b}{F/K_h b} \right| e^{i\varphi_h} = |H_{hF}|\, e^{i\varphi_h} \qquad (6\text{--}94)$$

The static value of this angle is, of course, 180°. We observe two effects of the airflow. It provides damping, rounding off the infinite resonance peaks of the system in vacuum; this damping is very small in the case of the peak near ω_h, however, because the wing mass is so large compared to the effective inertia of the air $(m/2\rho_\infty bS = 100)$. The second effect is a shift of the resonance peak locations, particularly noticeable in ω_2, as functions of Mach number and dynamic pressure. In the present example the peaks move apart at supersonic speeds, but more commonly they move toward one another. The implications of such behavior with regard to stability is taken up in Sec. 6–6. For some purposes it is convenient to present the mechanical admittance in complex polar form. Figure 6–10 shows such a graph for the case $M = 2$.

It bears repeating that, in principle, the availability of H_{hF} and $H_{\alpha F}$ implies the power to calculate response to any time-dependent $F(t)$ at any airspeed. For a periodic but not sinusoidal force, the Fourier series Eqs. 6–47 and 6–48 are valid. Good convergence should be obtained whenever $F(t)$ is not too jagged and the system is sufficiently stable. This technique is used in practice even for aperiodic inputs by applying the same force repeatedly, but at long enough intervals to allow the decay of transients. Alternatively, the aperiodic input may be handled by a

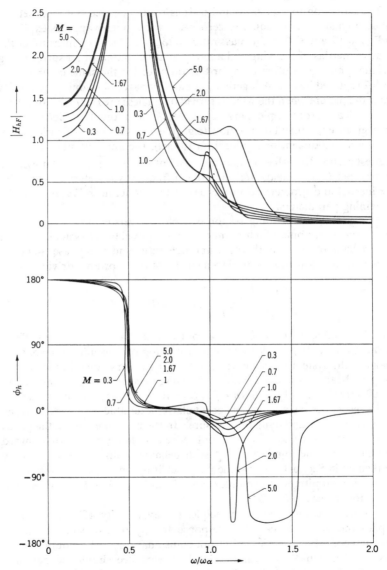

Fig. 6–9. Mechanical admittance $H_{hF}(i\omega, M)$ relating translational displacement to applied force on the typical section in an airstream at various Mach numbers. The following set of parameters was used in Eq. 6–92: $d/b = 0$, $x_\alpha = 0$, $r_\alpha = 0.5$, $\omega_h/\omega_\alpha = 0.5$, $m/2\rho_\infty bS = 100$, $a = -0.2$ (E.A. at 40% chord), $\omega_\alpha b/a_\infty = 1$ (needed to fix k). Note that the lack of inertia coupling implies $\omega_1 = \omega_h$ and $\omega_2 = \omega_\alpha$.

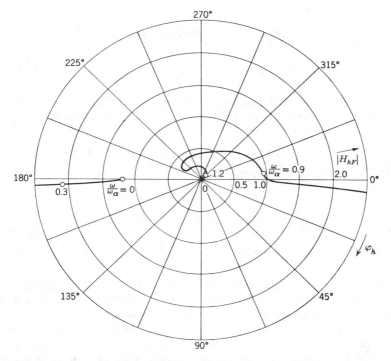

Fig. 6–10. Complex polar plot of H_{hF} from Fig. 6–9 for the case $M = 2$.

combination of formulas like 6–69b and 6–66. As will be seen soon, however, it is sometimes easier to determine indicial admittances directly rather than by numerical integration of sinusoidal data.

As a second example, suppose that the concentrated force is removed but that somehow we are able to generate in the airstream a sinusoidal gust with a vertical velocity distribution given by the real part of

$$w_G(x, t) = \bar{w}_G e^{i\omega[t - x/U]} \tag{6–95}$$

Since x is a downstream coordinate fixed relative to the test-section walls, this gust is embedded in the moving gas and, as seen by an observer on the wing, varies simple harmonically with circular frequency ω. Chapter 4 discusses the theory of two-dimensional airloads due to the sinusoidal gust and shows that the dimensionless lift and moment are functions of Mach number and reduced frequency based on ω. Hence we can write the disturbance loading

$$L^D(t) = qS \frac{\bar{w}_G}{U} e^{i\omega t} \bar{C}_{LG}(k, M) \qquad (6\text{-}96a)$$

$$M_y{}^D(t) = qS(2b) \frac{\bar{w}_G}{U} e^{i\omega t} \bar{C}_{MG}(k, M) \qquad (6\text{-}97a)$$

By reference to Eqs. 4–174, 4–175, 4–178, and 4–179, these coefficients can be cast in the following forms appropriate, for instance, to low subsonic or supersonic speeds:

$$\bar{C}_{LG}(k, M) = \begin{cases} 2\pi\{C(k)[J_0(k) - iJ_1(k)] + iJ_1(k)\}, & M \cong 0 \\ \dfrac{4e^{ik}}{\sqrt{M^2 - 1}} f_0(M, k), & M > 1 \end{cases} \qquad (6\text{-}96b)$$

$$\bar{C}_{MG}(k, M) = \begin{cases} \left[\dfrac{a}{2} + \dfrac{1}{4}\right] \bar{C}_{LG}(k, M = 0), & M \cong 0 \\ \dfrac{e^{ik}}{\sqrt{M^2 - 1}} [(2a + 1)f_0(M, k) - f_1(M, k)], & M > 1 \end{cases} \qquad (6\text{-}97b)$$

Here the origin of x is assumed to be at midchord; $C(k)$ and f_n are functions discussed in Chap. 4. It is of interest that the entire incompressible-flow lift acts at the $\frac{1}{4}$-chord A.C. ($a = -\frac{1}{2}$), regardless of the frequency.

After the disappearance of transients, it is clear that the foregoing gust will produce a steady-state response of frequency ω. Once again Eqs. 6–37 and 6–38 apply, and the airloads caused by the motion are given by Eqs. 6–88 and 6–89. The appropriate substitutions in the equations of motion, including a choice of supersonic coefficients, lead to Eqs. 6-98, as shown on page 225.

The solution of Eqs. 6–98 will not be written out, but it is obvious that there are two mechanical admittances for the gust input,

$$H_{hG}(i\omega, M) = \frac{\bar{h}/b}{\bar{w}_G/U} \qquad (6\text{-}99)$$

and

$$H_{\alpha G}(i\omega, M) = \frac{\bar{\alpha}}{\bar{w}_G/U} \qquad (6\text{-}100)$$

Except for the absence of the moment arm d/b, each of these depends on the same set of parameters as the admittances H_{hF} and $H_{\alpha F}$ in Eqs. 6–92 and 6–93. Plots like Figs. 6–9 and 6–10 could be constructed and would have the same general forms, since the only essential distinction between the concentrated force and the gust loading is that the latter is distributed along the wing chord. The techniques outlined in Secs. 6–3(a)

$$\begin{bmatrix} \{mb[\omega_h{}^2 - \omega^2] + 2\rho_\infty b^2 S\omega^2[L_1 + iL_2]\} & \{-S_\alpha\omega^2 + 2\rho_\infty b^2 S\omega[L_3 + iL_4]\} \\[2mm] \{-S_\alpha b\omega^2 + 2\rho_\infty b^3 S\omega^2[M_1 + iM_2]\} & \{I_\alpha[\omega_\alpha{}^2 - \omega^2] + 2\rho_\infty b^3 S\omega^2[M_3 + iM_4]\} \end{bmatrix} \begin{Bmatrix} \dfrac{h}{b} \\[2mm] \bar{\alpha} \end{Bmatrix}$$

$$= qS\,\frac{\bar{w}_G}{U}\begin{Bmatrix} -\bar{C}_{LG}(k, M) \\ 2b\bar{C}_{MG}(k, M) \end{Bmatrix} \qquad (6\text{-}98)$$

and (b) also provide a basis for calculating response to any sort of gust structure embedded in the airstream, once the mechanical admittances are known.

(b) Transient forcing

In view of the complexity of the aerodynamic lift and moment expressions (cf. Chap. 4 or Sec. 6–5 of Ref. 1–2), the problem of arbitrary time-dependent forcing will not be worked through for the full range of Mach numbers. As an introductory example, we return to the concentrated force of Fig. 6–5, initiating the motion from rest at $t = 0$, but we restrict the airspeed, U, so that incompressible-flow theory is valid. Under this limitation, the aerodynamic operators relating $L^M(t)$ and $M_y{}^M(t)$ to the displacements assume the relatively simple forms of Eqs. 4–170 and 4–171:

$$L^M(t) = \frac{\pi}{2} \rho_\infty bS[\ddot{h} + U\dot{\alpha} - ba\ddot{\alpha}] \qquad (6\text{--}101)$$

$$+ \pi\rho_\infty US \int_0^s \frac{d}{d\sigma} [h(\sigma) + U\alpha(\sigma) + b(\tfrac{1}{2} - a)\dot{\alpha}(\sigma)]\varphi(s - \sigma)\, d\sigma$$

$$M_y{}^M(t) = \frac{\pi}{2} \rho_\infty bS[ba\ddot{h} - Ub(\tfrac{1}{2} - a)\dot{\alpha} - b^2(\tfrac{1}{8} + a^2)\ddot{\alpha}]$$

$$+ \pi\rho_\infty UbS[\tfrac{1}{2} - a] \int_0^s \frac{d}{d\sigma} [h(\sigma) + U\alpha(\sigma) + b(\tfrac{1}{2} - a)\dot{\alpha}(\sigma)]$$

$$\times \; \varphi(s - \sigma)\, d\sigma \qquad (6\text{--}102)$$

Here $s = Ut/b$ is a dimensionless time measured in semichordlengths traveled after the start of motion; dots, however, indicate derivatives with respect to physical time. $\varphi(s)$ is the indicial admittance (of unit asymptote) which describes the circulatory lift build-up after a sudden change of incidence. This incompressible case is facilitated by the fact that only one such function appears, whereas generally four separate indicial admittances are needed for lift and moment due to vertical translation and pitching.

Turning to Eqs. 6–87, we set L^D and $M_y{}^D$ equal to zero, insert Eqs. 6–101 and 6–102, and take the Laplace transform *with respect to the dimensionless variable s*, viz.,

$$\bar{F}(\tilde{p}) = \int_0^\infty e^{-\tilde{p}s} F(s)\, ds \qquad (6\text{--}103)$$

using the tilde to distinguish from the dimensional p. The resulting matrix equation reads

$$
\begin{bmatrix}
\left\{\left[\dfrac{mU^2}{b} + \pi qS + 2\pi qS\bar{\varphi}(\tilde{p})\right]\tilde{p}^2 + mb\omega_h^2\right\} & \left\{\left[S_\alpha\dfrac{U^2}{b^2} - \pi qSa + 2\pi qS(\tfrac{1}{2}-a)\bar{\varphi}(\tilde{p})\right]\tilde{p}^2 \right. \\
& \left. + [\pi qS + 2\pi qS\bar{\varphi}(\tilde{p})]\tilde{p}\right\} \\[2ex]
\left\{\left[S_\alpha\dfrac{U^2}{b^2} - \pi qSba - 2\pi qSb(\tfrac{1}{2}+a)\bar{\varphi}(\tilde{p})\right]\tilde{p}^2\right\} & \left\{\left[I_\alpha\dfrac{U^2}{b^2} + \pi qSb(\tfrac{1}{8}+a^2)\right.\right. \\
& \left. - 2\pi qSb(\tfrac{1}{4}-a^2)\bar{\varphi}(\tilde{p})\right]\tilde{p}^2 \\
& \left. + [\pi qSb(\tfrac{1}{2}-a) - 2\pi qSb(\tfrac{1}{2}+a)\bar{\varphi}(\tilde{p})]\tilde{p} + I_\alpha\omega_\alpha^2\right\}
\end{bmatrix}
\begin{Bmatrix} \dfrac{h(\tilde{p})}{b} \\[2ex] \bar{\alpha}(\tilde{p}) \end{Bmatrix}
=
\begin{Bmatrix} -F(\tilde{p}) \\[1ex] F(\tilde{p})d \end{Bmatrix}
\tag{6-104}
$$

In terms of $k_\alpha = \omega_\alpha b/U$ and other previously defined parameters, Eqs. 6–104 have the following transformed solutions:

$$\frac{\bar{h}/b}{F(\bar{p})/K_h b} =$$

$$-k_\alpha^2 \left(\frac{\omega_h}{\omega_\alpha}\right)^2 \left(\frac{2m}{\pi\rho_\infty bS}\right) \left(\left\{\left[\left(\frac{2m}{\pi\rho_\infty bS}\right)r_\alpha^2 + (\tfrac{1}{8} + a^2) + (2a^2 - \tfrac{1}{2})\bar{\varphi}(\bar{p})\right]\bar{p}^2 \right.\right.$$

$$+ \left[(\tfrac{1}{2} - a) - (1 + 2a)\bar{\varphi}(\bar{p})\right]\bar{p} + \frac{2mr_\alpha^2 k_\alpha^2}{\pi\rho_\infty bS}\Bigg\}$$

$$\frac{\left. + \dfrac{d}{b}\left\{\left[\dfrac{2mx_\alpha}{\pi\rho_\infty bS} - a + (1 - 2a)\bar{\varphi}(\bar{p})\right]\bar{p}^2 + [1 + 2\bar{\varphi}(\bar{p})]\bar{p}\right\}\right)}{\Delta}$$

$$(6\text{–}105)$$

$$\frac{\bar{\alpha}(\bar{p})}{F(\bar{p})/K_h b} =$$

$$k_\alpha^2 \left(\frac{\omega_h}{\omega_\alpha}\right)^2 \left(\frac{2m}{\pi\rho_\infty bS}\right) \left(\frac{d}{b}\left\{\left[\frac{2m}{\pi\rho_\infty bS} + 1 + 2\bar{\varphi}(\bar{p})\right]\bar{p}^2 + k_\alpha^2\left(\frac{\omega_h}{\omega_\alpha}\right)^2\right.\right.$$

$$\frac{\left. \times \left(\frac{2m}{\pi\rho_\infty bS}\right)\right\} + \left\{\left[\left(\frac{2m}{\pi\rho_\infty bS}\right)x_\alpha - a - (1 + 2a)\bar{\varphi}(\bar{p})\right]\bar{p}^2\right\}\right)}{\Delta}$$

$$(6\text{–}106)$$

The common denominator determinant is

$$\Delta = \left\{\left[\frac{2m}{\pi\rho_\infty bS} + 1 + 2\bar{\varphi}(\bar{p})\right]\bar{p}^2 + k_\alpha^2\left(\frac{\omega_h}{\omega_\alpha}\right)^2 \frac{2m}{\pi\rho_\infty bS}\right\}$$

$$\times \left\{\left[\frac{2m}{\pi\rho_\infty bS}r_\alpha^2 + (\tfrac{1}{8} + a^2) + (2a^2 - \tfrac{1}{2})\bar{\varphi}(\bar{p})\right]\bar{p}^2\right.$$

$$+ \left[(\tfrac{1}{2} - a) - (1 + 2a)\bar{\varphi}(\bar{p})\right]\bar{p} + \frac{2m}{\pi\rho_\infty bS}r_\alpha^2 k_\alpha^2\Bigg\}$$

$$- \left\{\left[\frac{2m}{\pi\rho_\infty bS}x_\alpha - a - (1 + 2a)\bar{\varphi}(\bar{p})\right]\bar{p}^2\right\}$$

$$\times \left\{\left[\frac{2mx_\alpha}{\pi\rho_\infty bS} - a + (1 - 2a)\bar{\varphi}(\bar{p})\right]\bar{p}^2 + [1 + 2\bar{\varphi}(\bar{p})]\bar{p}\right\} \quad (6\text{–}107)$$

Obviously Eqs. 6–105 and 6–106 cannot be inverted except by numerical means, even when $F(s)$ has an elementary functional form. But the process is reducible to manageable proportions if one introduces an exponential approximation for $\varphi(s)$, of which

$$\varphi(s) \cong 1 - 0.165e^{-0.0455s} - 0.335e^{-0.3s} \qquad (4\text{–}172a)$$

is the best known. This has the transform

$$\bar{\varphi}(\tilde{p}) \cong \frac{0.5\tilde{p}^2 + 0.281\tilde{p} + 0.01365}{\tilde{p}(\tilde{p} + 0.3)(\tilde{p} + 0.0455)} \qquad (6\text{--}108)$$

and therefore leads to a ratio of polynomials in $\tilde{p}$ when inserted into Eqs. 6–105 through 6–107. Inversion is accomplished by the laborious procedure of factoring the tenth-degree denominator and applying the method of partial fractions (Chap. VI, Ref. 6–4) to construct the corresponding series of time functions. A step input $\bar{F}(\tilde{p}) \sim 1/\tilde{p}$ can thus be made to generate a pair of indicial admittances for h and α, whose usefulness in determining the response to more general transient forcing is implicit in Eq. 6–66.

We now examine a problem where the calculations are much less unwieldy. The torsional degree of freedom is suppressed by means of an infinitely rigid spring K_α, so that $\alpha = 0$. Regardless of the range of flight Mach number, we can then write (Eq. 4–156) for zero initial conditions

$$L^M(s) = 2\pi q S \int_0^s \frac{d^2}{d\sigma^2}\left[\frac{h(\sigma)}{b}\right] \varphi_c(s - \sigma)\, d\sigma \qquad (6\text{--}109)$$

The subscript c, meaning "compressible," on the dimensionless indicial function distinguishes it from its incompressible counterpart in Eqs. 6–101 and 6–102.

As an input, we create a transient gust structure $w_G(t - x/U)$ in the air stream, such that $w_G(t)$ [or $w_G(s)$] denotes the instantaneous upward velocity striking the leading edge at time t. According to Eq. 4–163, the gust-induced lift is

$$L_G(s) = 2\pi q S \int_0^s \frac{d}{d\sigma}\left[\frac{w_G(\sigma)}{U}\right] \psi_c(s - \sigma)\, d\sigma \qquad (6\text{--}110)$$

assuming the first encounter to occur at $s = t = 0$ and $w_G(0) = 0$. $\psi_c(s)$ is the indicial function for a step or sharp-edged gust.

Specializing Eq. 6–87a to the single degree of freedom, we get

$$m\left[\frac{U^2}{b^2}\frac{d^2h}{ds^2} + \omega_h{}^2 h\right] + 2\pi q S \int_0^s \frac{d^2}{d\sigma^2}\left[\frac{h(\sigma)}{b}\right] \varphi_c(s - \sigma)\, d\sigma$$

$$= -2\pi q S \int_0^s \frac{d}{d\sigma}\left[\frac{w_G(\sigma)}{U}\right] \psi_c(s - \sigma)\, d\sigma \qquad (6\text{--}111)$$

Clearly, the Laplace transform of the solution reads

$$\frac{\bar{h}(\tilde{p})}{b} = \frac{-\tilde{p}\bar{\psi}_c(\tilde{p})\bar{w}_G(\tilde{p})/U}{\left[\dfrac{m}{\pi\rho_\infty S b} + \bar{\varphi}_c(\tilde{p})\right]\tilde{p}^2 + \left(\dfrac{m}{\pi\rho_\infty S b}\right)k_h{}^2} \qquad (6\text{--}112)$$

where $k_h = \omega_h b / U$. Exponential approximations analogous to Eqs. 6–108 permit fairly straightforward inversions of Eq. 6–112.

An especially interesting case is that of high supersonic speed. Then the only aerodynamic force due to the motion is a lift directly opposite to the velocity h, which has the effect of putting a little "viscous" damping into the system. By Eqs. 4–183 and 4–184,

$$\varphi_c(s) = \frac{2}{\pi M} \tag{6–113}$$

$$\psi_c(s) = \begin{cases} \dfrac{s}{\pi M}, & 0 \le s \le 2 \\[2ex] \dfrac{2}{\pi M}, & 2 \le s \end{cases} \tag{6–114}$$

where the influence of airfoil thickness is neglected in Eq. 6–114 during the interval up to $s = 2$ while the gust front is enveloping the chord. Since $\varphi_c(s)$ is now just a step function, the denominator of Eq. 6–112 becomes a quadratic in $\tilde{p}$, leading to the familiar inversion

$$\mathscr{L}^{-1}\left\{ \frac{1}{\left(\dfrac{m}{\pi \rho_\infty S b}\right)\tilde{p}^2 + \dfrac{2}{\pi M}\tilde{p} + \left(\dfrac{m}{\pi \rho_\infty S b}\right)k_h^2} \right\}$$

$$= \frac{\exp\left[-\left(\dfrac{\rho_\infty S b}{m}\right)\dfrac{s}{M}\right] \sin\left(k_h s \sqrt{1 - \left(\dfrac{\rho_\infty S b}{m M k_h}\right)^2}\right)}{\sqrt{\left(\dfrac{m k_h}{\pi \rho_\infty S b}\right)^2 - \dfrac{1}{\pi^2 M^2}}} \tag{6–115}$$

Since $k_h s = \omega_h t$ and the radical in the argument of the sine is very nearly equal to unity, Eq. 6–115 shows that the typical section will execute damped vibrations at close to its natural bending frequency.

Inverting Eq. 6–112 by a double application of the convolution theorem, we obtain

$$\frac{h(s)}{b} = -\int_0^s \frac{\exp\left[-\left(\dfrac{\rho_\infty S b}{m}\right)\left(\dfrac{s-\sigma}{M}\right)\right] \sin\left(k_h(s-\sigma)\sqrt{1 - \left(\dfrac{\rho_\infty S b}{m M k_h}\right)^2}\right)}{\sqrt{\left(\dfrac{m k_h}{\pi \rho_\infty S b}\right)^2 - \dfrac{1}{\pi^2 M^2}}}$$

$$\times \left\{ \int_0^\sigma \frac{d}{d\sigma'}\left[\frac{w_G(\sigma')}{U}\right] \psi_c(\sigma - \sigma')\, d\sigma' \right\} d\sigma \tag{6–116}$$

In particular, for the response to a sharp-edged gust $w_G(s) = w_0 1(s)$,

$$\frac{h(s)}{b} \equiv A_{hG}(s)\frac{w_0}{U} = -\frac{w_0}{U}$$

$$\times \int_0^s \frac{\exp\left[-\left(\frac{\rho_\infty Sb}{m}\right)\left(\frac{s-\sigma}{M}\right)\right] \sin\left(k_h(s-\sigma)\sqrt{1 - \left(\frac{\rho_\infty Sb}{mMk_h}\right)^2}\right)\psi_c(\sigma)\,d\sigma}{\sqrt{\left(\frac{mk_h}{\pi\rho_\infty Sb}\right)^2 - \frac{1}{\pi^2 M^2}}} \tag{6-117}$$

Equation 6–117 can be integrated without difficulty when Eq. 6–114 is inserted. The very slowly decaying response to such a unit step at $M = 5$ is plotted in Fig. 6–11 for a typical section having $k_h = 0.1$ at a density ratio of 20.

(c) Random forcing

To illustrate the principles set forth in Sec. 6–3(c), we adopt the same physical system as in the last example (i.e., $\alpha = 0$) but expose it to a gust which is a stationary random function of time. In the buffeting problem, for instance, such a gust would represent the partially separated wake of a second lifting surface located upstream. Alternatively, it might be generated by a coarse screen or lattice hung across the test section. Especially in the latter case, the air motion would resemble so-called *homogeneous isotropic turbulence*, in which all three fluctuating velocity components, u, v, and w, are present and have similar statistical properties. If these components are generally small compared with U, as they must be to assure linearity, the u and v disturbances produce second-order airloads compared with $w = w_G$ and are neglected. We assume that the turbulence is embedded in the airstream and does not change appreciably in time during the interval needed to traverse one chordlength, so that $w_G = w_G\left(t - \frac{x}{U}\right)$. Finally, the turbulent structure is taken to be two-dimensional, with no variation of w_G in the spanwise direction; this is an unrealistic assumption, unless the wingspan is very small compared to the typical scale of turbulent eddies.

Under the prescribed conditions, Liepmann (Ref. 6–12) suggests that the following autocorrelation function and power spectral density represent rather well the turbulence which might be encountered:

$$\varphi_{GG}(\tau) = \frac{\overline{w_G^2}}{U^2} e^{\tau U/\lambda}[1 - (U/2\lambda)\tau], \quad (\tau \geq 0) \tag{6-118}$$

$$\Phi_{GG}(\omega) = \frac{\overline{w_G^2}}{U^2}\frac{\lambda}{\pi U}\frac{1 + 3(\omega\lambda/U)^2}{[1 + (\omega\lambda/U)^2]^2} \tag{6-119}$$

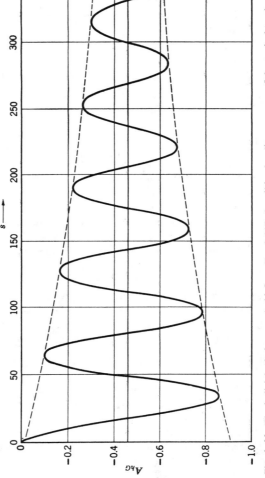

Fig. 6–11. The indicial admittance $A_{hG}(s)$ from Eq. 6–117, describing bending response of the typical section to a sharp-edged gust at high supersonic Mach number. System parameters are $M = 5$, $m/\pi\rho_\infty Sb = 20$, $k_h = \omega_h b/U = 0.1$.

Here $\overline{w_G^2}$ is the mean-square value; and λ is the *integral scale of turbulence,* which is a rough measure of the average size of eddies. In the earth's atmosphere, λ might be of the order of 500 to 1,000 feet, whereas wind-tunnel turbulence has a scale equal to or smaller than the vertical dimension of the test section. The reader can confirm that $\Phi_{GG}(\omega)$ satisfies Eq. 6–76.

Equation 6–83 furnishes the most convenient way of calculating statistical properties of the output, here chosen to be the bending displacement $h(t)$. The mechanical admittance relating w_G to h is found from the first member of Eq. 6–98, which reads, in the absence of twist,

$$\{mb(\omega_h^2 - \omega^2) + 2\rho_\infty b^2 S\omega^2[L_1 + iL_2]\}\frac{h}{b}$$
$$= -qS\bar{C}_{LG}(k, M)\frac{\bar{w}_G}{U} \quad (6\text{–}120)$$

Solving and reducing to dimensionless form, we obtain

$$\frac{h/b}{\bar{w}_G/U} = H_{hG}(i\omega, M)$$
$$= \frac{-\bar{C}_{LG}(k, M)}{\dfrac{4m}{\rho_\infty Sb}[k_h^2 - k^2] + 8k^2[L_1 + iL_2]} \quad (6\text{–}121)$$

By analogy with Eq. 6–83, the power spectral density of the bending motion is

$$\Phi_{hh}(\omega) = |H_{hG}(i\omega, M)|^2\Phi_{GG}(\omega)$$
$$= \left|\frac{\bar{C}_{LG}(k, M)}{\dfrac{4m}{\rho_\infty Sb}[k_h^2 - k^2] + 8k^2[L_1 + iL_2]}\right|^2 \frac{\overline{w_G^2}}{U^2}\frac{\lambda}{\pi U}\frac{1 + 3\left(\dfrac{k\lambda}{b}\right)^2}{\left[1 + \left(\dfrac{k\lambda}{b}\right)^2\right]^2} \quad (6\text{–}122)$$

The mean-square displacement is

$$\frac{\overline{h^2}}{b^2} = \int_0^\infty \Phi_{hh}(\omega)\, d\omega \quad (6\text{–}123)$$

and the autocorrelation function, if desired, can be calculated from a formula resembling Eq. 6–75.

The high-supersonic speed range again serves as a simple numerical

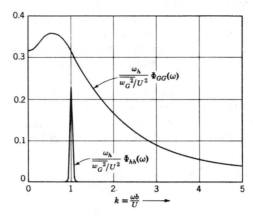

Fig. 6–12. Input and output power spectral densities for bending motion of the typical section exposed to isotropic turbulence. System parameters are $M = 5$, $k_h = 1$, $m/2\rho_\infty Sb = 40$, and $\lambda/b = 1$.

example. Piston theory, with thickness effects neglected, gives the following formulas for the aerodynamic coefficients:

$$\left.\begin{array}{c} L_1 = 0 \\ L_2 = 1/Mk \\ \bar{C}_{LG}(k, M) = \dfrac{8 \sin k}{Mk} \end{array}\right\} \qquad (6\text{-}124)$$

(The vanishing of $\bar{C}_{LG}$ for $k = n\pi$ occurs because, according to piston theory, the sinusoidal gust produces no net lift when the chord is an integral multiple of its wavelength.) After some algebra, we obtain

$$\Phi_{hh}(\omega) = \frac{\overline{w_G}^2}{U^2}\frac{\lambda}{\pi U}\left[\frac{1 + 3\left(\dfrac{k\lambda}{b}\right)^2}{\left(1 + \left(\dfrac{k\lambda}{b}\right)^2\right)^2}\right]$$

$$\times \frac{\sin^2 k}{M^2\left(\dfrac{m}{4\rho_\infty Sb}\right)^2 k^2\left[4(k^2 - k_h{}^2)^2 + \left(\dfrac{4\rho_\infty Sb}{mM}\right)^2 k^2\right]} \qquad (6\text{-}125)$$

Figure 6–12 illustrates the two power spectral densities for a particular set of system parameters. Some closely related examples, based on incompressible-flow aerodynamic coefficients, will be found worked out in Sec. 10–7 of Ref. 1–2.

6–5 EIGENVALUES; DYNAMIC AEROELASTIC INSTABILITY (FLUTTER) IN VARIOUS FLIGHT SPEED RANGES

The reader has no doubt already observed that the system consisting of a typical section in an airstream possesses dynamic eigenvalues. Mathematically they consist of values of the complex variable $\tilde{p}$ which cause the determinant of the square matrix in Eq. 6–104—or its compressible-flow counterpart—to vanish. Because $\bar{\varphi}(\tilde{p})$ and similar transformed indicial admittances are transcendental functions, there exists in theory an infinity of such roots. This should not surprise us, for the combined system, with air included, really has an infinite number of degrees of freedom. When computed in practice, the set of roots normally contains just two complex-conjugate pairs, the remainder being real and negative. These pairs are the ones of principal interest to the aeroelastician, since they represent coupled, damped (or divergent) oscillations of the wing acted on by airloads. In more complicated systems, we nearly always discover conjugate pairs up to the number of degrees of freedom of the elastic body; and the vibrations associated with them are called *aeroelastic modes*. In the case of the typical section, the two modes can be traced continuously as functions of airspeed down to $U = 0$, where they coalesce with the coupled natural modes whose frequencies are given by Eqs. 6–44.

The most logical way of studying the dynamic aeroelastic stability of a structure, and in particular of determining its margin of safety for any given flight condition, would seem to be to calculate a *roots locus* of $\tilde{p}$ as a function of airspeed and altitude. In engineering practice, however, this has not been the customary approach, for the very sound reason that more data are available on airloads resulting from simple harmonic motion. What is actually done is to establish a *stability boundary*, a curve or surface along which certain preassigned system parameters (either natural or artificially introduced for the purpose) vary in such a way that stability is neutral, that is, $\tilde{p} = ik$ for one of the roots. In the case of the typical section, for instance, maximum use would be made of the available aerodynamic operators by assuming sinusoidal motion in advance and finding roots of the determinant of the left-hand side in Eq. 6–90. Alternatively, one might apply the Nyquist criterion of stability (Ref. 6–5) to a series of response curves like the one in Fig. 6–10, as has been suggested by Dugundji (Ref. 6–13; see also Chap. 10 of Ref. 1–1).

For our discussion of the simple bending-torsion system, we propose to take up the less familiar roots locus first. It merits much more consideration for applications than it has received in the past, and some of the alternative methods may engender positively misleading conclusions.

There are two airspeed regimes in which the properties of the nonsinusoidal aeroelastic modes can be ascertained relatively easily: the low subsonic, where incompressible operators are valid, and the high supersonic. In the low subsonic case, the characteristic equation to be solved is stated

$$\Delta(\tilde{p}) = 0 \tag{6-126}$$

where Δ is the typical section determinant, made dimensionless by division with a nonzero factor and expanded as in Eq. 6-107. Each complex-conjugate pair of roots of Eq. 6-126 has the form

$$\tilde{p} = (p_R \pm i\omega)\frac{b}{U} = +\tilde{p}_R \pm ik \tag{6-127}$$

subscript R denoting the real part. The *damping ratio* of the associated mode, in the parlance of vibration engineering, is given by

$$\zeta = \frac{-p_R}{\sqrt{p_R{}^2 + \omega^2}} = \frac{-\tilde{p}_R}{\sqrt{\tilde{p}_R{}^2 + k^2}} \tag{6-128}$$

Provided no roots of Eq. 6-126 lie to the right of the imaginary axis in the complex $\tilde{p}$-plane, the linear system is known to be stable with respect to all small motions. One or more positive real parts—in particular, a negative value of ζ—imply instability. This situation is generally agreed to be inacceptable on a vehicle in flight, despite nonlinear effects which might limit the amplitude of the actual oscillation. The *critical flutter* condition is defined to occur at the lowest speed U_F, for fixed free-stream pressure and density, at which the damping ratio of any dynamic aeroelastic mode passes through zero.

Both conjugate pairs of eigenvalues have associated with them eigenfunctions of the form

$$\frac{h}{b} = \frac{h_0}{b} e^{p_R t} e^{i(\omega t + \varphi_h)} \tag{6-129a}$$

$$\alpha = \alpha_0 e^{p_R t} e^{i\omega t} \tag{6-129b}$$

when the complex representation is used for the vibration. Each of these is found, in the standard way, by substituting its eigenvalue into one or the other of the homogeneous Eqs. 6-104, obtained by setting $\bar{F}(\tilde{p}) = 0$, and solving for the amplitude ratio and phase angle between the two degrees of freedom. Since either α_0 or h_0 may be specified arbitrarily, the *mode shape* is completely defined by $\alpha_0 / \dfrac{h_0}{b}$ and φ_h. This information has only minor interest in stability studies, and modes are often not even calculated. Their variation with airspeed can throw some light on the

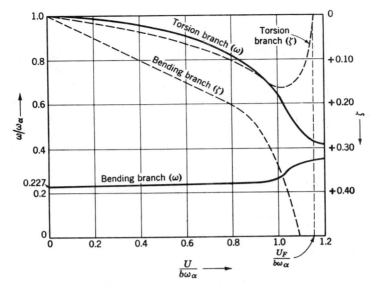

Fig. 6–13. Dimensionless frequencies ω/ω_α and damping ratios ζ of the aeroelastic modes of the typical section, plotted vs. airspeed parameter $1/k_\alpha \equiv U/b\omega_\alpha$ for incompressible flow. System parameters are $x_\alpha = 0.024$, $r_\alpha = 0.62$, $\omega_h/\omega_\alpha = 0.227$, $m/2\rho_\infty bS = 38$, $a = -0.36$. (Adapted from Ref. 6–14.)

physical nature of flutter, however, an interpretation we shall attempt in Sec. 6–6.

According to Eqs. 6–107 and 6–126, the aforementioned properties of the aeroelastic modes in incompressible flow depend on six dimensionless parameters characterizing the typical section and airstream: ω_h/ω_α, x_α, r_α, a, $2m/\pi\rho_\infty bS$, and $k_\alpha \equiv \omega_\alpha b/U$. The first four of these describe the wing and its supporting mechanism. The mass ratio* $(2m/\pi\rho_\infty bS)$ couples it to the fluid state through the relative densities. Finally $1/k_\alpha$ is a dimensionless speed that may be regarded as the independent variable parameter in a stability investigation. Another suitable combination for this purpose would be some multiple of $\pi\rho_\infty bS/2mk_\alpha^2 r_\alpha^2$, which is proportional to dynamic pressure q divided by K_α/l and is therefore a ratio between aerodynamic and structural stiffnesses.

As an illustration of how the eigenvalues vary with speed, we have adapted some calculations of Goland and Luke (Ref. 6–14) in Figs. 6–13 and 6–14. Although the objective of these authors was to analyze the

---

* The reader will recall that, in defining mass ratio, the $2/\pi$ is often replaced by $1/2$ or some other factor.

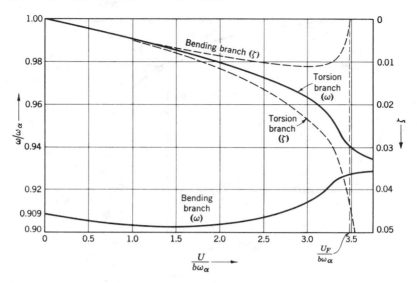

Fig. 6–14. Dimensionless frequencies ω/ω_α and damping ratios ζ of the aeroelastic modes of the typical section, plotted vs. airspeed parameter $1/k_\alpha \equiv U/b\omega_\alpha$ for incompressible flow. System parameters are $x_\alpha = 0.024$, $r_\alpha = 0.62$, $\omega_h/\omega_\alpha = 0.909$, $m/2\rho_\infty bS = 38$, $a = -0.36$. (Adapted from Ref. 6–14.)

stability of a uniform wing with concentrated masses at its midspan and tips, their system is mathematically equivalent to a typical section with certain values of the parameters listed in the previous paragraph. We have computed these as accurately as the original data (Ref. 6–15) permit, and they are enumerated in the figure captions. The scheme employed by Goland and Luke for finding the frequency and damping ratio is based on Eq. 6–126, combined with the approximation (6–108) for $\bar{\varphi}(\tilde{p})$. They devised an iteration procedure, starting with an assumed value of the root $\tilde{p}$ in $\bar{\varphi}(\tilde{p})$, factoring the quartic polynomial left after thus replacing $\bar{\varphi}(\tilde{p})$, using the roots so obtained to refine the first approximation, and repeating the steps to convergence. The final results are inexact, mainly* because of the small difference between Eq. 6–108 and its transcendental counterpart, which is

$$\bar{\varphi}(\tilde{p}) = \frac{K_1(\tilde{p})}{\tilde{p}[K_1(\tilde{p}) + K_0(\tilde{p})]} \qquad (6\text{--}130)$$

* As described in Ref. 6–14, this calculation appears to destroy the conjugate property of the roots by introducing complex coefficients into the characteristic polynomial through the approximation (6–108). The damping ratios are generally quite small, however, so the additional inaccuracy is probably negligible.

K_n being the modified Bessel function of the second kind. No significant error is suspected in the present case.

Figure 6–13 shows an example qualitatively resembling the larger-aspect-ratio cantilever wings and tail surfaces whose bending frequencies are rather small fractions of the torsional. The aeroelastic mode marked "torsion branch," the one which merges as $U \to 0$ with the higher-frequency coupled natural mode ω_2 involving mostly rotational vibration, exhibits a lower damping ratio than the "bending branch" at all airspeeds and proves to be the one which becomes unstable at a critical condition $\dfrac{U_F}{b\omega_\alpha} = 1.153$.

Figure 6–14 presents a case of nearly equal uncoupled frequencies such as might occur on an all-movable control surface, where ω_α refers to oscillation as a rigid body against whatever torsional restraint is attached to the axis. Here the "bending" aeroelastic mode shows the instability. In both figures, the damping, which is purely aerodynamic in origin, is seen to increase steadily with airspeed up to a point roughly 15% below critical. Then one of the modes reveals its tendency to turn unstable, while the other damping ratio simultaneously increases very rapidly. The frequencies of the aeroelastic modes approach each other gradually but do not become equal at flutter, as has sometimes been hypothesized. In fact, further work on this question indicates that, had the calculations been carried to higher values of $U/b\omega_\alpha$, they would converge onto the same horizontal asymptote from above and below.

In Fig. 6–15 is given the roots locus itself, on the complex $\tilde{p}$-plane, of the data from Fig. 6–13. Here the instability of the "torsion branch" is clearly indicated by its crossing the imaginary axis from negative to positive $\tilde{p}_R$. Incidentally, for any point on such a locus, the angle between the vertical and a line to that point from the origin has a sine equal* to ζ; this angle is often treated as a measure of the degree of stability. The reader will note that this system has such a small value of static unbalance x_α that ω_1 and ω_2 essentially coincide with ω_h and ω_α, respectively. It is also of interest to observe the damping ratio of the critical mode changing quite rapidly with $U/b\omega_\alpha$ in the neighborhood of neutral stability, suggesting that the onset of flutter would occur violently and any small excess of airspeed beyond U_F would cause certain destruction. This effect is more pronounced in Fig. 6–13 than Fig. 6–14, the latter wing being very lightly damped at all speeds below flutter.

At high supersonic Mach numbers ($M^2 \gg 1$, $M\delta \ll 1$), the piston theory

---

* This construction requires, of course, that the horizontal and vertical scales be equal, which is not the case in Fig. 6–15 where the abscissa has been expanded for clarity.

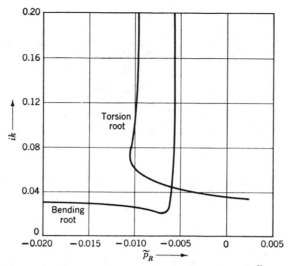

Fig. 6–15. Locus of conjugate roots of the characteristic equation $\Delta(\tilde{p}) = 0$ of the typical section, plotted on the upper half of the complex plane of $\tilde{p} = \tilde{p}_R + ik$, for the same parameters as in Fig. 6–13. $1/k_\alpha$ varies along each curve. (Note that horizontal and vertical scales are unequal.)

aerodynamic operators discussed in Secs. 4–5(d) and 4–6(d) are applicable, and the aeroelastic modes of the typical section can once more be determined with relative ease. It is a simple matter to adapt Eqs. 4–139 through 4–144 to arbitrary time-dependent motion, substitute for $L^M(t)$ and $M_y{}^M(t)$ in Eqs. 6–88,9, and arrive at the following dimensionless, homogeneous relations:

$$\left(\frac{m}{2\rho_\infty bS}\right)\frac{d^2}{ds^2}\left(\frac{h}{b}\right) + \frac{1}{M}\frac{d}{ds}\left(\frac{h}{b}\right) + \left(\frac{m}{2\rho_\infty bS}\right)k_\alpha{}^2\left(\frac{\omega_h}{\omega_\alpha}\right)^2\frac{h}{b}$$

$$+ \left(\frac{m}{2\rho_\infty bS}\right)x_\alpha\frac{d^2\alpha}{ds^2} - \frac{1}{M}\left[a + \left(\frac{\gamma+1}{4}\right)M\frac{A_w}{2b^2}\right]\frac{d\alpha}{ds} + \frac{1}{M}\alpha = 0 \quad (6\text{–}131a)$$

$$\left(\frac{m}{2\rho_\infty bS}\right)x_\alpha\frac{d^2}{ds^2}\left(\frac{h}{b}\right) - \frac{1}{M}\left[a + \left(\frac{\gamma+1}{4}\right)M\frac{A_w}{2b^2}\right]\frac{d}{ds}\left(\frac{h}{b}\right) + \left(\frac{m}{2\rho_\infty bS}\right)r_\alpha{}^2\frac{d^2\alpha}{ds^2}$$

$$+ \frac{1}{M}\left\{\frac{4}{3} - \left(\frac{\gamma+1}{4}\right)M\frac{M_w}{b^3} - (a+1)\left[1 - a - \left(\frac{\gamma+1}{4}\right)M\frac{A_w}{b^2}\right]\right\}\frac{d\alpha}{ds}$$

$$+ \left\{\left(\frac{m}{2\rho_\infty bS}\right)r_\alpha{}^2k_\alpha{}^2 - \frac{1}{M}\left[a + \left(\frac{\gamma+1}{4}\right)M\frac{A_w}{2b^2}\right]\right\}\alpha = 0 \quad (6\text{–}131b)$$

Here $s = Ut/b$ is the time variable from Sec. 6–4(b). Other quantities have been defined in this chapter, except for A_w and M_w, which are the cross-sectional area of the wing profile and its area moment about its leading edge, respectively. (The leading and trailing edges are assumed to be sharp-pointed.)

When the Laplace transformation is applied to Eqs. 6–131, we again get a characteristic equation like 6–126. Only a quartic in $\bar{p}$ now has to be factored for the exact eigenvalues, however, because all the coefficients in Eqs. 6–131 are constants. The process is sufficiently direct that it need not be elaborated. Clearly the eigenvalues and mode shapes are functions of the same basic set of dimensionless parameters as for incompressible flow, plus the Mach number M. When thickness effects are accounted for, two profile parameters MA_w/b^2 and MM_w/b^3 must be added; each of these will be directly proportional to $M\delta$ for a given geometry.

Recently Zisfein and Frueh presented an illuminating study of typical section flutter based on piston theory (Ref. 6–16). Figures 6–16 and 6–17 are replotted in dimensionless form from their report. The values of the parameters needed, after adopting their approximation that $\zeta \ll 1$, are listed in the captions. In both these examples, the aeroelastic mode corresponding to the "bending branch" goes unstable. The resemblance between the damping curves and those of Fig. 6–14 indicates that, for this simplified system, the dimensionless character of the modes is rather insensitive to the range of M wherein they are calculated. Although the large number of parameters makes it dangerous to attempt broad generalizations, experience does show that the form of these curves is principally controlled by the mechanical properties of the structure. Thus we find, in two-degree-of-freedom cases involving a trailing edge flap, that stability reappears at an airspeed higher than the minimum U_F. This leaves a finite range of flutter, something which is not observed in the bending-torsion case. When there are three degrees of freedom, often two aeroelastic modes become critical at different speeds. Even more complicated behavior is naturally observed on the elaborate structures met in practice, especially when large concentrated masses are attached to a lifting surface. A multitude of examples will be found in the published literature, and some are presented in subsequent chapters.

Zisfein and Frueh discover two interesting properties of curves of frequency versus speed like those in Figs. 6–16 and 6–17. At the lower end of the $1/k_\alpha$-scale, small ζ being assumed, the two branches merge with a so-called *base curve* formed by dropping first-derivative terms (i.e., eliminating all damping) from the equations of motion (6–131). When the C.G.

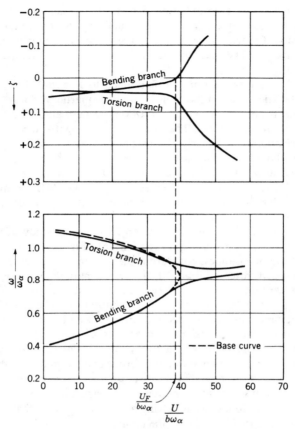

Fig. 6–16. Dimensionless frequencies ω/ω_α and damping ratios ζ of the aeroelastic modes of the typical section, plotted vs. $1/k_\alpha \equiv U/b\omega_\alpha$ for high supersonic M. System parameters are $x_\alpha = 0.2$, $r_\alpha = 0.5$, $\omega_h/\omega_\alpha = 0.4$, $m/2\rho_\infty bS = 50$, $a = -0.2$, thickness ratio $\delta = 0$, $a_\infty/b\omega_\alpha = 2.325$. $\left(\text{Note that } M = \dfrac{U}{b\omega_\alpha} \Big/ \dfrac{a_\infty}{b\omega_\alpha}.\right)$ (Adapted from Ref. 6–16.)

is at midchord ($x_\alpha = -a$), Ref. 6–16 derives a biquadratic relation for this base curve,

$$\frac{U}{b\omega_\alpha} = \frac{r_\alpha^2}{a}\left(\frac{\omega_\alpha}{\omega_h}\right)^2\left(\frac{\omega_\alpha b}{a_\infty}\right)\left(\frac{m}{2\rho_\infty bS}\right)\left[\frac{\omega^4}{\omega_\alpha^4}\left(1 - \frac{x_\alpha^2}{r_\alpha^2}\right) - \frac{\omega^2}{\omega_\alpha^2}\left(1 + \frac{\omega_h^2}{\omega_\alpha^2}\right) + \frac{\omega_h^2}{\omega_\alpha^2}\right]$$

$$= \frac{r_\alpha^2}{a}\left(\frac{\omega_\alpha}{\omega_h}\right)^2\left(\frac{\omega_\alpha b}{a_\infty}\right)\left(\frac{m}{2\rho_\infty bS}\right)\left[\left(\frac{\omega^2}{\omega_\alpha^2} - \frac{\omega_1^2}{\omega_\alpha^2}\right)\left(\frac{\omega^2}{\omega_\alpha^2} - \frac{\omega_2^2}{\omega_\alpha^2}\right)\right]\left(1 - \frac{x_\alpha^2}{r_\alpha^2}\right)$$

$$(6\text{–}132)$$

Equation 6–132 also assumes thickness ratio $\delta = 0$; it can be extended to other C.G. locations by replacing the factor a in the denominator with

$$a\left[1 - \frac{\omega^2}{\omega_h^2}\left(1 + \frac{x_\alpha}{a}\right)\right].$$

On the other hand, at high speeds beyond $U_F/b\omega_\alpha$, both ω/ω_α-branches are asymptotic to the single value

$$\frac{\omega_\infty}{\omega_\alpha} = \frac{\sqrt{1 + \dfrac{\left(1 + \dfrac{x_\alpha}{a}\right)\left(1 - \dfrac{\omega_\alpha^2}{\omega_h^2}\dfrac{x_\alpha}{a}\right)}{(1 - x_\alpha^2/r_\alpha^2)}} - 1}{\dfrac{\omega_\alpha^2}{\omega_h^2}\left(1 + \dfrac{x_\alpha}{a}\right)} \tag{6–133}$$

This expression simplifies considerably to

$$\omega_\infty = \sqrt{\frac{\omega_h^2 + \omega_\alpha^2}{2(1 - x_\alpha^2/r_\alpha^2)}} \tag{6–134}$$

for the case $x_\alpha = -a$ treated in Ref. (6–16).

Properties of the typical-section aeroelastic modes can, in principle, be computed for Mach numbers intermediate between the low subsonic of Figs. 6–13 through 6–15 and the high supersonic of Figs. 6–16 and 6–17. No exact results of this sort have been published. As discussed in Sec. 4–6, four indicial functions would be required for the characteristic equation in place of the single $\varphi(s)$ needed at $M = 0$. These functions are known only numerically when $M < 1$, while for supersonic speeds they reach fixed limits within a finite time and sometimes have multiple extrema, thus rendering curve-fittings like Eq. 4–172a unsuitable or, at least, very difficult to achieve accurately. The best prospect for future calculations on systems of all kinds probably lies in the use of the analogue computer, with aerodynamic elements designed to approximate the various indicial or mechanical admittances by electronic circuitry.

To find boundaries of neutral stability for the typical section by means of sinusoidal aerodynamic operators, we must in some way determine roots of the complex, algebraic equation*

$$\Delta(i\omega) = \left\{\frac{m}{2\rho_\infty bS}\left[1 + \left(\frac{\omega_\alpha}{i\omega}\right)^2\left(\frac{\omega_h}{\omega_\alpha}\right)^2\right] - [L_1 + iL_2]\right\}$$
$$\times \left\{\frac{m}{2\rho_\infty bS}r_\alpha^2\left[1 + \left(\frac{\omega_\alpha}{i\omega}\right)^2\right] - [M_3 + iM_4]\right\}$$
$$- \left\{\frac{mx_\alpha}{2\rho_\infty bS} - [L_3 + iL_4]\right\}\left\{\frac{mx_\alpha}{2\rho_\infty bS} - [M_1 + iM_2]\right\} = 0 \tag{6–135}$$

---

* Supersonic symbols are employed here for the lift and moment coefficients. The corresponding notation for subsonic speeds is discussed in Sec. 6–4 (a) and elsewhere in the book.

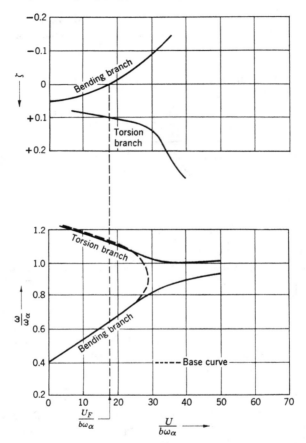

Fig. 6–17. Dimensionless frequency ratios ω/ω_α and damping ratios ζ of the aeroelastic modes of the typical section, plotted vs. $1/k_\alpha \equiv U/b\omega_\alpha$ for high supersonic M. System parameters are the same as in Fig. 6–16, except for lower radius of gyration, $r_\alpha = 0.354$. (Adapted from Ref. 6–16.)

which is obtained from Eq. 6–93a, the dimensionless denominator determinant of the homogeneous form of Eqs. 6–90. With airloads acting, the system is not conservative. Equation 6–135 constitutes an elementary example of a *complex eigenvalue problem*, a topic closely associated with the name of H. Wielandt (e.g., Ref. 3–8).

This characteristic equation is essentially a pair of real, implicit relations among the eight parameters k, M, ω/ω_α, ω_h/ω_α, x_α, r_α, a, and $(m/2\rho_\infty bS)$; A_w/b^2 and M_w/b^3 must be added to this list when thickness effects are introduced at supersonic speed. The quantities ω/ω_α, ω_h/ω_α, x_α, r_α, and $(m/2\rho_\infty bS)$ appear explicitly in Eq. 6–135, so that the linear or quadratic dependence of the real and imaginary parts of the equation upon each of them may be inferred by inspection. The E.A. location a enters linearly into $[L_3 + iL_4]$ and $[M_1 + iM_2]$, and quadratically into $[M_3 + iM_4]$; hence (6–135) is also quadratic in a. Closed-form solutions can be developed for any pair out of these six just mentioned. In contrast, all the aerodynamic coefficients are transcendental functions of k and M, except under the simplification of piston theory.

This mathematical situation is either convenient or very inconvenient, depending on whether one is making a parametric study of flutter or wishes to ascertain the critical speed of a particular wing in a particular flight condition. For the former case, k and M may be selected in advance, along with all but two of the other parameters, whereupon these two follow by elementary algebra from Eq. 6–135.

A common instance involves picking density ratio $m/2\rho_\infty bS$ and frequency ratio ω_α/ω as the dependent parameters. With other quantities fixed, the imaginary part of Eq. 6–135 furnishes a linear relation between them, which is used to eliminate one from the real part of Eq. 6–135, leaving a quadratic equation for the other. This process might be repeated at several values of k and M, yielding a neutral-stability surface on a three-dimensional plot of critical M_F versus $m/2\rho_\infty bS$ and $a_\infty/b\omega_\alpha$ (which equals $\omega/\omega_\alpha Mk$). Such a stability boundary has several uses. For example, a test of a certain wing model in a variable Mach-number wind tunnel is described by a line, which may penetrate the surface at one or more flutter points. The effect of varying altitude in the standard atmosphere on bending-torsion flutter is shown by a line, which is the intersection between the stability boundary and a cylindrical surface. The shape of this surface follows from the variation of ρ_∞ with $a_\infty = \sqrt{\gamma R T_\infty}$ in the atmosphere, and the surface generators are parallel to the M-axis.

A few results of such parametric studies and citations to many more in the literature are described in Subsecs. (a), (b), and (c) that follow presently.

To compute the critical speed and frequency of a typical section with given properties in air at an assigned thermodynamic state (ρ_∞ and a_∞ given), trial-and-error or implicit solution is unavoidable. The quantities k and M are the unknowns, with $\omega_\alpha b/a_\infty$, the mass ratio, and the structural and inertial parameters specified. Unfortunately, the complicated relationship between the aerodynamic coefficients k and M demands that these two be selected before Eq. 6–135 can be solved. Two of the other quantities must therefore be allowed to deviate artificially from their correct values, while a sufficient number of k-M combinations are tried to permit double interpolation to the true critical point. Since a curve of M_F versus altitude is often useful to the designer (at least those portions of it which do not turn out to be below sea level!), the most logical choice of artificial unknowns would seem to be the density ratio and frequency ratio ω_α/ω, leading to a stability surface like the one mentioned two paragraphs above. Past practice in the United States has leaned, however, toward the employment of a fictitious structural damping coefficient. This procedure is treated further during the discussion of distributed-property structures in Sec. 7–7, where more information is also supplied about the influence of internal friction on stability.

(a) Effects of some of the parameters in incompressible flow

This case constitutes a major simplification, because M is eliminated as an argument of the transcendental airloads. The many incompressible-flow studies that have been published retain more than historical importance, because of the light they throw on how aeroelastic stability is affected by the mechanical parameters at all airspeeds. Significant examples whose results may be applied to the typical section are the following: the extensive figures of Theodorsen and Garrick (Ref. 6–2), giving stability boundaries for bending-torsion, bending-flap, and torsion-flap flutter; a second paper by these authors (Ref. 6–17) on the same system with all three degrees of freedom coupled; Duncan and Griffith (Ref. 6–18); Duncan and Lyon (Ref. 6–19); van de Vooren and Greidanus (Refs. 6–20 and 6–21); and Falkner's work (Ref. 6–22) on the bending-flap case.

We shall present here a few curves adapted from Ref. 6–2. As in other early papers, Theodorsen and Garrick choose $U_F/b\omega_\alpha$ as the ordinate for their stability boundaries. On a plot of this quantity versus a second dimensionless parameter, with all other system properties held fixed, the area below the curve is identified as stable. This means that any combination of airspeed (for $M^2 \ll 1$), semichord, and uncoupled torsional frequency which ensures $U < U_F$ puts the wing in a flutter-free region, and conversely. Some writers now use $b\omega_\alpha/U_F$ as an incompressible-flow ordinate; this inversion simply makes stable conditions fall above the boundary, but it is more consistent with various ordinates proportional to $b\omega_\alpha/a_\infty$, which are widely used in subsonic and supersonic studies.

Figure 6–18 illustrates the influence of frequency ratio ω_h/ω_α for four values of the static unbalance. A little shading is added on the unstable side of each curve, as in other figures below. The most significant feature here—one which is very familiar to aeroelasticians and appears repeatedly on such plots—is the tendency for instability to be much more pronounced in the neighborhood of $\omega_h/\omega_\alpha = 1$, especially for moderately aft C.G. locations. When pointing out this so-called "resonance" phenomenon,

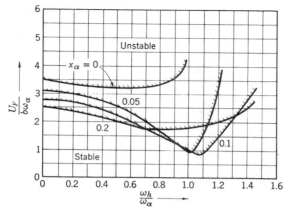

Fig. 6–18. Stability boundaries for the typical section in incompressible flow, shown on a plot of dimensionless flutter speed $U_F/b\omega_\alpha$ vs. frequency ratio ω_h/ω_α for four values of static unbalance. Other system parameters are $(2m/\pi\rho_\infty bS) = 20$, $a = -0.3$, $r_\alpha = 0.5$. (Reproduced from Ref. 6–2.)

we must add that reasonable amounts of internal friction, such as might occur in actual wing structures, tend to ameliorate, although not to eliminate, the more severe stability minima. Reference 6–2 and others demonstrate this, and they also show the much less dramatic stabilizing effects of structural damping at lower frequency ratios.

The curious dependence of $U_F/b\omega_\alpha$ on the relative density $(2m/\pi\rho_\infty bS)$ is exhibited in Fig. 6–19. Each curve is seen to fall almost linearly with decreasing $(2m/\pi\rho_\infty bS)$ until a minimum is reached somewhere between about 2 and 15 on the abscissa scale. This instability peak is followed rapidly by an asymptotic rise to infinity, predicting that a wing moving through a sufficiently dense medium should be free from flutter at all speeds. Although modern high-performance aircraft and missiles fly well out on the linear portions of boundaries like those in Fig. 6–19, some light personal airplanes, when at low altitudes, operate near the minima. Moreover, there exist flutter measurements in air and Freon-12 (Woolston and Castile, Ref. 6–23) that do not correlate well with these theoretical minima. They have recently given rise to interesting speculations (Refs. 6–24 and 6–25) on whether a practical design problem will ever arise from the *hydroelastic stability* of submerged lifting surfaces mounted on high-speed ships and submarines, which normally have small fractional values of $(2m/\pi\rho_\infty bS)$ and lie to the right of the theoretical asymptotes.*

As an indication of the combined effects of the E.A. and C.G. locations,

---

* See also the recent works of Herr (Ref. 6–51).

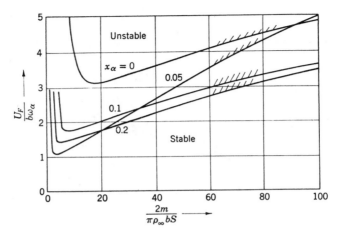

Fig. 6–19. Stability boundaries for the typical section in incompressible flow, shown on a plot of dimensionless flutter speed $U_F/b\omega_\alpha$ vs. relative density $(2m/\pi\rho_\infty bS)$ for four values of static unbalance. Other system parameters are $\omega_h/\omega_\alpha = 0.707$, $a = -0.3$, $r_\alpha = 0.5$. (Adapted from Ref. 6–2.)

we reproduce Figs. 6–20 and 6–21 from Ref. 6–2; the abscissa is $(\tfrac{1}{2} + a + x_\alpha)$, the dimensionless distance by which the C.G. lies behind the steady-flow aerodynamic center (A.C.) on the $\tfrac{1}{4}$-chord axis. Figure 6–20 contains points corresponding to four different E.A. positions. The fact that they fall very nearly on the same line proves, for the low values of ω_h/ω_α involved, that the instability is governed by the C.G.-A.C. offset. In fact, Theodorsen and Garrick suggest the empirical formula

$$\frac{U_F}{b\omega_\alpha} \cong \sqrt{\left(\frac{2m}{\pi\rho_\infty bS}\right)\frac{r_\alpha^2}{[1 + 2(a + x_\alpha)]}} \qquad (6\text{--}136)$$

which is also plotted on the figure. Using quasi-steady aerodynamic operators, Dugundji (Ref. 6–26) has provided a firmer theoretical foundation for Eq. 6–136 and also stated more precisely the circumstances under which it may be expected to hold; in particular, it may prove to be unconservative when the C.G. falls behind the 55 or 60% chord axis.

Figure 6–21 refers to a higher frequency ratio. A little examination will show that these curves, which correspond to six E.A. locations between the $\tfrac{1}{4}$- and $\tfrac{1}{2}$-chord lines, would draw much closer together if replotted versus x_α alone. Here it is the static unbalance itself that governs the instability. Larger positive values of x_α are unfavorable, with stability

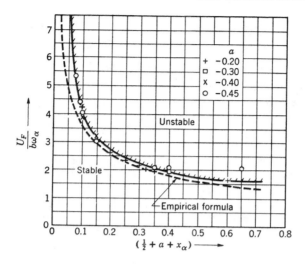

Fig. 6–20. Stability boundary for the typical section in incompressible flow, shown on a plot of dimensionless flutter speed $U_F/b\omega_\alpha$ vs. dimensionless A.C.–C.G. offset ($\frac{1}{2} + a + x_\alpha$) for four values of E.A. location. Other system parameters are $\omega_h/\omega_\alpha = 0$, $(2m/\pi\rho_\infty bS) = 10$, $r_\alpha = 0.5$. (Adapted from Ref. 6–2.)

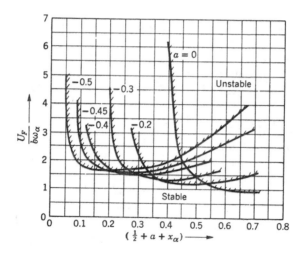

Fig. 6–21. Stability boundaries for the typical section in incompressible flow, shown on a plot of dimensionless flutter speed $U_F/b\omega_\alpha$ vs. dimensionless A.C.-C.G. offset ($\frac{1}{2} + a + x_\alpha$) for six values of E.A. location. Other system parameters are $\omega_h/\omega_\alpha = 0.707$, $(2m/\pi\rho_\infty bS) = 10$, $r_\alpha = 0.5$. (Adapted from Ref. 6–2.)

minima occurring near $x_\alpha = 0.2$. It is a simple illustration of the principle of mass-balancing that a C.G. ahead of the E.A. suppresses flutter completely, at least for the forward E.A. positions.

(b) Subsonic and supersonic compressible flow

Parametric studies which are applicable to typical section flutter in compressible flow are more restricted both in number and scope than those published for the low-Mach-number regime. Four important examples, all of them utilizing fully linearized aerodynamic operators, are the following: Garrick (Ref. 6–27) on the bending-torsion case at $M = 0.7$; Nelson and Berman (Ref. 6–28), who emphasize sonic speed but give a few curves for M's between 0 and 5; Garrick and Rubinow (Ref. 6–29) on the bending-torsion case in supersonic flow; and Woolston and Huckel (Ref. 6–30) on the bending-flap and torsion-flap cases between $M = 0$ and 2. Several other reports have been prepared, for limited distribution, by various industrial organizations, but these generally deal with finite wings and other systems having distributed properties.

Mach number is the primary independent variable used for presenting most parametric data. The practice of showing stability boundaries on a plot of $U_F/b\omega_\alpha$ versus M is regarded as undesirable, however, for the airspeed occurs in both ordinate and the abscissa. Following its introduction by Regier of NASA, the aforementioned ratio $b\omega_\alpha/a_\infty$, usually multiplied by some power of the relative density such as $\sqrt{2m/\pi\rho_\infty bS}$, is currently gaining wide acceptance as an ordinate. Figure 6–22, modified from Fig. 5 of Ref. 6–28, is such a graph. Here the boundaries are presented for a wing having a low frequency ratio and an E.A. at midchord. The wide range of relative densities represented by the four curves might be achieved either by varying the air density in a wind tunnel or with a series of different models.

The stable region lies above each boundary in Fig. 6–22, in the sense that a combination of wing characteristics and ambient state of the gas (i.e., test-section condition or altitude in the standard atmosphere) which yields a numerically higher ordinate, at a given M, assures stability. Thus, the torsional stiffness necessary to avoid flutter is measured by the value of ω_α required to bring a wing of given mass and geometry at least up to the boundary. Flight at a fixed ambient state corresponds to a certain horizontal straight line; no flutter is expected if this line falls entirely within the stable zone for all Mach numbers of interest.

The common tendency of lifting surfaces to be especially critical from the aeroelastic standpoint in the transonic range is typified by the sharp peak near $M = 1$ in the curve for the highest relative density. The behavior of the boundaries at the higher supersonic Mach numbers can be

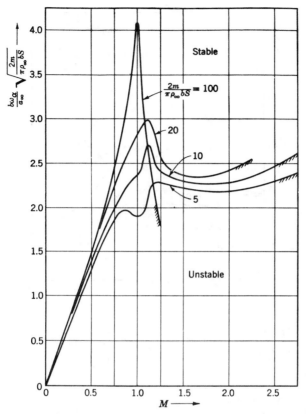

Fig. 6–22. Stability boundaries for the typical section in compressible flow, shown on a plot of $\dfrac{b\omega_\alpha}{a_\infty}\sqrt{\dfrac{2m}{\pi\rho_\infty bS}}$ vs. Mach number M for four values of relative density. Other system parameters are $\omega_h/\omega_\alpha = 0$, $a = 0$, $x_\alpha = 0.2$, $r_\alpha = 0.5$. (Adapted from Ref. 6–28.)

misleading, because of the destabilizing influence of profile thickness, which is not accounted for here but is discussed under Subsec. (c) presently. An incidental feature that renders plots like Fig. 6–22 helpful to designers is that flight of a particular wing at constant dynamic pressure q is described by a straight line through the origin of coordinates. The reader will have no trouble convincing himself that the slope of such a line is inversely proportional to $\sqrt{q}$.

Figure 6–23, also adapted from Ref. 6–28, shows the influence of frequency ratio ω_h/ω_α at five preassigned values of M between 0.7 and 5.

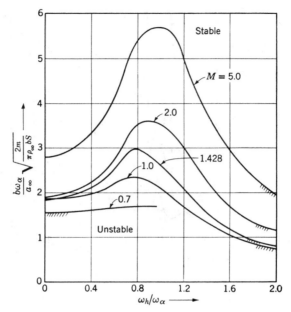

Fig. 6–23. Stability boundaries for the typical section in compressible flow, shown on a plot of $\dfrac{b\omega_\alpha}{a_\infty}\sqrt{\dfrac{2m}{\pi\rho_\infty bS}}$ vs. frequency ratio $\dfrac{\omega_h}{\omega_\alpha}$ for five values of Mach number M. Other system parameters are $(2m/\pi\rho_\infty bS) = 10$, $a = 0$, $x_\alpha = 0.2$, $r_\alpha = 0.5$.

As in Fig. 6–18, the undesirability of $\omega_h/\omega_\alpha \cong 1$ is evident especially from the high peak when $M = 5$.

(c) High supersonic speeds; the importance of thickness

The high-M simplification of piston theory has naturally produced a spate of analytical flutter studies, of which Refs. 6–31 through 6–34 are representative. Chawla (Ref. 6–31) and Morgan, Runyan, and Huckel (Ref. 6–32) treat bending-torsion instability of the typical section. The extensive tables of Weatherill and Zartarian (Ref. 6–33) cover all three binary cases associated with translation, rotation, and the trailing-edge flap, but most of their data also refer to the bending-torsion type; only a single profile shape—the symmetrical double wedge—is considered. Reference 6–34 undertakes to interpret the results of Ref. 6–33 and present them graphically. Elementary expressions for the stability boundary in many forms can be derived by manipulating the characteristic determinant of Eqs. 6–131, after it has been specialized to simple harmonic motion through the substitution $d/ds = ik$. The wing analyzed in Refs. 6–33 and 6–34 serves as a suitable illustration: for the double-wedge profile of thickness ratio δ, the cross-sectional area is $2b^2\delta$, and

$$M \frac{A_w}{2b^2} = M\delta \qquad (6\text{–}137)$$

$$M \frac{M_w}{b^3} = 2M\delta \qquad (6\text{–}138)$$

We then obtain the following formulas for dimensionless critical speed and frequency ratio:

$$\frac{U_F}{b\omega_\alpha}$$

$$= \frac{M\left(\dfrac{m}{2\rho_\infty bS}\right)}{\sqrt{\chi}}$$

$$\times \sqrt{\frac{x_\alpha{}^2 - r_\alpha{}^2[\chi - 1]\left[\left(\dfrac{\omega_h}{\omega_\alpha}\right)^2\chi - 1\right]}{M\left(\dfrac{m}{2\rho_\infty bS}\right)\left\{\left[-a - \left(\dfrac{\gamma + 1}{4}\right)M\delta\right]\right.}} }$$

$$\times \left[\left(\dfrac{\omega_h}{\omega_\alpha}\right)^2\chi - 1\right] + x_\alpha\Big\} + \left[\dfrac{\gamma + 1}{4}M\delta\right]^2 - \dfrac{1}{3} \qquad (6\text{–}139)$$

where

$$\chi \equiv \left(\frac{\omega_\alpha}{\omega}\right)^2$$

$$= \frac{r_\alpha{}^2 + 2x_\alpha\left[a + \left(\dfrac{\gamma + 1}{4}\right)M\delta\right] + \left[\dfrac{1}{3} + a^2 + 2a\left(\dfrac{\gamma + 1}{4}\right)M\delta\right]}{r_\alpha{}^2 + \left(\dfrac{\omega_h}{\omega_\alpha}\right)^2\left[\dfrac{1}{3} + a^2 + 2a\left(\dfrac{\gamma + 1}{4}\right)M\delta\right]} \qquad (6\text{–}140)$$

With thickness effects included, the eigenvalues are now seen to be functions of only six distinct parameters ω_h/ω_α, x_α, r_α, a, $M\delta$, and $M(m/2\rho_\infty bS)$, the Mach number and relative density having coalesced. While this is a great convenience, it renders inefficient Regier's parameter, which was used as the ordinate of Figs. 6–22 and 6–23. There is, however, a possibility of constructing another ratio with many of the same properties by inverting Eq. 6–139 and multiplying it through by $M(m/2\rho_\infty bS)$. This process yields

$$\frac{b\omega_\alpha}{a_\infty}\left(\frac{m}{2\rho_\infty bS}\right)$$

$$= \sqrt{\chi} \sqrt{\frac{M\left(\dfrac{m}{2\rho_\infty bS}\right)\left\{\left[-a - \left(\dfrac{\gamma+1}{4}\right)M\delta\right]\left[\left(\dfrac{\omega_h}{\omega_\alpha}\right)^2\chi - 1\right]\right.}{\left.+ x_\alpha\right\} + \left[\dfrac{\gamma+1}{4}M\delta\right]^2 - \dfrac{1}{3}}{x_\alpha^2 - r_\alpha^2\left[\left(\dfrac{\omega_h}{\omega_\alpha}\right)^2\chi - 1\right][\chi - 1]}} \qquad (6\text{–}141)$$

A considerable quantity of helpful information can be deduced by closely examining Eqs. 6–141, 6–139, or any of the simplified versions that are found by specializing the system. Since much of this is done in Refs. 6–31 and 6–34, we content ourselves here with reproducing a few figures that exhibit significant behavior of the bending-torsion typical section. When looking these over, the reader should bear in mind the limitation of piston theory to $M > 2.5$ or thereabouts for unswept wings; in Ref. 6–32 some discussion appears on how results like these can be extended to lower supersonic Mach numbers, using van Dyke's theory. On the plots taken from Ref. 6–34, the profile thickness is introduced through the parameter $\delta(2\rho_\infty bS/m)$, which is just the quotient of $M\delta$ and $M(m/2\rho_\infty bS)$ and is independent of the flight speed.

Figure 6–24 shows the influence of the basic Mach-number-density parameter at three values of thickness ratio. Increasing δ is definitely destabilizing to this system, whose properties are quite characteristic of the small, straight wings of certain high-speed vehicles (except that the C.G. lies somewhat farther aft than is normal). The ordinate varies in a roughly parabolic fashion with $M(m/2\rho_\infty bS)$, as can also be seen by studying Eq. 6–141. This fact means that an airplane or missile accelerating to higher M at a fixed altitude can be expected ultimately to encounter aeroelastic instability, even though its surfaces are stiff enough to survive in the dangerous transonic range.

Figure 6–25 presents the effects of frequency ratio, including the familiar peaking near $\omega_h/\omega_\alpha = 1$. A new phenomenon here is the crossing over of the boundaries for the three different thickness ratios just to the right of their maxima. While the larger δ is unfavorable at $\omega_h/\omega_\alpha < 1$, it has a small stabilizing tendency in the higher range, something of interest to the designer of all-movable controls, whose ω_α's are relatively low.

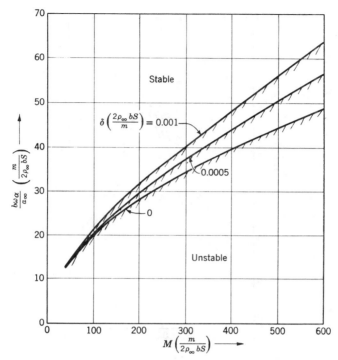

Fig. 6–24. Stability boundaries for the typical section with double-wedge profile according to piston theory, shown on a plot of $\dfrac{b\omega_\alpha}{a_\infty}\left(\dfrac{m}{2\rho_\infty bS}\right)$ vs. $M\left(\dfrac{m}{2\rho_\infty bS}\right)$ for three values of thickness-ratio parameter. Other system parameters are $\omega_h/\omega_\alpha = 0.7$, $a = 0$, $x_\alpha = 0.2$, $r_\alpha = 0.5$. (Adapted from Ref. 6–34.)

Figures 6–26 and 6–27 bear a close resemblance to Figs. 6–20 and 6–21, in that they summarize the roles of the relative chordwise locations of A.C., E.A., and C.G. for $\omega_h/\omega_\alpha \cong 0$ and for the less common higher $\omega_h/\omega_\alpha = 0.7$, respectively. Exactly as in incompressible flow, Fig. 6–26 presents a stability boundary which varies in direct proportion to the square root of the A.C.-C.G. offset for a variety of E.A. locations; but here the influence of thickness variations is also included. Since the A.C. is displaced forward by a distance proportional to $M\delta$, this phenomenon is the principal explanation of the much greater instability due to increased thickness ratio on wings which have both low ω_h/ω_α and C.G.'s in the vicinity of midchord. Incidentally, this A.C.-C.G. offset can be shown to affect flutter in generally the same way at all Mach numbers from near zero

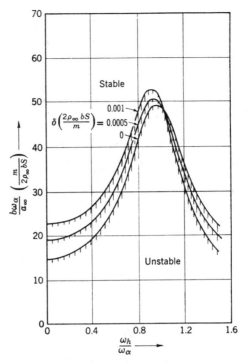

Fig. 6–25. Stability boundaries for the typical section with double-wedge profile according to piston theory, shown on a plot of $\dfrac{b\omega_\alpha}{a_\infty}\left(\dfrac{m}{2\rho_\infty bS}\right)$ vs. frequency ratio for three values of thickness-ratio parameter. Other system parameters are $M\left(\dfrac{m}{2\rho_\infty bS}\right) = 240$, $a = 0$, $x_\alpha = 0.2$, $r_\alpha = 0.4$. (Adapted from Ref. 6–34.)

to high supersonic. Figure 6–27 suggests that the static unbalance x_α tends to be the controlling factor at higher ω_h/ω_α. In fact, Ref. 6–34 indicates an extraordinary sensitivity of stability to this parameter when $\omega_h/\omega_\alpha = 1$ exactly, but the significance of this result is questionable because no account is taken of the presence of structural friction.

To emphasize the importance of thickness, we close this section by copying Fig. 6–28 from Ref. 6–32. Here the stability boundary is plotted versus frequency ratio for four airfoils: a 4% thick symmetrical double wedge, a 4% thick NACA 65A004 profile, a flat plate, and a single wedge with blunt trailing edge. Particularly at low ω_h/ω_α, the former two, which possess nearly coincident aerodynamic centers, have almost the same behavior and are much less stable than the latter two. The complete

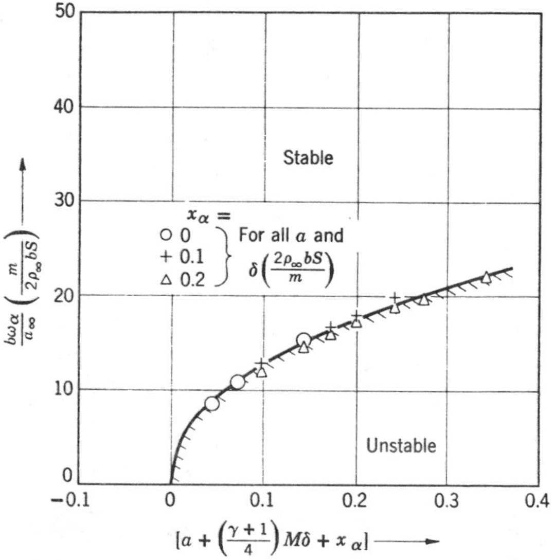

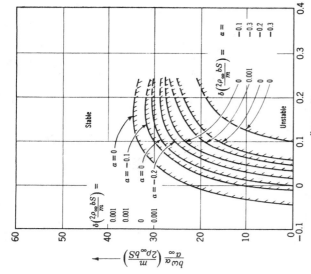

Fig. 6-27. Stability boundaries for the typical section with double-wedge profile according to piston theory, shown on a plot of $\frac{b\omega_\alpha}{a_\infty}\left(\frac{m}{2b_\infty\rho S}\right)$ vs. static unbalance parameter x_α for a variety of thickness ratios and E.A. positions. Other system parameters are $M\left(\frac{m}{2\rho_\infty bS}\right) = 240$, $\omega_h/\omega_\alpha = 0.7$, $r_\alpha = 0.5$. (Adapted from Ref. 6-34.)

Fig. 6-26. Stability boundary for the typical section with double-wedge profile according to piston theory, shown on a plot of $\frac{b\omega_\alpha}{a_\infty}\left(\frac{m}{2\rho_\infty bS}\right)$ vs. distance from A.C. to C.G. in semichords for a variety of thickness ratios and E.A. positions. Other system parameters are $M\left(\frac{m}{2\rho_\infty bS}\right) = 240$, $\omega_h/\omega_\alpha = 0$, $r_\alpha = 0.5$. (Adapted from Ref. 6-34.)

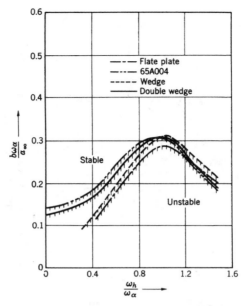

Fig. 6–28. Stability boundaries according to piston theory for the typical section with four different profile shapes, shown on a plot of $b\omega_\alpha/a_\infty$ vs. frequency ratio. For the single wedge and double wedge, $\delta = 0.04$. Other system parameters are $M = 5$, $(m/2\rho_\infty bS) = 250$, $a = -0.2$, $x_\alpha = 0.2$, $r_\alpha = 0.5$. (Reproduced from Ref. 6–32.)

stability of the flat plate and single wedge at $\omega_h/\omega_\alpha = 0$ is attributed to the fact that neither profile has any effect of thickness on A.C. position, the A.C.'s and C.G.'s falling together at midchord in both instances.

6–6 THE PHYSICAL EXPLANATION OF FLUTTER

It has been said that the development of unsteady airload theory for oscillating wings did more to promote misunderstanding of the flutter phenomenon than any other factor. While this is an exaggeration, it is true that the insights which aeroelasticians have—or think they have—about flutter are largely mathematical. An event of some importance is, therefore, the reappearance lately of papers such as those by Pines (Ref. 6–35), Dugundji (Ref. 6–26), and Crisp (Ref. 6–42), which try to contribute to the physical comprehension of this type of instability. One key step is the inspired return by these authors to quasi-steady aerodynamics when examining cases with more than one degree of freedom; this represents a more respectable approximation now that the success and accuracy of such theory at high supersonic speeds is recognized (cf. Ref. 6–16). Actually, quasi-steady theory was in common use prior to 1935. The reader will find many references and interesting examples of its application in the classical study by Frazer and Duncan (Ref. 6–36), and it underlies the discussions of flutter in more recent books by von Kármán and Biot (Ref. 1–11) and Rocard (Ref. 1–33).

Because the reduced frequencies are so high, neither steady-state nor quasi-steady aerodynamic operators are sufficiently accurate for most *quantitative* flutter predictions during the final design process, in the same sense that they serve for calculating dynamic stability of aircraft. When seeking physical explanations, however, things are different. For this purpose, we should like to distinguish between systems with one degree of freedom and those with two or more, and we hazard the assertion that unsteady effects are essential to the understanding of the former, but not of the latter. Let us try to clarify this point by treating two elementary examples in detail.

Consider, first, the purely rotational motion of the typical section (Fig. 2–3 or 6–5) obtained by restraining its axis with an infinitely stiff bending spring. Let the flow be incompressible. In the absence of any external driving force, the governing differential equation is

$$I_\alpha \ddot{\alpha} + K_\alpha \alpha = M_y \qquad (6\text{--}142)$$

The pitching moment M_y is given by Eq. 4–171 (with $h = 0$) for arbitrary small $\alpha(t)$, or by a complex function of $k = \omega b / U$ when the $e^{i\omega t}$ notation is used to describe a simple harmonic oscillation. In any event, the significant features of the problem are preserved by writing Eq. 6–142 in the approximate form

$$\left[I_\alpha + \frac{\pi}{2} \rho_\infty b^3 S \left(\frac{1}{8} + a^2 \right) \right] \ddot{\alpha} - \frac{\partial M_y}{\partial \dot{\alpha}} \dot{\alpha} + \left[K_\alpha - \frac{\partial M_y}{\partial \alpha} \right] \alpha = 0 \quad (6\text{--}143)$$

Here I_α is augmented by the virtual moment of inertia of air accelerated with the wing, which is usually small and does not affect stability. The derivatives $\partial M_y / \partial \dot{\alpha}$ and $\partial M_y / \partial \alpha$ are not constants but depend on the flow parameters and the time dependence of the rotation (although not its amplitude). If these parameters and the mode of motion are thought of as known, however, the derivatives take on fixed values; and Eq. 6–143 describes a second-order, lumped-parameter system with a single degree of freedom.

In terms of the Laplace-transform variable p, Eq. 6–143 would yield a characteristic polynomial $a_0 p^2 + a_1 p + a_2$. There are only two possible ways that instability can occur. One is by having the coefficient of α change from positive to negative, corresponding to the condition $a_2 \leq 0$ in Routh's criterion (Ref. 6–4). This is just torsional divergence, the state of static instability where the negative "aerodynamic spring" about the E.A. overpowers K_α; it is fully discussed in Sec. 6–2(a). The second possibility is that of negative damping, wherein $\partial M_y / \partial \dot{\alpha}$ changes from algebraically negative to positive ($a_1 \leq 0$ by Routh's criterion). It is the source of dynamic instability or flutter.

Since no mechanical friction has been included, the negative damping must be entirely aerodynamic in origin. It follows that the neutral stability boundary is defined, in part, by the necessary condition

$$\mathscr{I}m\{M_y\} = 0 \qquad (6\text{–}144)$$

The circumstances under which Eq. 6–144 can be fulfilled in incompressible, potential flow have been analyzed by Smilg (Ref. 6–37) and Fung (Ref. 1–1), who interpret the work of Greidanus (Ref. 6–38). These authors find that the rotational axis must lie ahead of the $\frac{1}{4}$-chord line ($a \leq -\frac{1}{2}$) and the reduced frequency must be sufficiently small (k less than an absolute maximum of 0.0435).

It is possible to explain qualitatively both the negative damping at low k and the existence of a minimum critical flutter speed. Regarding the former question, we imagine a slow, nearly sinusoidal variation of α, shown by Fig. 6–29 at four stages phased 90° apart. The principal part L_0 of the lift due to the incremental angle of attack is in phase with α and acts at the $\frac{1}{4}$-chord A.C., thus adding to the torsional spring K_α a further contribution proportional to the dynamic pressure q when $a < -\frac{1}{2}$. This inphase lift has associated with it a circulation Γ_0 bound to the airfoil, whose sense is indicated in each panel of the figure. Γ_0 is changing with time; and, because the total circulation in an incompressible, potential flow must remain constant, its alterations are reflected by the continual appearance of countervortices of equal strength. These countervortices are shed from the trailing edge and swept downstream at the airspeed U, giving rise to a wake vortex sheet, the near portions of which are pictured in Fig. 6–29. Now, an out-of-phase loading at low k is caused by the angle of attack (or upwash) induced along the airfoil chord by these wake vortices. This upwash resembles a gust, embedded in the air and going downstream with speed U, since the wake itself moves in this way. Hence, as shown for gusts in Sec. 4–6(a) and in the references listed there, the additional lift L_2 which it produces acts precisely at the $\frac{1}{4}$-chord. The induced upwash and L_2 are also sketched in Fig. 6–29 in the directions

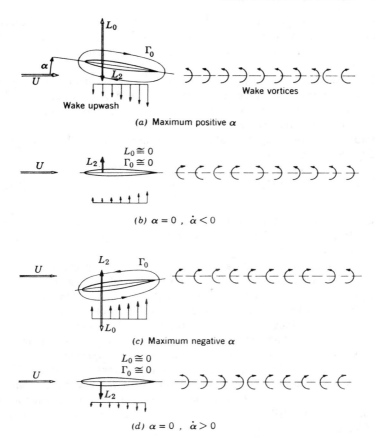

Fig. 6–29. Four positions of an airfoil oscillating about a fixed pitch axis in incompressible flow at low reduced frequency, showing the bound circulation Γ_0, wake vortices, upwash, and lift L_2 induced by the wake.

they would have at low reduced frequency, when the wake vortices mainly responsible for the upwash are those which left the trailing edge during the small fraction of a cycle just preceding the instant of observation, the remainder having been convected to large distances. The point of the discussion is as follows: when the rotation axis lies ahead of the $\frac{1}{4}$-chord, the moment due to L_2 is in the same sense as the angular velocity $\dot{\alpha}$ during more than half of each cycle of oscillation and therefore does a net positive work per cycle on the wing. This transfer of energy is the origin of the negative damping. At higher values of k, the damping becomes positive for a complex variety of reasons, among them the facts that more cycles of wake affect the upwash, the bound circulation lags behind the angular position, and the center of pressure of the resultant lift begins to oscillate.

The essence of the foregoing reasoning is the inclusion of unsteady effects. At least at subsonic speeds, quasi-steady theory is precisely what one arrives at when the influence of the wake is omitted entirely. To see how misleading are the results that one would obtain in the present example by overlooking the wake, we note that the incompressible quasi-steady damping derivative is

$$\left(\frac{\partial M_y}{\partial \dot{\alpha}}\right)_{QS} = \pi \rho_\infty U b^2 S [a(\tfrac{1}{2} - a)] \tag{6-145}$$

Equation 6–145 implies negative damping, independent of reduced frequency, about rotation axes between the $\frac{1}{2}$- and $\frac{3}{4}$-chord lines, at complete variance with the facts.

It is of interest to mention here the anomalous low-speed flutter discovered by Greidanus (Ref. 6–38), and Biot and Arnold (Ref. 6–39). Their discussions relate to bending-torsion instability with a vibration node near the $\frac{3}{4}$-chord axis. In the limit, however, such motion is indistinguishable from single-degree-of-freedom rotation about $a = \frac{1}{2}$. According to Eq. 4–171, the damping moment vanishes for just this particular value of a, so that the system is neutrally stable at all airspeeds.

Returning to the discussion of forward axis locations, the onset of flutter is easy to understand in terms of the variation of $\partial M_y/\partial \dot{\alpha}$ with k established above. At any values of U and q, the frequency ω of the aeroelastic mode is controlled primarily by the stiffness and inertia terms in Eq. 6–143. A good approximation is

$$\omega \cong \sqrt{\frac{K_\alpha - (\partial M_y/\partial \alpha)}{I_\alpha + \dfrac{\pi}{2} \rho_\infty b^3 S \left(\dfrac{1}{8} + a^2\right)}} \tag{6-146}$$

Since $\partial M_y/\partial \alpha$ is negative when $a < -\frac{1}{2}$ and proportional to q, ω starts from a value slightly below $\omega_\alpha = \sqrt{K_\alpha/I_\alpha}$ at $U = 0$ and rises slowly as the speed is increased. As a consequence, $k = \omega b/U$ falls steadily from its infinite still-air value. The damping stays positive until k reaches the boundary where $\partial M_y/\partial \dot{\alpha} = 0$, which defines the critical U_F. At higher U and lower k, the oscillation becomes and remains divergent.

Incidentally, a line of reasoning similar to the foregoing can be followed to prove that purely translational motion of this same system can never exhibit an instability, regardless of the Mach number range.

As a second attempt to visualize the flutter mechanism, we take up the two-degree-of-freedom typical section. For our aerodynamic operators, we adopt the same convention used in dynamic stability studies that the lift is determined by the instantaneous angle of attack $(\alpha + \dot{h}/U)$ on a steady-state basis and acts at the aerodynamic center. The equations of motion for any Mach number are then the following:

$$m[\ddot{h} + \omega_h^2 h] + S_\alpha \ddot{\alpha} = -qS \frac{\partial C_L}{\partial \alpha}\left[\alpha + \frac{\dot{h}}{U}\right] \qquad (6\text{--}147a)$$

$$S_\alpha \ddot{h} + I_\alpha[\ddot{\alpha} + \omega_\alpha^2 \alpha] = qS \frac{\partial C_L}{\partial \alpha} e\left[\alpha + \frac{\dot{h}}{U}\right] \qquad (6\text{--}147b)$$

All symbols in Eqs. 6–147 have been defined previously, including e, which is the distance by which the E.A. lies behind the A.C.; thus, $e = b[\frac{1}{2} + a]$ in subsonic flow, whereas according to piston theory with thickness effects $e = b\left[a + \left(\frac{\gamma + 1}{4}\right)M\frac{A_w}{2b^2}\right]$.

Avoiding unnecessary repetition, we merely observe that the eigenvalues and aeroelastic modes from Eqs. 6–147 are found by factoring a characteristic polynomial of fourth degree in the Laplace transform variable p. For a given wing, the coefficients are constants with a parametric dependence on q and possibly M. There are two oscillatory root pairs, each of the form $p = p_R \pm i\omega$. Choosing a representative set of subsonic parameters, we plot in Figs. 6–30 through 6–32 the dimensionless frequency and damping, amplitude ratio, and phase angle for each mode as functions of reduced speed $U/b\omega_\alpha$. The last two quantities are defined in Eqs. 6–129, the angle φ_h denoting the number of radians by which the downward displacement h leads the positive-nose-up rotation α.

The sequence of events by which instability develops in the higher-frequency torsional mode is as follows. From zero airspeed up to about half the critical value of the parameter $U/b\omega_\alpha$, the mode shape remains essentially the same as that for free vibration at the frequency ω_2, involving pure rotation about an axis slightly behind the C.G. This axis then begins to move forward, as indicated by the falling amplitude of bending accompanied by very little change in the 180° phase difference between the degrees of freedom. The gradual suppression of the h-motion is caused by the lift due to torsion. This lift, in phase with α, drives the bending freedom at a frequency ω greater than ω_h, so that the response to it has a maximum downward amplitude at the instant of maximum upward force and tends to cancel the bending component of the free-vibration mode.

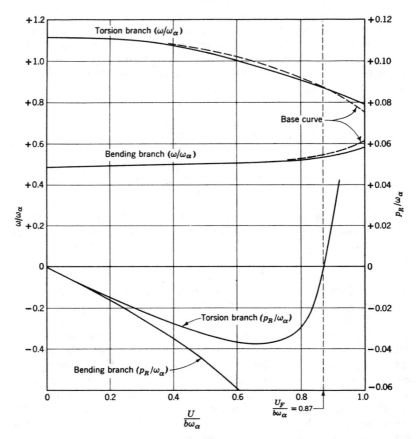

Fig. 6–30. Dimensionless frequency ω/ω_α and damping p_R/ω_α of the aeroelastic modes of the typical section, estimated using steady-state aerodynamic operators and plotted vs. reduced airspeed $U/b\omega_\alpha$. System parameters are $x_\alpha = 0.5$, $r_\alpha = 0.5$, $\omega_h/\omega_\alpha = 0.5$, $(2m'/\pi\rho_\infty bS) = 10$, $e/b = 0.4$, $\dfrac{\partial C_L}{\partial\alpha} = 2\pi$.

Simultaneously, ω drops, because the lift is applied at the $\frac{1}{4}$-chord and constitutes a negative "aerodynamic spring" on the torsional freedom with a "spring constant" proportional to the dynamic pressure. The small advance in φ_h is attributable to the lift due to h. Flutter sets in when the bending amplitude passes through zero, leaving a pure rotational oscillation around the E.A., on which no damping forces act according to the approximations underlying Eqs. 6–147.

The origin of instability becomes clearer when we examine the energy exchange between the airstream and the wing. The aerodynamic work done on the system per cycle of oscillation is

$$
W = -\int_{t=0}^{2\pi/\omega} \mathscr{R}e\,\{L\}\; \mathscr{R}e\left\{\frac{dh}{dt}\,dt\right\} + \int_{t=0}^{2\pi/\omega} \mathscr{R}e\,\{M_y\}\; \mathscr{R}e\left\{\frac{d\alpha}{dt}\,dt\right\}
$$
$$
\equiv W_h + W_\alpha \tag{6-148}
$$

Here we must take real parts because of the nonlinearity of this operation on quantities expressed in the complex notation. If we neglect the small deviations from simple harmonic motion, it is easy to show that the torsional work W_α is proportional to $\left|\dfrac{h_0}{b\alpha_0}\right|\dfrac{e}{b}\,qk\cos\varphi_h$ and is done entirely by the moment due to h. The lift work W_h is proportional to

$$
\left|\frac{h_0}{b\alpha_0}\right|q\left[\sin\varphi_h - k\left|\frac{h_0}{b\alpha_0}\right|\right],
$$

with the first term coming from the dephased α-lift and the second the pure damping experienced by any translational oscillation.

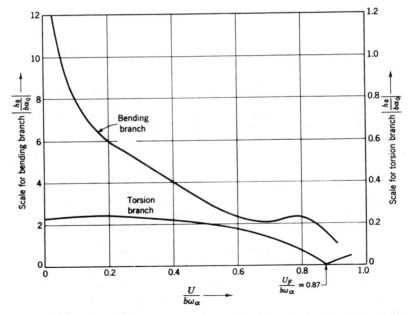

Fig. 6–31. Dimensionless amplitude ratio $|\,h_0/b\alpha_0\,|$ between the bending and torsional displacements of the aeroelastic modes of the system described in Fig. 6–30.

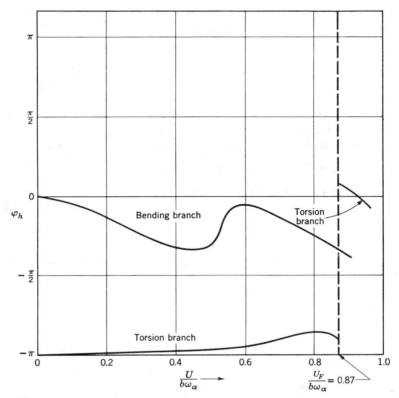

Fig. 6–32. Phase angle φ_h by which the bending leads the torsional displacement for the aeroelastic modes of the system described in Fig. 6–30.

The two contributions to W are plotted in dimensionless form in Fig. 6–33. As the airspeed is first brought up from zero, both W_h and W_α extract energy and assure stability, since the rotation about a rearward axis generates both a damping force and moment due to h. These negative works are roughly proportional to U at first

$$\left(\text{recall that } qk \equiv \left(\frac{\rho_\infty U^2}{2}\right)\left(\frac{\omega b}{U}\right) \sim U\right),$$

but they begin to drop off as the bending amplitude decreases and vanish simultaneously with it at flutter. For $U > U_F$, the effective rotation takes place about an axis ahead of the E.A., corresponding to a $180°$ shift in φ_h. The h-moment then does positive work and destabilizes the motion.

A similar analysis can be made of the behavior of the bending aeroelastic mode. At low speeds, this consists of rotation about a line far forward of the leading edge. Thus the stability is controlled by the translational damping in W_h, although there is a small positive work W_α due to the h-moment (see Fig. 6–34). This moment associated with the bending velocity also excites the torsional freedom, driving it at a frequency less than ω_α and therefore producing a rotation in phase with the torque itself. One can reason that this reduces $|h_0/b\alpha_0|$ and causes the increasingly negative, or lagging, φ_h. These phase and amplitude shifts are unable, however, completely to overpower the h-damping, so that this mode remains stable up to the flutter speed and beyond.

Not every bending-torsion instability comes about in the manner just described. In some cases, both the changes in mode shape and the variations of φ_h with airspeed play an important role. When unsteady aerodynamic effects are accounted for, we find that W_h and W_α do not simultaneously vanish at the critical $U_F/b\omega_\alpha$ but become equal in magnitude and opposite in sign. A qualitative description of one example of this type, in which the lower-frequency aeroelastic mode develops flutter because φ_h becomes positive and the torsional lift does work on the bending. has been attempted by Kassner and Fingado (Ref. 6–47; see also Ref. 6–48),

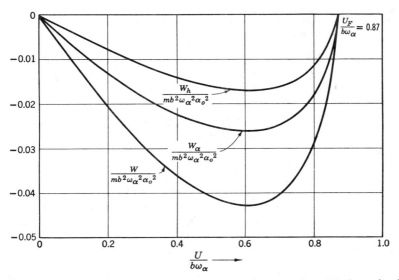

Fig. 6–33. Aerodynamic work done on the system of Fig. 6–30 during the cycle of oscillation commencing at $t = 0$, plotted vs. reduced airspeed for the aeroelastic mode associated with torsional frequency ω_2.

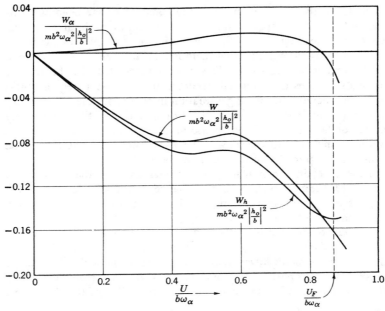

Fig. 6–34. Aerodynamic work done on the system of Fig. 6–30 during the cycle of oscillation commencing at $t = 0$, plotted vs. reduced airspeed for the aeroelastic mode associated with bending frequency ω_1.

The reader will find it an interesting exercise to undertake a physical explanation, along the foregoing lines, of flutter involving coupling between bending and rotation of a trailing edge flap. This reasoning proves somewhat less complicated, and, if careful attention is given to the effects of the aerodynamic hinge moments, one can demonstrate the stabilization attainable by mass-balancing the flap.

A significant conclusion which follows from all such discussions is that changes of the aeroelastic mode shapes have a major influence on how and where the instability of a multi-degree-of-freedom system sets in. We also believe that some light can be thrown on studies such as Ref. 6–49, wherein the alterations in the bending-torsion stability boundary are calculated which result from certain arbitrary modifications of various aerodynamic terms in the flutter equations. It is found, for instance, that the in-phase component of lift due to the torsional motion is often the airload to which stability is most sensitive. This discovery is not surprising in view of the example of the present section, where this lift causes the modal modification that leads to negative aerodynamic damping. Another verification is implied in the explanation given by Ref. 6–47.

(a) Instability and the merging of natural frequencies

Adopting even more extreme simplifications than those made in the foregoing discussion, Rocard (Ref. 1–33, Sec. 119) and Pines (Ref. 6–35) have suggested that flutter of a multi-degree-of-freedom system can be explained by discarding aerodynamic energy dissipation entirely. Although we feel that something essential is lost through eliminating the effects of continuous variations with airspeed of the phase angle φ_h, which is one consequence of this approximation, their conception is nevertheless intensely interesting.

Consider the equations of motion that are obtained by dropping the terms involving h/U from the right-hand sides of Eqs. 6–147. Only derivatives of even orders remain, and the resulting characteristic polynomial must be a biquadratic, $a_0 p^4 + a_2 p^2 + a_4$. In practice, the two p^2-roots turn out to be either real and negative, corresponding to undamped vibrations $p = \pm i\omega_1$, $p = \pm i\omega_2$, or complex conjugates. In fact, when plotted versus $U/b\omega_\alpha$, the p-roots form the "base curve" shown in Fig. 6–30 or Fig. 6–16. In Fig. 6–35 we sketch a typical roots locus on the upper half of the complex plane. Evidently the instability is associated with a merging of the frequencies of the aeroelastic modes (in Rocard's French, *auto-oscillation par confusion de deux fréquences propres*). This coalescence must occur at the critical speed U_F, because it follows from the realness of the coefficients a_0, a_2, a_4 that, at any $U > U_F$, there are two modes of equal frequency, one decaying and the other divergent.

Instability by merging of frequencies may also appear in an undamped system with more than two degrees of freedom. For example, exactly this phenomenon underlies Hedgepeth's very successful approximate scheme

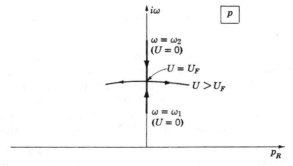

Fig. 6–35. Locus of roots of the characteristic polynomial for the typical section without aerodynamic damping; airspeed U increases in the direction of the arrows. (Lower half of the locus is symmetrical with respect to the p_R-axis.)

for predicting panel flutter at the higher supersonic Mach numbers (Ref. 6–40; see also Chap. 8 of Houbolt's thesis, Ref. 6–43). It is, of course, necessary that the system be unconservative and capable of drawing on external energy sources. The lack of energy conservation is always betrayed by *asymmetric coupling* between the degrees of freedom, such as can be seen in the aerodynamic terms of Eqs. 6–147. Indeed, when a system is conservative and the parameter which symmetrically couples two degrees of freedom is enlarged, the natural frequencies associated with them are forced apart and can never merge. This principle is demonstrated for free vibrations of the typical section by reference to the kinetic and potential energies, which are

$$T = \tfrac{1}{2}mh^2 + \tfrac{1}{2}I_\alpha\dot\alpha^2 + S_\alpha\dot\alpha h \qquad (6\text{–}149)$$

and

$$U = \tfrac{1}{2}K_h h^2 + \tfrac{1}{2}K_\alpha\alpha^2 \qquad (6\text{–}150)$$

When the structure is vibrating at either natural frequency ω_i, with modal amplitudes h_{0i} and α_{0i}, the equality between the maximum values of T and U leads to the following expression for ω_i^2:

$$\omega_i^2 = \frac{\tfrac{1}{2}K_h h_{0i}^2 + \tfrac{1}{2}K_\alpha\alpha_{0i}^2}{\tfrac{1}{2}mh_{0i}^2 + \tfrac{1}{2}I_\alpha\alpha_{0i}^2 + S_\alpha\alpha_{0i}h_{0i}}, \qquad (i = 1, 2) \qquad (6\text{–}151)$$

Imagining S_α to be increased, we find for the rate of change of frequency

$$\frac{\partial(\omega_i^2)}{\partial S_\alpha} = -\alpha_{0i}h_{0i}\frac{\tfrac{1}{2}K_h h_{0i}^2 + \tfrac{1}{2}K_\alpha\alpha_{0i}^2}{[\tfrac{1}{2}mh_{0i}^2 + \tfrac{1}{2}I_\alpha\alpha_{0i}^2 + S_\alpha\alpha_{0i}h_{0i}]^2} \qquad (6\text{–}152)$$

It is easily proved for an initially positive S_α that α_{01} and h_{01}, corresponding to the lower ω_1, are both of the same sign, whereas α_{02} and h_{02} are opposite. Therefore, added coupling tends to depress ω_1 and increase ω_2.*

(b) Pines' approximate rules for bending-torsion flutter

Many useful things can be learned from close examination of the idea of instability by frequency coalescence. For instance, Rocard is able to show that adding certain types of damping is always destabilizing, in the sense that it reduces the critical value of the coupling parameter (Ref. 1–33, Sec. 120; see also Salaün, Ref. 6–50). With such damping present, the frequencies do not merge by the time the stability boundary is reached. Putting the h-terms back into Eqs. 6–147 provides an illustration of this

---

* To complete the proof, we must point out that $\partial(\omega_i^2)/\partial h_{0i}$ and $\partial(\omega_i^2)/\partial\alpha_{0i}$ vanish, so that the frequency is stationary with respect to the small changes in mode shape that also occur. This is a special case of some important theorems of Rayleigh (Ref. 6–41) regarding the effects of system changes on natural frequencies. A partial account is given in Sec. 13–3 of Ref. 1–2.

surprising theorem, and it thus explains why flutter occurs to the left of the noses of the base curves in Figs. 6–16, 6–17, and 6–30. Very unfortunately, space limitations prevent our reproducing Rocard's demonstration in full. We also warn the reader that it is by no means universally applicable in aeroelasticity. Thus, Hedgepeth points out (Chaps. 7 and 8, Ref. 6–43) that damping of aerodynamic origin at high supersonic Mach numbers can increase the critical speed for a skin panel.

In Ref. 6–35, Pines deduces a set of approximate rules governing typical-section flutter by studying the characteristic polynomial of the undamped system. With the h-terms omitted from Eqs. 6–147 one finds that the (dimensionless) characteristic equation reads

$$\left(\frac{p}{\omega_\alpha}\right)^4\left[1 - \frac{x_\alpha^2}{r_\alpha^2}\right] + \left(\frac{p}{\omega_\alpha}\right)^2\left\{1 + \left(\frac{\omega_h}{\omega_\alpha}\right)^2 - \frac{qSb}{K_\alpha}\frac{\partial C_L}{\partial \alpha}\left[x_\alpha + \frac{e}{b}\right]\right\}$$

$$+ \left(\frac{\omega_h}{\omega_\alpha}\right)^2\left[1 - \frac{qSb}{K_\alpha}\frac{\partial C_L}{\partial \alpha}\frac{e}{b}\right] = 0 \quad (6\text{–}153)$$

It is easy to see that *static* instability comes about with the vanishing of the bracketed factor in the constant term,* which yields the dynamic pressure q_D of torsional divergence, as in Eq. 6–11.

Dynamic instability coincides with the merging of the natural frequencies, the test for this situation involving negative or zero values of the discriminant of Eq. 6–153, that is,

$$\left\{1 + \left(\frac{\omega_h}{\omega_\alpha}\right)^2 - \frac{qSb}{K_\alpha}\frac{\partial C_L}{\partial \alpha}\left[x_\alpha + \frac{e}{b}\right]\right\}^2$$

$$- 4\left[1 - \frac{x_\alpha^2}{r_\alpha^2}\right]\left(\frac{\omega_h}{\omega_\alpha}\right)^2\left[1 - \frac{qSb}{K_\alpha}\frac{\partial C_L}{\partial \alpha}\frac{e}{b}\right] \leq 0 \quad (6\text{–}154)$$

Pines identifies this as a limitation on the magnitude of the dynamic pressure parameter

$$Q \equiv \frac{qSb}{K_\alpha}\frac{\partial C_L}{\partial \alpha} \quad (6\text{–}155)$$

Equation 6–154 constitutes a quadratic expression in Q. Sufficient conditions for instability are that Q be real, positive, and lie between the roots of the quadratic,

$$Q_1 \leq Q \leq Q_2 \quad (6\text{–}156a)$$

---

* This is just Routh's condition $a_4 = 0$. Actually, the absence of a_1 and a_3 means that the system is never better than neutrally stable.

where

$$Q_{1,2} = \frac{1}{\left[x_\alpha + \dfrac{e}{b}\right]^2} \left\{ \left[1 + \left(\frac{\omega_h}{\omega_\alpha}\right)^2 \right]\left[x_\alpha + \frac{e}{b} \right] - 2\frac{e}{b}\left(\frac{\omega_h}{\omega_\alpha}\right)^2 \left[1 - \frac{x_\alpha{}^2}{r_\alpha{}^2} \right] \right.$$

$$\left. \mp 2\sqrt{ x_\alpha \left(\frac{\omega_h}{\omega_\alpha}\right)^2 \left[1 - \frac{x_\alpha{}^2}{r_\alpha{}^2} \right]\left[x_\alpha + \frac{e}{b} - \frac{e}{b}\left(\frac{\omega_h}{\omega_\alpha}\right)^2 \left(1 + \frac{e}{b}\frac{x_\alpha}{r_\alpha{}^2} \right) \right] } \right\}$$

$$(6\text{–}156b)$$

Equations 6–156 imply a range of dynamic pressure within which flutter might be observed. Furthermore, the coefficient of Q^2 and the constant term in Eq. 6–154 are both positive, so that Q_1 and Q_2, if real, must have the same sign. This sign will be positive provided the coefficient of the first power of Q is negative, which leads to the condition

$$x_\alpha + \frac{e}{b} + \left(\frac{\omega_h}{\omega_\alpha}\right)^2 \left[x_\alpha - \frac{e}{b} + 2\frac{e}{b}\frac{x_\alpha{}^2}{r_\alpha{}^2} \right] \geq 0 \qquad (6\text{–}157)$$

Additionally, Q_1 and Q_2 will be real only if the quantity under the radical in Eq. 6–156b is positive, that is,

$$x_\alpha \left[x_\alpha + \frac{e}{b} - \frac{e}{b}\left(\frac{\omega_h}{\omega_\alpha}\right)^2 \left(1 + \frac{e}{b}\frac{x_\alpha}{r_\alpha{}^2} \right) \right] \geq 0 \qquad (6\text{–}158)$$

Both of the inequalities (6–157) and (6–158) must be satisfied before dynamic instability can occur. These combined requirements generate the following conclusions (Ref. 6–35):

(1) No flutter exists for C.G. positions forward of the E.A. ($x_\alpha < 0$).

(2) Flutter at some speed is possible for all frequency ratios if the A.C. lies between the E.A. and the C.G. $\left(x_\alpha > 0, \dfrac{e}{b} < 0 \text{ and } |x_\alpha| > \left| \dfrac{e}{b} \right| \right)$.

(3) If the A.C. is forward of the E.A. $\left(\dfrac{e}{b} > 0, x_\alpha > 0 \right)$, flutter is possible only if

$$\left(\frac{\omega_h}{\omega_\alpha}\right)^2 \leq \frac{x_\alpha + (e/b)}{\dfrac{e}{b}\left[1 + \dfrac{e}{b}\dfrac{x_\alpha}{r_\alpha{}^2} \right]} \qquad (6\text{–}159)$$

(4) If the A.C. is aft of the C.G. $\left(x_\alpha > 0, \frac{e}{b} < 0, \left|\frac{e}{b}\right| > x_\alpha\right)$, flutter is possible only if

$$\left(\frac{\omega_h}{\omega_\alpha}\right)^2 > \frac{-x_\alpha + \left|\dfrac{e}{b}\right|}{\left|\dfrac{e}{b}\right|\left[1 + \dfrac{e}{b}\dfrac{x_\alpha}{r_\alpha^2}\right]} \qquad (6\text{-}160a)$$

and

$$r_\alpha^2 > x_\alpha \left|\frac{e}{b}\right| \qquad (6\text{-}160b)$$

It is not a difficult exercise to prove that these four statements follow from Eqs. 6–157 and 6–158, if account is taken of Eq. 6–45. Reference 6–35 illustrates them pictorially and makes some other interesting deductions. One case of practical importance is that of small $(\omega_h/\omega_\alpha)^2$. The limiting zero value of frequency ratio must be approached with care, because this removes an elastic constraint from the system and changes the total order of the differential equations. It is clear, however, that Eq. 6–156b can be approximated by

$$Q_{1,2} \cong \frac{1}{\left[x_\alpha + \dfrac{e}{b}\right]} \mp \frac{2\dfrac{\omega_h}{\omega_\alpha}\sqrt{x_\alpha\left[1 - \dfrac{x_\alpha^2}{r_\alpha^2}\right]}}{\left[x_\alpha + \dfrac{e}{b}\right]^{3/2}} \qquad (6\text{-}161)$$

For instability, $\left[x_\alpha + \dfrac{e}{b}\right]$ must be positive along with x_α itself, which means that the C.G. must fall behind the A.C. Flutter occurs only in the neighborhood of the dynamic pressure parameter

$$Q \equiv \frac{qSb}{K_\alpha}\frac{\partial C_L}{\partial \alpha} = \frac{1}{\left[x_\alpha + \dfrac{e}{b}\right]} \qquad (6\text{-}162)$$

When solved for the speed, Eq. 6–162 yields

$$\frac{U}{b\omega_\alpha} = \sqrt{\left(\frac{2m}{\rho_\infty bS}\right)\frac{r_\alpha^2}{\dfrac{\partial C_L}{\partial \alpha}\left[x_\alpha + \dfrac{e}{b}\right]}} \qquad (6\text{-}163)$$

which is reminiscent of the empirical formula (6–136) and shows the same effects of the various parameters. Since the flutter frequency can also be shown to vanish, Pines correctly identifies the limiting instability at

$\omega_h = 0$ as nonoscillatory dynamic divergence involving both translational acceleration and rotation about the C.G.

It must be realized that all the results just stated are based on rather extreme assumptions and cannot be accepted quantitatively. They are, however, very helpful when interpreting parametric data such as those of Refs. 6–2 and 6–34. Moreover, Ref. 6–35 points out how these ideas can be carried over to assist in dealing with much more complicated systems and states, "the problem of the flutter design engineer is to establish methods which will prevent regions of frequency coincidence from occurring within the flight regime." Based on considerable experience, Pines recommends procedures for assuring this prevention.

(c) Necessary conditions for instability based on aerodynamic energy input

As with any sort of instability, supercritical flutter exists because the total mechanical energy of the lifting surface is, on the average, increasing with time. When dissipative forces within the structure are negligible and the system is linear, this test is equivalent to the statement that the work per cycle of oscillation done by the aerodynamic loads on the surface is positive, as has already been discussed above in connection with the typical section. In the early works of Frazer and Duncan (Ref. 6–36), this idea was employed extensively for studying the stability of a particular type of semirigid wing with aileron. It has been elaborated in a number of more recent publications, among them those of Garrick (Ref. 6–44), Greidanus (Ref. 6–38 et al.), Rott (Ref. 6–45), and Duncan (Ref. 6–46, a very general and elegant investigation).

Crisp (Ref. 6–42) introduces an *ultimate stability criterion*, or relationship between mode shape and reduced frequency, which must be fulfilled in a given configuration before flutter can exist. Some of Crisp's results regarding energy transfer to a vibrating wing, which are essentially founded on the use of quasi-steady aerodynamic operators, contribute to the physical understanding we are seeking and will be reviewed here. Consider a linear system whose motion can be described in terms of a set of generalized coordinates $q_i(t)$ (e.g., $q_1 = \dfrac{h}{b}$ and $q_2 = \alpha$ for the typical section).

Crisp observes that, when the airloads involve only $q_i, \dot{q}_i,$ and $\ddot{q}_i$ multiplied by constant factors (this comprehends virtual inertias as well as the conventional quasi-steady effects), the equations of motion can be written in the matric notation

$$[A]\{\ddot{q}_i\} + [B]\{\dot{q}_i\} + [C]\{q_i\} = 0 \qquad (6\text{–}164)$$

The same matrices of coefficients appearing in Eqs. 6–164 form the basis of the following three quantities:

$$\tau = \tfrac{1}{2}\lfloor \dot{q}_i \rfloor [A]\{\dot{q}_i\} \qquad (6\text{–}165)$$

$$F = \tfrac{1}{2}\lfloor \dot{q}_i \rfloor [B]\{\dot{q}_i\} \qquad (6\text{–}166)$$

$$U = \tfrac{1}{2}\lfloor q_i \rfloor [C]\{q_i\} \qquad (6\text{–}167)$$

τ and U are generalized kinetic and potential energies, including aerodynamic inertias and stiffnesses, respectively. F is similar to the *dissipation function* of Rayleigh (Ref. 6–41). $[A]$, $[B]$, and $[C]$ are symmetrical when dealing with an elementary mechanical system, but the addition of the aerodynamic couplings spoils this symmetry in the case of the last two. Each of them can be decomposed into symmetric and skew-symmetric portions, for example,

$$[B] = [B_1] + [B_2] \qquad (6\text{–}168)$$

Moreover, only $[B_1]$ and $[C_1]$ contribute to F and U, in view of the property of skew-symmetric matrices exemplified by the equation $\lfloor q_i \rfloor [C_2]\{q_i\} = 0$.

Under conditions of uniform flow at infinity and in the absence of heat transfer, all three matrices are made up of constants; but Ref. 6–42 considers the possibility that they might be functions of time, thus allowing, on a quasi-steady basis, for variations in flight speed and altitude or for aerothermoelastic modifications to the flexibility of the structure. It can then be shown that the set of Eqs. 6–164 reads

$$[A(t)]\{\ddot{q}_i\} + ([\dot{A}(t)] + [B(t)])\{\dot{q}_i\} + [C(t)]\{q_i\} = 0 \qquad (6\text{–}169)$$

the dot over a matrix meaning that the time derivative of each element is implied.

The total energy of the system* is $(\tau + U) = E$. By taking the derivative of this quantity, using Eq. 6–169 to eliminate $\{\ddot{q}_i\}$, and applying the aforementioned property to drop out some skew-symmetric matrices, Crisp derives the following expression for the instantaneous rate at which work is being done by the air:

$$\frac{dE}{dt} = -\lfloor \dot{q}_i \rfloor ([B_1] + \tfrac{1}{2}[\dot{A}_1])\{\dot{q}_i\}$$

$$\qquad - \lfloor \dot{q}_i \rfloor [C_2]\{q_i\} + \tfrac{1}{2}\lfloor q_i \rfloor [\dot{C}_1]\{q_i\} \qquad (6\text{–}170)$$

The first term on the right of Eq. 6–170 is a generalized dissipation function, composed of the sum of $-2F$ and an obvious effect of the rate of change of

---

* Note that E is not exactly the mechanical energy of the solid material but includes certain aerodynamic inertias and "springs." It is nevertheless true that the stability is determined by the rate of change of E, with increasing E implying an unstable system.

inertia, it is nearly always negative and therefore stabilizing. The second term, containing the skew-symmetric part of the aerodynamic stiffness, is the one which has the major controlling influence on stability. Its contribution to dE/dt may be positive or negative, depending on the mode shape and dynamic pressure; but in any event $[C_2]$ is a large quantity in aeroelastic systems. For instance, on the typical section as described by Eqs. 6–147,

$$[C_2] = \begin{bmatrix} 0 & \frac{1}{2}qS\dfrac{\partial C_L}{\partial \alpha} \\[2ex] -\frac{1}{2}qS\dfrac{\partial C_L}{\partial \alpha} & 0 \end{bmatrix} \qquad (6\text{–}171)$$

The significant role of $[C_2]$ is even clearer in uniform flight, when the square matrices are independent of time and Eq. 6–170 reduces to

$$\frac{dE}{dt} = -(2F + \lfloor \dot{q}_i \rfloor [C_2]\{q_i\}) \qquad (6\text{–}172)$$

Several interesting conclusions regarding the influence of system properties on stability are deduced in Ref. 6–42 from Eqs. 6–170 and 6–172. The reader will find it instructive to examine more closely the application to the typical section, including the consequences of rapid variations in ambient gas state and airspeed, such as might be found in certain intermittent wind tunnels.

For a structure in vibration at a critical flutter boundary, one can state a *necessary condition* for equilibrium by requiring the net energy input per cycle to vanish:

$$E \Big|_{t=0}^{t=2\pi/\omega} = -\int_0^{2\pi/\omega} (2F + \lfloor \dot{q}_i \rfloor [C_2]\{q_i\})\, dt$$
$$= 0 \qquad (6\text{–}173)$$

The generalized coordinates here have the forms

$$q_i = Q_i \sin(\omega t + \varphi_i), \qquad (i = 1, 2, \cdots, n) \qquad (6\text{–}174)$$

Reference 6–42 shows that Eq. 6–173 is reducible to the dimensionless relation*

$$\sum_{i=1}^{n} \sum_{j=1}^{n} R_i R_j \left[b_{ij}^{(1)} \cos \lambda_{ij} - \frac{c_{ij}^{(2)}}{k} \sin \lambda_{ij} \right] = 0 \qquad (6\text{–}175)$$

where $R_i = Q_i/Q_R$, $\lambda_{ij} = (\varphi_i - \varphi_j)$, Q_R is the amplitude of some reference coordinate, and $b_{ij}^{(1)}$ and $c_{ij}^{(2)}$ are suitably dimensionless elements

---

* Rigorous unsteady aerodynamic operators can be employed in Eq. 6–173.

of the symmetric part of $[B]$ and the skew-symmetric part of $[C]$, respectively. When binary flutter is being analyzed ($n = 2$), Eq. 6–175 can be given the elementary graphical interpretation that all possible flutter conditions fall on a circle whose size and location depend on the aforementioned matrix elements. Flutter may be prevented, in theory, by reducing the radius of this circle to zero. This fact forms the basis of one of Crisp's stability criteria, and in certain systems no flutter is possible at reduced frequencies greater than some limiting value. Other criteria can be constructed by examining the energy input to individual degrees of freedom. Applications of some of the foregoing ideas to flutter of low-aspect-ratio wings will be found in Ref. 7–34.

REFERENCES

6–1.   Theodorsen, T., *General Theory of Aerodynamic Instability and the Mechanism of Flutter*, NACA Report 496, 1935.

6–2.   Theodorsen, T., and I. E. Garrick, *Mechanism of Flutter, a Theoretical and Experimental Investigation of the Flutter Problem*, NACA Report 685, 1940.

6–3.   Roxbee-Cox, H., and A. G. Pugsley, *Stability of Static Equilibrium of Elastic and Aerodynamic Actions on a Wing*, British A.R.C. Reports and Memoranda 1059, 1934.

6–4.   Gardner, M. F., and J. L. Barnes, *Transients in Linear Systems*, John Wiley and Sons, New York, 1942.

6–5.   Draper, C. S., W. McKay, and S. Lees, *Instrument Engineering*, Vol. 2, McGraw-Hill Book Company, New York, 1953.

6–6.   Sneddon, I. N., *Fourier Transforms*, McGraw-Hill Book Company, New York, 1951.

6–7.   Campbell, G. A., and R. M. Foster, *Fourier Integrals for Practical Applications*, Technical Publication, Bell Telephone System, 1931.

6–8.   Churchill, R. V., *Modern Operational Mathematics in Engineering*, McGraw-Hill Book Company, New York, 1944.

6–9.   Doetsch, G., *Theorie und Anwendung der Laplace-Transformation*, Dover Publications, New York, 1944.

6–10. Feller, W., *An Introduction to Probability Theory and Its Applications*, John Wiley and Sons, New York, 1950.

6–11. Laning, J. H., and R. H. Battin, *Random Processes in Automatic Control*, McGraw-Hill Book Company, New York, 1956.

6–12. Liepmann, H. W., "On the Application of Statistical Concepts to the Buffeting Problem," *J. Aero. Sciences*, Vol. 19, No. 12, Dec. 1952, pp. 793–800, 822.

6–13. Dugundji, J., "A Nyquist Approach to Flutter," *J. Aero. Sciences*, Readers' Forum, Vol. 19, No. 6, June 1952, pp. 422–423.

6–14. Goland, M., and Y. L. Luke, "A Study of the Bending-Torsion Aeroelastic Modes for Aircraft Wings," *J. Aero. Sciences*, Vol. 16, No. 7, July 1949, pp. 389–396.

6–15. Goland, M., and Y. L. Luke, "The Flutter of a Uniform Wing with Tip Weights," *J. Applied Mechanics*, Vol. 15, No. 1, March 1948, pp. 13–20.

6–16. Zisfein, M. B., and F. J. Frueh, *A Study of Velocity-Frequency-Damping Relationships for Wing and Panel Binary Systems in High Supersonic Flow*, USAF Office of Scientific Research Technical Note 59-969, October 1959.

6–17. Theodorsen, T., and I. E. Garrick, *Flutter Calculations in Three Degrees of Freedom*, NACA Report 741, 1942.

6–18. Duncan, W. J., and C. L. T. Griffith, *The Influence of Wing Taper on the Flutter of Cantilever Wings*, British A.R.C. Reports and Memoranda 1869, 1939.

6–19. Duncan, W. J., and H. M. Lyon, *Calculated Flexural-Torsional Flutter Characteristics of Some Typical Cantilever Wings*, British A.R.C. Reports and Memoranda 1782, 1937.

6–20. van de Vooren, A. I., and J. H. Greidanus, *Diagrams of Critical Flutter Speed for Wings of a Certain Standard Type*, Nationaal Luchtvaartlaboratorium, Amsterdam, Report V. 1297, 1946.

6–21. van de Vooren, A. I., *Diagrams of Flutter, Divergence and Aileron Reversal Speeds for Wings of a Certain Standard Type*, Nationaal Luchtvaartlaboratorium, Amsterdam, Report V. 1397, 1947.

6–22. Falkner, V. M., *Effect of Variation of Aileron Inertia and Damping on Flexural-Aileron Flutter of a Typical Cantilever Wing*, British A.R.C. Reports and Memoranda 1685, 1935.

6–23. Woolston, D. S., and G. E. Castile, *Some Effects of Variations in Several Parameters Including Fluid Density on the Flutter Speed of Light Uniform Cantilever Wings*, NACA Technical Note 2558, 1951.

6–24. Henry, C. J., Jr., J. Dugundji, and H. Ashley, *Aeroelastic Stability of Lifting Surfaces in High-Density Fluids*, USAF Office of Scientific Research Technical Note 58-626, June 1958.

6–25. Abramson, H. N., and W. H. Chu, *A Discussion of the Flutter of Submerged Hydrofoils*, Southwest Research Institute Technical Report 1, Contract No. Nonr 2470 (00), August 1958.

6–26. Dugundji, J., "Effect of Quasi-Steady Airforces on Incompressible Bending-Torsion Flutter," *J. Aero. Sciences*, Vol. 25, No. 2, February 1958, pp. 119–121.

6–27. Garrick, I. E., *Bending-Torsion Flutter Calculations Modified by Subsonic Compressibility Corrections*, NACA Report 836, 1948.

6–28. Nelson, H. C., and J. H. Berman, *Calculations on the Forces and Moments for an Oscillating Wing-Aileron Combination in Two-Dimensional Potential Flow at Sonic Speed*, NACA Report 1128, 1953.

6–29. Garrick, I. E., and S. I. Rubinow, *Flutter and Oscillating Airforce Calculations for an Airfoil in a Two-Dimensional Supersonic Flow*, NACA Report 846, 1946.

6–30. Woolston, D. S., and V. Huckel, *A Calculation Study of Wing-Aileron Flutter in Two Degrees of Freedom for Two-Dimensional Supersonic Flow*, NACA Technical Note 3160, 1954.

6–31. Chawla, J. P., "Aeroelastic Instability at High Mach Number," *J. Aero. Sciences*, Vol. 25, No. 4, April 1958, pp. 246–258.

6–32. Morgan, H. G., H. L. Runyan, and V. Huckel, "Theoretical Considerations of Flutter at High Mach Numbers," *J. Aero. Sciences*, Vol. 25, No. 6, June 1958, pp. 371–381.

6–33. Weatherill, W. H., and G. Zartarian, *Tabular Presentation of Supersonic Flutter Trends from Piston Theory Calculations*, USAF Wright Air Development Center Technical Note 57-310, October 1957.

6–34. Ashley, H., and G. Zartarian, *Supersonic Flutter Trends as Revealed by Piston*

Theory Calculations, USAF Wright Air Development Center Technical Report 58-74, May 1958.

6–35. Pines, S., "An Elementary Explanation of the Flutter Mechanism," *Proc.* Nat. Specialists Meeting on Dynamics and Aeroelasticity, Institute of the Aeronautical Sciences, Ft. Worth, Texas, November 1958, pp. 52–58.

6–36. Frazer, R. A., and W. J. Duncan, *The Flutter of Aeroplane Wings*, British A.R.C. Reports and Memoranda 1155, 1928.

6–37. Smilg, B., "The Instability of Pitching Oscillations of an Airfoil in Subsonic Incompressible Potential Flow," *J. Aero. Sciences*, Vol. 16, No. 11, November 1949, pp. 691–696.

6–38. Greidanus, J. H., "Low-Speed Flutter," *J. Aero. Sciences*, Readers' Forum, Vol. 16, No. 2, February 1949, pp. 127–128.

6–39. Biot, M. A., and L. Arnold, "Low Speed Flutter and Its Physical Interpretation," *J. Aero. Sciences*, Vol. 15, No. 4, April 1948, pp. 232–236.

6–40. Hedgepeth, J. M., "Flutter of Rectangular Simply Supported Panels at High Supersonic Speeds," *J. Aero. Sciences*, Vol. 24, No. 8, August 1957, pp. 563–573, 586.

6–41. Strutt, J. W. (Baron Rayleigh), *The Theory of Sound*, Vol. 1, 2nd ed., Dover Publications, New York, 1945.

6–42. Crisp, J. D. C., "The Equation of Energy Balance for Fluttering Systems with Some Applications in the Supersonic Regime," *J. Aero/Space Sciences*, Vol. 26, No. 11, November 1959, pp. 703–716, 738.

6–43. Houbolt, J. C., *A Study of Several Aerothermoelastic Problems of Aircraft Structures in High-Speed Flight*, Institut für Flugzeugstatik und Leichtbau, Eidgenössische Technische Hochschule, Zürich, Mitteilung Nr. 5, 1958.

6–44. Garrick, I. E., *Propulsion of a Flapping and Oscillating Airfoil*, NACA Report 567, 1936.

6–45. Rott, N., "Flügelschwingungsformen in ebener kompressibler Potentialströmung," *Z. angew. Math. Phys.*, Vol. I, fasc. 6, 1950, pp. 380–410.

6–46. Duncan, W. J., "Flutter of Systems of Many Degrees of Freedom," *Aero. Quarterly*, Vol. I, Part I, May 1949, pp. 59–76.

6–47. Kassner, R., and H. Fingado, "Flügelschwingungen," *Jahrbuch 1935 der Vereinigung für Luftfahrtforschung*, R. Oldenbourg, Munich, pp. 54–66.

6–48. Kassner, R., and H. Fingado, "The Two-Dimensional Problem of Wing Vibration," *J. Royal Aero. Soc.*, Vol. XLI, 1937, pp. 921–944 (Translated from Luftfahrtforschung, Vol. 13, No. 11, November 1936, pp. 374–387).

6–49. Williams, R. M., *Effect of Variations in Aerodynamic Coefficients on Flutter Speed Calculations at Mach 10/9*, Navy Bureau of Aeronautics Report 43, July 1954.

6–50. Salaün, P., *Influence de l'amortissement interne sur la vitesse critique de flottement*, La Recherche aéro., 61, November–December 1957, pp. 19–25.

6–51. Herr, R. W., *A Study of Flutter at Low Mass Ratios with Possible Applications to Hydrofoils*, NASA Technical Note D-831, 1961.

7

ONE-DIMENSIONAL
STRUCTURES

7-1 INTRODUCTION

For the purposes of this book, we define a one-dimensional structure to be any linear elastic system whose state of deformation can be adequately specified by a set of functions of a single space coordinate. Because of the presence of the additional independent variable time, we are still forced to deal with partial differential or integral equations when analyzing *dynamic* problems. But the conceptual and mathematical simplification that is accomplished by the one-dimensional approach is hard to over-emphasize; and for some years the practice of aeroelasticity was in-distinguishable from "beamology," as it is irreverently called. The many configurations which can be treated in this way—straight or swept wing and tail surfaces of large aspect ratio, elongated fuselages, struts, booms, and the like—will retain their importance simply because of the high flexibility which is associated with their shapes. More recently rockets, guided missiles, and slender-winged aircraft such as re-entry gliders have been added to the long list of approximately one-dimensional vehicles.

Consider an aggregate of elastic material clustered around a straight or curved line. If straight, let this line coincide with the (spanwise) y-axis of a rectangular Cartesian system; otherwise, let s denote distance along it measured from some reference point. Provided that cross sections of the structure taken normal to the coordinate line do not change their shapes or warp appreciably, any instantaneous small deformation from the un-strained position is fully specified by three orthogonal displacement components $u(y)$, $v(y)$, $w(y)$, [or $u(s)$, $\cdots$] of points on the line, plus the twist $\theta(y)$ of the cross sections. θ is taken positive in a right-hand sense about the direction of increasing y or s. The axial stretching $v(y)$ is neglected in aeronautical work. When dealing with lifting surfaces of small thickness ratio, the resistance to bending in the chordwise direction is usually so great that only the normal displacement $w(y)$ and the twist are significant. It is with such wing-like structures that we shall be mainly concerned in this chapter, although chordwise bending can easily be included when setting up equations of motion in terms of generalized coordinates. Both bending directions have comparable flexibility on fuselage and missile structures, but the two are usually uncoupled since they coincide with inertial and elastic principal axes.

From the discussion of aeroelastic equations in Chap. 3, we see that any system of loads acting on a one-dimensional wing-like configuration can be reduced to a running force $F_z(y, t)$ in the direction of w and a running torque $T_y(y, t)$ about the coordinate line. Each of these represents an aerodynamic, inertial, gravitational, etc. load *per unit distance* along y (or s). When studying the deformations caused by such forces and torques, it is convenient to distinguish between those situations where there is a straight elastic axis and all other cases. The concept of elastic axis, or line of shear centers, is a familiar one, fully treated in such places as Chap. 1 and Appendix I of Fung (Ref. 1–1). All that need be said here is that, when the axis is straight, bending and torsion are uncoupled in the sense that loads on the axis produce no twisting and couples about it produce no linear displacement. A line structure possessing this property is often called a *beam-rod* or simply a *beam*.

The elastic operators of the beam-rod were reviewed in Chap. 5. Its motion is most conveniently described by the differential equations of dynamic torsion and flexure, which are derived from considering the equilibrium of the segment of length dy pictured in Fig. 7–1. Here the displacement w of the segment's center from its unstrained position is composed of portions α due to bending strains and β due to shearing strains:

$$w(y, t) = \alpha(y, t) + \beta(y, t) \tag{7-1}$$

With the arbitrary infinitesimal dy cancelled, the conditions of force and moment balance read, respectively,

$$m\ddot{w} = F_z + \frac{\partial S_e}{\partial y} \tag{7-2}$$

$$\mu\frac{\partial}{\partial y}\ddot{\alpha} = S_e + \frac{\partial M_e}{\partial y} \tag{7-3}$$

For the moment, the C.G.'s are assumed to lie along the E.A.; $m(y)$ is the mass per unit length; $\mu(y)\, dy$ is the mass moment of inertia (*rotary*

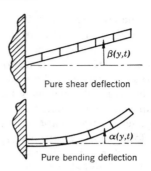

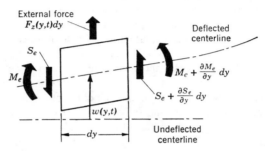

Fig. 7–1. Normal forces and moments acting on a segment dy of a beam-rod. All quantities are shown in their positive senses.

inertia) of the segment about an x-axis through its center; S_e and M_e are the elastic shear and bending moment, the subscript e being used to distinguish them from aerodynamic symbols. Dots indicate partial differentiations with respect to t. In view of the elastic operators connecting S_e and M_e with the corresponding deflections,

$$GK \frac{\partial \beta}{\partial y} = S_e \qquad (7-4)$$

$$EI \frac{\partial^2 \alpha}{\partial y^2} = M_e \qquad (7-5)$$

Eqs. 7–2 and 7–3 can be transformed to the following force-deflection relations:

$$\frac{\partial^2}{\partial y^2}\left[EI \frac{\partial^2 \alpha}{\partial y^2} \right] - \frac{\partial}{\partial y}\left[\mu \frac{\partial \ddot{\alpha}}{\partial y} \right] + m\ddot{w} = F_z \qquad (7-6)$$

$$GK \frac{\partial \beta}{\partial y} + \frac{\partial}{\partial y}\left[EI \frac{\partial^2 \alpha}{\partial y^2} \right] - \mu \frac{\partial \ddot{\alpha}}{\partial y} = 0 \qquad (7-7)$$

The bending and shearing stiffnesses $EI(y)$ and $GK(y)$ conform with the notation of Chap. 5.

When the properties of the beam vary along its span, Eqs. 7–1, 7–6, and 7–7 must be solved simultaneously for w, α, and β, subject to geometrical boundary conditions on β, α, and $\partial\alpha/\partial y$, or elastic boundary conditions on S_e and M_e at the ends. For dynamic problems, initial conditions on α, $\dot{\alpha}$, β, and $\dot{\beta}$ must be supplied at $t = 0$.

The case of a uniform beam is a fair approximation to some aircraft structures and serves even better for promoting the understanding of more complex problems. The coefficients EI, GK, m, and μ are defined to be constants. It then proves possible, by successively eliminating α, β, M_e, and S_e among Eqs. 7–4 through 7–7, to construct a partial differential equation for the single variable $w(y, t)$:

$$EI \frac{\partial^4 w}{\partial y^4} - \left[\frac{mEI}{GK} + \mu\right] \frac{\partial^2 \ddot{w}}{\partial y^2} + m\ddot{w} + \frac{\mu m}{GK} \overset{....}{w} = F_z \qquad (7\text{–}8)$$

Similar equations can be derived for α and β, except for different terms on the right.

Equations 7–6 through 7–8 have been derived principally for the purpose of discussing the importance of shear deformation and rotary inertia in aeroelastic applications. Let us temporarily define dimensionless variables $\tilde{w} = w/l$, $\tilde{y} = y/l$, and $\tilde{t} = \omega t$, where l is the beam's length and ω a circular frequency at which it might be oscillating. Using these quantities, the left-hand member of Eq. 7–8 can be recast as follows:

$$\frac{l^3}{EI} \left\{ EI \frac{\partial^4 w}{\partial y^4} - \left[\frac{mEI}{GK} + \mu\right] \frac{\partial^2 \ddot{w}}{\partial y^2} + m\ddot{w} + \frac{\mu m}{GK} \overset{....}{w} \right\}$$

$$= \frac{\partial^4 \tilde{w}}{\partial \tilde{y}^4} + \frac{ml^4\omega^2}{EI} \left\{ \frac{\partial^2 \tilde{w}}{\partial \tilde{t}^2} - \left[\frac{EI}{GKl^2} + \frac{\mu}{ml^2}\right] \frac{\partial^4 \tilde{w}}{\partial \tilde{y}^2 \partial \tilde{t}^2} + \frac{\mu\omega^2}{GK} \frac{\partial^4 \tilde{w}}{\partial \tilde{t}^4} \right\} \qquad (7\text{–}9)$$

During the oscillation which we envision, the four derivatives with respect to $\tilde{y}$ and $\tilde{t}$ all maintain the same order of magnitude as $\tilde{w}$ itself. Hence the sizes of their coefficients measure, at least qualitatively, their importance in determining the characteristics of the motion. The first two terms on the right form the basis of what we shall call *slender beam* theory. They evidently predominate over the effects of shearing when

$$\frac{EI}{GKl^2} \ll 1 \qquad (7\text{–}10a)$$

and over those of rotary inertia when

$$\frac{\mu}{ml^2} \ll 1 \qquad \text{and} \qquad \frac{\mu\omega^2}{GK} \ll 1 \qquad (7\text{–}10b, c)$$

In essence, the inequalities (7–10a) and (7–10b) require that the ratio of depth to length of the beam be small. Equation 7–10c calls for a frequency which is not too large, that is, for motions confined to the lower end of the vibration spectrum. Now, if we leave aside low-aspect-ratio plates or shells and occasional impulsive inputs, the latter condition is generally met in aeroelastic problems. On the other hand, the slenderness condition is no longer fulfilled so universally today as it used to be. In our forthcoming applications of the bending differential equation, we shall be neglecting shear deformation, not because it is always insignificant but because it complicates the mathematical manipulations. In any event, shear, rotary inertia, and several other effects are automatically accounted for in certain alternative techniques of solution, such as those based on integral equations or modal superposition.

Accordingly, we set $\beta = 0$ and adopt the slender-beam differential equation,

$$\frac{\partial^2}{\partial y^2} \left[EI \frac{\partial^2 w}{\partial y^2} \right] + m\ddot{w} = F_z \tag{7–11}$$

for the examples of this chapter. Confirmation of the validity of the limitations (7–10) is provided by analyses of free vibrations of beams which do include shearing and rotary inertia. References 7–1 and 7–2 are two excellent papers on the subject. Although the techniques are straightforward, not much has been done toward working out exact solutions of Eq. 7–8 for cases of forced motion. (Leonard does provide some useful examples in Ref. 7–2.) The reader might find it an interesting exercise to try to extend some of the uniform beam results given below, particularly to the study of shear deformation of swept wings under aerodynamic loading.

Turning to torsional motion of the beam-rod, we first generalize in the following way the elastic operator which determines the internal torque T_e associated with the twist θ:

$$T_e = GJ \frac{\partial \theta}{\partial y} - \frac{\partial}{\partial y} \left[\sigma \frac{\partial^2 \theta}{\partial y^2} \right] \tag{7–12}$$

$GJ(y)$ is the familiar torsional stiffness that is characteristic of solid sections, closed structural boxes, and the like. The second term on the right is contributed by differential bending, which is the resistance to twisting provided by upward and downward bending of a pair (or more) of spanwise longerons or spars acting in opposition, as described, for example, in Sec. 12–3 of Ref. 1–2. $\sigma(y)$ represents the second moment of the bending stiffnesses of all such spars; in the case of front and rear spars separated a distance D and having individual stiffnesses EI_f and EI_r, respectively,

$$\sigma = \frac{D^2 EI_r EI_f}{EI_r + EI_f} \qquad (7\text{--}13)$$

Equilibrating the applied and inertia couples on a segment dy of the beam-rod, we find

$$I_y \ddot{\theta} = T_y + \frac{\partial T_e}{\partial y} \qquad (7\text{--}14)$$

where T_y is the running external torque and I_y the moment of inertia per unit length about the E.A. When the internal torque is eliminated between Eqs. 7–12 and 7–14,

$$\frac{\partial^2}{\partial y^2}\left[\sigma \frac{\partial^2 \theta}{\partial y^2}\right] - \frac{\partial}{\partial y}\left[GJ \frac{\partial \theta}{\partial y}\right] + I_y \ddot{\theta} = T_y \qquad (7\text{--}15)$$

When analyzing aeroelastic problems by means of differential equations in this chapter, we shall generally neglect the differential bending term in Eq. 7–15 for the same reasons that we choose to emphasize slender beams. Moreover, most of the torsion material in current wing structures is furnished by thick metal skins, which tend to minimize the differential bending parameter σ/GJl^2 (analogous to EI/GKl^2 in Eq. 7–9). Its greatest interest may be in connection with dynamic model wings having multi-spar, sectional construction, as discussed in Ref. 1–2.

It is presumed that the reader is familiar with the uncoupled free-oscillation characteristics of slender, uniform beam-rods under various combinations of boundary conditions, since these are covered in so many books on elasticity and vibrations. The torsional mode shapes are, of course, simple sine curves. Valuable information on bending eigenvalues ·and modes will be found in two University of Texas publications (Refs. 7–3 and 7–4).

For some applications, especially those concerning nonuniform or nonslender beam-rods, it is more convenient to cast the equations of motion in integral form. We then need two flexibility influence functions [cf. Sec. 5–2(a)]: $C^{zz}(y, \eta)$, the normal deflection w at station y due to unit concentrated normal force at station η; and $C^{\theta\theta}(y, \eta)$, the twist at y due to unit concentrated torque about the elastic axis at η. These functions imply within themselves all structural boundary conditions, and they lend themselves to experimental determination in this and more complicated situations. Since the net external load and torque on a strip

$d\eta$ at station η are $[-m(\eta)\ddot{w}(\eta, t) + F_z(\eta, t)] \, d\eta$ and $[-I_y(\eta)\ddot{\theta}(\eta, t) + T_y(\eta, t)] \, d\eta$, the integral relations equivalent to Eqs. 7–15 and 7–11 (or possibly to Eqs. 7–1, 7–6, 7–7) are

$$w(y, t) = \int_0^l C^{zz}(y, \eta)[-m(\eta)\ddot{w}(\eta, t) + F_z(\eta, t)] \, d\eta \qquad (7\text{–}16)$$

$$\theta(y, t) = \int_0^l C^{\theta\theta}(y, \eta)[-I_y(\eta)\ddot{\theta}(\eta, t) + T_y(\eta, t)] \, d\eta \qquad (7\text{–}17)$$

The structure here extends from $y = 0$ to $y = l$. Two important assumptions are implicit in Eqs. 7–16 and 7–17. First, the beam-rod is fully restrained and has no rigid-body degrees of freedom; the removal of this constraint is taken up in Chap. 9. Second, the chordwise centers of gravity of all sections lie on the E.A.; if there is actually a static unbalance $S_y(\eta)$ per unit span about this axis, the terms $S_y(\eta)\ddot{\theta}(\eta, t)$ and $S_y(\eta)\ddot{w}(\eta, t)$, respectively, should be added inside the integrand brackets of Eqs. 7–16 and 7–17.

Motions of a one-dimensional structure without an E.A. can be conceived of in terms of integral equations which resemble Eqs. 7–16 and 7–17. We consider the situation illustrated in Fig. 7–2, where there is a curved *structural reference line* so chosen that sections normal to it experience negligible warping or in-plane deformation during bending and twisting. The displacements are $w(s, t)$ and $\bar{\theta}(s, t)$, the latter representable by a vector locally tangent to the line. Since it is no longer true that forces produce only pure linear deflections and torques, pure twist, we need four influence functions: $C^{zz}(s, \sigma)$ and $C^{\theta\theta}(s, \sigma)$, defined analogously to

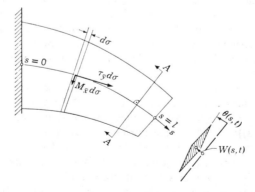

Fig. 7–2. One-dimensional structure with coupling between bending and twist, showing the running moments acting at station σ and the elastic displacements of a typical cross section.

$C^{zz}(y, \eta)$ and $C^{\theta\theta}(y, \eta)$ except that torques are vectors tangent to the structural reference line; $C^{z\theta}(s, \sigma)$, the normal deflection at s due to unit concentrated torque at σ; and $C^{\theta z}(s, \sigma)$, the twist at s due to unit concentrated normal force at σ. A new type of loading often found on curved or swept wings is the distributed external bending moment, a couple with its vector normal to the s-axis. We use symbol $M_{\bar{x}}(s, t)$ for such a moment per unit length, positive in a sense to compress the upper fibers. It is not difficult (cf. Sec. 4–4 of Ref. 1–1) to treat $M_{\bar{x}}$ as a limit of equal and opposite Z-forces, and thus to show that $\partial C^{zz}/\partial\sigma$ and $\partial C^{\theta z}/\partial\sigma$ are the influence functions giving the normal deflection and twist, respectively, due to unit concentrated bending moment at σ.

As in Ref. 1–1 or 1–2, we can synthesize integral equations for $w(s, t)$ and $\bar{\theta}(s, t)$ from the information just set down. For the sake of brevity, let the inertia forces and moments be temporarily included in the definitions of F_z, $M_{\bar{x}}$, and the torque $T_{\bar{y}}$ about the s-axis. The desired equations will then read

$$w(s, t) = \int_0^l C^{zz}(s, \sigma)F_z(\sigma, t)\,d\sigma$$
$$+ \int_0^l C^{z\theta}(s, \sigma)T_{\bar{y}}(\sigma, t)\,d\sigma + \int_0^l \frac{\partial C^{zz}(s, \sigma)}{\partial\sigma} M_{\bar{x}}(\sigma, t)\,d\sigma \quad (7\text{–}18)$$

$$\bar{\theta}(s, t) = \int_0^l C^{\theta z}(s, \sigma)F_z(\sigma, t)\,d\sigma$$
$$+ \int_0^l C^{\theta\theta}(s, \sigma)T_{\bar{y}}(\sigma, t)\,d\sigma + \int_0^l \frac{\partial C^{\theta z}(s, \sigma)}{\partial\sigma} M_{\bar{x}}(\sigma, t)\,d\sigma \quad (7\text{–}19)$$

We close this section with a few words about aerodynamic operators. The equations of motion of one-dimensional structures contain the airloads in the form of running lifts and moments, which are chordwise integrals of the overall load distribution. It is, therefore, natural to consider two successive degrees of refinement in the theoretical technique of relating lift and moment to the displacements. The simpler of these is strip theory, wherein the local loads at a particular spanwise station are assumed to depend only on the local motion or angle of attack there, with perhaps a time-independent multiplying factor to approximate the effects of three-dimensional flow. The more refined method is lifting-line theory, wherein spanwise induction between stations is accounted for. On very slender configurations which are one-dimensional in the chordwise direction, the various forms of slender-body theory are counterparts of the lifting line. As a general rule, it is inconsistent to connect lifting-surface aerodynamic theory with line-type structures, and this we shall not do. Indeed, when analyzing the motion of beam-rods by the differential equation approach, we shall not go beyond strip theory, for span-wise induction is much more compatible with integral forms of the structural and aerodynamic operators.

7-2 STATIC PROBLEMS OF LARGE-ASPECT-RATIO STRAIGHT WINGS

In this section we study the cantilever lifting surface of large aspect ratio which behaves structurally as a beam-rod with an unswept E.A. This system has, in common with the typical section of Sec. 6–2, a fully de-coupled bending freedom. The complete aerodynamic loading can be determined from the torsional equilibrium, after which bending may be calculated, if desired, with the already known spanwise lift distribution as an applied force. The concentration here is on the torsion, since it con-stitutes a true aeroelastic problem. Lacking space for exhaustive coverage of the available methods of solution, we examine representative cases of both symmetrical and antisymmetrical loading.

The system with its steady-state loads per unit span is pictured in Fig. 7–3. Allowance is made for an upward normal acceleration $gN(y)$, which produces a running downward force mNg along the line of C.G.'s, located at a distance $d(y)$ ahead of the E.A.

(a) Symmetrical load distribution and divergence

Consider first a *uniform* cantilever wing or tail. Adopting the differ-ential equation approach and neglecting differential bending, we rewrite Eq. 7–15

$$\frac{d^2\theta}{d\tilde{y}^2} = -\frac{l^2}{GJ}\,T_v = -\frac{l^2}{GJ}\,[eL(\tilde{y}) + M_{AC}(\tilde{y}) - mNgd]   (7\text{–}20)$$

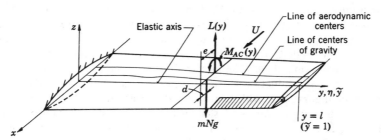

Fig. 7–3. A straight, cantilever lifting surface subjected to steady-state aerodynamic loads and an upward maneuvering acceleration $gN(y)$.

The boundary conditions are

$$\theta(0) = 0 \quad \text{and} \quad \frac{d\theta(1)}{d\tilde{y}} = 0 \qquad (7\text{–}21a, b)$$

in terms of the dimensionless spanwise variable $\tilde{y} = y/l$. If the undeformed surface has an initial angle of attack α_0, measured from the zero-lift attitudes of individual sections, the resultant incidence is $(\alpha_0 + \theta)$. Section 4–3 furnishes the strip-theory aerodynamic operators for running lift and moment about the A.C.,

$$L(\tilde{y}) = qcc_l = qc \frac{\partial c_l}{\partial \alpha}(\alpha_0 + \theta) \qquad (7\text{–}22)$$

$$M_{AC}(\tilde{y}) = qc^2 c_{mAC} \qquad (7\text{–}23)$$

Lower-case symbols are employed for aerodynamic coefficients per unit span; in this immediate example α_0, $\partial c_l / \partial \alpha$, c_{mAC}, and N are assumed constants, but they may in other applications be dependent on y, allowing for variable camber, aerodynamic twist, or a spanwise factor to adjust approximately for three-dimensional effects.

Combining Eqs. 7–20, 7–22, and 7–23 yields

$$\frac{d^2\theta}{d\tilde{y}^2} + \lambda^2 \theta = K \qquad (7\text{–}24)$$

where

$$\lambda^2 = \frac{qcel^2}{GJ} \frac{\partial c_l}{\partial \alpha} \qquad (7\text{–}25)$$

and

$$K = \frac{mNgdl^2}{GJ} - \frac{qcl^2}{GJ}\left[e \frac{\partial c_l}{\partial \alpha} \alpha_0 + cc_{mAC} \right] \qquad (7\text{–}26)$$

The homogeneous and particular solutions to Eq. 7–24 add up to

$$\theta(\tilde{y}) = A \sin \lambda \tilde{y} + B \cos \lambda \tilde{y} + \frac{K}{\lambda^2} \qquad (7\text{–}27)$$

When applying boundary conditions to determine the free constants A and B in Eq. 7–27, imagine that the surface is a half-wing *encastré* at $\tilde{y} = 0$ in the body of an aircraft or missile. A symmetrical flight condition is assumed. There are then two ways of looking at the problem.

First, the rigid angle of attack α_0 might be a given constant. K becomes, in turn, a known quantity, and Eqs. 7–21 are evidently satisfied by

$$\theta(\tilde{y}) = \frac{K}{\lambda^2}\left[1 - \tan \lambda \sin \lambda \tilde{y} - \cos \lambda \tilde{y} \right] \qquad (7\text{–}28)$$

The total lift on two halves of the wing is

$$C_L = \frac{2qcl}{qS}\int_0^1 c_l \, d\tilde{y} = 2\frac{cl}{S}\frac{\partial c_l}{\partial \alpha}\int_0^1 (\alpha_0 + \theta) \, d\tilde{y}$$

$$= \frac{2cl}{S}\left\{\frac{\tan \lambda}{\lambda}\frac{\partial c_l}{\partial \alpha}\alpha_0 + \left[1 - \frac{\tan \lambda}{\lambda}\right]\left[\frac{mNgd}{qce} - \frac{c}{e}c_{mAC}\right]\right\} \quad (7\text{--}29)$$

(Note that the factor multiplying the braces approaches unity if the body width is negligible compared to the wingspan. This result also reduces to what would be expected in the absence of elastic twist when the dynamic pressure is small and $\tan \lambda/\lambda \cong 1$.) Equations 7–28 and 7–29 may be used directly for any situation where N is given, such as a wind-tunnel test of a flexible model with a preset root angle α_0 ($N = 1$ because of the weight). For an aircraft in free flight one would have to solve for N by equating the total lift from Eq. 7–29 to NW, W being the vehicle weight, as set forth in Sec. 8–3 of Ref. 1–2.

As a second interpretation of Eqs. 7–27 and 7–28, we might consider a typical structural design condition, where the total wing lift $C_L qS$ is the prescribed quantity. Now N is known, because, to an excellent degree of approximation on most large-aspect-ratio airplanes, $C_L qS = NW$. The unknown quantity is α_0, for which we solve by multiplying Eq. 7–29 by qS and equating the result to NW,

$$\alpha_0 = \frac{\lambda}{\tan \lambda}\frac{NW}{2qcl\,\partial c_l/\partial \alpha}$$

$$+ \left[1 - \frac{\lambda}{\tan \lambda}\right]\left[\frac{mNgd}{qce\,\partial c_l/\partial \alpha} - \frac{c}{e}\frac{c_{mAC}}{\partial c_l/\partial \alpha}\right] \quad (7\text{--}30)$$

The first term on the right here is the rigid-wing value when $\lambda/\tan \lambda = 1$. The elastic deformation can be computed by inserting Eq. 7–30 into Eq. 7–26 and using this value of K in Eq. 7–28. It is interesting to examine the modification of spanwise lift *distribution* due to the flexibility. For an untwisted, rigid surface pulling $N\,g$'s, one has

$$c_l{}^r = \frac{NW}{2qcl} \quad (7\text{--}31)$$

Some algebraic manipulation yields the running lift on the flexible wing at the same load factor

$$\frac{c_l{}^e}{c_l{}^r} = \frac{\lambda \cos \lambda\tilde{y}}{\tan \lambda} + \lambda \sin \lambda\tilde{y}$$

$$+ \left[\frac{2mg\,dl}{We} - \frac{2qc^2l}{NWe}c_{mAC}\right]\left[1 - \lambda \sin \lambda\tilde{y} - \frac{\lambda \cos \lambda\tilde{y}}{\tan \lambda}\right] \quad (7\text{--}32)$$

Plots of Eq. 7–32 for large-aspect-ratio surfaces in subsonic flight show appreciable distortion even at rather low q. For example, on an uncambered wing with $mg\ dl/We \cong 0$, the overload ratio at the tip is

$$\frac{c_l^e}{c_l^r}\bigg|_{\tilde{y}=1} = \frac{\lambda}{\sin \lambda} \tag{7–33}$$

This rises from 1 to $\pi/2$ before the structure becomes unstable.

The mention of instability calls to mind that there is an eigenvalue problem associated with Eqs. 7–24 and 7–21. This can be seen from the blowing up of solutions like Eqs. 7–28 and 7–29 at certain critical values of the parameter λ. But perhaps the most important fact is that, when the forcing terms on the right-hand side of Eq. 7–24 are equated to zero, we obtain an elementary example of a *Sturm-Liouville problem* (cf. Sec. 6–3 of Ref. 7–5). It can be proved that there is a denumerable infinity of distinct, real, positive eigenvalues, and, that the corresponding eigenfunctions form a *complete set*, that is, they can be used to represent to an arbitrary degree of accuracy any sufficiently well-behaved function in the interval $0 \le \tilde{y} \le 1$, by an extension of the familiar concept of Fourier series.

For twisting of the uniform beam-rod, the characteristic values and functions are

$$\left.\begin{aligned} \lambda_n &= (2n+1)\frac{\pi}{2} \\ \varphi_n(\tilde{y}) &= \sin \lambda_n \tilde{y} \end{aligned}\right\}, \quad (n = 0, 1, 2, \cdots) \tag{7–34a, b}$$

Each such combination represents a dynamic pressure and mode of deformation in which the wing (starting from zero initial incidence, camber, and normal load factor N) can assume an equilibrated twisted position of arbitrary amplitude. The fundamental eigenvalue $\lambda_0 = \pi/2$ has the most direct physical interest. It corresponds to the lowest dynamic pressure of torsional divergence, which Eq. 7–25 shows to be

$$q_D = \frac{\pi^2}{4l^2} \frac{GJ}{ce\ \partial c_l/\partial \alpha} \tag{7–35}$$

Significant comparisons can be made with Eq. 6–11 for the typical section. Once more Eq. 7–35 is merely an implicit formula unless e and $\partial c_l/\partial \alpha$ are independent of flight Mach number. The inverse proportionality to l^2 reveals the major influence of wingspan or aspect ratio, but even this difference from Eq. 6–11 disappears if we regard GJ/l as an effective torsional spring stiffness K_α and also replace lc by the wing area S.

In common with buckling of columns and flutter, the divergence problem has a special emphasis on the lowest eigenvalue. It is therefore worth mentioning, in passing, the utility of Rayleigh's energy method for estimating λ_0 for systems either with constant or variable spanwise properties. The twisting wing is neutrally stable in any mode $\Theta(y)$ provided the potential strain energy stored in the elastic deformation just equals the work done by the deforming aerodynamic torques. According to strip theory, these torques grow in direct proportion to the local θ; so one obtains the following formula:

$$\int_0^l \tfrac{1}{2} GJ \left(\frac{d\Theta}{dy} \right)^2 dy = q_D \int_0^l \tfrac{1}{2} ce \frac{\partial c_l}{\partial \alpha} \Theta^2 \, dy \qquad (7\text{-}36)$$

If the divergence mode is known, it can be substituted for $\Theta(y)$ in Eq. 7-36, leading to the exact q_D. However, any reasonable function which satisfies the structural boundary conditions will yield a close approximate q_D, whether or not $\Theta(y)$ can be found analytically. At worst, the two simple integrals in Eq. 7-36 have to be evaluated numerically. The estimated dynamic pressure always exceeds the exact, since the true potential energy is a minimum relative to any approximation which applies artificial constraints. If we use the uniform cantilever as an example, but replace the true mode $\sin(\pi\tilde{y}/2)$ with $\Theta = [2\tilde{y} - \tilde{y}^2]$, Rayleigh's method proceeds as follows:

$$q_D \cong \frac{\dfrac{GJ}{2l} \displaystyle\int_0^1 \left[\frac{d\Theta}{d\tilde{y}} \right]^2 d\tilde{y}}{\dfrac{l}{2} ce \dfrac{\partial c_l}{\partial \alpha} \displaystyle\int_0^1 \Theta^2 \, d\tilde{y}}$$

$$= \frac{GJ}{l^2 ce \dfrac{\partial c_l}{\partial \alpha}} \frac{\displaystyle\int_0^1 [2 - 2\tilde{y}]^2 \, d\tilde{y}}{\displaystyle\int_0^1 [2\tilde{y} - \tilde{y}^2]^2 \, d\tilde{y}} = \frac{5}{2l^2} \frac{GJ}{ce \dfrac{\partial c_l}{\partial \alpha}} \qquad (7\text{-}37)$$

This result is greater than Eq. 7-34 by only 1.4%. Incidentally, improved aerodynamic theory of the lifting-line type can be introduced by replacing $\dfrac{\partial c_l}{\partial \alpha} \Theta$ with $c_l \Theta$ in the right-hand member of Eq. 7-36 and solving for $c_l(\tilde{y})$ corresponding to the preassigned $\Theta(\tilde{y})$ before evaluating the work integral.

A well-known by-product of Sturm-Liouville theory is the powerful way in which the eigenfunctions of the homogeneous differential equation are adapted to solving nonhomogeneous problems. We illustrate this technique by returning to Eq. 7-24. λ^2 must still be constant, but K may be a function of $\tilde{y}$, thus allowing for aerodynamic twist, variations in C.G. position, concentrated masses like external stores, or variable normal load factor. The solution $\theta(\tilde{y})$ is constructed by assuming that both it and $K(\tilde{y})$ can be expanded in Fourier series of the functions $\varphi_n(\tilde{y})$ from Eq. 7–34b,

$$\theta(\tilde{y}) = \sum_{n=0}^{\infty} a_n \varphi_n(\tilde{y}) \tag{7–38}$$

$$K(\tilde{y}) = \sum_{n=0}^{\infty} A_n \varphi_n(\tilde{y}) \tag{7–39}$$

Since K is treated as a given quantity, the constants A_n are efficiently calculated by resort to the orthogonality of the eigenfunctions,

$$\int_0^1 \varphi_n(\tilde{y}) \varphi_m(\tilde{y})\, d\tilde{y} = \int_0^1 \sin{(2n+1)}\frac{\pi \tilde{y}}{2} \sin{(2m+1)}\frac{\pi \tilde{y}}{2}\, d\tilde{y}$$

$$= \begin{cases} 0, & n \ne m \\ \tfrac{1}{2}, & n = m \end{cases} \tag{7–40}$$

With Eq. 7–40 established, Eq. 7–39 is multiplied through by a particular $\varphi_m(\tilde{y})$ and integrated over the span to produce

$$A_m = 2\int_0^1 K(\tilde{y}) \varphi_m(\tilde{y})\, d\tilde{y}, \qquad (m = 0, 1, 2, \cdots) \tag{7–41}$$

To determine the constants a_n in Eq. 7–38, we insert both series into Eq. 7–24 and recall that $\varphi_n(\tilde{y})$ satisfies the homogeneous differential equation when $\lambda = \lambda_n$, that is,

$$\frac{d^2 \varphi_n}{d\tilde{y}^2} + \lambda^2 \varphi_n = (\lambda^2 - \lambda_n{}^2)\varphi_n \tag{7–42}$$

These steps lead to the relation

$$a_n = \frac{A_n}{\lambda^2 - \lambda_n{}^2} \tag{7–43}$$

and to the final solution

$$\theta(\tilde{y}) = -\frac{2}{\lambda^2} \sum_{n=0}^{\infty} \frac{\displaystyle\int_0^1 K(\tilde{y})\varphi_n(\tilde{y})\, d\tilde{y}}{\left(\dfrac{\lambda_n{}^2}{\lambda^2} - 1\right)} \varphi_n(\tilde{y}) \tag{7–44}$$

Good convergence is assured, in practice, by the fact that the denominator quantities $\lambda_n{}^2/\lambda^2$ increase from term to term in proportion to the squares of the odd integers 1, 9, 25, 49, $\cdots$. For any realizable flight condition, $\lambda < \lambda_0$; the blowing up of $\theta(\tilde{y})$ as torsional divergence is approached shows clearly in Eq. 7–44.

One other class of straight, cantilever wings for which Sturm-Liouville methods have genuine practical merit consists of the generalized tapered planforms first identified by H. Reissner (for a full treatment, see the paper by F. B. Hildebrand and E. Reissner, Ref. 7–6). The chord, A.C.–E.A. offset, and torsional stiffness are described by the functions

$$\frac{c(\tilde{y})}{c_R} = \frac{e(\tilde{y})}{e_R} = [1 - \chi\tilde{y}]^{\gamma_1} = y_1^{\gamma_1} \tag{7-45}$$

$$\frac{GJ(\tilde{y})}{GJ_R} = [1 - \chi\tilde{y}]^{\gamma_2} = y_1^{\gamma_2} \tag{7-46}$$

Here subscript R refers to the wing-root section, and y_1 is a dimensionless variable running from $(1 - \chi)$ at the tip to 1 at the root. With the steady-state loads illustrated in Fig. 7–3, the differential equation (7–15) can be written in terms of y_1 for the system under consideration,

$$\frac{d}{dy_1}\left(y_1^{\gamma_2}\frac{d\theta}{dy_1}\right) + \left(\frac{\lambda_R}{\chi}\right)^2 y_1^{2\gamma_1}\theta = K_1(y_1) \tag{7-47}$$

The modified parameters are

$$\lambda_R^2 = \frac{qc_R e_R l^2}{GJ_R}\frac{\partial c_l}{\partial\alpha} \tag{7-48}$$

$$K_1(y_1) = \frac{mNg\,dl^2}{GJ_R} - \frac{qcl^2}{GJ_R}\left[e\frac{\partial c_l}{\partial\alpha}\alpha_0 + cc_{mAC}\right] \tag{7-49}$$

The section lift-curve slope must be a constant. As long as $\chi < 1$, the boundary conditions are Eqs. 7–21.

When developing solutions for Eq. 7–47, it is convenient to distinguish between two sub-cases.

(1) $\gamma_2 \neq 2(\gamma_1 + 1)$. Under this restriction, any extensive treatment of Bessel functions (e.g., Chap. 2 of von Kármán and Biot, Ref. 7–7) shows that the homogeneous solution reads

$$\theta_H(y_1) = y_1^{\nu/\delta}\left[AJ_\nu\left(\frac{\lambda_R\delta}{\chi}y_1^{1/\delta}\right) + BY_\nu\left(\frac{\lambda_R\delta}{\chi}y_1^{1/\delta}\right)\right] \tag{7-50}$$

where

$$\delta = \frac{2}{2\gamma_1 - \gamma_2 + 2}, \qquad \nu = \frac{1 - \gamma_2}{2\gamma_1 - \gamma_2 + 2} \tag{7-51a, b}$$

J_ν and Y_ν are Bessel functions of the first and second kinds and order ν; except when ν is an integer, $Y_\nu \equiv J_{-\nu}$. The eigenvalues and eigenfunctions are established by applying conditions (7–21), which process yields transcendental relations whose roots must be determined for each distinct combination of γ_1, γ_2, and χ. Once these characteristic quantities are available, $K_1(y_1)$ and $\theta(y_1)$ may be expanded as in Eqs. 7–38 and 7–39, leading to systematic and rapidly convergent series. When computing the coefficients A_n, we make use of the orthogonality of eigenfunctions with respect to the weight $y_1^{2\gamma_1}$.

(2) $\gamma_2 = 2(\gamma_1 + 1)$. In this important special case, the transformation $z_1 = \ln y_1$ leads to the following homogeneous solution, which also satisfies the boundary condition at the root:

$$\theta_H(y_1) = \frac{A \sin (\mu \ln y_1)}{y_1^{[(1+2\gamma_1)/2]}} \tag{7-52}$$

where

$$\mu = \sqrt{\left(\frac{\lambda_R}{\chi}\right)^2 - \left(\frac{1 + 2\gamma_1}{2}\right)^2} \tag{7-53}$$

Substituting Eq. 7–52 into Eq. 7–21b, we get a transcendental characteristic equation

$$\tan (\mu \ln (1 - \chi)) = \frac{2\mu}{1 + 2\gamma_1} \tag{7-54}$$

whence the sequence of positive eigenvalues λ_{Rn} can be computed by numerical or graphical means. The eigenfunctions $\varphi_n(y_1)$ then follow from Eqs. 7–52 and 7–53. The nonhomogeneous solution is finally found as in the first sub-case. Particular examples of torsional divergence will be found worked out in Ref. 7–6 and Sec. 8–3 of Ref. 1–2.

Reissner's results have received less attention in engineering applications than they deserve. The reader will be able to find an excellent approximation in the form of Eqs. 7–45 and 7–46 for almost any beam-rod cantilever which has no discontinuities. Sub-case (2) has been elaborated because it encompasses $\gamma_1 = 1$, $\gamma_2 = 4$, an exact description of a straight-tapered surface with geometrically similar cross sections at all spanwise stations, fabricated from the same structural material.

(b) An elementary example of aerodynamic heating effects

The study of uniform wings by means of the torsional differential equation provides an opportunity for us to illustrate the role of aerodynamic heating in aeroelasticity. Although they are taken up to some extent in Chaps. 2 and 5, the full presentation of boundary-layer heat transfer, temperature distribution in solid structures, and generation of thermal stresses would require another book larger than the present one. Accordingly, we must accept without proof certain essential results and emphasize those consequences of the thermal environment which have primary importance for the aeroelastician, namely, the losses of stiffness due to thermal stresses and to deterioration of the elastic constants.

Budiansky and Mayers (Ref. 7–8) have analyzed an unswept lifting surface of constant cross section, which has a large enough aspect ratio to permit the examination of aerodynamic heating and its effects on a two-dimensional basis. If a particular section, of area A, has an instantaneous distribution of temperature T throughout the material, this will give rise to a varying spanwise normal stress σ_y (positive in tension). It can be reasoned, either by rigorous application of the theory of thermoelasticity, or by observing that the stresses σ_y alter their directions along with the structural fibers to which they are attached, that a twist rate $\partial\theta/\partial y$ causes the following internal torque to appear across the section *because of thermal stresses alone:*

$$T_\sigma = \frac{\partial\theta}{\partial y} \iint\limits_A \sigma_y r^2 \, dA \qquad (7\text{–}55)$$

r is the distance from the E.A. to the area element dA at which σ_y acts. Unless there is a net spanwise tensile force, which would be present only in whirling beams such as rotors or propeller blades, the integral of σ_y over A equals zero, and T_σ is a pure couple. To get the total internal torque T_e we must add the familiar contribution $GJ \, \partial\theta/\partial y$ of the shear stresses, that is,

$$T_e = \left[GJ + \iint\limits_A \sigma_y r^2 \, dA \right] \frac{\partial\theta}{\partial y} \qquad (7\text{–}56)$$

Since this relation replaces Eq. 7–12, we see that there is an *effective torsional stiffness*

$$\frac{GJ_{\text{eff.}}}{GJ} = 1 + \frac{1}{GJ} \iint\limits_A \sigma_y r^2 \, dA \qquad (7\text{–}57)$$

to be inserted in place of GJ in all of the beam-rod formulas.

The foregoing development presupposes that the shear modulus G is unaffected by heating. Should there be appreciable reductions of G in the hot zones, the shear equilibrium of the cross section must be re-examined to find a reduced value of the coefficient GJ for Eqs. 7–56 and 7–57. This adjustment is usually much smaller than the one associated with thermal stress, because strength losses force the designer to turn to different, more heat-resistant materials before temperatures are reached where E and G drop significantly. Moreover, the elastic constants which control dynamic phenomena, such as vibrations and transient response, turn out to be nearer their "cold" values than those which apparently must be used when dealing with static problems, because the latter are always affected by a certain amount of high-temperature creep (Ref. 7-9). It is also of interest that self-equilibrated spanwise thermal stresses do not alter the bending stiffness at all, within the framework of slender beam-rod theory.

The implications of Eq. 7-57 are clarified by studying what happens following a rapid, large increase or decrease in the forward speed of the vehicle mounting the lifting surface. If we adopt the assumptions of Budiansky and Mayers that recovery temperature and heat-transfer coefficient are independent of location along the airfoil chord and that radiative heat loss is negligible, the entire structure starts the maneuver at the uniform *adiabatic wall temperature*

$$(T_{aw})_0 = T_\infty \left[1 + \eta_r \frac{\gamma - 1}{2} M_0{}^2 \right] \tag{7-58}$$

associated with the initial flight Mach number M_0 and ambient temperature T_∞. The *recovery factor* η_r is a number of order 0.8–0.9 which adjusts for the drop of T_{aw} below stagnation temperature achieved by convection of some heat away in the viscous boundary layer. If the vehicle accelerates, heat is fed in at a rate roughly proportional to the defect of instantaneous skin temperature below a new T_{aw} appropriate to the higher flight Mach number. Since there is relatively little time for this heat to be conducted chordwise along the slender airfoil section, it penetrates largely through the depth of the wing at the place where it is received. T therefore increases faster at the leading and trailing edges, where there is relatively small heat capacity compared with the thicker center section. Trying to expand excessively in the spanwise direction, these edges find themselves restrained by the cooler center and develop compressive thermal stresses (negative σ_y). A simultaneous positive σ_y appears around midchord to preserve equilibrium. Since r is larger where σ_y is negative, the integral in Eqs. 7-55 through 7-57 has a minus value, and the effective torsional stiffness is reduced. Severe, sudden heat inputs, especially to structures having thick or solid cross sections, may even lead to negative $GJ_{\text{eff.}}$ and thermal buckling. By the converse of the foregoing reasoning, deceleration produces a torsional stiffening and is favorable from the standpoint of elastic and aeroelastic stability. In either event, the modification to GJ is a transient effect, which disappears once thermal equilibrium is attained.

In their applications of the theory, Budiansky and Mayers adopt the general assumptions discussed above, going so far as to neglect chordwise heat conduction completely and to treat T as constant through the thickness at any point on the wing. For solid cross sections, the temperature is then calculated from a simple first-order differential equation and is related to the thermal stress as follows:

$$\sigma_y = E\alpha_T \left\{ \frac{1}{A} \int_{\text{chord}} [T - (T_{aw})_0] z_t \, dx + [T - (T_{aw})_0] \right\} \quad (7\text{-}59)$$

where α_T is the coefficient of linear thermal expansion and $z_t(x)$ the local depth of the section. Based on Eq. 7–59, effective torsional stiffness can be determined in terms of tabulated functions of a modified time variable when the profile has a simple analytical shape. Most numerical results in Ref. 7–8 refer to a 3% thick symmetrical double-wedge airfoil with a 3-ft chord, fabricated of solid steel. Figure 7–4 reproduces their calculations for a sudden acceleration from low flight speed to a final Mach number M_f at an altitude of 50,000 ft in the standard atmosphere. For $M_f = 3$, there is a peak loss of all but 22.5% of the torsional stiffness of the cold wing, occurring about $1\frac{1}{4}$ min after the speed-up. An appreciable interval of thermal buckling appears when $M_f = 4$.

Reference 7–8 investigates the relieving effects of finite acceleration and of the presence of a free wingtip where $\sigma_y = 0$. The numbers show, for the particular solid wing considered, that the minimum value of transient $GJ_{\text{eff.}}$ is relatively insensitive to the rate at which the velocity is increased over a range that might be regarded as including any full-power maneuver of a supersonic fighter or missile (e.g., for a time of less than 5 min to reach $M_f = 3$). As measured by the reduction in fundamental torsional frequency squared, ω_x^2, on a uniform, free-free wing of span B, Budiansky and Mayers find only very small deviations from two-dimensional results when B/c exceeds roughly 2. Thus one may conclude that simplified estimates of $GJ_{\text{eff.}}$ are adequate at least for preliminary design purposes on unswept lifting surfaces having continuous, thick-skinned structures, but that stability margins will always contain an element of conservatism.

The effective torsional stiffness may be inserted directly into formulas like Eqs. 7–28, 7–32, and 7–35 for static aeroelastic analyses. Since high supersonic speed is involved, we must take care to include the influence of profile shape and thickness on the A.C. location and $c_{m,AC}$. Piston theory will normally yield aerodynamic operators of sufficient accuracy. For example, it gives the following lift-curve slope and A.C.–E.A. offset in the case of a symmetrical double-wedge airfoil of thickness ratio δ:

$$\frac{\partial c_l}{\partial \alpha} = \frac{4}{M} \quad (7\text{-}60)$$

$$e = \frac{c}{2} \left[a + \left(\frac{\gamma + 1}{4} \right) M\delta \right] \quad (7\text{-}61)$$

Here a is the distance in semichords by which the E.A. lies aft of the midchord line. When Eqs. 7–60 and 7–61 are substituted into Eq. 7–35, we obtain the implicit divergence formula ($S_s = lc$ is semispan area)

$$q_D = \frac{\pi^2 M_D GJ_{\text{eff.}}}{8 S_s^{\,2} \left[a + \left(\frac{\gamma + 1}{4} \right) M_D \delta \right]} \tag{7-62}$$

This can be solved as a quadratic for M_D. Typically, the E.A. of such a wing would be close to midchord ($a = 0$); and the Mach number then becomes

$$M_D = \frac{\pi}{S_s} \sqrt{\frac{GJ_{\text{eff.}}}{\gamma(\gamma + 1)p_\infty \delta}} \tag{7-63}$$

Equation 7–63 reveals another undesirable effect of airfoil thickness at high speeds: forward displacement of the A.C., occurring in direct proportion to $M\delta$, increases the moment arm of the lift and augments torsional instability. It is clear from Fig. 7–4 how thermal effects can temporarily reduce the rigidity below a safe level and cause dangerous deformations or divergence on a surface which otherwise possesses a perfectly adequate margin. Typical calculations of this sort will be found

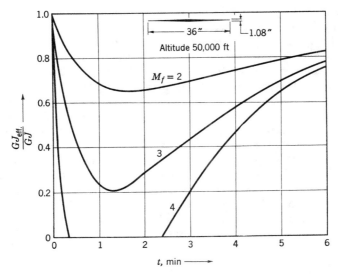

Fig. 7–4. Time history of effective torsional stiffness for the double-wedge steel wing shown, following a sudden acceleration to M_f. (Taken from Budiansky and Mayers, Ref. 7–8.)

in Ref. 7–11. On the other hand, a decelerated maneuver might be able to pass successfully through a zone of "cold" instability.

Since $GJ_{\text{eff.}}$ is a time-dependent quantity, its variation might conceivably generate dynamic responses on a loaded wing in steady flight undergoing thermal transients. The characteristic times of temperature rise and fall in the structure are so long compared with a typical period of vibration, however, that a quasi-static approach always seems to be justified in aeronautical practice. We recall, nevertheless, that quenching following heat treatment sometimes causes metal objects to sing.

More refined torsional stiffness calculations of the sort discussed above will be found in Refs. 7–10 through 7–13, along with further aeroelastic applications. One especially significant error may arise from the neglect of midplane stretching where the thermal stresses become so large that the simplified theory predicts buckling. When this stretching and initial imperfections are considered in a large-deflection theory, $GJ_{\text{eff.}}$ will always remain greater than zero, and the curve in Fig. 7–4 for $M_f = 4$ is therefore unrealistic.

(c) Antisymmetrical load distribution by integral equation methods; control effectiveness

In Sec. 7–2(a) we analyzed the effects of torsional deformation on a surface attached either to a fixed cantilever mount or to an airborne vehicle executing a symmetrical maneuver. We now look at antisymmetrical maneuvers and, in view of the general restriction to one-dimensional structures, consider only loadings produced by a flapped control of the aileron or elevon type. Linear theory permits the portions of the total load distribution which are even and odd functions of the spanwise variable y to be examined separately. Accordingly, the lifts and moments needed for the present treatment arise from only three effects: control surface rotation $\delta(y)$, antisymmetrical twist $\theta(y)$, and angle of attack changes due to motions in the lateral degrees of freedom of the vehicle. Angles are taken to be positive when in the nose-up sense on the right wing or stabilizer. Among the lateral freedoms, only the rolling velocity p (positive in a sense to swing the right wing upward) will be included, because aileron effectiveness is normally measured with yaw and sideslip trimmed to zero while performing some sort of helical motion about a fixed longitudinal axis. This situation is achieved exactly in wind-tunnel tests where the model is constrained to roll about a spindle, as in the experiments described by Refs. 7–14 and 7–15. A more complete discussion of aeroelastic effects on lateral stability and control of free-flying aircraft appears in Chap. 9.

The system shown in Fig. 7–3, with its straight E.A., still serves our purposes except for the added freedom in roll. To provide variety in the exposition of methods of solution, however, we set out from the integral equation (7–17) and from lifting-line aerodynamic operators. The introduction of approximate integration methods will then lead us to a matrix formulation of the sort commonly used in the aircraft industry for nonuniform one-dimensional structures. It is consistent with the lifting-line approach to write the antisymmetrical angle of attack distribution

$$\alpha(y) = \theta(y) + \frac{\partial \alpha}{\partial \delta} \delta(y) - \frac{py}{U} \qquad (7\text{–}64)$$

where py/U is the (small) angle of attack produced by rolling and $\partial \alpha/\partial \delta$, which may also vary across the span, is the wing incidence equivalent to unit control rotation (cf. Eq. 6–17b). As discussed in Sec. 4–4, the running lift is given in linear operator form as follows:

$$\frac{L(y)}{qc} \equiv c_l(y) = \mathscr{A}\left\{\theta(y) + \frac{\partial \alpha}{\partial \delta} \delta(y) - \frac{py}{U}\right\}$$

$$= c_l{}^e(y) + \frac{\partial c_l(y)}{\partial \delta} \delta_R + \frac{\partial c_l(y)}{\partial \left(\dfrac{pl}{U}\right)} \frac{pl}{U} \qquad (7\text{–}65)$$

δ_R is the angle at some reference station, and this representation is satisfactory as long as either δ is a constant or $\delta(y)$ is known in advance; if the control surface is permitted to twist under the action of aerodynamic hinge moments, the term $\delta_R \, \partial c_l(y)/\partial \delta$ must be replaced by a $c_l{}^\delta(y)$ which is computed in connection with the hinge-moment equilibrium equation. Otherwise the lift influence functions $\partial c_l(y)/\partial \delta$ and $\partial c_l(y)/\partial \left(\dfrac{pl}{U}\right)$ can be determined by measurement or theory independently of the aeroelastic analysis.

The lift in Eq. 7–65 acts along the line of aerodynamic centers. The only antisymmetrical pitching moment accompanying it is that due to δ, which can be written

$$\frac{M_{AC}(y)}{qc^2} \equiv c_{mAC}(y) = \frac{\partial c_{mAC}(y)}{\partial \delta} \delta_R \qquad (6\text{–}66)$$

the right-hand member being useful when twisting of the control surface is neglected. A final contribution to the torque in Eq. 7–17 might come from the inertia force $(-m\ddot{p}y)$ caused by roll acceleration, acting with its arm d about the E.A. Such a time-dependent maneuver can properly be considered if its rise time is long compared to the fundamental structural frequency; elastic deformation and airloads may then be calculated on a quasi-static basis, as is sometimes also true of longitudinal maneuvers.

Making the appropriate substitutions into Eq. 7–17 and dropping $\ddot{\theta}$, we obtain

$$\theta(y) = q \int_0^l C^{\theta\theta}(y, \eta) e c c_l{}^e \, d\eta + f_a(y) \qquad (7\text{-}67a)$$

where

$$f_a(y) = \int_0^l C^{\theta\theta}(y, \eta) \left[qec \left(\frac{\partial c_l}{\partial \delta} \delta_R + \frac{\partial c_l}{\partial \left(\frac{pl}{U}\right)} \frac{pl}{U} \right) + qc^2 \frac{\partial c_{mAC}}{\partial \delta} \delta_R - m\dot{p}\eta d \right] d\eta$$

$$(7\text{-}67b)$$

In these relations, as in Eq. 7–64, $y = 0$ may be regarded as the vehicle centerline if the torsional influence function and the aerodynamic operators are suitably defined.

No exact solution of Eqs. 7–67 or their differential counterpart can be expected unless the geometrical and structural properties are given as elementary functions. Hence we turn to quadrature. As discussed in Sec. 3–5(a) or any book on numerical processes (Ref. 7–16 is an excellent example), the introduction of Simpson's rule, the trapezoidal rule, Gauss' formula, or any equivalent device permits a typical integral to be written for a set of spanwise stations y_i on the right wing,

$$\left\{ \int_0^l C^{\theta\theta}(y_i, \eta) e c c_l{}^e \, d\eta \right\} \cong \left[C^{\theta\theta} \right] \left[\overline{W} \right] \left[e \right] \left[c \right] \{c_l{}^e\} \qquad (7\text{-}68)$$

where $\left[\overline{W} \right]$ is a diagonal matrix of weighting numbers, and the definitions of the other matrices are obvious. To construct algebraic equations in terms of the single unknown quantity $c_l{}^e(y_i)$, we use any standard approximate scheme for solving the lifting-line aerodynamic problem to recast the required terms in Eq. 7–65,

$$\left\{ \frac{c}{c_R} c_l{}^e \right\} = [\mathscr{A}^a] \{\theta\} \qquad (7\text{-}69)$$

Here c_R is the chord at any convenient reference station. If possible, Eq. 7–69 involves the same locations y_i chosen for Eq. 7–68, taking advantage of antisymmetry to work on only half the wingspan; otherwise, an interpolating matrix must be employed to make the two sets compatible.

Inverting Eq. 7–69 and carrying out a series of operations similar to Eq. 7–68, we find it possible to rewrite Eqs. 7–67

$$[\mathscr{A}^a]^{-1}\left\{\frac{cc_l{}^e}{c_R}\right\} = qc_R[E]\left\{\frac{cc_l{}^e}{c_R}\right\} + \{f_a\} \tag{7-70a}$$

$$\{f_a\} = q\left[E\right]\left[\diagdown_c\right]\left(\left\{\frac{\partial c_l}{\partial \delta}\right\}\delta_R + \left\{\frac{\partial c_l}{\partial\left(\frac{pl}{U}\right)}\right\}\frac{pl}{U}\right)$$

$$+ q[F]\left\{\frac{\partial c_{mAC}}{\partial \delta}\right\}\delta_R - \left[G\right]\left[\diagdown_y\right]\{m\}\dot{p} \tag{7-70b}$$

The three square matrices on the right-hand sides are abbreviations for the following:

$$\left[E\right] = \left[C^{\theta\theta}\right]\left[\overline{W}\right]\left[\diagdown_e\right] \tag{7-70c}$$

$$\left[F\right] = \left[C^{\theta\theta}\right]\left[\overline{W}\right]\left[\diagdown_{c^2}\right] \tag{7-70d}$$

$$\left[G\right] = \left[C^{\theta\theta}\right]\left[\overline{W}\right]\left[\diagdown_d\right] \tag{7-70e}$$

Treating δ_R, p, and $\dot{p}$ for the moment as known coefficients, the antisymmetrical load distributions are found by elementary matric manipulations,

$$\left\{\frac{cc_l{}^e}{c_R}\right\} = ([\mathscr{A}^a]^{-1} - qc_R[E])^{-1}\{f_a\} \tag{7-71}$$

$$\left\{\frac{cc_l}{c_R}\right\} = \left\{\frac{cc_l{}^e}{c_R}\right\} + \left[\diagdown\frac{c}{c_R}\right]\left(\left\{\frac{\partial c_l}{\partial \delta}\right\}\delta_R + \left\{\frac{\partial c_l}{\partial\left(\frac{pl}{U}\right)}\right\}\frac{pl}{U}\right) \tag{7-72}$$

The total rolling moment exerted by the lifting surface is

$$l_R = 2q\int_0^l cc_l y\,dy \cong 2qc_R\lfloor H\rfloor\left\{\frac{cc_l}{c_R}\right\} \tag{7-73}$$

where the row matrix resulting from spanwise numerical integration is

$$\lfloor H \rfloor = \lfloor 1 \rfloor\left[\overline{W}\right]\left[\diagdown_y\right] \tag{7-74}$$

Two maneuvers of practical importance to the handling qualities of aircraft will serve to illustrate how the foregoing solution is used. The first of these is the response to a sudden fixed displacement δ_0 of the ailerons,

$$\delta_R = \delta_0 1(t) \tag{7-75}$$

With all but the rolling degree of freedom suppressed, this motion is governed by the differential equation

$$I_x \dot{p} = l_R \tag{7-76}$$

and the initial condition $p(0) = 0$. I_x is the mass moment of inertia of the vehicle about its roll-axis, a quantity which is practically unaffected by wing twist. When Eqs. 7-70b, 7-71, 7-72, and 7-73 are combined with Eq. 7-77, and the roles of input and output are specifically identified, we obtain the following:

$$A_1 \dot{p} + B_1 \frac{pl}{U} = C_1 \delta_0 \tag{7-77}$$

Here the constant coefficients are the rather elaborate expressions

$$A_1 = \frac{I_x}{2qc_R} + \lfloor H \rfloor \left([\mathscr{A}^a]^{-1} - qc_R[E] \right)^{-1} [G] \lceil \mathbf{y} \rfloor \{m\} \tag{7-78a}$$

$$B_1 = -\lfloor H \rfloor \left\langle \left\lceil \frac{c}{c_R} \right\rfloor + q\left([\mathscr{A}^a]^{-1} - qc_R[E] \right)^{-1} [E] \lceil c_l \rfloor \right\rangle \left\{ \frac{\partial c_l}{\partial \left(\frac{pl}{U} \right)} \right\} \tag{7-78b}$$

$$C_1 = q\lfloor H \rfloor \left([\mathscr{A}^a]^{-1} - qc_R[E] \right)^{-1} [F] \left\{ \frac{\partial c_{mAC}}{\partial \delta} \right\}$$
$$+ \lfloor H \rfloor \left\langle \left\lceil \frac{c}{c_R} \right\rfloor + q\left([\mathscr{A}^a]^{-1} - qc_R[E] \right)^{-1} [E] \lceil c_l \rfloor \right\rangle \left\{ \frac{\partial c_l}{\partial \delta} \right\} \tag{7-78c}$$

Equation 7-78a is solved by an exponential expression typical of first-order systems,

$$\frac{pl}{U} = \left[1 - e^{-(B_1 l/A_1 U)t} \right] \frac{C_1}{B_1} \delta_0 \tag{7-79}$$

We observe how the motion starts out with a maximum roll acceleration

$$\dot{p}(0) = \frac{C_1}{A_1} \delta_0 \tag{7-80}$$

and speeds up smoothly until the moment due to rolling velocity (related to the *damping in roll*) just balances the effect of the ailerons.

The second maneuver to be considered is steady rolling, which is described by the asymptotic form of Eq. 7-79.

$$\left. \frac{pl}{U} \right|_{t \to \infty} = \frac{C_1}{B_1} \delta_0 \tag{7-81}$$

When divided by δ_0, Eq. 7–81 yields the aileron effectiveness $\partial\left(\frac{pl}{U}\right)\Big/\partial\delta$ or rolling power of the elastic wing. After substituting Eqs. 7–78b and 7–78c into Eq. 7–81, we can divide numerator and denominator by the wing area S and the ratio l/c_R. We then recognize terms in C_1 and B_1 which are, respectively, the aileron derivative and damping in roll of the rigid wing:

$$C_{l_\delta} = \frac{1}{qS(2l)}\, 2q\lfloor\, H\, \rfloor\lceil c\rfloor\left\{\frac{\partial c_l}{\partial\delta}\right\} \tag{7–82}$$

$$C_{l_p} = \frac{1}{qS(2l)}\, 2q\lfloor\, H\, \rfloor\lceil c\rfloor\left\{\frac{\partial c_l}{\partial\left(\frac{pl}{U}\right)}\right\} \tag{7–83}$$

With these defined, the aileron effectiveness becomes

$$\frac{\partial\left(\frac{pl}{U}\right)}{\partial\delta} = -\frac{C_{l_\delta} + \frac{qc_R}{Sl}\lfloor H\rfloor([\mathscr{A}^a]^{-1} - qc_R[E])^{-1} \times \left([E]\lceil c\rfloor\left\{\frac{\partial c_l}{\partial\delta}\right\} + [F]\left\{\frac{\partial c_{mAC}}{\partial\delta}\right\}\right)}{C_{l_p} + \frac{qc_R}{Sl}\lfloor H\rfloor([\mathscr{A}^a]^{-1} - qc_R[E])^{-1}[E]\lceil c\rfloor\left\{\frac{\partial c_l}{\partial\left(\frac{pl}{U}\right)}\right\}} \tag{7–84a}$$

From the corresponding rigid-wing performance

$$\left(\frac{\partial\left(\frac{pl}{U}\right)}{\partial\delta}\right)^r = -\frac{C_{l_\delta}}{C_{l_p}} \tag{7–84b}$$

we can construct a ratio somewhat analogous to Eq. 6–17a for the typical section.

One significance of results like Eq. 7–84a is that rigid-wing derivatives are available from the wind tunnel or from accurate aerodynamic calculations, and the aeroelastic corrections can often be made on a modified strip-theory basis with considerable economy and little loss of over-all precision. Since C_{l_p} is negative, Eq. 7–84a yields a numerically positive result that decreases with increasing dynamic pressure because of the influence of the principal numerator term, which is the one containing the negative quantity $\partial c_{mAC}/\partial\delta$. As in the case of the typical section treated in Sec. 6-2(b), an aileron reversal condition $\partial\left(\frac{pl}{U}\right)\Big/\partial\delta = 0$ is reached at $q = q_R$, given implicitly by the equation

$$q_R = -\frac{\dfrac{Sl}{c_R}C_{l_\delta}}{\lfloor H \rfloor([\mathscr{A}^a]^{-1} - q_R c_R[E])^{-1}} \times \left([F]\left\{\frac{\partial c_{mAC}}{\partial\delta}\right\} + [E]\lceil c\rfloor\left\{\frac{\partial c_l}{\partial\delta}\right\}\right) \tag{7-85}$$

Since the aerodynamic quantities vary with flight Mach number, it is usually easiest to find q_R by calculating aileron effectiveness over the speed range of interest and observing if and where it vanishes.

While discussing reversal, it is natural to bring up the eigenvalue problem for the integral equation of the elastic wing. We note, for example, that Eq. 7-70a has a nonzero homogeneous solution of the divergence type at the dynamic pressure q_D for which the determinant

$$|[\mathscr{A}^a]^{-1} - q_D c_R[E]| = 0 \tag{7-86}$$

When $[\mathscr{A}^a]$ is independent of Mach number, Eq. 7-86 expands into a polynomial in q_D, whose degree is one less than the number of spanwise stations and each of whose real positive roots identifies a possible anti-symmetrical divergence. Only the lowest root is practically important; and even it serves mainly to measure the level of rigidity, because $q_R < q_D$ on straight wings with trailing-edge controls.

This is a good example of a problem where the technique of matrix iteration can be efficiently applied. We rewrite Eq. 7-70a with $f_a = 0$ as follows:

$$\left\{\frac{cc_l^e}{c_R}\right\} = q_D c_R[\mathscr{A}^a][E]\left\{\frac{cc_l^e}{c_R}\right\} \tag{7-87}$$

A trial solution for $c_l(y_i)$ is inserted on the right, and a preliminary estimate for q_D is found by requiring the two sides of Eq. 7-87 to be equal, either at a chosen spanwise station or in some average sense. The result of the calculation $[\mathscr{A}^a][E]\left\{\frac{cc_l^e}{c_R}\right\}$ is chosen for the second approximation to $\left\{\frac{cc_l^e}{c_R}\right\}$, and the process is repeated until convergence occurs in the sequence of q_D values. The final column of lift coefficients also yields the divergence mode of twist through the operation $\{\theta\} = [\mathscr{A}^a]^{-1}\left\{\frac{cc_l^e}{c_R}\right\}$.

If desired, the aerodynamic matrix can be adjusted after each step to reflect the improved estimate of Mach number. This procedure is much more efficient than factoring a polynomial, when a realistic number of stations is employed. It can be rigorously proved to converge onto the lowest positive eigenvalue if the problem is self-adjoint, that is, if the square matrix is symmetrical (Ref. 7–17). This condition is fulfilled for straight wings when strip theory is used, and satisfactory convergence occurs in practice even with much more complicated aerodynamic operators.

Incidentally, the matrix solution for *symmetrical* divergence resembles the foregoing, except that a different aerodynamic matrix $[\mathscr{A}^s]$ and possibly another influence function $C^{\theta\theta}$ appropriate to the other type of wing twist must be used. In some simple examples, the symmetrical q_D has turned out slightly lower than the antisymmetrical, because the evenly twisted finite wing is a more efficient generator of lift and torque.

Another iteration technique can be adopted to compute reversal from Eq. 7–85. A trial number is substituted for q_R in the denominator on the right, the calculation being repeated until the two sides of the equation and the Mach number of the aerodynamic quantities all agree (Ref. 7–18).

An extended presentation of various methods for finding the antisymmetrical load distribution and aileron effectiveness of straight wings will be found in Sec. 8–3 of Ref. 1–2. It is pointed out, for example, that many of the foregoing results can be closely approximated by assuming the E.A. and the line of A.C.'s to coincide, so that elastic terms containing $\partial c_l/\partial \delta$ and $\partial c_l/\partial \left(\dfrac{pl}{U}\right)$ may be discarded and only the deformation due to aileron twisting moment $\partial c_{mAC}/\partial \delta$ retained. Strip-theory solutions are derived for the uniform straight lifting surface, which parallel those developed in Sec. 7–2(a). Using the notation defined therein, the aileron effectiveness can be expressed as follows (Eq. 8–110 of Ref. 1–2):

$$\frac{\partial \left(\dfrac{pl}{U}\right)}{\partial \delta} = \frac{\left[\dfrac{\cos \lambda_1}{\cos \lambda} - 1\right]\dfrac{\partial c_l}{\partial \delta} + \left[\dfrac{\cos \lambda_1}{\cos \lambda} - 1 - \dfrac{\lambda^2 - \lambda_1{}^2}{2}\right]\dfrac{c}{e}\dfrac{\partial c_{mAC}}{\partial \delta}}{\dfrac{\partial c_l}{\partial \alpha}\left[\dfrac{\tan \lambda}{\lambda} - 1\right]} \quad (7\text{--}88)$$

The control surface envisioned in Eq. 7–88 is one of constant chord, extending from $y = l_1$ to the tip $y = l$. λ_1 is the dynamic pressure parameter

$$\lambda_1{}^2 = \frac{qcel_1{}^2}{GJ}\frac{\partial c_l}{\partial \alpha} \quad (7\text{--}89)$$

associated with the inboard end. Reversal occurs at the vanishing of the

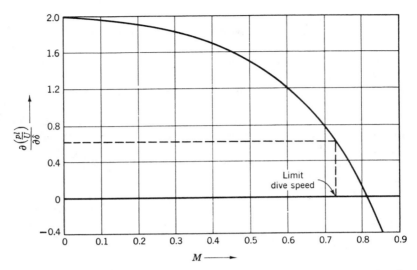

Fig. 7–5. Aileron effectiveness plotted vs. flight Mach number at sea level for the straight wing of a subsonic monoplane. (Taken from Chap. 8 of Bisplinghoff, Ashley, and Halfman, Ref. 1–2.)

numerator in Eq. 7–88, but the presence of transcendental functions of λ and λ_1 prevents the writing of an explicit formula for q_R, even when the aerodynamic derivatives are independent of M.

As a warning, it must be stated that uncorrected strip theory is less accurate for the quantitative estimation of aileron effects than for symmetrical loadings. Except at high supersonic and hypersonic speeds, the aerodynamic induction is much stronger in connection with angle-of-attack discontinuities like those at the control-surface extremities than it is when the spanwise incidence variation is continuous. Special care must therefore be taken in evaluating $\partial c_{mAC}/\partial \delta$, $\partial c_l/\partial \delta$, and $C_{l\delta}$.

Despite the difficulty of determining absolute values of $\partial \left(\dfrac{pl}{U}\right) \Big/ \partial \delta$ and q_R, it turns out on most straight wings that the linear variation of effectiveness with q/q_R found for the typical section in Sec. 6–2(b) is a close approximation of the true behavior. This point is well illustrated by Example 8–4 of Ref. 1–2, from which we reproduce Fig. 7–5. The system considered is a tapered wing of semispan $l = 500$ in. ($Æ = 6.15$), root and tip chords 225 in. and 100 in., respectively. The A.C. is at 25% chord behind the leading edge, and there is a straight E.A. normal to the flight direction at 35% chord. The spanwise stations y_i are chosen, in connection with a

7-point Gauss quadrature formula, to be equally spaced in an angle variable defined by $y = l \cos \theta$, so that on the semispan*

$$\lceil y \rfloor = \begin{bmatrix} 461.94 & & & \\ & 353.55 & & \\ & & 191.34 & \\ & & & 0 \end{bmatrix} \text{ in.} \qquad (7\text{-}90)$$

Other necessary matrices of wing properties are

$$\lceil e \rfloor = \begin{bmatrix} 10.95 & & & \\ & 13.66 & & \\ & & 17.72 & \\ & & & 22.50 \end{bmatrix} \text{ in.} \qquad (7\text{-}91a)$$

$$\lceil c \rfloor = \begin{bmatrix} 109.515 & & & \\ & 136.612 & & \\ & & 177.165 & \\ & & & 225 \end{bmatrix} \text{ in.} \qquad (7\text{-}91b)$$

$$[C^{\theta\theta}] = \begin{bmatrix} 424.3 & 424.3 & 424.3 & 0 \\ 424.3 & 186.6 & 186.6 & 0 \\ 424.3 & 186.6 & 78.45 & 0 \\ 0 & 0 & 0 & 0 \end{bmatrix} \times 10^{-10} \text{ rad/in.-lb} \qquad (7\text{-}91c)$$

There is an aileron lying between $l_1 = 370$ in. and $l_2 = 487$ in., which is assumed to have constant sectional derivatives $\partial\alpha/\partial\delta = 0.4$ and $\partial c_{mAC}/\partial\delta = -0.45$. The data in Fig. 7-5 were computed from Eq. 7-84a for sea-level flight, using strip theory throughout, neglecting the twist due to the A.C.–E.A. offset, and adjusting all aerodynamic coefficients by a factor $1/\sqrt{1 - M^2}$ for the (subsonic) influence of compressibility. A generally parabolic variation of effectiveness with M is evident, corresponding to

* To be precise, these Gaussian (or Multhopp) stations are used in Ref. 1–2 for divergence and symmetrical loading calculations, while a different set of y_i is employed in Example 8–4, concentrating on the portion of the span covered by the aileron. Equations 7–90 and 7–91 serve, however, to give a description of the system adequate for our purposes here.

the aforementioned linear dependence on q/q_R. A hypothetical limit dive speed of 480 knots at sea level is indicated on the figure. At this limit 70% of the rolling power has already been lost to aeroelastic effects. Reversal occurs at 534 knots.

In Examples 8–1 and 8–3 of Ref. 1–2, the symmetrical influence of twist on load distribution and the divergence eigenvalue are computed for this same wing from the integral equation. To permit comparison with U_R, we list the following estimates of sea-level U_D, found by using three different forms of the aerodynamic operator: 1317 knots, by strip theory with $\partial c_l/\partial \alpha = 5.5$; 1583 knots, for symmetrical divergence by lifting-line theory; 1659 knots, for antisymmetrical divergence by lifting-line theory. No attempt was made at a Mach-number correction, since these speeds exceed that of sound and are therefore artificial. In actuality, this system would never encounter torsional instability, because the A.C. would be shifted behind the E.A. in supersonic flight.

With the background provided by this relatively brief treatment of the integral-equation-matrix approach to the aeroelastic analysis of straight, one-dimensional structures, the reader will be able to fill many gaps left open because of space limitations. For instance, if the trailing-edge control surface is replaced with a spoiler, we would commence by substituting $\Delta c_l(\delta_s)$ and $\Delta c_{mAC}(\delta_s)$ for the terms $\delta_R \partial c_l/\partial \delta$ and $\delta_R \partial c_{mAC}/\partial \delta$ in Eqs. 7–67. Even though these coefficients vary nonlinearly with spoiler deflection δ_s, it is possible to calculate the rolling power at a given value of δ_s, just as was done in Sec. 6–2(b). Because of the relative ineffectiveness of this type of control as a generator of unfavorable twisting moments, aeroelastic problems are less important. No practical case of spoiler reversal on a straight or sweptback wing is known to the authors.

Another more critical situation concerns the "wind-up" or "blowback" of the aileron under aerodynamic hinge moments. This can be treated by introducing first a command aileron rotation δ_0, which corresponds to the lateral displacement of the control column and is the input to a series of flexible links such as control cables, actuators, etc. It is then necessary to compute the true aileron position $\delta(y)$, to be used in Eqs. 7–67 by replacing $\delta_R \partial c_l/\partial \delta$ with $c_l{}^\delta(y)$. This is done by writing an integral equation for the equilibrium of elastic and aerodynamic torques about the hinge, involving an influence function $C^{\delta\delta}(y, \eta)$, which equals the twist from the undeformed position δ_0 at station y due to unit hinge moment concentrated at station η. The influence function can be arranged to include contributions from a control cable or actuator and back-up structure. The use of numerical integration formulas will lead to algebraic equations, which can be solved in combination with a modified version of Eqs. 7–70 to find aileron effectiveness per unit δ_0 and other interesting results.

7-3 STATIC PROBLEMS OF STRAIGHT AND SWEPT WINGS SOLVED BY APPROXIMATE METHODS

Turning to lifting surfaces without clearly defined elastic axes or with appreciable sweep, we are confronted by an extensive literature, containing a bewildering array of techniques for predicting their static aeroelastic properties. The objective of what follows is merely to illustrate a few of the more powerful and broadly useful methods, furnishing as we go some information on the influence of the principal parameters.

Appearing at the close of World War II, the swept wing offered to the aeroelastician his first real novelty since the cantilever monoplane. It is characterized by a *structural aspect ratio* (ratio of the swept span to the chordwise dimension taken normal to the swept axis) which exceeds the aerodynamic aspect ratio in rough proportion to $1/\cos^2 \Lambda$. Λ denotes the sweep angle, positive for sweepback. Thus the deformations are generally larger, but not necessarily more undesirable, than those of comparable straight wings. Indeed, the most interesting feature of sweep is the aerodynamic coupling between flexure and torsion. It is easy to see that a small upward bending slope $\partial w/\partial \bar{y}$, $\bar{y}$ denoting a coordinate measured out along the swept structural reference axis, is equivalent to an incidence $-\sin \Lambda \; \partial w/\partial \bar{y}$ in a plane parallel to the flight direction. A positive lift increment on a flexible, sweptback wing usually unloads the tip region and displaces the center of pressure inboard and forward. Such a wing is incapable of diverging, since added torque due to positive twisting is more than counteracted by the negative effect of the bending slope.

Even 5 or 10 degrees of sweep angle can cause q_D to go to infinity, the asymptotic value depending mainly on the GJ/EI ratio. This observation led Hill in England to the concept of the *aero-isoclinic wing* (Ref. 7–19), which has a combination of parameters that makes q_D just infinite; the spanwise load distribution is unaffected by aeroelasticity, with bending and torsion cancelling each other identically.

Sweepback obviously detracts from lateral control effectiveness, because upward bending and nose-down twisting both work against a positively deflected aileron. As a result, most large-aspect-ratio swept wings are fitted with spoilers. Sweptforward surfaces are fine from the control standpoint, but their very low divergence speeds seem to have contributed heavily toward their abandonment in practice.

The often predominant influence of bending is illuminated by the elementary example of a uniform, slender swept wing whose line of A.C.'s coincides with its E.A. In the absence of camber, there is no torsional deformation, and the flexural differential equation (7–11), for the steady-state deflection $w(\bar{y})$ under aerodynamic loading, reads

$$EI \frac{d^4 w}{d\bar{y}^4} = [L^e + L^r] \cos \Lambda \qquad (7\text{--}92)$$

Here L is lift force *per unit distance normal to the flight direction*, as it would be given by a theory based on chordwise sections taken this way. Superscripts r and e refer, respectively, to loads due to initial incidence of the rigid airfoil and to those added by bending itself. We assume that $L^r(\bar{y}) \equiv qcc_l{}^r(\bar{y})$ is known and that $L^e(\bar{y})$ can be estimated on a strip basis from the aforementioned angle of attack due to bending slope,

$$L^e(\bar{y}) = qc \frac{\partial c_l}{\partial \alpha} \cos \Lambda \left[- \frac{dw}{d\bar{y}} \sin \Lambda \right] \qquad (7\text{--}93)$$

The derivative $\partial c_l / \partial \alpha$ represents lift-curve slope for a straight wing whose profile is the same as a section normal to the $\bar{y}$-axis, with $\cos \Lambda$ providing the adjustment for sweep effect discussed in Sec. 3–4(e). Cantilever boundary conditions require w and $dw/d\bar{y}$ to vanish at the root $\bar{y} = 0$, while the second and third derivatives are zero at the free tip $\bar{y} = l$.

Solution of Eqs. 7–92 and 7–93 is facilitated by adopting

$$\Gamma \equiv \frac{dw}{d\bar{y}} \qquad (7\text{--}94a)$$

as the principal unknown and introducing dimensionless quantities,

$$\bar{\eta} = 1 - \frac{\bar{y}}{l} \qquad (7\text{--}94b)$$

$$\bar{a} = \frac{qc\bar{l}^3 \sin \Lambda \cos^2 \Lambda}{EI} \frac{\partial c_l}{\partial \alpha} \qquad (7\text{--}94c)$$

$$\bar{b} = \frac{qc\bar{l}^3 \cos \Lambda}{EI} \qquad (7\text{--}94d)$$

$\bar{a}$ and $\bar{b}$ are constants on the uniform wing. In these terms, the differential equation and remaining boundary conditions are

$$\frac{d^3 \Gamma}{d\bar{\eta}^3} - \bar{a}\Gamma = -\bar{b}c_l{}^r(\bar{\eta}) \qquad (7\text{--}95)$$

$$\Gamma(1) = 0, \qquad \frac{d\Gamma(0)}{d\bar{\eta}} = 0, \qquad \frac{d^2\Gamma(0)}{d\bar{\eta}^2} = 0 \qquad (7\text{--}96a, b, c)$$

The $w = 0$ root condition is applied when integrating Γ to find w.

Our principal interest here is in the homogeneous problem obtained by equating the right-hand side of Eq. 7–95 to zero. Despite its simple appearance, this is no longer a Sturm-Liouville problem, because of the presence of the third derivative, and it would be incorrect to conclude that the eigenvalues $\bar{a}$ are all real and positive. The homogeneous solution has the form

$$\Gamma = A_1 e^{r_1 \bar{\eta}} + A_2 e^{r_2 \bar{\eta}} + A_3 e^{r_3 \bar{\eta}} \tag{7-97}$$

where r_i are the roots of the equation

$$r^3 - \bar{a} = 0 \tag{7-98}$$

When Eq. 7–97 is substituted into the three conditions (7–96), the requirement for nonzero coefficients A_i turns out to be the following transcendental equation:

$$1 + \frac{r_1}{r_2}\left[\frac{r_1 - r_3}{r_3 - r_2}\right] e^{r_2 - r_1} + \frac{r_1}{r_3}\left[\frac{r_2 - r_1}{r_3 - r_2}\right] e^{r_3 - r_1} = 0 \tag{7-99}$$

After the three well-known roots $r_1 = \sqrt[3]{\bar{a}}$, $r_2 = \frac{1}{2}(-1 + i\sqrt{3})\sqrt[3]{\bar{a}}$, $r_3 = \frac{1}{2}(-1 - i\sqrt{3})\sqrt[3]{\bar{a}}$ are inserted into Eq. 7–99, it simplifies to

$$e^{(3/2)\sqrt[3]{\bar{a}}} + 2\cos\left(\frac{\sqrt{3}}{2}\sqrt[3]{\bar{a}}\right) = 0 \tag{7-100}$$

Equation 7–100 is not satisfied by any real, positive $\bar{a}$, because the exponential term is always too large to be cancelled by the cosine; but there is a denumerably infinite sequence of negative eigenvalues. The first of these,

$$\bar{a}_0 = -6.33 \tag{7-101a}$$

must be found by trial and error, while the rest are closely approximated by the vanishing of the cosine term,

$$\bar{a}_n = -\frac{(2n + 1)^3 \pi^3}{3\sqrt{3}}, \qquad (n = 1, 2, 3, \cdots) \tag{7-101b}$$

By reference to Eq. 7–94c, we note the physical interpretation of these results that only sweptforward wings can undergo pure bending divergence. Corresponding to $\bar{a}_0$, the dynamic pressure of the fundamental mode is

$$q_D = \frac{6.33EI}{cl^3 |\sin \Lambda| \cos^2 \Lambda\, \partial c_l/\partial\alpha}$$

$$= \frac{6.33EI}{\bar{c}l^3 |\sin \Lambda| \cos \Lambda\, \partial c_l/\partial\alpha}, \qquad (\Lambda < 0) \tag{7-102}$$

where $\bar{c} = c \cos \Lambda$ is the chord measured normal to the E.A. For the sweptback case, the bending divergence speed is always imaginary.

By way of application, the last member of Eq. 7–102 refers to a cantilever wing of fixed planform dimensions, which can be rotated about its root

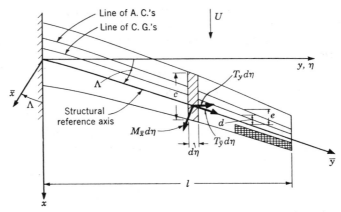

Fig. 7–6. Swept wing of arbitrary planform and stiffness, showing aerodynamic moments acting on a section dy parallel to the centerline.

from $\Lambda = 0°$ to $90°$. Neglecting variations of $\partial c_l/\partial\alpha$ with Mach number, q_D approaches infinity as $1/|\sin \Lambda|$ when the sweep is brought back toward zero. It is also infinite when the wing is swung all the way forward, because of the theoretically complete loss of aerodynamic efficiency. Between these limits q_D has a minimum value of $12.66EI/\bar{c}\bar{l}^3\ \partial c_l/\partial\alpha$ at $\Lambda = -45°$.

Nonhomogeneous solutions of Eq. 7–80 describe the bending of the preloaded wing, either sweptback or sweptforward. They are found by adding Eq. 7–97 to the appropriate particular solution. For arbitrary $c_l{}^r(\bar{\eta})$, the latter is best developed from the standard integral formulas (pp. 529–530 of Ref. 7–5). If $c_l{}^r$ is constant, the particular solution is just $\bar{b}c_l{}^r/\bar{a}$, and we can easily work out the complete result by applying Eqs. 7–96:

$$\Gamma(\bar{\eta}) = \frac{\bar{b}}{\bar{a}}c_l{}^r\left\{ 1 - \frac{e^{\bar{\eta}\sqrt[3]{\bar{a}}} + 2e^{-\frac{1}{2}\bar{\eta}\sqrt[3]{\bar{a}}}\cos\left(\dfrac{3\bar{\eta}\sqrt[3]{\bar{a}}}{2\sqrt{3}}\right)}{e^{\sqrt[3]{\bar{a}}} + 2e^{-\frac{1}{2}\sqrt[3]{\bar{a}}}\cos\left(\dfrac{3\sqrt[3]{\bar{a}}}{2\sqrt{3}}\right)} \right\} \qquad (7\text{–}103)$$

In the analytical presentations comprising the remainder of this section, we shall focus on one-dimensional structures having swept, rectilinear reference axes but with arbitrary spanwise variations of all properties. For use with modern digital computers, the most convenient mathematical approach relies on numerical approximation of the integral equations. As pointed out in Sec. 7–2, aerodynamic operators of lifting-line type fit in naturally. We introduce the airloads in connection with strips parallel to the vehicle centerline (Fig. 7–6), whose cross sections are assumed to remain undeformed, because such a viewpoint is especially compatible with Weissinger's theory of spanwise loading in subsonic flight.

This choice of system is not intended to reflect critically upon the many published treatments based either on the beam-rod differential equations or on integral equations with sections normal to the reference axis. In fact we should like to call particular attention to two examples of the former which are just as useful today, for the continuous structures to which they apply, as they were when proposed a decade ago. The first is the method of bending-torsion divergence calculation of Diederich and Budiansky (Ref. 7–20). Using strip theory modified by an over-all aspect-ratio correction, these authors have solved the cantilever beam-rod equations for two important cases: the wing with uniform properties, and the linearly tapered wing with EI and GJ proportional to the fourth power of the chord [cf. H. Reissner's solution in Sec. 7–2(a) with $\gamma_1 = 1$, $\gamma_2 = 4$]. Their solutions show the divergence eigenvalues to be governed primarily by the ratio of two stiffness parameters

$$\frac{q\bar{e}_R\bar{c}_R\bar{l}^2\cos^2\Lambda}{GJ_R}\frac{\partial c_l}{\partial\alpha} \quad \text{and} \quad \frac{q\bar{c}_R\bar{l}^3\cos^2\Lambda\tan\Lambda}{EI_R}\frac{\partial c_l}{\partial\alpha}$$

where R identifies properties of the root station. We reproduce Fig. 7–7, which plots both theoretical and experimental results on a uniform, sweptforward surface from Ref. 7–20 and points up the controlling influence of Λ.

A second procedure deserving special mention is Diederich's numerical integration of the differential equations for a cantilever with arbitrary spanwise stiffness distributions (Refs. 7–21 and 7–22). Reference 7–22 contains charts and approximate formulas from which such things as spanwise load distribution, lift-curve slope, aerodynamic center location, and damping in roll can be corrected for aeroelastic effects.

In order to synthesize integral equations for the deformations of the surface in Fig. 7–6, we must carefully distinguish between barred quantities, associated with cross sections taken perpendicular to the structural reference axis, and unbarred quantities, related to streamwise segments. Either y or $\bar{y}$ is suitable as a spanwise coordinate, but we choose the former in conformity with our emphasis on sections normal to it. Then it is clear, for example, that the bending deflection $w(y)$ and twist $\bar{\theta}(y)$ about the axis combine to produce a (small) rotation $\theta(y)$ of streamwise segments, as follows:

$$\theta(y) = \bar{\theta}(y)\cos\Lambda - \frac{dw}{d\bar{y}}\sin\Lambda$$

$$= \bar{\theta}(y)\cos\Lambda - \frac{dw}{dy}\sin\Lambda\cos\Lambda \tag{7–104}$$

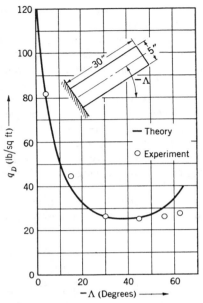

Fig. 7-7. Measured divergence dynamic pressure q_D on a uniform, sweptforward wing compared with modified strip-theory calculations as a function of sweep angle Λ_0. (Taken from Diederich and Budiansky, Ref. 7-20.)

To modify the general equations (7-18 and 7-19) into a system containing w and θ as dependent variables, we let

$$F_z(\eta) = L(\eta) - m(\eta)N(\eta)g \qquad (7\text{-}105)$$

stand for resultant upward force per unit distance normal to the flight direction, acting at $y = \eta$. The total aerodynamic moment on the strip $d\eta$, which is the resultant of $T_{\bar{y}} \, d\eta$ and $M_{\bar{x}} \, d\eta$, is called $T_y(\eta) \, d\eta$, and to it must be added the inertia torque $-m(\eta)N(\eta)g \, d(\eta) \, d\eta$. Thus we are led to the following formulation:

$$w(y) = \int_0^l C^{zz}(y, \eta)[L(\eta) - mNg] \, d\eta$$
$$+ \int_0^l C^{z\theta}(y, \eta)[T_y(\eta) - mNgd] \, d\eta \qquad (7\text{-}106)$$

$$\theta(y) = \int_0^l C^{\theta z}(y, \eta)[L(\eta) - mNg] \, d\eta$$
$$+ \int_0^l C^{\theta\theta}(y, \eta)[T_y(\eta) - mNgd] \, d\eta \qquad (7\text{-}107)$$

The definitions of the structural influence functions in Eqs. 7–106 and 7–107 are clear from the manner in which they appear. When lifting-line or streamwise-strip aerodynamic operators are employed, this general approach gives the impression of having decoupled bending from torsion, because L and T_y depend only on the spanwise distribution of θ. This is somewhat misleading, since bending-slope contributions are implicit in $C^{\theta z}$, $C^{\theta\theta}$, and θ itself. Nevertheless, it is a great convenience to carry out aeroelastic calculations with the single Eq. 7–107 and to determine the flexural deformations afterward from Eq. 7–106. Any advantage is lost when we seek information on shears, bending moments, torques, and stresses.

The assumption, implicit in the foregoing, that streamwise segments are wholly free from camber bending is fulfilled best when the ribs are oriented parallel to the vehicle centerline. Since most large-aspect-ratio swept wings are laid out with ribs normal to some swept axis, it is reassuring to know that experience proves their camber bending to have very little influence on aeroelastic phenomena. Any lifting surface which deforms in this way to a significant degree is probably best treated by the methods of Chap. 8.

(a) Symmetrical load distribution and divergence

For the symmetrically loaded case, N is a constant, and the angle of attack can be separated into portions due to twist and initial incidence of the undeformed surface,

$$\alpha(y) = \theta(y) + \alpha_0(y) \qquad (7\text{–}108)$$

The generic lifting-line operator (cf. Eq. 4–88), relates α to the dimensionless running force,

$$\frac{L(y)}{qc} \equiv c_l(y) = \overline{\mathscr{A}^s}\{\theta(y) + \alpha_0(y)\}$$
$$= c_l^e(y) + c_l^r(y) \qquad (7\text{–}109)$$

This is presumed to act along some line of aerodynamic centers, whose location may be influenced by the finite span, and it is accompanied by a moment about the A.C. Therefore, the combined torque must be

$$T_y(y) = e(y)L(y) + M_{AC}(y)$$
$$= qcec_l(y) + qc^2 c_{mAC}(y) \qquad (7\text{–}110)$$

When both airload expressions are inserted into Eq. 7–107 and the terms describing the effect of flexibility are separated out, we obtain

$$\theta(y) = q\int_0^l \bar{C}(y, \eta)cc_l^e \, d\eta + \bar{f}_s(y) \qquad (7\text{–}111a)$$

where

$$f_s(y) = q \int_0^l \bar{C}(y, \eta) c c_l^r \, d\eta + q \int_0^l C^{\theta\theta}(y, \eta) c^2 c_{mAC} \, d\eta$$
$$- g \int_0^l [C^{\theta z}(y, \eta) + C^{\theta\theta}(y, \eta) d(\eta)] mN \, d\eta \qquad (7\text{-}111b)$$

The quantity

$$\bar{C}(y, \eta) = C^{\theta z}(y, \eta) + C^{\theta\theta}(y, \eta) e(\eta) \qquad (7\text{-}112)$$

is a modified torsional influence function, which plays a role analogous to that of $C^{\theta\theta}$ on a straight wing with a true E.A. $\bar{C}$ is evidently not symmetrical under an interchange of y and η.

We introduce numerical integrations with a weighting matrix $\lceil \overline{W} \rfloor$ and approximate Eqs. 7-111 through a series of operations similar to those in Sec. 7-2(c). Assuming that the simultaneous equations

$$\left\{ \frac{c}{c_R} c_l \right\} = [\overline{\mathscr{A}^s}]\{\alpha\} \qquad (7\text{-}113)$$

defined by the form of the aerodynamic operator, can be set up at or transformed to the same stations used in the structural analysis, we finally obtain

$$[\overline{\mathscr{A}^s}]^{-1} \left\{ \frac{c c_l^e}{c_R} \right\} = q c_R [\bar{E}] \left\{ \frac{c c_l^e}{c_R} \right\} + \{f_s\} \qquad (7\text{-}114a)$$

The various matrices abbreviated in Eq. 7-114a are the following:

$$\{f_s\} = q c_R [\bar{E}] \left\{ \frac{c c_l^r}{c_R} \right\} + q[F]\{c_{mAC}\} - Ng[\bar{G}]\{m\} \qquad (7\text{-}114b)$$

$$[\bar{E}] = ([C^{\theta z}] + [C^{\theta\theta}]\lceil e \rfloor)\lceil \overline{W} \rfloor \qquad (7\text{-}114c)$$

$$[F] = [C^{\theta\theta}]\lceil c^2 \rfloor \lceil \overline{W} \rfloor \qquad (7\text{-}114d)$$

$$[\bar{G}] = ([C^{\theta z}] + [C^{\theta\theta}]\lceil d \rfloor)\lceil \overline{W} \rfloor \qquad (7\text{-}114e)$$

As discussed in connection with the straight-wing differential equation in Sec. 7-2(a), we can conceive of solving Eqs. 7-114 either for a given incidence at the vehicle's centerline or for a given normal load factor N. The former case is quite straightforward because $\alpha_0(y)$, $c c_l^r(y)$, and $c_{mAC}(y)$ are known quantities, the rigid load distribution often being available from wind-tunnel tests, if not theoretically. Equation 7-114a is manipulated to give the elastic loading

$$\left\{ \frac{c c_l^e}{c_R} \right\} = ([\overline{\mathscr{A}^s}]^{-1} - q c_R [\bar{E}])^{-1} \{f_s\} \qquad (7\text{-}115)$$

whence the resultant lift on the full span and its symmetrical spanwise distribution are

$$\left\{\frac{cc_l}{c_R}\right\} = \left\{\frac{cc_l{}^e}{c_R}\right\} + \left\{\frac{cc_l{}^r}{c_R}\right\} \tag{7-116}$$

$$C_L = \frac{2c_R}{S} \lfloor\, 1 \,\rfloor [\overline{W}] \left\{\frac{cc_l}{c_R}\right\} \tag{7-117}$$

If Eqs. 7-115 through 7-117 refer to a free-flight condition rather than a restrained wing, the final term in $\{f_s\}$ contains the unknown factor N, which must be eliminated by equating total lift to N times the vehicle weight W. (This involves neglecting the tail lift, but it can be introduced by the techniques appropriate to unrestrained elastic aircraft presented in Chap 9.) The most direct way of handling the problem of free flight is to replace $cc_l{}^e$ by $(cc_l - cc_l{}^r)$ in Eq. 7-114a and make the substitution

$$[\overline{\mathscr{A}^s}]^{-1}\left\{\frac{cc_l{}^r}{c_R}\right\} = \{\alpha_0\} \tag{7-118}$$

The resulting relation is

$$[\overline{\mathscr{A}^s}]^{-1}\left\{\frac{cc_l}{c_R}\right\} = qc_R[\bar{E}]\left\{\frac{cc_l}{c_R}\right\} + \{\alpha_0\} + q[\bar{F}]\{c_{mAC}\} - Ng[\bar{G}]\{m\} \tag{7-119}$$

Eliminating N by means of the balance condition $qSC_L = NW$, we are led to the following solution for the lift distribution:

$$\left\{\frac{cc_l}{c_R}\right\} = \left([\overline{\mathscr{A}^s}]^{-1} - qc_R[\bar{E}] + \frac{2gqc_R}{W}[\bar{G}]\{m\}\lfloor\, 1 \,\rfloor[\overline{W}]\right)^{-1}$$
$$\times (\{\alpha_0\} + q[\bar{F}]\{c_{mAC}\}) \tag{7-120}$$

Total lift and pitching moments are determined by operations like Eq. 7-117.

One useful application of Eqs. 7-115 and 7-120 is to compute wing aeroelastic effects on stability derivatives. Consider, for instance, the lift- and moment-curve slopes, which are found by adding a fixed increment $\Delta\alpha_0$ to the rigid-wing angle of attack and are unaffected by M_{AC}. Two situations are possible: the process might be carried out on a restrained elastic model, as if mounted on a wind-tunnel balance, in which case $N = 1$ and does not influence the incremental load Δcc_l; or it might be done for free flight, in which case the airplane must be balanced, and Eq. 7-120 applies. The solution of the former problem results from replacing $\{\bar{f}_s\}$ by $\Delta\alpha_0 qc_R[\bar{E}][\overline{\mathscr{A}^s}]\{1\}$ in Eqs. 7-115 and 7-116:

$$\left\{\frac{\Delta cc_l}{c_R}\right\} = \left\{\frac{\Delta cc_l{}^r}{c_R}\right\}$$
$$+ \Delta\alpha_0 qc_R([\overline{\mathscr{A}^s}]^{-1} - qc_R[\bar{E}])^{-1}[\bar{E}][\overline{\mathscr{A}^s}]\{1\} \tag{7-121}$$

The changes in total load coefficients are

$$\Delta C_L = \frac{2c_R}{S} \lfloor 1 \rfloor \left[\overline{W}\right] \left\{\frac{\Delta c c_l}{c_R}\right\} \qquad (7\text{–}122)$$

and

$$\Delta C_M = -\frac{2c_R \tan \Lambda}{(MAC)S} \lfloor y \rfloor \left[\overline{W}\right] \left\{\frac{\Delta c c_l}{c_R}\right\} \qquad (7\text{–}123)$$

Here pitching moment is taken about the y-axis in Fig. 7–6,[*] for which the lift at station y has an arm $-x = -y \tan \Lambda$; (MAC) is the mean aerodynamic chord, used as a reference length. The actual derivatives are

$$\frac{\partial C_L}{\partial \alpha} = \frac{\Delta C_L}{\Delta \alpha_0} = \left(\frac{\partial C_L}{\partial \alpha}\right)^r$$

$$+ \frac{2q c_R^2}{S} \lfloor 1 \rfloor \left[\overline{W}\right] \left([\overline{\mathscr{A}^s}]^{-1} - q c_R [\bar{E}]\right)^{-1} [\bar{E}][\overline{\mathscr{A}^s}]\{1\}$$

and $(7\text{–}124a)$

$$\frac{\partial C_M}{\partial \alpha} = \frac{\Delta C_M}{\Delta \alpha_0} = \left(\frac{\partial C_M}{\partial \alpha}\right)^r$$

$$- \frac{2q c_R^2 \tan \Lambda}{(MAC)S} \lfloor y \rfloor \left[\overline{W}\right] \left([\overline{\mathscr{A}^s}]^{-1} - q c_R [\bar{E}]\right)^{-1} [\bar{E}][\overline{\mathscr{A}^s}]\{1\}$$

$$(7\text{–}124b)$$

where the rigid-wing contributions, which correspond to the first term on the right of Eq. 7–121, have been separately identified. The static stability parameter about this moment axis is

$$\frac{\partial C_M}{\partial C_L} = \frac{\partial C_M / \partial \alpha}{\partial C_L / \partial \alpha} \qquad (7\text{–}124c)$$

For the somewhat more realistic example of free flight, Eqs. 7–120, 7–122, and 7–123 yield the following derivative expressions:

$$\frac{\partial C_L}{\partial \alpha} = \frac{2c_R}{S} \lfloor 1 \rfloor \left[\overline{W}\right] \left([\overline{\mathscr{A}^s}]^{-1} - q c_R [\bar{E}]\right.$$

$$+ \frac{2gq c_R}{W}[\bar{G}]\{m\} \lfloor 1 \rfloor \left[\overline{W}\right]\Big)^{-1} \{1\} \qquad (7\text{–}125a)$$

$$\frac{\partial C_M}{\partial \alpha} = -\frac{2c_R \tan \Lambda}{(MAC)S} \lfloor y \rfloor \left[\overline{W}\right] \left([\overline{\mathscr{A}^s}]^{-1} - q c_R [\bar{E}]\right.$$

$$+ \frac{2gq c_R}{W}[\bar{G}]\{m\} \lfloor 1 \rfloor \left[\overline{W}\right]\Big)^{-1} \{1\} \qquad (7\text{–}125b)$$

---

[*] The moment can be transferred to any other spanwise axis by the standard method.

The reader will have no difficulty in reorganizing Eqs. 7–125 so that the rigid-wing derivatives appear explicitly, if he will first separate out the elastic part from Eq. 7–120 before carrying out the total lift and moment integrations.

On a typical *sweptback* wing, the bending deformations under load overpower the twist, tending to reduce the lift-curve slope and the static stability (i.e., rendering $\partial C_M/\partial \alpha$ and $\partial C_M/\partial C_L$ more positive). This fact is evident from Eqs. 7–124, if we observe that the principal contributions to matrix $[\bar{E}]$ are from $[C^{\theta z}]$, whose elements are obviously negative. The converse is true in the sweptforward case. It is interesting that the inertia loads resulting from a symmetrical flight maneuver partially counteract these static aeroelastic effects, regardless of the sense of the sweep, as can be seen from the opposite signs of the terms containing $[\bar{E}]$ and $[\bar{G}]$ in Eqs. 7–125. Pai and Sears seem to be the first to have analyzed this phenomenon quantitatively (Ref. 7–23). Numerical examples computed by the integral-equation-matrix approach will be found in their paper, as well as in Sec. 8–4 of Ref. 1–2.

Nothing has been done in the foregoing about the case of flight at a prescribed value of the normal load factor. This situation can be handled, however, by substituting the known N and an unknown centerline angle of attack $\alpha_0(0)$ into Eqs. 7–114. Equation 7–117 is then employed to compute $\alpha_0(0)$ from the force-balance on the vehicle. The very comprehensive report by Pian and Lin (Ref. 7–24) contains illustrative calculations of this sort on the XB–47 airplane. Figure 7–8 is adapted from them.* It shows the rigid- and elastic-wing spanwise lift distributions during a $3g$ pull-out at $M = 0.8$ and 27,000 ft in the standard atmosphere. Gross weight is taken to be 125,000 lb, and other parameters can be obtained from Ref. 7–24. These authors employ the lifting-line theory of Weissinger in their development, choosing the spanwise stations called for in the $m = 7$ Weissinger aerodynamic matrix [cf. Sec. 4–4(a)] when setting up all the numerical integrations. They consider the effects on wing deformation of lift, aerodynamic pitching moment, torques caused by offset engine thrust lines, aerodynamic moments due to external stores, concentrated or distributed inertia forces, and drag loads; they demonstrate how all these various contributions may be separately determined and superimposed. It is of interest that the linearized consequence of a thrust or drag force is apparently to modify the structural influence functions; this alteration turns out to be negligibly small on the XB–47. In Ref. 7–24, an approximate procedure based on strip theory is also presented and shown to

---

* This calculation also forms the basis of Example 8–7 of Ref. 1–2, but the airplane could not be identified at the time for security reasons.

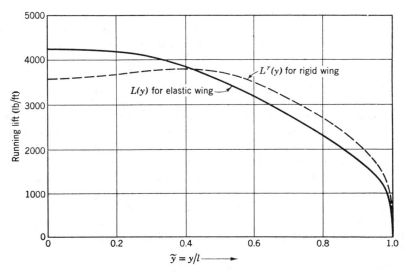

Fig. 7–8. Spanwise load distribution, with and without aeroelastic effects, on the right wing of the XB-47 in a 3*g* pull-out at $M = 0.8$ and 27,000-ft altitude. (Taken from Pian and Lin, Ref. 7–24.)

provide fairly good agreement with the more precise computations mentioned above.

The divergence problem for swept wings and one-dimensional lifting surfaces without clearly defined E.A.'s, which form the principal subject matter of this section, is commonly analyzed by "brute-force" numerical calculation of the eigenvalues of the integral equation. Although the distinction was not brought out in Sec. 7–2, there are actually two different types of symmetrical bending-torsion divergence. The more familiar static divergence coincides with the blowing up of the elastic loading or twist described by Eq. 7–115, that is, at the roots of the determinantal equation

$$\left| \, [\overline{\mathscr{A}^s}]^{-1} - q_R c_R [\bar{E}] \, \right| = 0 \qquad (7\text{--}126)$$

The lowest positive value of q_R, if any, defines an aeroelastic stability boundary for a wing that is either restrained or mounted on a vehicle flown at a fixed normal load factor.

In uncontrolled free flight, however, Eq. 7–120 applies, and the eigenvalues must be found from

$$\left| \, [\overline{\mathscr{A}^s}]^{-1} - q_R{}' \left(c_R [\bar{E}] - \frac{2gc_R}{W} [\bar{G}]\{m\} \lfloor \, \overline{W} \, \rfloor \right) \right| = 0 \qquad (7\text{--}127)$$

This second instability, associated with dynamic pressure $q_R{}'$, is sometimes called *dynamic divergence*. Its onset would be accompanied by a looping maneuver which went into an ever-tightening spiral as the increasing structural deformations augmented the acceleration Ng.

Both q_R and $q_R{}'$ are likely to be negative on swept-back surfaces. But in the sweptforward case the problem is a severe and real one. As may be guessed from the opposite signs of the two terms in parentheses in Eq. 7–127, it usually turns out that $q_R{}' > q_R$. During the approach to dynamic divergence, the downward inertia forces tend, in part, to counteract the positive bending caused by upward airloads when $\Lambda < 0$.

With high-speed computing machinery, there are many ways of determining the roots of Eqs. 7–126 and 7–127; these will not be discussed here. A trial-and-error procedure is generally indicated, since Mach-number effects must be included in the aerodynamic matrix. Primarily because $[\bar{E}]$ and $[\bar{G}]$ are not symmetrical, matrix iteration is, strictly speaking, not applicable. Nevertheless, satisfactory convergence will usually be obtained in cases of practical significance, such as swept-forward wings, where the eigenvalues are positive.

Divergence and loading of large-aspect-ratio swept wings have special theoretical interest because they involve the solution of adjoint differential and integral equations. General information on the theory will be found in Ref. 7–5 (Chap. 5 and Sec. 8.2). We mention here one simple illustration, which can be given a physical interpretation. In the case of instability at a constant normal load factor, the homogeneous form of Eq. 7–111a applies,

$$\theta(y) = q \int_0^l \bar{C}(y, \eta) c c_l{}^e \, d\eta \qquad (7\text{–}128)$$

For the moment, suppose that we neglect twisting and adopt the strip-theory aerodynamic operator, Eq. 4–110, to rewrite this in terms of the single variable θ.

$$\theta(y) = q \frac{\partial c_l}{\partial \alpha} \cos \Lambda \int_0^l C^{\theta z}(y, \eta) c \theta \, d\eta \qquad (7\text{–}129)$$

Equation 7–129 is self-adjoint only if $C^{\theta z}(y, \eta) = C^{\theta z}(\eta, y)$, but this is not true of swept structures. The adjoint integral equation is, therefore, obtained by interchanging the variables in the influence function,

$$w(y) = q \frac{\partial c_l}{\partial \alpha} \cos \Lambda \int_0^l C^{\theta z}(\eta, y) c w \, d\eta \qquad (7\text{–}130a)$$

The reason for choosing the symbol w in Eq. 7–130a becomes apparent when we observe that, by the reciprocity relation for linear elastic systems, $C^{\theta z}(\eta, y) = C^{z\theta}(y, \eta)$, and

$$w(y) = q \frac{\partial c_l}{\partial \alpha} \cos \Lambda \int_0^l C^{z\theta}(y, \eta) cw \, d\eta \qquad (7\text{-}130b)$$

Since $C^{z\theta}(y, \eta)$ represents the bending displacement at y due to unit concentrated streamwise torque at η, Eq. 7–130b describes the bending of a wing which is somehow loaded with distributed torques everywhere proportional to cw. Thus there is a meaning, albeit somewhat artificial, to the adjoint problem of Eq. 7–129, which itself describes streamwise twisting of the same structure under normal forces proportional to $c\theta$.

As proved in books on integral equations, the eigenvalues of Eqs. 7–129 and 7–130 are equal, if real, and conjugate, if complex. Moreover, the eigenfunctions corresponding to distinct eigenvalues are orthogonal with respect to the weighting factor c. That is, if $\Theta_i(y)$, q_i and $W_j(y)$, q_j are solutions of Eq. 7–129 and Eq. 7–130b, respectively,

$$\int_0^l c\Theta_i W_j \, dy = 0, \qquad (i \neq j) \qquad (7\text{-}131)$$

This result has particular importance in connection with methods of approximate solution based on superposition of eigenfunctions, because it is definitely not true that the integrals of $c\Theta_i\Theta_j$ and cW_iW_j vanish.

All the foregoing ideas are readily extended to the more general integral equations of swept wings and to the matrix relations which are obtained by numerical integration thereof. For example, the adjoint of Eq. 7–128 has the kernel function $\bar{C}(\eta, y)$, when aerodynamic strip theory is used. The identity of eigenvalues and orthogonality of adjoint eigenfunctions (or eigenvectors) is preserved, but the possibility of simple physical interpretation is lost.

The practical value of *biorthogonal eigenfunctions*, as the sets of adjoint solutions are sometimes called (Refs. 7–5, 7–26), is brought out when we introduce the idea of generalized coordinates for solving the nonhomogeneous integral equations. A variety of such procedures is discussed in Chap. 3, the two in most common use being those of Rayleigh-Ritz and Galerkin. In self-adjoint problems, Rayleigh-Ritz consists simply of substituting a series of approximating functions into the variational statement of the principle of minimum potential energy. It can also be adapted to swept surfaces (cf. Flax, Ref. 7–25) but will lose its elementary physical interpretation.

We illustrate here the Galerkin method, applying it to Eq. 7–111a.

Let us assume the following representations for the twist and elastic lift distributions:

$$\theta(y) = \sum_{i=1}^{n} \Theta_i(y)\bar{q}_i \qquad (7\text{–}132a)$$

$$c_l{}^e(y) = \sum_{i=1}^{n} c_{l_i}(y)\bar{q}_i \qquad (7\text{–}132b)$$

Here the $\Theta_i(y)$ satisfy the boundary conditions $\theta(0) = d\theta(l)/dy = 0$. Each member of the lift coefficient series is calculated from the corresponding $\Theta_i(y)$ by whatever aerodynamic operator has been chosen,

$$c_{l_i}(y) = \overline{\mathscr{A}^s\{\Theta_i(y)\}} \qquad (7\text{–}133)$$

When substituted into Eq. 7–111a, Eqs. 7–132 yield

$$\sum_{i=1}^{n} \left\{ \Theta_i(y) - q \int_0^l \bar{C}(y, \eta) c c_{l_i} \, d\eta \right\} \bar{q}_i = \bar{f}_s(y) \qquad (7\text{–}134)$$

One version of Galerkin's scheme is to solve for the coordinates $\bar{q}_i$ by requiring the two sides of Eq. 7–134 to be equal, in the mean, when weighted in turn with each of the assumed twist functions. Multiplying by $c(y)\Theta_j(y)$ and integrating over the span, we obtain

$$\sum_{i=1}^{n} A_{ij}\bar{q}_i = B_j, \qquad (j = 1, 2, \cdots, n) \qquad (7\text{–}135)$$

where

$$A_{ij} = \int_0^l c\Theta_i\Theta_j \, dy - q \int_0^l c\Theta_j \int_0^l \bar{C}(y, \eta) c c_{l_i} \, d\eta \, dy \qquad (7\text{–}136)$$

and

$$B_j = \int_0^l c\Theta_j \bar{f}_s \, dy \qquad (7\text{–}137)$$

Both the twist and lift distributions are calculated by solving Eqs. 7–135 for the $\bar{q}_i$, then inserting these constants into the series 7–132a and 7–132b.

A more efficient variant of the foregoing is to choose for $\Theta_i(y)$ the eigenfunctions of the homogeneous integral equation (7–128) and use the adjoint eigenfunctions $W_j(y)$ in the weighting process. It is then known that

$$\int_0^l \bar{C}(y, \eta) c c_{l_i} \, d\eta = \frac{1}{q_{D_i}} \Theta_i(y) \qquad (7\text{–}138)$$

q_{D_i} being the associated eigenvalue, and the coefficients A_{ij} in Eq. 7–136 are replaced by

$$A_{ij} = \int_0^l c\Theta_i W_j \, dy - q \int_0^l c W_j \int_0^l \bar{C}(y, \eta) c c_{l_i} \, d\eta \, dy$$

$$= \left[1 - \frac{q}{q_{D_i}} \right] \int_0^l c\Theta_i W_j \, dy \qquad (7\text{–}139a)$$

Hence,

$$A_{ij} = \begin{cases} 0, & i \neq j \\ \left[1 - \dfrac{q}{q_{D_i}}\right]\displaystyle\int_0^l c\Theta_i W_i \, dy, & (i = j) \end{cases} \qquad (7\text{-}139b)$$

Since they have been decoupled by the orthogonality, Eqs. 7–135 can now be solved independently,

$$\bar{q}_j = \frac{B_j}{A_{jj}} = \frac{\displaystyle\int_0^l cW_j \bar{f}_s \, dy}{\left[1 - \dfrac{q}{q_{D_j}}\right]\displaystyle\int_0^l c\Theta_j W_j \, dy} \qquad (7\text{-}140)$$

Thus one laborious step in the mathematical process is avoided in cases where the biorthogonal eigenfunctions are known.

On sweptforward wings, for which the q_{D_j} are usually real and positive, Eqs. 7–132 and 7–140 indicate clearly the divergence of the solution to infinity at each critical dynamic pressure. Sweptback wings often have only real, negative q_{D_j}'s, so that the bracketed factor in the denominator of Eq. 7–140 represents an attenuation of the deformation—the familiar influence of the bending slope. Indeed, in some simple examples, such as a sweptback surface which bends without twisting, $|q_{D_j}|$ can be proved to be the jth divergence dynamic pressure of the same structure sweptforward at the angle $(-\Lambda)$.

For a more thorough and rigorous treatment of the solution of swept-wing aeroelastic problems by means of biorthogonal eigenfunctions, the mathematically-minded reader is directed to Seifert's article (Ref. 7–26).

(b) Antisymmetrical load distribution; control effectiveness

When we consider the lifting surface in Fig. 7–6 to be rolling about the x-axis with angular velocity p (positive to swing the right tip upward), the purely antisymmetrical load distribution arises from only three effects: antisymmetrical twist $\theta(y)$, deflection $\delta(y)$ of a lateral control device, and p itself. Therefore, the development followed in Sec. 7–2(c) can be adapted for our purposes here, with a slight redefinition of symbols. Equations 7–64, 7–65, and 7–66 are unchanged, except that the aerodynamic operator $\mathscr{A}$ may have to include sweep effects. The normal load factor $N(y)$ is determined by the rolling acceleration, according to

$$Ng = \dot{p}y \qquad (7\text{-}141)$$

Turning to Eq. 7–107, we make a series of substitutions similar to those which produced Eqs. 7–111 and are led to

$$\theta(y) = q\int_0^l \bar{C}(y, \eta)cc_i^e \, d\eta + \bar{f}_a(y) \qquad (7\text{-}142a)$$

where

$$
\bar{f}_a(y) = q \int_0^l \bar{C}(y, \eta) \left[c \frac{\partial c_l}{\partial \delta} \delta_R + c \frac{\partial c_l}{\partial \left(\dfrac{pl}{U} \right)} \frac{pl}{U} \right] d\eta
$$

$$
+ q \int_0^l C^{\theta\theta}(y, \eta) c^2 \frac{\partial c_{mAC}}{\partial \delta} \delta_R \, d\eta
$$

$$
- \dot{p} \int_0^l [C^{\theta z}(y, \eta) + C^{\theta\theta}(y, \eta) \, d(\eta)] m\eta \, d\eta \qquad (7\text{-}142b)
$$

δ_R is the control rotation at some reference station, and $\bar{C}$ is given once more by Eq. 7–112.

Introducing a numerical integration formula with the weighting matrix $\lceil \overline{W} \rfloor$, along with the antisymmetrical aerodynamic matrix $[\overline{\mathscr{A}^a}]$, we derive the following system of simultaneous equations for the running lift at the y_i-stations:

$$
[\overline{\mathscr{A}^a}]^{-1} \left\{ \frac{cc_l^e}{c_R} \right\} = qc_R[\bar{E}] \left\{ \frac{cc_l^e}{c_R} \right\} + \{\bar{f}_a\} \qquad (7\text{-}143a)
$$

The column matrix of "known" quantities on the right is evidently

$$
\{\bar{f}_a\} = q[\bar{E}] \lceil c \rfloor \left(\left\{ \frac{\partial c_l}{\partial \delta} \right\} \delta_R + \left\{ \frac{\partial c_l}{\partial \left(\dfrac{pl}{U} \right)} \right\} \frac{pl}{U} \right)
$$

$$
+ q[F] \left\{ \frac{\partial c_{mAC}}{\partial \delta} \right\} \delta_R - [\bar{G}] \lceil y \rfloor \{m\} \dot{p} \qquad (7\text{-}143b)
$$

with the three elastic-geometric matrices $[\bar{E}]$, $[F]$, and $[\bar{G}]$ still those defined in Eqs. 7–114c, d, e.

By direct analogy with Sec. 7–2(c), a number of conclusions can now be reached without recourse to detailed derivations. The motion of the vehicle in response to a sudden deflection $\delta_R = \delta_0 1(t)$ of the roll controls is given by

$$
\frac{pl}{U} = \left[1 - \exp \left(-\frac{\bar{B}_1 l}{\bar{A}_1 U} t \right) \right] \frac{\bar{C}_1}{\bar{B}_1} \delta_0 \qquad (7\text{-}144)
$$

where $\bar{A}_1$, $\bar{B}_1$, and $\bar{C}_1$ are the constants from Eqs. 7–78b, c, d with bars added over the square matrices $[\mathscr{A}^a]$, $[E]$, $[F]$, and $[G]$. The asymptotic form of Eq. 7–144 corresponds to the steady-state rolling power of the elastic swept wing and can be written

$$
\frac{\partial\left(\frac{pl}{U}\right)}{\partial\delta}
$$

$$
= -\frac{C_{l\delta} + \frac{qc_R}{Sl}\lfloor H \rfloor([\overline{\mathscr{A}^a}]^{-1} - qc_R[\bar{E}])^{-1} \times \left([\bar{E}][^c{}_\diagdown]\left\{\frac{\partial c_l}{\partial\delta}\right\} + [F]\left\{\frac{\partial c_{mAC}}{\partial\delta}\right\}\right)}{C_{l_p} + \frac{qc_R}{Sl}\lfloor H \rfloor([\overline{\mathscr{A}^a}]^{-1} - qc_R[\bar{E}])^{-1}[\bar{E}][^c{}_\diagdown]\left\{\frac{\partial c_l}{\partial\left(\frac{pl}{U}\right)}\right\}} \quad (7\text{--}145)
$$

Here $C_{l\delta}$ and C_{l_p} are the control-surface derivative and damping in roll of the rigid wing. The dynamic pressure for reversal occurs at the vanishing of the numerator in Eq. 7–145, expressed by the implicit formula

$$
q_R = \frac{\frac{Sl}{c_R}C_{l\delta}}{-\lfloor H \rfloor([\overline{\mathscr{A}^a}]^{-1} - q_R c_R[\bar{E}])^{-1}\left([\bar{E}][^c{}_\diagdown]\left\{\frac{\partial c_l}{\partial\delta}\right\} + [F]\left\{\frac{\partial c_{mAC}}{\partial\delta}\right\}\right)}
$$

$$(7\text{--}146)$$

Antisymmetrical bending-torsion divergence is a constant load-factor phenomenon and therefore takes place at the fundamental positive eigenvalue of Eq. 7–126, $[\mathscr{A}^s]$ being replaced by $[\mathscr{A}^a]$ and adjustments made to the structural influence coefficients for the effects of antisymmetrical loading, if they are required.

To illustrate the severe deleterious influence of sweepback on the operation of trailing-edge controls, we reproduce the results of two calculations on the XB-47 airplane. Figure 7–9 shows the antisymmetrical portion of the loading on the right wing, both flexible and infinitely rigid, at the start of a maneuver consisting of a sudden $10°$ aileron deflection during a $2g$ pull-out. Flight Mach number is 0.72 at an altitude of 5,000 ft in the standard atmosphere, and the gross weight is 125,000 lb. Geometry of the controls and other details of this calculation, which is based on the same integral equation studied in the present section, appear in the paper by Pian and Lin (Ref. 7–24), from which Fig. 7–9 is adopted. Our second example, Fig. 7–10, is taken from Sec. 8–4 of Ref. 1–2. It shows how the aileron rolling power varies with M for the XB-47 flying at sea level and

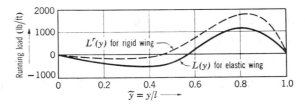

Fig. 7–9. Antisymmetrical load distribution, with and without aeroelastic effects, due to sudden 10° aileron deflection on the right wing of the XB-47 in a 2g pull-out at $M = 0.72$ and 5000-ft altitude. (Taken from Pian and Lin, Ref. 7–24.)

indicates reversal at a speed not far from the maximum performance of the bomber. Needless to say, results like these were influential in leading to the adoption of spoilers for the B-47 and B-52. Perhaps as undesirable as the low reversal condition is the manner in which the effectiveness drops off more rapidly with M/M_R than the parabolic dependence observed on straight wings. A comparison between Figs. 7–5 and 7–10 illuminates this point.

In addition to the publications cited above, the reader will find extensive information on antisymmetrical loading of swept wings in Refs. 7–27, 7–28, and 7–29.

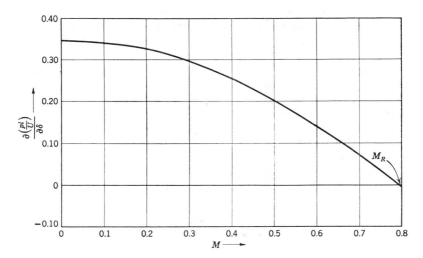

Fig. 7–10. Aileron effectiveness as a function of flight Mach number for the XB-47 flying at sea level. (Taken from Example 8-9 of Bisplinghoff, Ashley, and Halfman, Ref. 1–2.)

7-4   CHORDWISE DEFORMATIONS AND
VERY SLENDER CONFIGURATIONS

Sections 7-2 and 7-3 were concerned with the relatively familiar and long-understood static aeroelastic properties of large-aspect-ratio wings and tails. To the list of one-dimensional problems there have been added more recently those where the deformation depends principally on the single chordwise coordinate x. The attainment of supersonic flight has augmented in a number of ways both the catalogue and the severity of such problems. For one thing, lifting surfaces of small to fractional aspect ratio are the optimum solution for many high-speed design requirements. The majority of these must be treated by two-dimensional structural methods, but some are slender enough so that the spanwise variations of their bending deflections are negligible. Most missiles also fall in this category of slender configurations.

Another well-known consequence of supersonic Mach numbers is that the local aerodynamic loading often grows in nearly direct proportion to the local surface angle of attack or chordwise slope. This behavior, which differs entirely from the subsonic regime except when the aspect ratio is very low, implies the possibility of a new type of divergence.

Moreover, we hardly need mention the enormously larger dynamic pressures that usually go with supersonic speeds. We are now entering an era when rapidly accelerated missiles and entry vehicles will encounter as great dynamic pressures as will probably ever be attained by controlled flying machines within the earth's atmosphere. Somewhat later, high-performance submarines will be setting similar records for liquid media, albeit not supersonic speeds. The aeroelastic problems which arise, accompanied by intense aerodynamic heating in the case of entry, may cause our earlier experiences to look like child's play.

The differential and integral equations which govern the chordwise deformations differ from those presented in Sec. 7-1 mainly by a change of independent variable from y (or s or $\bar{y}$) to x. When analyzing simple bending, for example, we overlook rotary inertia and shear deflections, rewriting Eq. 7-11

$$\frac{\partial^2}{\partial x^2}\left[EI\,\frac{\partial^2 w}{\partial x^2}\right] + m(x)\ddot{w}(x,\,t) = F_z(x,\,t) \qquad (7\text{-}147)$$

where $F_z(x,\,t)$ is the running external load and $m(x)$ the mass per unit chordwise distance. There are two possible interpretations of Eq. 7-147, corresponding to the two systems which we shall examine in Secs. 7-4(a) and (b) below. Either we can consider chordwise (camber) bending of a uniform, relatively large aspect-ratio surface on a structurally one-dimensional (plane-strain) basis, or we may be dealing with a genuinely slender configuration. In the former instance, $m(x)$ represents mass per unit area, and $EI(x)$ is the component of flexural rigidity of a plate resisting curvature about a spanwise axis.

Antisymmetrical deformation of slender wings and bodies consists of torsion about the longitudinal axis and side bending. Torsion is subject to a differential equation resembling Eq. 7–15, with the differential bending term omitted,

$$\frac{\partial}{\partial x}\left[GJ_x\frac{\partial\theta}{\partial x}\right] - I_x(x)\ddot{\theta}(x, t) = -T_x(x, t) \qquad (7\text{–}148)$$

The meaning of these symbols is obvious. It should be noted that the torsional problem appears less important in practice than bending. In fact, there are situations involving high longitudinal accelerations (e.g., the anti-missile missile), where time-dependent axial deformation $u(x, t)$ may be much more significant than torsion. The appropriate differential equation reads

$$\frac{\partial}{\partial x}\left[EA\frac{\partial u}{\partial x}\right] - m(x)\ddot{u}(x, t) = -F_x(x, t) \qquad (7\text{–}149)$$

where $A(x)$ is the cross-sectional area of structural material and F_x is a running longitudinal force. Equation 7–149 has homogeneous solutions of the longitudinal-vibration type, which usually occur at high frequencies compared to lateral bending but may be violently excited by certain impulsive loads, such as the starting thrust of a rocket engine. For missile structures composed of thin cylindrical skins, Eq. 7–149 is oversimplified, and the problem must be approached by means of shell theory.

An integral equation for chordwise bending, in direct analogy with Eq. 7–16, can be written

$$w(x, t) = \int_0^l C^{zz}(x, \xi)Z(\xi, t)\, d\xi \qquad (7\text{–}150a)$$

where Z is the resultant running load, inertia forces included. Since slender configurations are usually encountered in the free-free condition, $w(x, t)$ is then the elastic displacement relative to a reference surface tangent to the structure at a given station. $\ddot{w}$ is not the absolute normal acceleration, because it also receives contributions from the rigid-body accelerations in vertical translation and pitching. Sometimes the slope

$\alpha_e = -\partial w/\partial x$ represents a more useful variable. In such cases, the derivative of Eq. 7–150a with respect to x may be used.

$$\alpha_e(x, t) = \int_0^l C^{\alpha z}(x, \xi)Z(\xi, t)\,d\xi \qquad (7\text{–}150b)$$

with

$$C^{\alpha z}(x, \xi) = -\frac{\partial}{\partial x}\,C^{zz}(x, \xi) \qquad (7\text{–}150c)$$

The remainder of this section discusses static problems which illustrate what can be done with a one-dimensional chordwise model. The first of these relates to the curling up and divergence of thin leading edges in supersonic flow, emphasizing elementary profile shapes that are amenable to differential equation solutions. Next, both the differential and integral-equation approaches are exemplified in connection with the chordwise stability of slender vehicles.

(a) Chordwise bending at supersonic speeds

We consider a thin, uniform structure with a chordwise cross section of the form shown in Fig. 7–11. The intention is to represent the forward half of the profile of a straight wing in a supersonic airstream, replacing the connection to the rearward half by a cantilever base. The section is solid and has a slightly blunted leading edge, because, as we shall see, chordwise stability is strongly affected by the closeness of approach to mathematical sharpness. There is a mean incidence α_0 at midchord. The aspect ratio and structural arrangement are assumed to be such that the

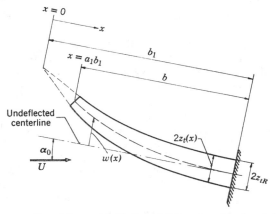

Fig. 7–11. Cross section of forward half of an airfoil undergoing chordwise bending in a two-dimensional supersonic airstream.

flow can be approximated two-dimensionally and the mean-surface bending one-dimensionally according to Eq. 7–147. All angles are small, Eliminating time-dependent quantities from Eq. 7–147, we obtain

$$\frac{d^2}{dx^2}\left[E_1 I \frac{d^2 w}{dx^2}\right] = \Delta p_a(x) \tag{7-151}$$

Here $E_1 = E/(1 - \nu^2)$ is a reduced modulus of elasticity for the assumed state of plane strain, ν being Poisson's ratio. Since the equation refers to a unit spanwise segment,

$$I = \tfrac{2}{3}z_t^3 \tag{7-152}$$

in dimensions of feet[4] per foot. For convenience in a case of linear chordwise taper to be treated below, the origin of x in Fig. 7–11 is placed at the vertex of a hypothetical point added onto the leading edge; the actual chord is $c = 2b$.

We are able to include in the analysis the important second-order aerodynamic effects of profile thickness, which yield a linear relation between upward force per unit area and the slope $(dw/dx - \alpha_0)$, containing a nonlinear factor dependent on the thickness distribution $z_t(x)$. Busemann's theory (Ref. 7–30) would provide the rigorously correct form of the operator; but an excellent approximation for all but the lowest supersonic speeds is found by replacing M with

$$\beta = \sqrt{M^2 - 1} \tag{7-153}$$

in the denominator of the piston theory operator, Eq. 4–70. This step gives*

$$\Delta p_a(x) \cong \frac{4q}{\beta}\left[1 + \left(\frac{\gamma + 1}{2}\right)M \frac{dz_t}{dx}\right]\left[-\frac{dw}{dx} + \alpha_0\right] \tag{7-154}$$

Since w itself will not appear in the differential equation, we work in terms of the bending slope

$$\alpha_e(x) = -\frac{dw}{dx} \tag{7-155}$$

and also employ a dimensionless independent variable

$$\tilde{x} = \frac{x}{b_1} \tag{7-156}$$

---

* It should be noted that the general property of supersonic flow involving no upstream influence on aerodynamic loading provides an excellent justification for analyzing deformations of the forward half of the profile without regard for what occurs to the rear.

Inserting Eqs. 7–152 through 7–156 into Eq. 7–151, the following dimensionless relation is obtained:

$$\frac{d^2}{d\tilde{x}^2}\left[\frac{z_t^3}{z_{tR}^3}\frac{d\alpha_e}{d\tilde{x}}\right] + k_1\alpha_e = -k_1\alpha_0 \qquad (7\text{–}157)$$

where

$$k_1 = \frac{6q}{\beta E_1}\left(\frac{b_1}{z_{tR}}\right)^3\left[1 + \left(\frac{\gamma+1}{2}\right)M\frac{dz_t}{dx}\right] \qquad (7\text{–}158)$$

and $z_{tR} = z_t(1)$ is the semi-thickness at midchord. Boundary conditions for the free edge at $\tilde{x} = a_1$ and clamped midchord line at $\tilde{x} = 1$ read

$$\alpha_e(1) = \left(\frac{z_t}{z_{tR}}\right)^3\frac{d\alpha_e}{d\tilde{x}}\bigg|_{\tilde{x}=a_1} = \frac{d}{d\tilde{x}}\left[\left(\frac{z_t}{z_{tR}}\right)^3\frac{d\alpha_e}{d\tilde{x}}\right]\bigg|_{\tilde{x}=a_1} = 0 \qquad (7\text{–}159)$$

Equations 7–157 and 7–159 define a nonself-adjoint mathematical problem reminiscent of the pure bending of the swept wing, discussed in the introduction to Sec. 7–3. One achieves the exact analogue of a uniform sweptforward wing by specializing to the trivial case of a flat-plate airfoil ($z_t = z_{tR}$, $dz_t/dx = 0$) and setting the origin at the leading edge ($a_1 = 0$). Then, k_1 becomes identical with the parameter $(-\bar{a})$ in Eq. 7–95. From Eqs. 7–101, the eigenvalues of the homogeneous form of Eq. 7–157 are known to be

$$k_{1_n} = \begin{cases} 6.33, & (n = 0) \\ \dfrac{(2n+1)^3\pi^3}{3\sqrt{3}}, & (n = 1, 2, 3, \cdots) \end{cases} \qquad (7\text{–}160)$$

Each of these corresponds to a possible mode of chordwise bending divergence of the flat-slab wing.

According to Eq. 7–158, k_{1_0} furnishes the lowest dynamic pressure at which static instability is possible

$$q_D = 1.055\beta E_1\left(\frac{z_t}{b}\right)^3 \qquad (7\text{–}161a)$$

In view of the relation $q = \frac{\gamma}{2}p_\infty M^2$, this implicit representation can be solved for the divergence Mach number

$$M_D = \sqrt{\frac{K^2}{2} \pm \sqrt{\frac{K^4}{4} - K^2}} \qquad (7\text{–}161b)$$

where

$$K = \frac{2.11E_1}{\gamma p_\infty}\left(\frac{z_t}{b}\right)^3 \qquad (7\text{–}161c)$$

(One limitation is $K > 2$, as it ordinarily will be because of the magnitude of E_1. The root with the minus sign in Eq. 7-161b is likely to be physically meaningless; it occurs near $M_D = 1$ because of the smallness of β in the denominator of Eq. 7-54, and this is a range where the aerodynamic theory is invalid.) Finally, Eq. 7-103 gives us, for the deformation of the plate under load at $q < q_D$,

$$\alpha_e(\tilde{x}) = \alpha_0 \left\{ \frac{e^{\tilde{x}\sqrt[3]{-k_1}} + 2e^{-\frac{1}{2}\tilde{x}\sqrt[3]{-k_1}} \cos\left(\dfrac{3\tilde{x}\sqrt[3]{-k_1}}{2\sqrt{3}}\right)}{e^{\sqrt[3]{-k_1}} + 2e^{-\frac{1}{2}\sqrt[3]{-k_1}} \cos\left(\dfrac{3\sqrt[3]{-k_1}}{2\sqrt{3}}\right)} - 1 \right\} \qquad (7\text{-}162)$$

Equation 7-161a reveals the proportionality between dynamic pressure and elastic modulus which characterizes all divergence formulas for distributed-parameter systems. Somewhat more unusual is the very strong dependence on the profile thickness ratio z_t/b.

In his exhaustive studies of the chordwise-deformation problem (e.g., Refs. 7-31, 7-32, and 7-33), Biot has solved Eqs. 7-157 through 7-159 for a truncated, tapered wedge having

$$z_t = z_{tR} \frac{x}{b_1} = z_{tR}\tilde{x} \qquad (7\text{-}163)$$

In this case dz_t/dx is a constant related to the thickness ratio of the profile, so that k_1 is independent of x. The particular integral of Eq. 7-157 is simply

$$\alpha_{e_p} = -\alpha_0 \qquad (7\text{-}164)$$

and Biot therefore concentrates on the homogeneous solution and the related question of stability.

Equation 7-163 leads to the following restatement of the eigenvalue problem:

$$\frac{d^2}{d\tilde{x}^2}\left[\tilde{x}^3 \frac{d\alpha_e}{d\tilde{x}}\right] + k_1\alpha_e = 0 \qquad (7\text{-}165a)$$

$$\alpha_e(1) = \tilde{x}^3 \frac{d\alpha_e}{d\tilde{x}}\bigg|_{\tilde{x}=a_1} = \frac{d}{d\tilde{x}}\left[\tilde{x}^3 \frac{d\alpha_e}{d\tilde{x}}\right]\bigg|_{\tilde{x}=a_1} = 0 \qquad (7\text{-}165b)$$

This differential equation is of the equidimensional type, and we can easily show that it is solved by

$$\alpha_e = \sum_{n=1,2,3} C_n \tilde{x}^{m_n} \qquad (7\text{-}166)$$

where $m_n = z_n - 1$ and the z_n are the three roots of

$$z(z^2 - 1) + k_1 = 0 \qquad (7\text{-}167)$$

Since $k_1 > 0$, Eq. 7–167 must have at least one real root, z_1, which is less than -1. z_1 can be plotted versus k_1 by substituting a series of values of z_1 into Eq. 7–167, and it varies from -1 when $k_1 = 0$ to $-\sqrt[3]{k_1}$ when k_1 is large. It follows from the conditions $z_1 + z_2 + z_3 = 0$ and $z_1 z_2 + z_1 z_3 + z_2 z_3 = -1$ that the other two roots may be written

$$z_2, z_3 = -\frac{z_1}{2} \pm \tfrac{1}{2}\sqrt{4 - 3z_1^2} \qquad (7\text{–}168a, b)$$

z_2 and z_3 are real or complex-conjugate, depending on whether z_1 is algebraically greater or less, respectively, than $-2/\sqrt{3} \equiv -1.1548$. Through Eq. 7–167, this condition corresponds to k_1 being less or greater than $2/(3\sqrt{3}) \equiv 0.3849$.

By applying the boundary conditions (7–165b) to the solution, Eq. 7–166, Biot (Ref. 7–32) constructs the eigenvalue problem for k_1 and proves that, when $0 < a_1 < 1$, no real k_1 are possible for real z_2 and z_3. Hence the fundamental k_{1_0} exceeds 0.3849 always. With the latter condition fulfilled, the characteristic equation becomes

$$\frac{D z_1[z_1 + 1]}{z_1 - 1}\left(\frac{1}{a_1}\right)^{\frac{3z_1}{2}} + \frac{2 + 3z_1}{2}\sin\left(D \ln \frac{1}{a_1}\right)$$
$$+ D[2z_1 + 1]\cos\left(D \ln \frac{1}{a_1}\right) = 0 \quad (7\text{–}169a)$$

where

$$D = \tfrac{1}{2}\sqrt{3z_1^2 - 4} \qquad (7\text{–}169b)$$

Having found z_1 as a function of a_1 from Eqs. 7–169, we can compute the eigenvalues by substituting z_1 into Eq. 7–167. There is, of course, an infinite number of roots for each bluntness parameter a_1; but only the fundamental is of interest. Biot defines a second characteristic parameter, based on the actual semichord,

$$k = \frac{6q}{\beta E_1}\left(\frac{b}{z_{tR}}\right)^3\left[1 + \left(\frac{\gamma + 1}{2}\right)M\left(\frac{z_{tR}}{b}\right)\right]$$
$$= k_1[1 - a_1]^3 \qquad (7\text{–}170)$$

and gives the following table of k_D, corresponding to the minimum divergence dynamic pressure. Critical dynamic pressure or Mach number can be computed from Eq. 7–170 as in the case of the slab with constant thickness.

TABLE 7–I

Critical value of the stability parameter for chordwise divergence of a two-dimensional straight wedge, vs. bluntness parameter a_1

a_1	k_D
0	0.3849
0.000076	0.528
0.0204	1.04
0.069	1.51
0.1055	1.79
0.177	2.25
0.241	2.63
0.460	3.78
0.582	4.39
0.660	4.74
0.712	4.99
0.823	5.51
0.981	6.27
1.000	6.33

(b) Slender aircraft and missiles treated as one-dimensional structures

Figure 7–12 depicts a generalized flight vehicle of the type we shall examine next. Since dynamic phenomena are touched on in Sec. 7–7, we restrict ourselves here to the static case and drop terms containing $\ddot{w}$ from equations of motion such as Eqs. 7–147 and 7–150. If the product of Mach number by the maximum local angle between the flight direction and planes tangent to the wing or fuselage surfaces is small enough compared to unity,* the slender-body aerodynamic operator (4–97) furnishes the required running load per unit x-distance. Differential equation (7–147) then becomes

$$\frac{d^2}{dx^2}\left[EI(x)\frac{d^2w}{dx^2}\right] = 2q\frac{d}{dx}\left\{S(x)\left[\alpha_0 - \frac{dw}{dx}\right]\right\} \qquad (7\text{–}171)$$

The cross-sectional virtual mass $\rho_\infty S(x)$ can be taken from one of Eqs. 4–100 through 4–102 in most practical applications; for instance, Eq. 4–101 yields

$$S(x) = \pi[s^2(x) - R^2(x) + R^4(x)/s^2(x)]$$

when symmetrical deformation of the midwing-body-of-revolution in Fig. 7–12 is analyzed.

* The greatest local slope usually occurs along the wing leading edge.

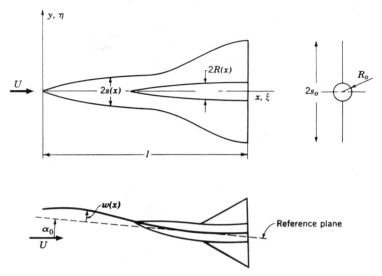

Fig. 7–12. Three views of a slender configuration, consisting of a body of revolution centered in a thin wing and undergoing small symmetrical bending deflections $w(x)$ relative to a plane tangent at some reference station.

Among the several published treatments of the static problem, the reader will find those by Dugundji and Crisp (Ref. 7–34) and Martin and Watkins (Ref. 7–35) especially illuminating. Reference 7–35 presents a scheme for numerical integration of Eq. 7–171 that can be adapted to arbitrary chordwise distributions of bending stiffness and to aerodynamic operators more general than slender-body theory. For flat, triangular planforms having $I(x)$ proportional to any power of x, Martin and Watkins also furnish closed Bessel-function solutions to Eq. 7–171 with $\alpha_0 = 0$ and calculate chordwise-divergence eigenvalues corresponding to boundary conditions of clamping or pitch-spring restraint at the trailing edge $x = l$. Because their procedure has certain parallels to Biot's work discussed in Sec. 7–4(a), we shall not go over it in detail but rather turn to a modal analysis of divergence in free flight (Ref. 7–34).

As pointed out in Chap. 3, a very natural choice of assumed shape (such as the γ_i in Eq. 3–103) to approximate the deformations of a one-dimensional or two-dimensional structure consists of its normal modes of vibration *in vacuo*. This is true even for static problems, because normal modes still satisfy the elastic boundary conditions and their orthogonality eliminates off-diagonal terms from the $[\mathscr{S}]$ matrix. The authors of Ref. 7–34 consider both flutter and divergence of the slender, plane triangular (delta) wing which has uniform mass and stiffness distributions, so that both $m(x)$ and $I(x)$ vary directly with x. Thus we can write

$$EI = EI_0\xi \qquad (7\text{-}172a)$$

$$m = m_0\xi \qquad (7\text{-}172b)$$

where subscript zero refers to the trailing-edge station, and $\xi = x/l$. Free vibration in the ith normal mode $\varphi_i(\xi)$, defined according to

$$w(\xi, t) = l\varphi_i(\xi)e^{i\omega_i t} \qquad (7\text{-}173)$$

is governed by the homogeneous form of Eq. 7–147, which is easily reduced to

$$\frac{d^2}{d\xi^2}\left[\xi\frac{d^2\varphi_i}{d\xi^2}\right] - \theta_i^2\xi\varphi_i = 0 \qquad (7\text{-}174)$$

The characteristic parameter is

$$\theta_i = \omega_i l^2\sqrt{m_0/EI_0} \qquad (7\text{-}175)$$

Dugundji and Crisp solve Eq. 7–174 subject to the free-flight boundary conditions

$$\left.\begin{array}{l} \text{Moment} \sim \xi\dfrac{d^2\varphi_i}{d\xi^2} = 0 \\[3mm] \text{Shear} \sim \dfrac{d}{d\xi}\left(\xi\dfrac{d^2\varphi_i}{d\xi^2}\right) = 0 \end{array}\right\}, \qquad (\xi = 0, 1) \qquad (7\text{-}176)$$

There are, of course, two rigid-body degrees of freedom associated with the zero-frequency normalized modes $\varphi_1 = 1$, $\varphi_2 = [1 - \tfrac{3}{2}\xi]$. The elastic modes are determined by truncating a power-series solution, and each has the form

$$\varphi_i(\xi) = 1 + \alpha_4\xi^4 + \alpha_8\xi^8 + \alpha_{12}\xi^{12} + \cdots + \alpha_1\xi + \alpha_5\xi^5 + \alpha_9\xi^9 + \cdots \qquad (7\text{-}177)$$

The α_n-coefficients tabulated in Appendix D, Ref. 7–34, for $i = 3, 4$, and 5, show satisfactory convergence. Eigenvalues corresponding to them are $\theta_3 = 28.80992$, $\theta_4 = 73.21273$, $\theta_5 = 137.4223$. Orthogonality requires that

$$\int_0^1 \xi\varphi_i(\xi)\varphi_j(\xi)\,d\xi = 0, \qquad (i \neq j) \qquad (7\text{-}178)$$

Given the φ_i as computed above, we represent the total z-displacement relative to an inertial coordinate system in the manner already exemplified by Eqs. 3–96, 3–103, 3–111, and 3–127:

$$w_{\text{TOT}}(\xi) = l \sum_{i=1} \varphi_i(\xi) q_i \qquad (7\text{--}179)$$

The q_i are independent, dimensionless generalized coordinates. Therefore, Lagrange's equations provide a description of the aerodynamically loaded system (cf. Sec. 2–6). By applying Eq. 7–178, together with the fact that the maximum kinetic and potential energies are equal during free vibration, we calculate for the potential function U

$$U = \tfrac{1}{2} m_0 l^3 \sum_{i=3} q_i^2 \omega_i^2 \int_0^1 \xi \varphi_i^2 \, d\xi \qquad (7\text{--}180)$$

The generalized force exerted by the running lift on the ith degree of freedom is

$$\begin{aligned}
Q_i &= l^2 \int_0^1 L(\xi) \varphi_i(\xi) \, d\xi \\
&= -2ql \sum_{j=1} q_j \int_0^1 \varphi_i(\xi) \frac{d}{d\xi} \left[S(\xi) \frac{d\varphi_j}{d\xi} \right] d\xi \qquad (7\text{--}181)
\end{aligned}$$

where the operator (4–97) has been substituted. In the last member of Eq. 7–181, only loading due to the displacements themselves is shown, although an additional "disturbing" load can easily be added. From Eq. 4–100,

$$S(\xi) = \pi s^2(\xi) = \pi l^2 \xi^2 \tan^2 \Lambda' \qquad (7\text{--}182)$$

Λ' being the semi-vertex angle of the delta planform (the complement of the conventional sweep angle Λ). By substituting Eqs. 7–180 through 7–182 into the general Lagrange equation (2–67) and eliminating dimensional factors, we derive the homogeneous equations

$$([\mathscr{S}] - [\mathscr{A}])\{q_i\} = 0 \qquad (7\text{--}183)$$

Here the individual elements of the aerodynamic matrix are

$$A_{ij} = -\tfrac{1}{2} \varphi_j{}'(1)\varphi_i(1) + \tfrac{1}{2} \int_0^1 \xi^2 \varphi_j{}'(\xi)\varphi_i{}'(\xi) \, d\xi \qquad (7\text{--}184)$$

while those of the structural matrix may be written

$$\begin{aligned}
S_{ij} &= \left(\frac{m_0 \omega_j{}^2}{4q\pi \tan^2 \Lambda'} \right) \int_0^1 \xi \varphi_j(\xi)\varphi_i(\xi) \, d\xi \\
&= \begin{cases} \left(\dfrac{EI_0}{ql^4} \right) \left(\dfrac{\theta_j{}^2}{4\pi \tan^2 \Lambda'} \right) \displaystyle\int_0^1 \xi \varphi_j{}^2(\xi) \, d\xi, & (i = j) \\[4pt] 0, & (i = 1, 2 \quad \text{or} \quad i \neq j) \end{cases} \qquad (7\text{--}185)
\end{aligned}$$

Truncating Eqs. 7–183 at $i = 5$, Dugundji and Crisp have calculated the following numerical quantities:

$$A_{ij} = \begin{bmatrix} 0 & 0.75 & -0.85297 & 1.0504 & -1.2095 \\ 0 & 0 & 0.21096 & -0.35281 & 0.45685 \\ 0 & 0 & 0.10305 & 0.07130 & -0.18634 \\ 0 & 0 & -0.03450 & 0.19524 & 0.00279 \\ 0 & 0 & -0.00697 & -0.06808 & 0.28415 \end{bmatrix} \quad (7\text{–}186a)$$

$$S_{ij} = \frac{EI_0}{4\pi q l^4 \tan^2 \Lambda'} \begin{bmatrix} 0 & 0 & 0 & 0 & 0 \\ 0 & 0 & 0 & 0 & 0 \\ 0 & 0 & 0.02065\theta_3^2 & 0 & 0 \\ 0 & 0 & 0 & 0.01321\theta_4^2 & 0 \\ 0 & 0 & 0 & 0 & 0.00969\theta_5^2 \end{bmatrix} \quad (7\text{–}186b)$$

The zeroes in Eq. 7–186a can be interpreted as meaning that, although steady-state displacements of the elastic degrees of freedom generate forces tending to excite the rigid-body freedoms, the converse is not true. (This statement is obvious in the case of the translation φ_1, which causes no static airloads at all.)

Regarding the phenomenon of divergence, it is seen to be controlled entirely by the elastic freedoms, provided some mechanism is assumed to act which preserves static equilibrium or "trimmed flight," so that a dynamic divergence of the sort discussed in Sec. 7–3(a) does not take place. Divergence can occur at each of the eigenvalues of the determinant of Eq. 7–183,

$$|[\mathscr{S}] - [\mathscr{A}]| = 0 \qquad (7\text{–}187)$$

After the zero roots are cancelled, this evidently reduces itself to third order in some quantity proportional to $\dfrac{EI_0}{q l^4}\left(\dfrac{\theta_3^2}{4\pi \tan^2 \Lambda'}\right)$. Reference 7–34 solves the resulting cubic equation and gives three real eigenvalues, choosing to present them in the form of reduced divergence speeds

$$\frac{U_D}{l\omega_3\sqrt{(m_0/2\pi\rho_\infty l^2 \tan^2 \Lambda')}} \equiv \sqrt{\frac{q_D l^4}{EI_0}\frac{4\pi \tan^2 \Lambda'}{\theta_3^2}}$$

$$= 0.494, \qquad 0.541, \qquad 0.973 \quad (7\text{–}188)$$

The lowest figure yields, of course, an upper bound on the speeds or dynamic pressures at which the wing can safely fly.

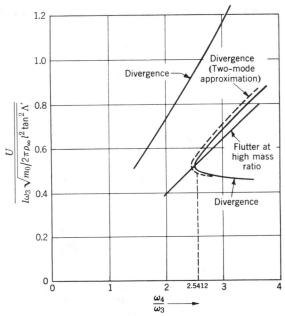

Fig. 7–13. Divergence and flutter characteristics of a free-flying, flat-plate delta wing according to slender-body aerodynamic theory. (Adapted from Ref. 7–34.)

Since elementary slender-body theory has been used, Eq. 7–188 shows that the dimensionless divergence properties of this particular configuration are independent of flight Mach number and are fixed by a single combination of system parameters. Thus, we can say that $U/l\omega_3$ is determined entirely by the mass ratio

$$(m_0/\pi \rho_\infty l^2 \tan^2 \Lambda'), \quad \text{or} \quad q_D l^4/EI_0 \quad \text{by} \quad (4\pi \tan^2 \Lambda'/\theta_3{}^2).$$

To shed more light on the couplings involved, Dugundji and Crisp have computed the influence on these eigenvalues of changing the frequency ratios ω_4/ω_3 and ω_5/ω_3. For instance, Fig. 7–13 shows divergence speed vs. ω_4/ω_3, with the mode shapes $\varphi_i(\xi)$ kept unaltered and ω_5 adjusted according to

$$\frac{\omega_5}{\omega_3} = 2.4461 \frac{\omega_4}{\omega_3} - 1.4461 \qquad (7\text{–}189)$$

(Equation 7–189 causes the three frequencies to spread in linear proportion as ω_4 is increased.) It is of interest that, just below the frequency ratio $(\omega_4/\omega_3) = 2.5412$ which characterizes the actual flat-plate delta, the divergence speed jumps from a value associated with a mode that couples principally the φ_3 and φ_4 degrees of freedom to a much higher value associated with the φ_5 normal mode. If the φ_5 freedom is dropped from Eqs. 7–183, we obtain the closed-form solution

$$\frac{U_D}{l\omega_3\sqrt{(m_0/2\pi\rho_\infty l^2 \tan^2 \Lambda')}} \cong \sqrt{\frac{(\omega_4/\omega_3)^2}{7.390 + 2.495\left(\frac{\omega_4}{\omega_3}\right)^2 \pm \sqrt{6.225\left(\frac{\omega_4}{\omega_3}\right)^4 - 45.891\left(\frac{\omega_4}{\omega_3}\right)^2 + 54.612}}}$$

(7–190)

Equation 7–190 is plotted as the dashed curve on Fig. 7–13.

A great deal more information on the delta wing as a one-dimensional structure will be found in Ref. 7–34. Thus we have added on Fig. 7–13 a curve marked "flutter speed," which is an excellent approximation to their slender-body result for high mass ratios $(m_0/\pi\rho_\infty l^2 \tan^2 \Lambda') > 25$. At the true frequency ratio $(\omega_4/\omega_3) = 2.5412$, the static instability is seen to occur at a slightly lower speed than the dynamic. In contrast, in Sec. 9–60 there are presented similar computations for flutter speed based on the piston-theory aerodynamic operator, whose use is limited to higher Mach numbers and $M \sin \Lambda' > 1$.

As an indication of the success with which slender-body theory predicts divergence of triangular wings, we reproduce Fig. 7–14 from Martin and

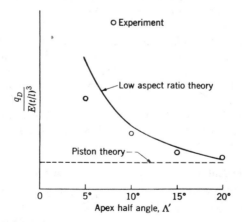

Fig. 7–14. Experimental and theoretical divergence characteristics at Mach number 2.0 of flat-plate delta wings cantilevered from the trailing edge. The plate thickness is t. (Reproduced from Ref. 7–35.)

Watkins (Ref. 7–35). These data refer to a flat-plate delta supported as a cantilever from the rear. The writers suggest that one possible explanation for the experimental points' falling below theory at very low vertex angles is nonlinear aerodynamic effects; on such wings, it is well known that vortex separation from the leading edge above rather small angles of attack causes the lift to build up more rapidly than estimated by slender-body methods.

Among other studies on the static aeroelasticity of low-aspect-ratio wings, we cite the papers by Hedgepeth and Waner (Ref. 7–36) and Hancock (Ref. 7–37). These references go somewhat beyond the limitation to one-dimensional deformations made in the present chapter. For example, Hedgepeth and Waner treat a rectangular planform supported along a chordwise line at midspan; they assume deflections which are quadratic in the spanwise coordinate y, but make no restrictions on the x-dependence.

7–5 PRESCRIBED TIME-DEPENDENT INPUTS

Superposition of normal free-vibration modes, which has been applied to a static problem in the preceding section, also serves as a very efficient method for finding the dynamic response to an input whose form can be prescribed independently of the system's motion. These situations arise during the ground operation of aircraft, such as when conducting vibration tests, braking or taxiing over rough surfaces, catapulting, and analyzing the effects of certain weapons. One might also mention the static firing of rocket engines and transient loads associated with missile launching, especially when this takes place from a hardened site. A related class of problems consists of those where the aerodynamics due to the response have a negligible influence; landing loads and vibrations induced in a missile or other flexible structure exposed to the wind usually fall into this category.

In the present section, we illustrate the analysis of forced transients with examples involving a uniform cantilever beam and the landing impact of an airplane.

(a) Time-dependent forcing of a uniform slender wing

The mass of any lifting surface is usually a rather small fraction of the total mass of the fuselage or missile body to which it is attached. Provided the uncoupled vibration frequencies of the body and the wing aspect ratio are both high enough, its bending motion can then often be simulated by that of a cantilever beam built in at some station $y = 0$ between the body's centerline and side wall.* We adopt this approximation and also assume that the bending stiffness EI and running mass m are independent of spanwise distance. With the applied load $F_z(y, t)$ regarded as the given input, Eq. 7–11 becomes

$$EI \frac{\partial^4 w}{\partial y^4} + m\ddot{w} = F_z(y, t) \tag{7-191}$$

The convenient transformation of variables $\tilde{y} = y/l$, $\tilde{t} = (t/l^2)\sqrt{EI/m}$, $\tilde{w} = w/l$, $\tilde{f} = F_z l^3/EI$ eliminates all coefficients from Eq. 7–191, leaving

$$\frac{\partial^4 \tilde{w}}{\partial \tilde{y}^4} + \frac{\partial^2 \tilde{w}}{\partial \tilde{t}^2} = \tilde{f}(\tilde{y}, \tilde{t}) \tag{7-192}$$

The cantilever boundary conditions call for an *encastré* root and zero shear and bending moment at the tip:

$$\tilde{w}(0, \tilde{t}) = \frac{\partial \tilde{w}}{\partial \tilde{y}}(0, \tilde{t}) = \frac{\partial^2 \tilde{w}}{\partial \tilde{y}^2}(1, \tilde{t}) = \frac{\partial^3 \tilde{w}}{\partial \tilde{y}^3}(1, \tilde{t}) = 0 \tag{7-193}$$

Any modern book on vibrations develops the solutions to the homogeneous (i.e., $\tilde{f} = 0$) form of Eqs. 7–192 and 7–193. The normal mode shapes are composed of trigonometric and hyperbolic functions in the combinations

$$\varphi_i(\tilde{y}) = \pm \tfrac{1}{2}\big[(\cosh \sqrt{\theta_i}\tilde{y} - \cos \sqrt{\theta_i}\tilde{y})$$
$$- \left(\frac{\sinh \sqrt{\theta_i} - \sin \sqrt{\theta_i}}{\cosh \sqrt{\theta_i} + \cos \sqrt{\theta_i}}\right)(\sinh \sqrt{\theta_i}\tilde{y} - \sin \sqrt{\theta_i}\tilde{y})\big] \tag{7-194}$$

where the eigenvalues θ_i and corresponding natural frequencies

$$\omega_i = (\theta_i/l^2)\sqrt{EI/m}$$

are roots of the characteristic equation

$$1 + \cosh \sqrt{\theta_i} \cos \sqrt{\theta_i} = 0 \tag{7-195}$$

Section 3–2 of Ref. 1–2 lists

$$\left.\begin{array}{l} \theta_1 = (0.597\pi)^2 \\[4pt] \theta_2 = (1.49\pi)^2 \\[4pt] \theta_i = \left(\dfrac{2i-1}{2}\,\pi\right)^2, \qquad (i = 3, 4, \cdots) \end{array}\right\} \tag{7-196}$$

* A similar structural configuration occurs with a vertically erected missile clamped firmly to its launching pad, as is done prior to firing Atlas.

We have chosen the constant outside the brackets in Eq. 7–194 so as to make the tip amplitude $\varphi_i(1)$ equal to unity; the sign should be taken the same as that of $\sin \sqrt{\theta_i}$ and therefore alternates between plus and minus with increasing i. As shown, for example, on pp. 327 ff. of Ref. 1–32, the orthogonality condition reduces to

$$\int_0^1 \varphi_i(\bar{y})\varphi_j(\bar{y})\,d\bar{y} = \begin{cases} \frac{1}{4}\varphi_i^2(1) \equiv \frac{1}{4}, & (i = j) \\ 0, & (i \ne j) \end{cases} \qquad (7\text{–}197)$$

Two different schemes are in common use for computing the forced transient response from Eq. 7–192. The classical procedure, which need not be reproduced in detail, represents the entire deflection as discussed in Sec. 2–5(c):

$$\bar{w}(\bar{y}, \bar{t}) = \sum_{i=1} \varphi_i(\bar{y})q_i(\bar{t}) \qquad (7\text{–}198)$$

When Eq. 7–198 is substituted, multiplication by $\varphi_j(\bar{y})$, integration across the span, and application of known properties of the normal modes $\left(\text{e.g.,}\ \dfrac{d^4\varphi_i}{d\bar{y}^4} = \theta_i^2\varphi_i\right)$ lead to

$$\ddot{q}_j + \theta_j^2 q_j = 4\int_0^1 \bar{f}(\bar{y}, \bar{t})\varphi_j(\bar{y})\,d\bar{y} \equiv 4Q_j(\bar{t}), \quad (j = 1, 2, 3, \cdots) \quad (7\text{–}199)$$

Dots here indicate dimensionless time-differentiation. The uncoupled equations (7–199) can readily be solved by Laplace transformation for the individual generalized coordinates. For instance, we might consider motion starting from rest at $\bar{t} = 0$, whence

$$\bar{w}(\bar{y}, 0) = \dot{\bar{w}}(\bar{y}, 0) = 0 \qquad (7\text{–}200)$$

corresponding to

$$q_j(0) = \dot{q}_j(0) = 0 \qquad (7\text{–}201)$$

The response to the jth generalized force then becomes

$$q_j(\bar{t}) = \frac{4}{\theta_j}\int_0^{\bar{t}} Q_j(\tau)\sin\theta_j(\bar{t} - \tau)\,d\tau \qquad (7\text{–}202)$$

Returning to Eq. 7–198 for the complete response,

$$\bar{w}(\bar{y}, \bar{t}) = 4\sum_{i=1} \frac{\varphi_i(\bar{y})}{\theta_i}\int_0^{\bar{t}} Q_i(\bar{t})\sin\theta_i(\bar{t} - \tau)\,d\tau$$

$$= 4\int_0^{\bar{t}}\int_0^1\left[\bar{f}(\eta, \tau)\sum_{i=1}\frac{\varphi_i(\bar{y})\varphi_i(\eta)}{\theta_i}\sin\theta_i(\bar{t} - \tau)\right]d\eta\,d\tau \qquad (7\text{–}203)$$

A less familiar technique, but one which has many practical advantages in aeronautical problems, is that originally proposed by Williams (Ref. 7–38). The deflection is written as

$$\bar{w}(\bar{y}, \bar{t}) = \bar{w}_s(\bar{y}, \bar{t}) + \sum_{i=1} \varphi_i(\bar{y}) r_i(\bar{t}) \tag{7-204}$$

$\bar{w}_s(\bar{y}, \bar{t})$ being the *quasi-steady displacement* due to the instantaneous load $\bar{f}(\bar{y}, \bar{t})$ under the constraint that all vibratory motion is suppressed. By direct integration of the static equation

$$\frac{d^4 \bar{w}_s}{d\bar{y}^4} = \bar{f}(\bar{y}, \bar{t}) \tag{7-205}$$

treating time as a parameter and employing the boundary conditions (7–193), we find

$$\bar{w}_s(\bar{y}, \bar{t}) = \int_0^{\bar{y}} \int_0^{\bar{y}} \int_{\bar{y}}^1 \int_{\bar{y}}^1 \bar{f}(\bar{y}, \bar{t})(d\bar{y})^4 \tag{7-206}$$

(The reader must recognize the presence of four dummy integration variables in Eq. 7–206, each distinct from the upper limit of the outer integral, which gives the actual dependence of $\bar{w}_s$ on the spanwise coordinate.)

To derive the dynamic correction to $\bar{w}(\bar{y}, \bar{t})$, we once more insert Eq. 7–204 into Eq. 7–192, use Eq. 7–205 to cancel $\bar{f}$ against the fourth derivative of $\bar{w}_s$, and observe that $\dfrac{d^4 \varphi_i}{d\bar{y}^4} = \theta_i^2 \varphi_i$.

$$\sum_{i=1} [\ddot{r}_i + \theta_i^2 r_i] \varphi_i(\bar{y}) = -\ddot{\bar{w}}_s(\bar{y}, \bar{t}) \tag{7-207}$$

Thus we see that this correction may be interpreted as the beam's dynamic response to the inertia loading due to the actual time-dependence of $\bar{w}_s$. Multiplication by $\varphi_j(\bar{y})$ and integration across the span yield the uncoupled equation for the jth generalized coordinate.

$$\ddot{r}_j + \theta_j^2 r_j = -4 \int_0^1 \ddot{\bar{w}}_s(\bar{y}, \bar{t}) \varphi_j(\bar{y}) \, d\bar{y}$$
$$\equiv -4 \ddot{R}_j(\bar{t}), \qquad (j = 1, 2, 3, \cdots) \tag{7-208}$$

The expression for the "generalized force" on $r_j(\bar{t})$ is capable of simplification. Thus, we can replace $\varphi_j(\bar{y})$ in terms of its fourth derivative and integrate by parts four times, using the structural boundary conditions to eliminate integrated portions, as follows:

$$R_j(\tilde{t}) \equiv \int_0^1 \tilde{w}_s(\tilde{y}, \tilde{t})\varphi_j(\tilde{y})\, d\tilde{y}$$

$$= \frac{1}{\theta_j^2} \int_0^1 \tilde{w}_s(\tilde{y}, \tilde{t})\frac{d^4\varphi_j}{d\tilde{y}^4}\, d\tilde{y} = \frac{1}{\theta_j^2}\int_0^1 \frac{d^4\tilde{w}_s}{d\tilde{y}^4}\, \varphi_j(\tilde{y})\, d\tilde{y} \quad (7\text{-}209)$$

In view of Eq. 7–205, however,

$$R_j(\tilde{t}) = \frac{1}{\theta_j^2}\int_0^1 \tilde{f}(\tilde{y}, \tilde{t})\varphi_j(\tilde{y})\, d\tilde{y}$$

$$= Q_j(\tilde{t})/\theta_j^2 \qquad (7\text{-}210)$$

Equations 7–208 through 7–210 require initial conditions. For instance, the system might be at rest at $\tilde{t} = 0$, as described by Eqs. 7–200. If Eq. 7–204 is substituted into 7–200, then the usual multiplication by $\varphi_j(\tilde{y})$ and spanwise integration lead to

$$\left.\begin{aligned} r_j(0) &= -4R_j(0) \\ \dot{r}_j(0) &= -4\dot{R}_j(0) \end{aligned}\right\} \qquad (7\text{-}211)$$

The solution to Eqs. 7–208 through 7–211 reads

$$r_j(\tilde{t}) = -4R_j(\tilde{t}) + 4\theta_j\int_0^{\tilde{t}} R_j(\tau)\sin\theta_j(\tilde{t} - \tau)\, d\tau$$

$$= -4R_j(0)\cos\theta_j\tilde{t} - 4\int_0^{\tilde{t}} \dot{R}_j(\tau)\cos\theta_j(\tilde{t} - \tau)\, d\tau \quad (7\text{-}212)$$

the latter form being preferable because of improved convergence and the vanishing of $R_j(0)$ whenever the force F_z has no step function at $\tilde{t} = 0$. Returning to Eq. 7–204 for the complete response,

$$\tilde{w}(\tilde{y}, \tilde{t}) = \tilde{w}_s(\tilde{y}, \tilde{t})$$

$$- 4\sum_{i=1}\varphi_i(\tilde{y})\left[R_i(0)\cos\theta_i\tilde{t} + \int_0^{\tilde{t}}\dot{R}_i(\tau)\cos\theta_i(\tilde{t} - \tau)\, d\tau\right]$$

$$= \tilde{w}_s(\tilde{y}, \tilde{t}) - 4\int_0^1\tilde{f}(\eta, 0)\sum_{i=1}\left[\frac{\varphi_i(\tilde{y})\varphi_i(\eta)}{\theta_i^2}\cos\theta_i\tilde{t}\right]d\eta$$

$$- 4\int_0^{\tilde{t}}\int_0^1\left[\dot{\tilde{f}}(\eta, \tau)\sum_{i=1}\frac{\varphi_i(\tilde{y})\varphi_i(\eta)}{\theta_i^2}\cos\theta_i(\tilde{t} - \tau)\right]d\eta\, d\tau \quad (7\text{-}213)$$

$\tilde{w}_s(\tilde{y}, \tilde{t})$ is, of course, calculated by Eq. 7–206 and involves no summing of infinite series. If $\tilde{f}(\eta, 0) = 0$, we can compare the series in Eqs. 7–213 and 7–203 and discover why good convergence of practical calculations is to be expected from the Williams solution, in view of the fast-growing factor θ_i^2 in the denominator.

When making vibration tests on a structure, we apply distributed or concentrated *sinusoidal* forces. Therefore $\tilde{f}(\tilde{y}, \tilde{t})$ is given by the real part of

$$\tilde{f}(\tilde{y}, \tilde{t}) = F(\tilde{y})e^{i\omega t} = F(\tilde{y})e^{i\theta l} \qquad (7\text{-}214)$$

$\theta = \omega l^2 \sqrt{m/EI}$ being the dimensionless driving frequency. (A concentrated force is treated by including in $F(\tilde{y})$ a suitable Dirac delta function centered on its point of action.) In this case, the alternative forms (7–203) and (7–213) for the system response must be modified to yield the steady oscillation which is all that remains after the input has acted for any length of time. The reader will have no difficulty in showing that the two results now read, respectively,

$$\tilde{w}(\tilde{y}, \tilde{t}) = 4e^{i\theta l} \sum_{i=1}^{\infty} \frac{\varphi_i(\tilde{y}) \int_0^1 F(\eta)\varphi_i(\eta)\, d\eta}{(\theta_i^2 - \theta^2)} \qquad (7\text{-}215)$$

$$\tilde{w}(\tilde{y}, \tilde{t}) = W_s(\tilde{y})e^{i\theta l} + 4e^{i\theta l} \sum_{i=1}^{\infty} \left(\frac{\theta^2}{\theta_i^2}\right) \frac{\varphi_i(\tilde{y}) \int_0^1 F(\eta)\varphi_i(\eta)\, d\eta}{(\theta_i^2 - \theta^2)} \qquad (7\text{-}216)$$

where

$$W_s(\tilde{y}) = \int_0^{\tilde{y}} \int_0^{\tilde{y}} \int_{\tilde{y}}^1 \int_{\tilde{y}}^1 F(\tilde{y})(d\tilde{y})^4 \qquad (7\text{-}217)$$

In particular, a constant loading amplitude $F(\tilde{y}) = F_0$ gives

$$W_s(\tilde{y}) = F_0 \left[\frac{\tilde{y}^4}{24} - \frac{\tilde{y}^3}{6} + \frac{\tilde{y}^2}{4} \right] \qquad (7\text{-}218)$$

whereas a concentrated force P_0 at $\tilde{y} = \tilde{y}_0$, which is described by

$$F(\tilde{y}) = P_0 \delta(y - y_0) \qquad (7\text{-}219)$$

will be found to produce a somewhat more complicated W_s, reflecting the shear discontinuity at $\tilde{y}_0$.

The faster convergence of the series in Eq. 7–216 is again apparent. Other interesting deductions may be made from these sinusiodal results. We note the familiar resonance condition that occurs whenever the driving frequency passes through any natural frequency of the beam. The theoretically infinite resonance peaks would in reality be reduced to large finite size by the inevitable presence of friction in the structure. Such damping has been neglected in the foregoing development; although this is an acceptable approximation when analyzing many forced motions, it is wholly unsatisfactory for the resonance phenomenon.

Equation 7–215 reveals that the key requirement in an experiment designed to measure properties of one particular normal mode is to have the force-amplitude distribution $F(\eta)$ as nearly as possible orthogonal to all other mode shapes of the system. This requirement becomes progressively more stringent the higher the order of the mode one wishes to isolate, because of the decreasing relative size of the coefficients $1/(\theta_i{}^2 - \theta^2)$ whenever the driving frequency does not coincide exactly with the desired ω_j. In this connection, many excellent discussions of vibration testing will be found in the literature; the papers by Mazet (Ref. 7–39) and Lewis and Wrisley (Ref. 7–40) are representative.

The two procedures for transient-response calculation have as their counterparts the *mode-displacement* and *mode-acceleration* methods for finding dynamic stresses in aircraft structures (cf. Sec. 10–3(b) of Ref. 1–2). The idea is that local stress, shear, bending moment, etc., may at any instant be written as linear sums of contributions from each normal mode. The advantage of the mode-acceleration scheme is that the principal part of the stress distribution is the static one, obtained from the instantaneous loading $F_z(y, t)$ without any reference to its time history whatever. This can be done by the tried, well-developed rules of the stress analyst and can include allowance for types of structural nonlinearity that are not encompassed by ordinary modal superposition. Thus the mode-acceleration design stress at a station y is expressible as

$$\sigma_y(y, t) = \sigma_{ys}(y, t) - \sum_{i=1}^{\ } \frac{A_y^{(i)}}{\omega_i^2} \ddot{q}_i \qquad (7\text{–}220)$$

where, as shown in Ref. 1–2, $A_y^{(i)}$ denotes the maximum stress at y due to unit displacement of the ith normal mode. For a given accuracy in a practical computation, the series in Eq. 7–220 can be truncated much lower than it should be if the mode displacement method were being applied. Incidentally, an ingenious variant of mode acceleration, for situations with external loads due to the system's motion, has been derived by Mar, Pian, and Calligeros (Ref. 7–41).

We close this discussion on transients in uniform cantilevers by directing the reader to many other examples of this sort which will be found in Ref. 7–2 and papers listed therein. Thus, Leonard tabulates the exact Williams solutions for cantilever and simply supported configurations and for free-free beams with a central concentrated mass. Results are given for both the elementary beam and the "Timoshenko beam" having finite shear flexibility and rotary inertia. The importance of these latter effects is assessed by means of illustrative computations, in which exact responses are compared with approximate numerical estimates found by the method of characteristics and by a stepwise-integration scheme adapted from Houbolt (Ref. 7–42).

(b) The landing problem

When predicting accelerations and stresses due to ground landing impact, dynamics engineers agree that the lift and moment on the airplane can generally be assumed constant during the critical phase. Hence one rather difficult part of the loading due to motion is eliminated from this type of transient analysis. What remains is not wholly straightforward, however, because of nonlinearities in the shock-absorbing strut, landing-gear structure, and tire force-deflection characteristic. For a period of a few tenths of a second following contact, these elements build up a system of vertical and horizontal forces which depend in a complicated way on the heights of the two landing-gear attachment points above the runway surface and on the rates of change of these heights. Simultaneously, the rest of the elastic vehicle vibrates in response to these forces. Both symmetrical and antisymmetrical modes are observed, except in the rare event that the angle of bank is exactly zero at touchdown.

If the linear techniques of aeroelasticity are to be used to compute landing loads, it seems necessary to be able to find the time history of the gear forces independently of vibrations in the airframe. Fortunately, several investigations (e.g., Ramberg and McPherson, Ref. 7-43, Pian and Flomenhoft, Ref. 7-44) indicate that this is the case to an acceptable degree of accuracy. By adopting their approximation and considering only one-dimensional wing structures, we bring the problem within the purview of the present section.

Let us examine the vertical translation and wing bending produced by a pair of landing forces $F_{z_R}(t)$ and $F_{z_L}(t)$ acting upward along the elastic axis at distances $y = \pm y_G$ from the centerline. Typical time variations of F_{z_R} and F_{z_L} are given in Refs. 7-43, 7-44, and elsewhere; Fung (Ref. 7-45) has published illuminating data on their statistical properties. Regarded as portions of a distributed spanwise load, they would be expressed

$$F_z(y, t) = F_{z_R}(t)\, \delta(y - y_G) + F_{z_L}(t)\, \delta(y + y_G) \qquad (7\text{-}221)$$

where the delta functions contribute unit area to any integral passing through their respective stations, in the usual manner.

Almost universally today, this problem is treated by superimposing normal modes. We therefore write the vertical position of the elastic axis relative to some inertial frame as

$$w(y, t) = l\left\{ q_0(t) + \varphi_0(t)\tilde{y} + \sum_{i=1}^{\infty} \varphi_i(\tilde{y})q_i(t) \right\} \qquad (7\text{-}222)$$

Here $2l$ is the wingspan and $\tilde{y} = y/l$. The normal mode shapes $\varphi_i(\tilde{y})$ of the free-free structure are presumed known; they are orthogonal with respect to the spanwise distribution of running mass $m(\tilde{y})$. Two rigid-body degrees of freedom are allowed: (1) vertical translation, in which $lq_0(t)$ denotes the position of the C.G. of the *instantaneously deformed* system; and (2) roll, in which $\varphi_0(t)$ denotes the angle of bank about the fore-and-aft principal axis. Pitching is excluded, as are chordwise deflections and fuselage bending, but it is an easy matter to add any of them to such an analysis whenever their importance is suspected.

As outlined in the preceding subsection or in Chap. 3, we apply the Lagrange technique and develop the following equations of motion:

$$M_0 \ddot{q}_0 = \int_{-1}^{1} F_z(\tilde{y}, t) \, d\tilde{y} \qquad (7\text{-}223a)$$

$$\frac{I_x}{l^2} \ddot{\varphi}_0 = \int_{-1}^{1} F_z(\tilde{y}, t) \tilde{y} \, d\tilde{y} \qquad (7\text{-}223b)$$

$$M_i[\ddot{q}_i + 2\zeta_i\omega_i\dot{q}_i + \omega_i^2 q_i]$$
$$= \int_{-1}^{1} F_z(\tilde{y}, t)\varphi_i(\tilde{y}) \, d\tilde{y}, \qquad (i = 1, 2, 3, \cdots) \quad (7\text{-}223c)$$

M_0 and I_x are the total mass and roll moment of inertia of the vehicle. M_i is the generalized mass of the ith elastic degree of freedom,

$$M_i = l \int_{-1}^{1} m(\tilde{y})\varphi_i^2(\tilde{y}) \, d\tilde{y} \qquad (7\text{-}224)$$

and is often made equal to M_0 by appropriately choosing the reference amplitude of the corresponding mode shape. [Note that the masses of nacelles, central body, etc., must be included in $m(\tilde{y})$.] For variety, we allow here in an approximate way for energy dissipation in the structural vibration. The symbol ζ_i represents the damping ratio of free oscillation in the ith normal mode. A viscous type of dissipation mechanism is hypothesized, and no consideration is given to generalized forces on other modes due to frictional forces caused by any one of them (zero "damping coupling"). The peak response to certain landing loads may not occur until several cycles of significant higher modes have gone by, and in such a circumstance the natural decay of these modes might affect the design. It is nearly always conservative to omit damping effects.

When Eq. 7–221 is substituted into Eqs. 7–223 and we take account of symmetry, a further separation into motions which are symmetrical or antisymmetrical with respect to the centerline occurs. The former are described by

$$M_0 \ddot{q}_0 = F_{z_R}(t) + F_{z_L}(t) \qquad (7\text{-}225a)$$

$$M_i[\ddot{q}_i + 2\zeta_i\omega_i\dot{q}_i + \omega_i^2 q_i]$$
$$= [F_{z_R}(t) + F_{z_L}(t)] \, \varphi_i(\tilde{y}_G), \qquad (i \text{ for symm. modes}) \quad (7\text{-}225b)$$

whereas the latter obey

$$I_x \ddot{\varphi}_0 = L_R(t) \qquad (7\text{-}226a)$$

$$M_i[\ddot{q}_i + 2\zeta_i \omega_i \dot{q}_i + \omega_i^2 q_i]$$

$$= [F_{z_R}(t) - F_{z_L}(t)]\, \varphi_i(\tilde{y}_G), \qquad (i \text{ for antisymm. modes}) \quad (7\text{-}226b)$$

In Eq. 7–226a, the generalized force is the rolling moment about the airplane centerline,

$$L_R(t) = y_G F_{z_R}(t) - y_G F_{z_L}(t) \qquad (7\text{-}227)$$

Clearly there will be no antisymmetrical response at all in a landing with $F_{z_R} = F_{z_L}$.

The solution of the foregoing equations can be illustrated by reference to the symmetrical case. Typical initial conditions would involve a starting rate of descent v_0 and undisturbed normal modes,

$$q_0(0) = 0, \qquad \dot{q}_0(0) = -v_0 \qquad (7\text{-}228)$$

$$q_i(0) = \dot{q}_i(0) = 0 \qquad (7\text{-}229)$$

Equations 7–225a through 7–228 are solved for the rigid-body response

$$\dot{q}_0(t) = -v_0 + \frac{1}{M_0} \int_0^t [F_{z_R}(\tau) + F_{z_L}(\tau)]\, d\tau \qquad (7\text{-}230a)$$

$$q_0(t) = -v_0 t + \frac{1}{M_0} \int_0^t \int_0^\tau [F_{z_R}(\tau') + F_{z_L}(\tau')]\, d\tau'\, d\tau \quad (7\text{-}230b)$$

Since the vertical velocity must drop to zero during landing, we observe that the total impulse

$$\int_0^\infty [F_{z_R}(\tau) + F_{z_L}(\tau)]\, d\tau$$

must ultimately settle down to equal the momentum change $M_0 v_0$. The vibratory responses from Eqs. 7–225b through 7–229 may be calculated by Laplace transformation or by the indicial admittance method [cf. Sec. 6–3(b)]. In general, they will contain terms with both sines and cosines of $\omega_i t$, but a very satisfactory approximation, in view of the smallness of the ζ_i, reads

$$q_i(t) \cong \frac{\varphi_i(\tilde{y}_G)}{M_i \omega_i} \int_0^t e^{-\zeta_i \omega_i (t-\tau)} \sin \omega_i(t-\tau)[F_{z_R}(\tau) + F_{z_L}(\tau)]\, d\tau,$$

$$(i \text{ for symm. modes}) \quad (7\text{-}231)$$

The resultant motion is described by the symmetrical terms in Eq. 7–222.

Local accelerations, bending moments, stresses, and the like can be computed efficiently by adjusting the mode acceleration procedure to include the small influence of damping. Since the various contributing modes are going at different frequencies, it may not be an easy task to determine the instant at which a particular one of these quantities reaches its peak value. One suggestion (Biot and Bisplinghoff, Ref. 7–46) is to add the maxima from the individual modes in scalar fashion, without regard for their phase differences; but this may be an oversimplification when the structure is a complex one.

Limitations of space prevent more than this rather sketchy review of what is often a severe design problem, so we must call attention to the many thorough treatments now available in the literature (e.g., Refs. 7–34 through 7–49). Two further points should be made. The first is that even greater complexity enters the picture when we deal with rough-water landing of seaplanes and flying boats. There is considerably more doubt as to whether the mutual interaction between vehicle motion and hull forces can be neglected, and a statistical approach to estimating loads is certainly indicated (Refs. 7–45, 7–50). Finally it must be observed that the fascination of the touchdown phenomenon sometimes obscures the fact that other aspects of ground operation, such as taxiing and instabilities in the braking system, often affect the design of landing-gear structure to an equal or greater extent (McBrearty, Ref. 7–51).

7–6 FORCING IN THE PRESENCE OF EXTERNAL LOADS DEPENDENT ON THE MOTION

In connection with the forced, time-dependent behavior of systems with a finite number of degrees of freedom, Secs. 6–3 and 6–4 devoted considerable attention to the concepts of *mechanical admittance, indicial admittance*, and *unit impulse response*. It should be evident to the reader that these ideas can be carried over to more complex linear systems, including the continuous, one-dimensional variety covered in the present chapter. Moreover, techniques such as Fourier series, Fourier integrals, and Duhamel integrals for calculating the response to arbitrary inputs also retain their usefulness. Provided the spatial variations of the driving force are known in advance, the position coordinates play the role of parameters. Simple harmonic excitation of the uniform cantilever, as described by Eqs. 7–215 through 7–217, furnishes an illustration of this point. Thus, one may divide out the common time function $e^{i\omega t} \equiv e^{i\theta \bar{t}}$ and normalize the force amplitude in some such manner as

$$F(\bar{y}) = F_0 \Phi(\bar{y}) \tag{7–232}$$

Equation 7-215 is then manipulated to yield the following admittance, relating bending amplitude as the output to F_0 as input:

$$\frac{W(\tilde{y})}{F_0} \equiv H_{WF}(\theta, \tilde{y})$$

$$= 4 \sum_{i=1} \frac{\varphi_i(\tilde{y}) \int_0^1 \Phi(\eta)\varphi_i(\eta)\,d\eta}{(\theta_i^2 - \theta^2)} \qquad (7\text{-}233)$$

Similar admittance functions exist for acceleration, shear, bending stress, and the like.

With these introductory observations, we undertake to review the analysis of three representative situations wherein the vehicle is driven by a prescribed external agency, while its response is significantly affected by airloads due to transient deformations of the lifting surfaces. The first is a case of steady-state sinusoidal forcing, and we have chosen the Holzer-Myklestad procedure of dividing the structure into a number of concentrated segments. The second example deals with discrete gust or blast loading of a swept wing and typifies the indicial approach. The third involves the (more realistic) statistical treatment of the effects of continuous atmospheric turbulence.

(a) A straight wing under sinusoidal forcing; in-flight vibration testing; the Holzer-Myklestad approach

It is a routine part of the testing of most large aircraft to excite the aeroelastic modes and observe their frequency and damping characteristics throughout a series of flight conditions. The airborne excitation may take the form of rapid control-surface displacements, impulsive loads caused by small shaped charges, or sinusoidal forcing from an electro-mechanical, hydraulic, or aerodynamic shaker. Although the impulsive type of excitation is the more economical in time, experience seems to prove that the greatest amount of useful information—especially regarding the nearness to a critical flutter boundary—can be had from carefully interpreted sinusoidal measurements over a range of frequencies.*

To provide an analytical description of a flight vibration test, let us study a straight-tapered wing with a rectilinear elastic axis normal to the body centerline. Following a procedure suggested by Myklestad (Ref. 1-9) in connection with free vibration and flutter prediction on beam-rod structures, we split the wing up into several bays and concentrate the mass

* For confirmation of this statement, the reader is referred to the consensus of papers collected in Ref. 7-52.

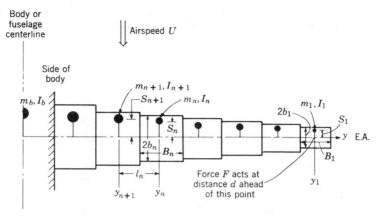

Fig. 7–15. Lumped-mass equivalent for the right half of a tapered wing with straight elastic axis normal to the fuselage centerline.

m_n and moment of inertia I_n from each bay at a suitably located central point y_n. As shown in Fig. 7–15, the center of gravity of each lump lies a distance S_n ahead of the E.A. It is connected to its neighbors by weightless flexures and torsion members. Each segment is assigned constant chord $2b_n$ and span B_n for aerodynamic purposes; the total lift and pitching moment act at station y_n.

We remark that the process of constructing this lumped equivalent for any given configuration is an art, beyond the scope of the present discussion. Such approximate representations have their greatest value today when analog computation is applied to aeroelastic problems.

The wing of Fig. 7–15 is assumed to be driven by a pair of shakers, each acting d feet ahead of the E.A. and aligned with the outermost mass m_1 on its half of the span. The forces are simple harmonic and positive upward,

$$F = \bar{F}e^{i\omega t} \qquad (7\text{–}234)$$

Let $\bar{w}_i e^{i\omega t}$ (positive upward), $\bar{\alpha}_i e^{i\omega t}$, and $\bar{\theta}_i e^{i\omega t}$ (positive leading-edge-up) denote the bending displacement, bending slope, and twist angle, respectively, at station y_i. The aerodynamic operators (4–129) and (4–130) give for the upward lift and pitching moment at y_i

$$\bar{L}e^{i\omega t} = 4\rho_\infty b_i^3 B_i \omega^2 \left\{ -[L_1 + iL_2]\frac{\bar{w}_i}{b_i} + [L_3 + iL_4]\bar{\theta}_i \right\} e^{i\omega t} \qquad (7\text{–}235)$$

$$\bar{M}_y e^{i\omega t} = -4\rho_\infty b_i^4 B_i \omega^2 \left\{ -[M_1 + iM_2]\frac{\bar{w}_i}{b_i} + [M_3 + iM_4]\bar{\theta}_i \right\} e^{i\omega t} \qquad (7\text{–}236)$$

It is necessary here to use strip theory, because the Holzer-Myklestad procedure makes no allowance for aerodynamic induction between the various stations. With minor exceptions to date, this same limitation affects any analog prediction of lifting-surface transients. Factors independent of the mode of motion can, of course, be introduced at each station to adjust roughly for span effects.

For arbitrary n, we are now in a position to write down the complex amplitudes of shear $\bar{S}_n$, bending moment $\bar{M}_n$, and torque $\bar{T}_n$ carried by the structure just to the left of y_n. The first two quantities are chosen positive in the senses of Fig. 7–1. Torque is positive when the outer side of a cross section exerts a nose-up twisting action on the inner side. Consolidating the inertia forces and torques at their respective stations and cancelling $e^{i\omega t}$ throughout, we obtain

$$\bar{S}_n = \sum_{i=1}^{n} \omega^2 \{[m_i - 4\rho_\infty b_i^2 B_i(L_1 + iL_2)]\bar{w}_i$$
$$+ [m_i s_i + 4\rho_\infty b_i^3 B_i(L_3 + iL_4)]\bar{\theta}_i\} + \bar{F} \qquad (7\text{–}237)$$

$$\bar{M}_n = \sum_{i=1}^{n-1} \omega^2 \{[m_i - 4\rho_\infty b_i^2 B_i(L_1 + iL_2)]\bar{w}_i$$
$$+ [m_i s_i + 4\rho_\infty b_i^3 B_i(L_3 + iL_4)]\bar{\theta}_i\}[y_i - y_n] + \bar{F}[y_1 - y_n] \quad (7\text{–}238)$$

$$\bar{T}_n = \sum_{i=1}^{n} \omega^2 \{[m_i s_i + 4\rho_\infty b_i^3 B_i(M_1 + iM_2)]\bar{w}_i$$
$$+ [I_i - 4\rho_\infty b_i^4 B_i(M_3 + iM_4)]\bar{\theta}_i\} + \bar{F}d \qquad (7\text{–}239)$$

The four bracketed factors appearing in Eqs. 7–237 through 7–239 are abbreviated as follows:

$$\left.\begin{aligned}
[m_i - 4\rho_\infty b_i^2 B_i(L_1 + iL_2)] &\equiv \bar{P}_i \\
[m_i s_i + 4\rho_\infty b_i^3 B_i(L_3 + iL_4)] &\equiv \bar{P}_i' \\
[m_i s_i + 4\rho_\infty b_i^3 B_i(M_1 + iM_2)] &\equiv \bar{Q}_i \\
[I_i - 4\rho_\infty b_i^4 B_i(M_3 + iM_4)] &\equiv \bar{Q}_i'
\end{aligned}\right\} \qquad (7\text{–}240)$$

These quantities depend on the properties of the ith section, on the air density, and (through the dimensionless aerodynamic operators) on the E.A. location, flight Mach number, and reduced frequency $\omega b_i/U$. All this information is available for a specified flight-test condition.

Using Myklestad's notation, we next introduce five influence coefficients which characterize completely the flexibility of the structure between station y_n and station y_{n+1}:

v_{T_n} = Difference between angles of twist at y_{n+1} and y_n when unit torque acts at y_n.

v_{F_n} = Difference between bending slopes at y_{n+1} and y_n when unit shear (and no bending moment) acts at y_n.

v_{M_n} = Difference between bending slopes at y_{n+1} and y_n when unit bending couple (and no shear) acts at y_n.

d_{F_n} = Bending displacement of y_n relative to y_{n+1} when y_{n+1} is held as a cantilever and unit shear (but no bending moment) acts at y_n.

d_{M_n} = Bending displacement of y_n relative to y_{n+1} when y_{n+1} is held as a cantilever and unit bending couple (but no shear) acts at y_n.

All of these are positive as defined.* They are illustrated pictorially in Ref. 1–9, and their calculation from the known flexural, shear, and torsional stiffness distributions of the wing is discussed. Because the structure is linear, the deflection and slope changes over the length l_n can be expressed as follows:

$$\bar{\alpha}_{n+1} = \bar{\alpha}_n - v_{F_n}\bar{S}_n - v_{M_n}\bar{M}_n$$

$$= \bar{\alpha}_n - v_{F_n}\omega^2\sum_{i=1}^{n}[\bar{P}_i\bar{w}_i + \bar{P}_i'\bar{\theta}_i] - v_{F_n}\bar{F}$$

$$- v_{M_n}\omega^2\sum_{i=1}^{n-1}[\bar{P}_i\bar{w}_i + \bar{P}_i'\bar{\theta}_i][y_i - y_n]$$

$$- v_{M_n}\bar{F}[y_1 - y_n] \tag{7-241}$$

$$\bar{w}_{n+1} = \bar{w}_n - l_n\bar{\alpha}_{n+1} - d_{F_n}\bar{S}_n - d_{M_n}\bar{M}_n$$

$$= \bar{w}_n - l_n\bar{\alpha}_{n+1} - d_{F_n}\omega^2\sum_{i=1}^{n}[\bar{P}_i\bar{w}_i + \bar{P}_i'\bar{\theta}_i]$$

$$- d_{F_n}\bar{F} - d_{M_n}\omega^2\sum_{i=1}^{n-1}[\bar{P}_i\bar{w}_i + \bar{P}_i'\bar{\theta}_i][y_i - y_n]$$

$$- d_{M_n}\bar{F}[y_1 - y_n] \tag{7-242}$$

$$\bar{\theta}_{n+1} = \bar{\theta}_n - v_{T_n}\bar{T}_n$$

$$= \bar{\theta}_n - v_{T_n}\omega^2\sum_{i=1}^{n}[\bar{Q}_i\bar{w}_i - \bar{Q}_i'\bar{\theta}_i] - v_{T_n}\bar{F}d \tag{7-243}$$

Here Eqs. 7–237 through 7–240 have been combined into each third member.

* Note that the structure need not be assumed a simple beam-rod. It is also not difficult to account for the effect of rotary inertia by adding a concentrated bending couple at y_n proportional to $\ddot{\alpha}_n$.

The computation proceeds inward from the tip station, where the slopes and deflection have the still-unknown values

$$\left.\begin{aligned} \bar{\alpha}_1 &\equiv \bar{\varphi} \\ \bar{w}_1 &\equiv \bar{\delta} \\ \bar{\theta}_1 &\equiv \bar{\psi} \end{aligned}\right\} \tag{7-244}$$

The linear character of the problem also permits these same quantities at any inboard station to be written

$$\left.\begin{aligned} \bar{\alpha}_n &= \bar{f}_{\varphi_n}\bar{\varphi} - \bar{f}_{\delta_n}\bar{\delta} - \bar{f}_{\psi_n}\bar{\psi} - \bar{f}_{F_n}\bar{F} \\ \bar{w}_n &= -\bar{g}_{\varphi_n}\bar{\varphi} + \bar{g}_{\delta_n}\bar{\delta} + \bar{g}_{\psi_n}\bar{\psi} - \bar{g}_{F_n}\bar{F} \\ \bar{\theta}_n &= \bar{h}_{\varphi_n}\bar{\varphi} - \bar{h}_{\delta_n}\bar{\delta} + \bar{h}_{\psi_n}\bar{\psi} - \bar{h}_{F_n}\bar{F} \end{aligned}\right\} \tag{7-245}$$

where the *amplitudes coefficients* $\bar{f}_{\varphi_n}, \cdots, \bar{h}_{F_n}$ are independent of $\bar{\varphi}$, $\bar{\delta}$, $\bar{\psi}$, and $\bar{F}$. (It is evident that $\bar{f}_{\varphi_1} = \bar{g}_{\delta_1} = \bar{h}_{\psi_1} = 1$, whereas the remaining coefficients vanish for $n = 1$.) Equations 7–245 are substituted into Eqs. 7–241 through 7–243, replacing n by $n + 1$ where appropriate. We thus arrive at three relations of the form

$$A\bar{\varphi} + B\bar{\delta} + C\bar{\psi} + D\bar{F} = 0 \tag{7-246}$$

Since $\bar{\varphi}$, $\bar{\delta}$, $\bar{\psi}$, and $\bar{F}$ are, as yet, arbitrary and independent, their coefficients must separately equal zero. Hence one is led to twelve equations among the amplitudes coefficients, which may be abbreviated as follows:

$$\left.\begin{aligned} \bar{f}_{\varphi_{n+1}} &= \bar{f}_{\varphi_n} + v_{F_n}\bar{G}_{\varphi_n} + v_{M_n}\bar{G}_{\varphi_n}{}' \\ \bar{f}_{\delta_{n+1}} &= \bar{f}_{\delta_n} + v_{F_n}\bar{G}_{\delta_n} + v_{M_n}\bar{G}_{\delta_n}{}' \\ \bar{f}_{\psi_{n+1}} &= \bar{f}_{\psi_n} + v_{F_n}\bar{G}_{\psi_n} + v_{M_n}\bar{G}_{\psi_n}{}' \\ \bar{f}_{F_{n+1}} &= \bar{f}_{F_n} + v_{F_n} + v_{M_n}\sum_{i=1}^{n-1} l_i \\ \bar{g}_{\varphi_{n+1}} &= \bar{g}_{\varphi_n} + l_n\bar{f}_{\varphi_{n+1}} - d_{F_n}\bar{G}_{\varphi_n} - d_{M_n}\bar{G}_{\varphi_n}{}' \\ \bar{g}_{\delta_{n+1}} &= \bar{g}_{\delta_n} + l_n\bar{f}_{\delta_{n+1}} - d_{F_n}\bar{G}_{\delta_n} - d_{M_n}\bar{G}_{\delta_n}{}' \\ \bar{g}_{\psi_{n+1}} &= \bar{g}_{\psi_n} + l_n\bar{f}_{\psi_{n+1}} - d_{F_n}\bar{G}_{\psi_n} - d_{M_n}\bar{G}_{\psi_n}{}' \\ \bar{g}_{F_{n+1}} &= \bar{g}_{F_n} - l_n\bar{f}_{F_{n+1}} + d_{F_n} + d_{M_n}\sum_{i=1}^{n-1} l_i \\ \bar{h}_{\varphi_{n+1}} &= \bar{h}_{\varphi_n} + v_{T_n}\bar{H}_{\varphi_n} \\ \bar{h}_{\delta_{n+1}} &= \bar{h}_{\delta_n} + v_{T_n}\bar{H}_{\delta_n} \\ \bar{h}_{\psi_{n+1}} &= \bar{h}_{\psi_n} + v_{T_n}\bar{H}_{\psi_n} \\ \bar{h}_{F_{n+1}} &= \bar{h}_{F_n} + v_{T_n}d \end{aligned}\right\} \tag{7-247}$$

The nine capital-letter factors in Eqs. 7–247 stand for

$$
\left.
\begin{aligned}
\bar{G}_{\varphi_n} &= \omega^2 \sum_{i=1}^{n} [\bar{P}_i \bar{g}_{\varphi_i} - \bar{P}_i' \bar{h}_{\varphi_i}] \\
\bar{G}_{\delta_n} &= \omega^2 \sum_{i=1}^{n} [\bar{P}_i \bar{g}_{\delta_i} - \bar{P}_i' \bar{h}_{\delta_i}] \\
\bar{G}_{\psi_n} &= \omega^2 \sum_{i=1}^{n} [\bar{P}_i \bar{g}_{\psi_i} + \bar{P}_i' \bar{h}_{\psi_i}] \\
\bar{H}_{\varphi_n} &= \omega^2 \sum_{i=1}^{n} [\bar{Q}_i \bar{g}_{\varphi_i} - \bar{Q}_i' \bar{h}_{\varphi_i}] \\
\bar{H}_{\delta_n} &= \omega^2 \sum_{i=1}^{n} [\bar{Q}_i \bar{g}_{\delta_i} - \bar{Q}_i' \bar{h}_{\delta_i}] \\
\bar{H}_{\psi_n} &= \omega^2 \sum_{i=1}^{n} [\bar{Q}_i \bar{g}_{\psi_i} + \bar{Q}_i' \bar{h}_{\psi_i}] \\
\bar{G}_{\varphi_n}' &= \sum_{i=1}^{n-1} l_i \bar{G}_{\varphi_i} \\
\bar{G}_{\delta_n}' &= \sum_{i=1}^{n-1} l_i \bar{G}_{\delta_i} \\
\bar{G}_{\psi_n}' &= \sum_{i=1}^{n-1} l_i \bar{G}_{\psi_i}
\end{aligned}
\right\} \quad (7\text{–}248)
$$

It is a simple task to show that the shear, moment, and torque amplitudes just to the left of y_n can be expressed in terms of the tip motion as follows:

$$\bar{S}_n = -\bar{G}_{\varphi_n}\bar{\varphi} + \bar{G}_{\delta_n}\bar{\delta} + \bar{G}_{\psi_n}\bar{\psi} + \bar{G}_{F_n}\bar{F} \quad (7\text{–}249)$$

$$\bar{M}_n = -\bar{G}_{\varphi_n}'\bar{\varphi} + \bar{G}_{\delta_n}'\bar{\delta} + \bar{G}_{\psi_n}'\bar{\psi} + \bar{G}_{F_n}'\bar{F} \quad (7\text{–}250)$$

$$\bar{T}_n = -\bar{H}_{\varphi_n}\bar{\varphi} + \bar{H}_{\delta_n}\bar{\delta} + \bar{H}_{\psi_n}\bar{\psi} + \bar{H}_{F_n}\bar{F} \quad (7\text{–}251)$$

where

$$
\left.
\begin{aligned}
\bar{G}_{F_n} &= 1 - \omega^2 \sum_{i=1}^{n} [\bar{P}_i \bar{g}_{F_i} + \bar{P}_i' \bar{h}_{F_i}] \\
\bar{G}_{F_n}' &= \sum_{i=1}^{n-1} l_i \bar{G}_{F_i} \\
\bar{H}_{F_n} &= d - \omega^2 \sum_{i=1}^{n} [\bar{Q}_i \bar{g}_{F_i} + \bar{Q}_i' \bar{h}_{F_i}]
\end{aligned}
\right\} \quad (7\text{–}252)
$$

For each flight altitude, Mach number, and forcing frequency, the computation implicit in Eqs. 7–247 and 7–248 can be started at $n = 2$ and repeated for each successive spanwise station until the centerline $n = b$ is reached. Reference 1–9 offers many illustrations of well-organized tabular procedures for this purpose. Since the aircraft is in the free-free condition, m_b and I_b represent one-half the mass and pitching moment of inertia of the central body. When the excitation is symmetrical at the two wingtips, the centerline boundary conditions require the vanishing of bending slope $\bar{\alpha}_b$, shear $\bar{S}_b$, and torque $\bar{T}_b$. In the foregoing notation, we therefore get the three equations

$$\bar{f}_{\varphi_b}\bar{\varphi} - \bar{f}_{\delta_b}\bar{\delta} - \bar{f}_{\psi_b}\bar{\psi} = \bar{f}_{F_b}F \qquad (7\text{-}253)$$

$$-\bar{G}_{\varphi_b}\bar{\varphi} + \bar{G}_{\delta_b}\bar{\delta} + \bar{G}_{\psi_b}\bar{\psi} = -\bar{G}_{F_b}F \qquad (7\text{-}254)$$

$$-\bar{H}_{\varphi_b}\bar{\varphi} + \bar{H}_{\delta_b}\bar{\delta} + \bar{H}_{\psi_b}\bar{\psi} = -\bar{H}_{F_b}F \qquad (7\text{-}255)$$

These constitute a simultaneous system with complex coefficients, which is easily solved for the tip motion amplitudes $\bar{\varphi}$, $\bar{\delta}$, and $\bar{\psi}$ per unit amplitude of F. $\bar{\varphi}/F$, $\bar{\delta}/F$, etc., are typical mechanical admittances or transfer functions for the wing; they depend on frequency ω, flight altitude, and Mach number. Through the extensive collection of relations derived above, we can obtain a mechanical admittance connecting any other load, stress, or property of the motion to F. These computations are well adapted for high-speed digital machinery.

One can also treat a case of antisymmetrical forcing, in which the wingtip shakers are operated 180° out of phase. The calculation then proceeds inward along the right wing just as before, but the boundary conditions consist of

$$\bar{w}_b = \bar{\theta}_b = \bar{M}_b = 0 \qquad (7\text{-}256)$$

These again provide the three relations needed to specify $\bar{\varphi}/F$, $\bar{\delta}/F$, and $\bar{\psi}/F$. Other types of root and tip restraint are studied in the various examples of Ref. 1-9. The majority of these deal with free vibration but are readily modified to include airloads.

(b) The discrete gust problem for a swept wing

Disturbances due to natural or artificial nonuniformity of the wind and pressure fields are a frequent occurrence during regular operation of airplanes and missiles. Although natural atmospheric turbulence can be fully described only by statistical means, it is a common design practice to compute the response to discrete gusts of preassigned shape and size. Such results are often very useful, when properly interpreted in the light of the designer's experience. Moreover, discrete-gust analyses apply directly to encounters with blast waves from large explosions such as bomb bursts. For these, considerable data on the structure, decay, and propagation rates are now available, so that accurate load calculations are a fairly straightforward matter (Ref. 7-53).

We can gain a crude idea of the relative severity of the gust and blast problems between one airplane and another by reference to the familiar *sharp-edged gust formula*, which estimates the normal acceleration Δn due to a sudden encounter with a wind component w_G normal to the wing planform:

$$\frac{\Delta n}{g} = \frac{\rho_\infty U(\partial C_L/\partial\alpha)w_G}{2W/S} \tag{7-257}$$

Here g is the gravitational constant and W/S the wing loading based on the area for which the lift-curve slope $\partial C_L/\partial\alpha$ is defined. Equation 7-257 assumes that the vehicle's structure is rigid, that gust envelopment takes place instantaneously, and that the incremental lift build-up is not delayed by unsteady flow effects. To improve upon these approximations, all of which are deviated from significantly in practice, aeroelasticians devised the concept of *alleviation factor*. This is a number by which the result of Eq. 7-257 or a similar formula for bending moment, stress, etc., is multiplied to yield the actual maximum of the quantity during the time-dependent response. The name is somewhat misleading, because the factor often exceeds unity on account of dynamic overshoot.

Wings with appreciable sweepback have a special interest, because their alleviation factors turn out smaller than those for straight surfaces with the same general stiffness level, aspect ratio, and wing loading. The reduction is due principally to the longer envelopment interval and to the unloading influence of negative angles of attack produced by bending deformations, the latter being the more important effect. In this section, we sketch how we would go about computing the response and alleviation on a large-aspect-ratio swept wing meeting the elementary "one-minus-cosine" gust, which forms the basis of certain military specifications. Our modal approach is suggested by the work of Pian, Lin, and co-authors (Refs. 7-54 and 7-55); but it also resembles most other treatments in the recent literature.

Consider the configuration of which the right half is pictured in Fig. 7-16. The vertical gust velocity which first strikes the vertex of the leading edge at $t = 0$ is described by

$$w_G(s_0) = \begin{cases} \frac{1}{2}w_0\left[1 - \cos\frac{\pi s_0}{s_G}\right], & (0 \leq s_0 \leq 2s_G) \\ 0, & (2s_G \leq s_0) \end{cases} \tag{7-258}$$

where $s_0 \equiv Ut/b_0$ is a dimensionless time based on midspan semichord, and s_G is a dimensionless half wavelength. Only bending deflections are permitted, through a set of mutually-orthogonal cantilever modes $\gamma_i(y)$. Freedom in rigid-body translation is introduced by means of the vertical coordinate $q_0(t)$ at the centerline, but the pitching moment of inertia is so large as to suppress any rotation about y. Hence the upward displacement of the elastic axis in Fig. 7–16 becomes

$$w(y,\,t) = q_0(t) + \sum_{i=1} \gamma_i(y)q_i(t) \qquad (7\text{–}259)$$

Only symmetrical modes are included because of the obvious symmetry of the input.

Following the Lagrangian approach of Secs. 2–6 and 3–4, we derive for the instantaneous kinetic energy of the right half of the system

$$\tau = \tfrac{1}{2}M_0\dot{q}_0{}^2 + \dot{q}_0\sum_{i=1}\dot{q}_i\int_0^l m(y)\gamma_i(y)\,dy + \tfrac{1}{2}\sum_{i=1}\dot{q}_i{}^2\int_0^l m(y)\gamma_i^2(y)\,dy \qquad (7\text{–}260)$$

Here $m(y)$ is the mass per unit distance perpendicular to the centerline, and M_0 the total mass. In terms of the natural frequencies ω_i of cantilever bending vibration, the potential energy is

$$U = \tfrac{1}{2}\sum_{i=1}q_i{}^2\omega_i{}^2\int_0^l m(y)\gamma_i^2(y)\,dy \qquad (7\text{–}261)$$

All generalized forces can be built up from the running lift $L(y,\,t)\,dy$ on the typical chordwise strip, shown shaded in Fig. 7–16. In the manner of Sec. 6–4, L is separated into a portion L_G due to the gust and a portion L^M caused by the dynamic response itself. Thus we find

$$Q_0 \equiv Q_0{}^M + Q_0{}^D = \int_0^l L^M(y,\,t)\,dy + \int_0^l L_G(y,\,t)\,dy \qquad (7\text{–}262)$$

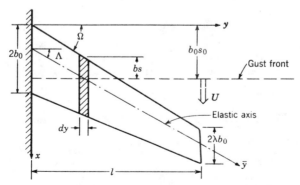

Fig. 7–16. Half span of a sweptback wing with rectilinear elastic axis, encountering a two-dimensional gust front normal to the flight direction. The taper is linear.

and

$$Q_i \equiv Q_i{}^M + Q_i{}^D = \int_0^l L^M(y, t)\gamma_i(y)\, dy + \int_0^l L_G(y, t)\gamma_i(y)\, dy,$$

$$(i = 1, 2, 3, \cdots) \tag{7–263}$$

Applying Eqs. 2–67 and substituting s_0 for physical time t, we obtain

$$M_0\left(\frac{U}{b_0}\right)^2 q_0{}'' + \left(\frac{U}{b_0}\right)^2 \sum_{j=1} q_j{}'' \int_0^l m(y)\gamma_j(y)\, dy = Q_0{}^M + Q_0{}^D \tag{7–264}$$

$$q_0{}''\left(\frac{U}{b_0}\right)^2 \int_0^l m(y)\gamma_i(y)\, dy + q_i{}''\left(\frac{U}{b_0}\right)^2 \int_0^l m(y)\gamma_i{}^2(y)\, dy$$

$$+ q_i\omega_i^2 \int_0^l m(y)\gamma_i{}^2(y)\, dy = Q_i{}^M + Q_i{}^D, \qquad (i = 1, 2, 3, \cdots) \tag{7–265}$$

primes denote differentiation with respect to s_0.

A unique aspect of the present development comes in the form of the disturbing airloads $Q_0{}^D$ and $Q_i{}^D$. According to Eq. 4–160, the aerodynamic operator which gives running lift due to a unit step gust striking the leading edge of any strip of the wing at $t = 0$ is

$$L_G = q(2b)a_\Lambda \frac{1}{U}\, \psi(s) \tag{7–266}$$

Here $s = Ut/b$, and a_Λ is a generalized lift-curve slope chosen, in this example, so as to make the total wing lift approach some preassigned value as $s \to \infty$, $\psi(s)$ being defined such that $\psi(\infty) = 1$. For instance, in incompressible flow over a swept wing of very large span and constant chord, $\psi(s)$ would be Küssner's function described in Sec. 4–6(a), and $a_\Lambda = 2\pi \cos \Lambda$.

The local semichord on the wing of Fig. 7–16 is

$$b(\tilde{y}) = b_0[1 - \delta\tilde{y}] \tag{7–267}$$

where $\tilde{y} \equiv y/l$, $\delta \equiv 1 - \lambda$, and λ is the taper ratio. A gust front which meets the vertex at $t = 0$ would begin to envelop station y at $t = \dfrac{y \tan \Omega}{U}$.

Hence s in Eq. 7–266 must be replaced by a delayed dimensionless time variable

$$s = \frac{U\left[t - \dfrac{y \tan \Omega}{U}\right]}{b} = \frac{s_0 - \beta\tilde{y}}{1 - \delta\tilde{y}} \tag{7–268}$$

where $\beta \equiv l \tan \Omega/b_0$. Leading-edge sweep is called Ω to distinguish it from Λ when the chord is not constant. Substituting for dynamic pressure, local semichord, and s in Eq. 7–266, we get

$$L_G = \rho_\infty U b_0 a_\Lambda [1 - \delta \tilde{y}] \psi \left(\frac{s_0 - \beta \tilde{y}}{1 - \delta \tilde{y}} \right) \tag{7–269}$$

Neglecting spanwise induction effects, the build-up of total lift on half the wing due to the step gust would therefore be

$$\begin{aligned}
(L_G)_{\text{TOT}} &= l \int_0^{\tilde{y}_1} L_G(\tilde{y}, s_0) \, d\tilde{y} \\
&= \rho_\infty U b_0 a_\Lambda l \int_0^{\tilde{y}_1} [1 - \delta \tilde{y}] \psi \left(\frac{s_0 - \beta \tilde{y}}{1 - \delta \tilde{y}} \right) d\tilde{y} \\
&\equiv \rho_\infty U b_0 a_\Lambda l \psi_0(s_0) \tag{7–270}
\end{aligned}$$

$\psi_0(s_0)$ is an indicial function for gust loading of the finite planform; the upper limit $\tilde{y}_1$ is used to distinguish whether the gust front has reached the tip station $y = l$

$$\tilde{y}_1 = \begin{cases} s_0/\beta, & (0 \le s_0 \le \beta) \\ 1, & (\beta \le s_0) \end{cases} \tag{7–271}$$

In the manner exemplified by Eq. 4–163, it is now possible to write the generalized lift force $Q_0{}^D$ associated with an arbitrary gust structure $w_G(s_0)$, which starts from zero at $s_0 = 0$, as follows:

$$\begin{aligned}
Q_0{}^D &= \rho_\infty U b_0 a_\Lambda l \int_0^{s_0} \frac{dw_G}{d\sigma} \int_0^{\tilde{y}_1} [1 - \delta \tilde{y}] \psi \left(\frac{s_0 - \sigma - \beta \tilde{y}}{1 - \delta \tilde{y}} \right) d\tilde{y} \, d\sigma \\
&= \rho_\infty U b_0 a_\Lambda l \int_0^{s_0} \frac{dw_G}{d\sigma} \psi_0(s_0 - \sigma) \, d\sigma \tag{7–272}
\end{aligned}$$

Here σ is a dummy variable, replacing s_0 in the convolution. Similarly,

$$\begin{aligned}
Q_i{}^D &= \rho_\infty U b_0 a_\Lambda l \int_0^{s_0} \frac{dw_G}{d\sigma} \int_0^{\tilde{y}_1} \gamma_i(\tilde{y}) [1 - \delta \tilde{y}] \psi \left(\frac{s_0 - \sigma - \beta \tilde{y}}{1 - \delta \tilde{y}} \right) d\tilde{y} \, d\sigma \\
&= \rho_\infty U b_0 a_\Lambda l \int_0^{s_0} \frac{dw_G}{d\sigma} \psi_i(s_0 - \sigma) \, d\sigma \tag{7–273}
\end{aligned}$$

where

$$\psi_i(s_0) = \int_0^{\tilde{y}_1} \gamma_i(\tilde{y}) [1 - \delta \tilde{y}] \psi \left(\frac{s_0 - \beta \tilde{y}}{1 - \delta \tilde{y}} \right) d\tilde{y} \tag{7–274}$$

Equation 7–258 can be substituted directly when that particular gust model is employed.

In Refs. 7–54 and 7–55, $\psi_0(s_0)$ and $\psi_1(s_0)$ (for $\gamma_1 = \tilde{y}^2$) are evaluated in closed form, using the rational-fraction and exponential approximations to the incompressible flow Küssner function, Eqs. 4–173b and 4–173a, respectively. The two approaches have comparable accuracies in practice, but the former is simpler and much better suited for analytical work. As an illustration, Ref. 7–54 gives

$$
\psi_0 \cong [s_0\delta - \beta]^2\left[\frac{0.09984}{(\beta + 0.32\delta)^3}\ln\left(\frac{0.32(\beta - s_0\delta)}{\beta(s_0 + 0.32)}\right)\right.
$$

$$
\left. + \frac{1.72}{(\beta + 2.5\delta)^3}\ln\left(\frac{2.5(\beta - s_0\delta)}{\beta(s_0 + 2.5)}\right)\right]
$$

$$
+ 0.312\frac{\dfrac{s_0}{\beta}\left[\beta^2 - \dfrac{\beta\delta}{2}(\beta - 0.32\delta)\dfrac{s_0}{\beta}\right]}{(\beta + 0.32\delta)^2}
$$

$$
+ 0.688\frac{\dfrac{s_0}{\beta}\left[\beta^2 - \dfrac{\beta\delta}{2}(\beta - 2.5\delta)\dfrac{s_0}{\beta}\right]}{(\beta + 2.5\delta)^2} \tag{7–275}
$$

for the range $0 \le s_0 \le \beta$ prior to when the front reaches the wingtip.

The generalized forces due to the motion follow a similar pattern, originating with the modified Eq. 4–150,

$$
L^M = q(2b)a_\Lambda\varphi(s) \tag{7–276}
$$

which represents running lift after unit step change of streamwise incidence at $s = 0$. (Once more $\varphi(s)$ is to be chosen so that $\varphi(\infty) = 1$.) The bending response of the wing will commence at the instant the gust hits the vertex. Therefore, $s = Ut/b$ in Eq. 7–276, and s can be expressed in terms of the reference time parameter s_0 by

$$
s = \frac{s_0}{1 - \delta\tilde{y}} \tag{7–277}
$$

Referred to properties of the midspan station,

$$
L^M = q(2b_0)a_\Lambda[1 - \delta\tilde{y}]\varphi\left(\frac{s_0}{1 - \delta\tilde{y}}\right) \tag{7–278}
$$

The total lift on the half-wing resulting from the unit step of angle of attack is

$$
(L^M)_{\text{TOT}} = l\int_0^1 L^M\,d\tilde{y} = q(2b_0)a_\Lambda l\int_0^1 [1 - \delta\tilde{y}]\varphi\left(\frac{s_0}{1 - \delta\tilde{y}}\right)d\tilde{y} \tag{7–279}
$$

Hence, for arbitrary angle-of-attack history, $\alpha(\tilde{y}, s_0)$, we get from Duhamel's principle

$$(L^M)_{\text{TOT}} = q(2b_0)a_\Lambda l \int_0^{s_0} \int_0^1 \frac{\partial \alpha_0 (\tilde{y}, \sigma)}{\partial \sigma} [1 - \delta \tilde{y}] \varphi\left(\frac{s_0 - \sigma}{1 - \delta \tilde{y}}\right) d\tilde{y} \, d\sigma$$

(7–280)

The instantaneous angle on the flexible wing receives contributions from the bending slope and vertical velocity. Thus, introducing Eq. 7–259 and noting that $\tilde{y} = (\bar{y}/l) \cos \Lambda$,

$$\alpha(\tilde{y}, t) = -\frac{1}{U}\frac{\partial w}{\partial t} - \sin \Lambda \frac{\partial w}{\partial \bar{y}}$$

$$= -\frac{\dot{q}_0}{U} - \frac{1}{U}\sum_{i=1}\gamma_i(\tilde{y})\dot{q}_i - \frac{\sin \Lambda \cos \Lambda}{l}\sum_{i=1}\frac{\partial \gamma_i}{\partial \tilde{y}}q_i \quad (7\text{–}281)$$

In terms of dimensionless time,

$$\alpha(\tilde{y}, s_0) = -\frac{q_0'}{b_0} - \frac{1}{b_0}\sum_{i=1}\gamma_i(\tilde{y})q_i' - \frac{\sin \Lambda \cos \Lambda}{l}\sum_{i=1}\frac{\partial \gamma_i}{\partial \tilde{y}}q_i(s_0) \quad (7\text{–}282)$$

Combining Eqs. 7–280 and 7–282, we get for the generalized force on the rigid-body degree of freedom

$$Q_0^M = -2qa_\Lambda l \int_0^{s_0} \int_0^1 \left[q_0''(\sigma) + \sum_{j=1}\gamma_j(\tilde{y})q_j''(\sigma)\right.$$

$$\left. + \sin \Lambda \cos \Lambda \frac{b_0}{l}\sum_{j=1}\frac{d\gamma_j}{d\tilde{y}}q_j'(\sigma)\right]$$

$$\times [1 - \delta \tilde{y}]\varphi\left(\frac{s_0 - \sigma}{1 - \delta \tilde{y}}\right) d\tilde{y} \, d\sigma \qquad (7\text{–}283)$$

In a similar fashion, the force on the ith bending degree of freedom must read

$$Q_i^M = -2qa_\Lambda l \int_0^{s_0} \int_0^1 \left[q_0''(\sigma) + \sum_{j=1}\gamma_j(\tilde{y})q_j''(\sigma)\right.$$

$$\left. + \sin \Lambda \cos \Lambda \frac{b_0}{l}\sum_{j=1}\frac{d\gamma_j}{d\tilde{y}}q_j'(\sigma)\right]$$

$$\times \gamma_i(\tilde{y})[1 - \delta \tilde{y}]\varphi\left(\frac{s_0 - \sigma}{1 - \delta \tilde{y}}\right) d\tilde{y} \, d\sigma \qquad (7\text{–}284)$$

When Eqs. 7–272, 7–273, 7–283, and 7–284 are substituted into Eqs. 7–264 and 7–265, we are able to construct the final set of equations of motion. These can be made dimensionless by division with $(2qa_\Lambda lb_0)$, whereupon the response to a given gust is found to depend in an important way on the mass ratio $(M_0/\rho_\infty b_0{}^2 la_\Lambda)$ and the mass distribution

$$m(\tilde{y})/\rho_\infty b_0{}^2$$

After the number and form of the modes are selected, we are confronted with the problem of solving a linear set of integrodifferential equations with s_0 as the independent variable. This may be done, in principle, by means of Laplace transformation, but Ref. 7–54 suggests the impracticability of such a procedure. In preference, these authors turn to stepwise schemes of numerical integration due to Houbolt (cf. Ref. 7–42).

Without going into full details, we reproduce as Fig. 7–17 the results of a typical computation of this type from Ref. 7–54. The actual mass distribution of the airplane is simplified to one concentrated mass $m_0 = 56$ slugs at each wingtip plus a mass $2(M_0 - m_0) = 3886.8$ slugs at the

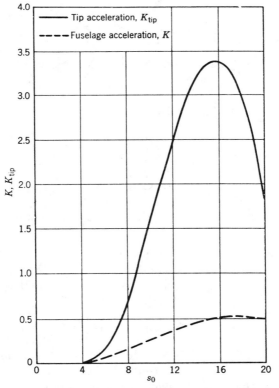

Fig. 7–17. Time histories of acceleration ratios at centerline and wingtip for a simplified sweptback wing. Details of the configuration are listed in the text.

fuselage centerline. A single parabolic bending mode $\gamma_1(\tilde{y}) = \tilde{y}^2$ was used. Other significant parameters are

$l = 58$ ft	$a_\Lambda = 6$
$\lambda = 0.418$	$\omega_1 = 7.55$ radians/sec
$\Omega = 36°\,40'$	Altitude $= 11,000$ ft
$\Lambda = 34°$	E.A.S. $= 460$ mph
$S = 1428$ ft^2	$s_G = 15$
$b_0 = 8.7$ ft	

Plotted in the figure versus s_0 are the acceleration ratios or alleviation factors K and K_{tip}. These are the ratios of the instantaneous centerline and wingtip accelerations, respectively, to the reference acceleration from Eq. 7–257 for a sharp-edged gust of amplitude equal to that of the one-minus-cosine gust used in the actual analysis. The peak value of K is 0.517, which clearly illustrates the ameliorative effect of sweep; however, the tip motion is quite violent in this case. Many other similar examples of swept-wing gust response will be found in Ref. 7–56.

(c) Dynamic response to continuous turbulence

To arrive at a more sophisticated formulation of the gust-loads problem than the foregoing, we must seek to generalize the ideas set forth in Secs. 6–3(c) and 6–4(c) to distributed-parameter systems. The foundations of such procedures are given in a paper by Liepmann (Ref. 7–57) and elaborated by Diederich (Ref. 7–58), Ribner (Ref. 7–59), Foss and McCabe (Ref. 7–60), and a host of other authors cited, *inter alia*, in the last reference. Perhaps the two most significant generalizations concern the influences of finite wingspan and three-dimensionality of atmospheric turbulence.

First, we need a statistical description of the turbulence itself. It has, of course, three Cartesian velocity components u_G, v_G, w_G, which are at any instant random functions of a position vector $\mathbf{r} = x\mathbf{i} + y\mathbf{j} + z\mathbf{k}$ in a coordinate system fixed relative to the mean motion of the air. Each component has a mean-square value; and they are further specified by a *correlation tensor*, composed of nine averaged products like $\overline{u_G(\mathbf{r}_1)u_G(\mathbf{r}_2)}$, $\overline{u_G(\mathbf{r}_1)v_G(\mathbf{r}_2)}$, $\cdots$. In the most general sense, the mean squares depend on position and direction, while the correlation tensor elements vary with the locations $\mathbf{r}_1$ and $\mathbf{r}_2$ of the individual points for which the observations are averaged. In the commonly assumed *homogeneous, isotropic* turbulence, however, $\overline{u_G{}^2} = \overline{v_G{}^2} = \overline{w_G{}^2} = $ const., and the correlations depend only on relative location $\Delta \mathbf{r} = \mathbf{r}_1 - \mathbf{r}_2$; moreover, it can be shown that only two correlations are distinct.

When linearized theory is applied to the present problem, the component w_G normal to the plane of the lifting surface is the only one producing first-order loads, and the history of these loads is determined by the variation of w_G in the x-y-plane. A partial statistical description of the dimensionless w_G/U is provided by the autocorrelation function $\varphi_{GG}(\tau, y_1 - y_2)$, which generalizes the $\varphi_{GG}(\tau)$ defined in Sec. 6–4(c) to account for the spanwise y-direction. Suppose we adopt Liepmann's symbol

$$R_{33}(x_1 - x_2, y_1 - y_2, z_1 - z_2) \equiv \overline{w_G(\mathbf{r}_1)w_G(\mathbf{r}_2)} \tag{7–285}$$

for the z-z-element of the correlation tensor. Then, if the turbulence embedded in the airstream U flowing past the flight vehicle undergoes no significant change during the brief period required to move one chord-length, it is evident that

$$\varphi_{GG}(\tau, y_1 - y_2) = R_{33}(x_1 - x_2, y_1 - y_2, 0) \tag{7–286}$$

Here

$$\tau = t_1 - t_2 = \frac{x_1 - x_2}{U} \tag{7–287}$$

is the time interval corresponding to the correlation distance $(x_1 - x_2)$ in the flight direction.

Associated with φ_{GG} and R_{33} are power spectra, which play an important role in practical computations. To understand their definitions it is helpful to start from the concept of the power spectrum of the correlation tensor. Since there are three coordinate directions, we expect that there should be three wavelengths or frequency variables appearing in this spectrum. They are defined in terms of the *wave numbers* k_1, k_2, and k_3, giving the number of waves per unit x-, y-, and z-distances, respectively, in a particular harmonic component of the turbulent structure. In a manner quite analogous to Eq. 6–75, we therefore define the z-z-element Φ_{33} of the power spectrum tensor by means of

$$R_{33}(x_1 - x_2, y_1 - y_2, z_1 - z_2) = \int_{-\infty}^{\infty} \int_{-\infty}^{\infty} \int_{-\infty}^{\infty} \Phi_{33}(k_1, k_2, k_3)$$
$$\times \exp\left[i\{k_1(x_1 - x_2) + k_2(y_1 - y_2) + k_3(z_1 - z_2)\}\right] dk_1\, dk_2\, dk_3 \tag{7–288}$$

The dependence of Φ_{33} on k_3 is of no interest in the present discussion and may be integrated out. Thus, the combination of Eqs. 7–286 and 7–288 yields

$$\varphi_{GG}(\tau, y_1 - y_2) = \int_{-\infty}^{\infty} \int_{-\infty}^{\infty} \Phi_{GG}(k_1, k_2)$$

$$\times \exp\{i[k_1(x_1 - x_2) + k_2(y_1 - y_2)]\}\, dk_1\, dk_2 \quad (7\text{–}289)$$

where

$$\Phi_{GG}(k_1, k_2) \equiv \int_{-\infty}^{\infty} \Phi_{33}(k_1, k_2, k_3)\, dk_3 \quad (7\text{–}290)$$

The mean-square gust velocity is

$$\frac{\overline{w_G^2}}{U^2} = \varphi_{GG}(0, 0) = \int_{-\infty}^{\infty} \int_{-\infty}^{\infty} \Phi_{GG}(k_1, k_2)\, dk_1\, dk_2 \quad (7\text{–}291)$$

Reference 7–57 suggests the following approximate formula for describing isotropic atmospheric turbulence:

$$\Phi_{33}(k_1, k_2, k_3) = \frac{2}{\pi} \frac{\overline{w_G^2}}{U^2} \lambda^5 \left\{ \frac{k_1^2 + k_2^2}{[1 + \lambda^2(k_1^2 + k_2^2 + k_3^2)]^3} \right\} \quad (7\text{–}292)$$

Here, as in Chap. 6, λ is the integral scale of turbulence. The integration in Eq. 7–290 leads to

$$\Phi_{GG}(k_1, k_2) = \frac{3}{4\pi} \frac{\overline{w_G^2}}{U^2} \lambda^4 \left\{ \frac{k_1^2 + k_2^2}{[1 + \lambda^2(k_1^2 + k_2^2)]^{5/2}} \right\} \quad (7\text{–}293)$$

(A second spanwise integration gives a result equivalent to Eq. 6–119.) Many studies (e.g., Ref. 7–62) suggest that, while Eqs. 7–292 and 7–293 may be inadequate for representing all the gustiness encountered over a long history of flight operations, the superposition of a few such structures with different intensities, scales, and durations will probably yield satisfactory design-load estimates.

As in the case of two-dimensional flow, spectral techniques greatly reduce the actual computing labor. We illustrate this point first with Liepmann's simple example of the gust-induced total lift $L(t)$ experienced by a wing of infinite mass and rigidity but arbitrary planform. The wing is exposed to a gust field $w_G(t, y)$, where y is now a coordinate normal to the flight direction and fixed to the aircraft so as to pass through the foremost point on the leading edge (cf. Fig. 7–16). Then $L(t)$ can be expressed in terms of a unit impulse response $h_{LG}(t, y)$ as in Eq. 6–77,

$$L(t) = \int_{-l}^{l} \int_{-\infty}^{t} \frac{w_G(\tau', y)}{U} h_{LG}(t - \tau', y)\, d\tau'\, dy \quad (7\text{–}294)$$

The wingspan is $2l$. Diederich's discussion in Ref. 7–58 of the exact meaning of $h_{LG}(t, y)$ and similar response functions will be found helpful; it represents the lift generated over the entire wing when a unit Dirac-delta function* of w_G/U sweeps past the y-axis at spanwise station y and time $t = 0$. The roles of t and x are interchangeable, since $x = x_0 + Ut$ for any gust element.

The upper time limit in Eq. 7–294 may be replaced by $+\infty$. The transformation $\tau = (t - \tau')$ thereupon gives an equation paralleling 6–78,

$$L(t) = \int_{-l}^{l} \int_{-\infty}^{\infty} \frac{w_G(t - \tau, y)}{U} h_{LG}(\tau, y) \, d\tau \, dy \tag{7-295}$$

Proceeding as in Eq. 6–79, it is easy to show that the autocorrelation function for lift is related to that for the gust as follows:

$$\varphi_{LL}(\tau) \equiv \lim_{T \to \infty} \frac{1}{2T} \int_{-T}^{T} L(t)L(t + \tau) \, dt$$

$$= \int_{-l}^{l} \int_{-l}^{l} \int_{-\infty}^{\infty} \int_{-\infty}^{\infty} h_{LG}(\tau_1, y_1)h_{LG}(\tau_2, y_2)$$

$$\times \varphi_{GG}(\tau + \tau_1 - \tau_2, y_1 - y_2) \, d\tau_1 \, d\tau_2 \, dy_1 \, dy_2 \tag{7-296}$$

The power spectral density of the lift is defined, somewhat as before, to be the Fourier transform of φ_{LL},

$$\varphi_{LL}(\tau) = \int_{-\infty}^{\infty} e^{i\omega\tau} \Phi_{LL}(\omega) \, d\omega \tag{7-297}$$

We introduce Eqs. 7–297 and 7–289 into Eq. 7–296, noting that it is possible to make the substitution

$$\omega(\tau + \tau_1 - \tau_2) = \omega\tau + k_1(x_1 - x_2) \tag{7-298}$$

where $k_1 = \omega/U$ is the wave number connected with the frequency ω of sinusoidal-gust passage. After breaking up the six integrals on the right by appropriately distributing the five factors in the exponential, we are led to

$$\int_{-\infty}^{\infty} e^{i\omega\tau} \Phi_{LL}(\omega) \, d\omega = \int_{-\infty}^{\infty} e^{i\omega\tau} \int_{-\infty}^{\infty} \Phi_{GG}(k_1, k_2) \Bigg\{ \left[\int_{-l}^{l} \int_{-\infty}^{\infty} h_{LG}(\tau_1 \, y_1) \right.$$

$$\times \exp\left[i(k_1 x_1 + k_2 y_1) \right] d\tau_1 \, dy_1 \bigg] \cdot \left[\int_{-l}^{l} \int_{-\infty}^{\infty} h_{LG}(\tau_2, y_2) \right.$$

$$\times \exp\left[-i(k_1 x_2 + k_2 y_2) \right] d\tau_2 \, dy_2 \bigg] \Bigg\} dk_2 \, dk_1 \tag{7-299}$$

---

* That is, unit volume under a plot of w_G/U vs. x, y when the volume integral is carried out over a very small x-y area in the vicinity of where the impulse is located.

The reasoning which leads from Eq. 6–80 to Eq. 6–83 can now be generalized. It is not difficult to prove that

$$H_{LG}(k_1, k_2) = \int_{-l}^{l} \int_{-\infty}^{\infty} h_{LG}(\tau_2, y_2) \exp\{-i[k_1 x_2 + k_2 y_2]\}\, d\tau_2\, dy_2$$

$$= \frac{1}{U} \int_{-l}^{l} \int_{-\infty}^{\infty} h_{LG}\left(\frac{x}{U}, y\right) \exp\{-i[k_1 x + k_2 y]\}\, dx\, dy$$

$$(7\text{–}300)$$

represents the complex amplitude of the simple harmonic lift force produced when the wing flies through a steady sinusoidal gust having wave numbers k_1 in the x-direction and $(-k_2)$ in the y-direction. (The crests of these gust waves would be oriented at an angle $\tan^{-1}(k_2/k_1)$ to the y-axis, and their wavelength is $2\pi/\sqrt{k_1^2 + k_2^2}$.) Moreover, the product of the two double integrals in brackets in Eq. 7–299 is the square of the magnitude of this mechanical admittance, $|H_{LG}(k_1, k_2)|^2$. Thus we come to the conclusion that the spectra of the lift and the gust are connected by the relatively simple relation

$$\Phi_{LL}(\omega) = \frac{1}{U} \int_{-\infty}^{\infty} \Phi_{GG}(k_1, k_2) |H_{LG}(k_1, k_2)|^2\, dk_2 \qquad (7\text{–}301)$$

where $k_1 = \omega/U$. The mean-square lift is

$$\overline{L^2} = \int_{-\infty}^{\infty} \Phi_{LL}(\omega)\, d\omega$$

$$= U\int_{-\infty}^{\infty} \Phi_{LL}\, dk_1 \int_{-\infty}^{\infty} \int_{-\infty}^{\infty} \Phi_{GG}(k_1, k_2) |H_{LG}(k_1, k_2)|^2\, dk_1\, dk_2 \quad (7\text{–}302)$$

Several interesting specializations of Eq. 7–302 are worked out in Ref. 7–57.

Probably the most valuable aspect of the foregoing development is its generality. In exactly the same sense that Eq. 6–83 relates any input-output pair in a lumped, linear system under stationary random forcing, so we can say that any time-dependent characteristic of the random-gust response of a rigid or flexible flight vehicle may be computed from the turbulence spectrum as in Eqs. 7–301 and 7–302. For instance, Foss and McCabe (Ref. 7–60) write the mean-square stress at any point in the structure as

$$\overline{\sigma^2} = \int_{-\infty}^{\infty} \int_{-\infty}^{\infty} \Phi_{GG}(k_1, k_2) |H_{\sigma G}(k_1, k_2)|^2\, dk_1\, dk_2 \qquad (7\text{–}303)$$

Their paper provides considerable detail on the practical problem of calculating $\overline{\sigma^2}$ and other properties of the response to the spectrum in Eq.

7-293.   A rigid airplane with freedom in vertical translation and an airplane with a large-aspect-ratio wing permitted only bending deformations are used as examples. Peak stresses and accelerations from discrete-gust analyses are compared with statistical results in the important special case where the scale λ is quite large relative to the wingspan. It is found that similar conclusions with respect to the size of design gust loads are reached from both computations in many practical cases, but that the discrete-gust approach may be seriously unconservative when the inherent aerodynamic damping on the airplane is low, as may occur in flight at very high altitudes.

As mentioned above, there now exist numerous studies of the statistical treatment of loading due to gusts, buffeting, ground operation, and other repetitive but individually unpredictable phenomena. The case for the rationality of this approach and a strong plea for its continued development will be found in Ref. 7-61. Several excellent illustrations, including comparisons with flight measurements, have been published, among them Ref. 7-63. It is a matter of distress to the present authors that space limitations preclude their providing a more extended treatment of this singularly important aspect of the influence of aeroelasticity on design.

Finally, the remark made in Chaps. 4 and 6 bears repeating, that unsteady aerodynamic operators are much more fully developed for sinusoidal motion than for other types of transients. This is even more true when three-dimensional flow and finite-span effects are being considered, as in the foregoing section. It is exactly quantities like the $H_{LG}(k_1, k_2)$ and $H_{\sigma G}(k_1, k_2)$ in Eqs. 7-301 through 7-303 that are the easiest to determine from the aerodynamic standpoint.

7-7   EIGENVALUES; FLUTTER OF ONE-DIMENSIONAL STRUCTURES

(a) The aeroelastic modes of a uniform straight wing

Most of the conclusions reached in Secs. 6-5 and 6-6 regarding the influence of various parameters on the aeroelastic modes of the typical section can be carried over, at least qualitatively, to the bending-torsion flutter of unswept, one-dimensional lifting surfaces. In view of this fact and of the prominence which flutter has received in the aeroelastic literature, we confine ourselves in this chapter to summarizing a number of interesting special results. The first of these concerns the method (originally suggested by Goland, Ref. 7-64, and refined by Runyan and Watkins in Ref. 7-66) for solving exactly the homogeneous differential equations of the uniform beam-rod in an airstream. So universal in practice is the use of generalized coordinates for analyzing flutter that engineers may overlook this "exact solution," its value as a standard of comparison for approximate calculations, and even its direct applicability to some simple wing and tail designs.

Consider a rectangular surface with straight elastic axis, whose stiffnesses EI, GJ, running inertias m, I_α, and running unbalance S_α (about the E.A., positive when C.G. is aft of E.A.) are independent of the spanwise coordinate y. Adapting the differential structural operators of the uniform beam and rod from Chap. 5, we derive for the general equations of motion

$$m \frac{\partial^2 w}{\partial t^2} - S_\alpha \frac{\partial^2 \theta}{\partial t^2} + EI \frac{\partial^4 w}{\partial y^4} = L(y, t) \qquad (7\text{--}304a)$$

$$I_\alpha \frac{\partial^2 \theta}{\partial t^2} - S_\alpha \frac{\partial^2 w}{\partial t^2} - GJ \frac{\partial^2 \theta}{\partial y^2} = M_y(y, t) \qquad (7\text{--}304b)$$

Representative of the boundary conditions which might be associated with Eqs. 7–304 are those for the cantilever,

$$w(0, t) = \frac{\partial w (0, t)}{\partial y} = \theta(0, t) = \frac{\partial^2 w (l, t)}{\partial y^2} = \frac{\partial^3 w (l, t)}{\partial y^3} = \frac{\partial \theta (l, t)}{\partial y} = 0$$
$$(7\text{--}305)$$

A characteristic-value problem is obtained when we replace the running lift L and moment M_y by means of the strip-theory equations (4–156) and (4–157), then substitute the motion coordinates w, θ themselves for $(-h)$ and α. If we examine the case of unsteady flow at arbitrary flight Mach number, the indicial functions φ, φ_q, etc., are so complicated that there is no hope of solving Eqs. 7–304 and 7–305 and in this way determining the frequencies, decay rates, and shapes of the coupled aeroelastic modes. Under the approximation of quasi-steady aerodynamics or of piston-theory operators (e.g., Eq. 4–183), however, we can reduce Eqs. 7–304 to a linear set with constant coefficients. Although the total order is six in y and four in t, a solution can be worked out—for example, by Laplace transforming on both the independent variables. It will consist of products of exponential functions of y and t, the complete set being denumerably infinite in correspondence with the infinite number of complex-frequency eigenvalues. The details have apparently never been carried through.

A less unwieldy problem, and one in which the exact forms of the unsteady operators may be retained, consists in seeking the stability boundary where a particular mode is simple harmonic,

$$w(y, t) = \bar{w}(y)e^{i\omega t} \qquad (7\text{--}306)$$

$$\theta(y, t) = \bar{\theta}(y)e^{i\omega t} \qquad (7\text{--}307)$$

When the supersonic airload notation, Eqs. 4–129 and 4–130, is inserted into Eqs. 7–304 along with Eqs. 7–306 and 7–307, we can cancel the common factor $e^{i\omega t}$ to obtain

$$EI \frac{d^4\bar{w}}{dy^4} - [m - 4\rho_\infty b^2(L_1 + iL_2)]\omega^2\bar{w}$$
$$- [-S_\alpha + 4\rho_\infty b^3(L_3 + iL_4)]\omega^2\bar{\theta} = 0 \quad (7\text{–}308a)$$

$$GJ \frac{d^2\bar{\theta}}{dy^2} + [-S_\alpha + 4\rho_\infty b^3(M_1 + iM_2)]\omega^2\bar{w}$$
$$+ [I_\alpha - 4\rho_\infty b^4(M_3 + iM_4)]\omega^2\bar{\theta} = 0 \quad (7\text{–}308b)$$

The pattern of Eqs. 7–308 is as follows:

$$\frac{d^4\bar{w}}{dy^4} - \alpha\bar{w} - \beta\bar{\theta} = 0 \quad (7\text{–}309a)$$

$$\frac{d^2\bar{\theta}}{dy^2} + \gamma\bar{w} + \delta\bar{\theta} = 0 \quad (7\text{–}309b)$$

Here

$$\alpha \equiv \frac{[m - 4\rho_\infty b^2(L_1 + iL_2)]}{EI}\omega^2 \quad (7\text{–}310)$$

and the remaining coefficients have similar expressions that are evident from the two sets of equations. The cantilever boundary conditions (7–305) become

$$\bar{w}(0) = \bar{w}'(0) = \bar{\theta}(0) = \bar{w}''(l) = \bar{w}'''(l) = \bar{\theta}'(l) = 0 \quad (7\text{–}311)$$

Equations 7–309 and 7–311 may formally be solved by means of a series of terms like $A_n e^{a_n y}$ (Ref. 7–64); but it is more efficient to adopt Laplace transformation on y,

$$\bar{b}(p) \equiv \int_0^\infty e^{-py} b(y)\, dy \quad (7\text{–}312)$$

Applying this operation to Eqs. 7–309 and using the boundary conditions at $y = 0$ to eliminate three of the residual terms arising from $d^4\bar{w}/dy^4$ and $d^2\bar{\theta}/dy^2$, we get

$$p^4\bar{\bar{w}}(p) - \alpha\bar{\bar{w}}(p) - \beta\bar{\bar{\theta}}(p) = pW_2 + W_3 \quad (7\text{–}313a)$$

$$p^2\bar{\bar{\theta}}(p) + \gamma\bar{\bar{w}}(p) + \delta\bar{\bar{\theta}}(p) = \Theta_1 \quad (7\text{–}313b)$$

The following abbreviations are employed on the right-hand sides:

$$\left.\begin{aligned} W_2 &\equiv \bar{w}''(0) \\ W_3 &\equiv \bar{w}'''(0) \\ \Theta_1 &\equiv \bar{\theta}'(0) \end{aligned}\right\} \quad (7\text{–}314)$$

The algebraic solution of Eqs. 7–313 reads

$$\bar{\bar{w}}(p) = \frac{[p^3 + \delta p]W_2 + [p^2 + \delta]W_3 + \beta\Theta_1}{\Delta(p)} \qquad (7\text{–}315a)$$

$$\bar{\bar{\theta}}(p) = \frac{[p^4 - \alpha]\Theta_1 - p\gamma W_2 - \gamma W_3}{\Delta(p)} \qquad (7\text{–}315b)$$

the denominator polynomial being

$$\Delta(p) = p^6 + \delta p^4 - \alpha p^2 + (\beta\gamma - \delta\alpha) \qquad (7\text{–}316)$$

The standard inversion of Eqs. 7–315 requires finding the roots of the bicubic $\Delta(p)$ and then expressing $\bar{w}(y)$ and $\bar{\theta}(y)$ as series of exponentials in y by means of the familiar technique for rational fractions (Ref. 6–4, Chap. 6). To avoid the excessive labor thus encountered, Runyan and Watkins (Ref. 7–66) proposed to take advantage of the relationship between the power series expansion of a function in y and its transform in $(1/p)$. They write

$$\frac{1}{\Delta(p)} = \frac{1}{p^6} \sum_{n=0}^{\infty} \frac{T_n}{p^{2n}} \qquad (7\text{–}317)$$

whence there is no difficulty in showing that

$$\left.\begin{array}{l} T_0 = 1 \\ T_1 = -\delta \\ T_2 = \alpha + \delta^2 \\ T_3 = -\delta^3 - \delta\alpha - \beta\gamma \\ T_n = -\delta T_{n-1} + \alpha T_{n-2} + (\alpha\delta - \beta\gamma)T_{n-3}, \qquad (n \geq 3) \end{array}\right\} \qquad (7\text{–}318)$$

Because of rapid convergence, very few of the T_n need to be computed in practice.

Once Eq. 7–317 is substituted into Eqs. 7–315, all terms in both formulas are of the form A_m/p^m and can be inverted through

$$\mathscr{L}^{-1}\left\{\frac{1}{p^m}\right\} = \frac{1}{(m-1)!} y^{m-1} \qquad (7\text{–}319)$$

We are thus led to something like the following:

$$\bar{w}(y) = h_1(y)W_2 + h_2(y)W_3 + h_3(y)\Theta_1 \qquad (7\text{–}320a)$$

$$\bar{\theta}(y) = g_1(y)W_2 + g_2(y)W_3 + g_3(y)\Theta_1 \qquad (7\text{–}320b)$$

where, for example,

$$h_1(y) = \sum_{n=0}^{\infty} \frac{T_n y^{2n+2}}{(2n+2)!} + \delta \sum_{n=0}^{\infty} \frac{T_n y^{2n+4}}{(2n+4)!} \qquad (7\text{–}321)$$

The remaining polynomials are listed in Ref. 7–66; their convergence is obvious.

The eigenvalues of the problem can be constructed by resort to the boundary conditions at the wingtip (the last three members of Eq. 7–311). They produce the homogeneous equations

$$\left.\begin{array}{c} h_1''(l)W_2 + h_2''(l)W_3 + h_3''(l)\Theta_1 = 0 \\ h_1'''(l)W_2 + h_2'''(l)W_3 + h_3'''(l)\Theta_1 = 0 \\ g_1'(l)W_2 + g_2'(l)W_3 + g_3'(l)\Theta_1 = 0 \end{array}\right\} \quad (7\text{–}322)$$

Clearly, if the solution is to be nontrivial, the denominator determinant of Eq. 7–322 must vanish,

$$\begin{vmatrix} h_1''(l) & h_2''(l) & h_3''(l) \\ h_1'''(l) & h_2'''(l) & h_3'''(l) \\ g_1'(l) & g_2'(l) & g_3'(l) \end{vmatrix} = 0 \quad (7\text{–}323)$$

Equation 7–323 is nothing but the flutter determinant. It is of interest to observe that its order is just one-half that of the original system of ordinary differential equations, whereas when modal methods are employed the determinant has the same order as the preselected number of generalized coordinates. In terms of the flutter eigenvalues, Eq. 7–323 has a much higher degree, controlled by where the series (7–317) is truncated. Indeed, factorization is generally impossible, since the T_n (through α, β, γ, and δ) are complex, transcendental functions of the reduced frequency $k \equiv \omega b/U$.

A trial and error process must be used in practice to ascertain flutter speeds and frequencies. Given a set of wing parameters, flight Mach number, and ambient density, for instance, one can compute the real and imaginary parts $\mathscr{R}e$ and $\mathscr{I}m$ of the determinant (7–323) for several values of ω and k. Plotting $\mathscr{R}e$ and $\mathscr{I}m$ versus ω for each k, we find a set of ω-k combinations at which $\mathscr{R}e$ and $\mathscr{I}m$ vanish separately. A cross plot of $\mathscr{R}e = 0$ and $\mathscr{I}m = 0$ on the ω-k-plane finally yields one or more intersections, each corresponding to a speed and frequency of neutral stability. If desired, the aeroelastic mode shape at flutter can be expressed as two complex power series in y by substituting into Eqs. 7–320 the values of T_n calculated from the known eigenvalues. It should be evident to the reader that the generality of these results can be increased, when making parametric studies and the like, by recasting Eqs. 7–309 and 7–311 in dimensionless terms before proceeding.

A very important extension of the exact solution has been presented (Refs. 7–65, 7–66), which accounts for a concentrated mass attached at an arbitrary spanwise and chordwise location on the uniform wing. Such a mass causes discontinuities in the vibratory shear and torque, but these are efficiently handled by the Laplace transformation and do not change the order of the flutter determinant. The special significance of this development lies in the fact that the exact method is quite successful in predicting measured flutter boundaries (Ref. 7–66) even for a rather large mass placed forward along the chord at outboard stations, a situation where the conventional modal solution may be seriously unconservative (Ref. 7–67). On the other hand, modal calculations on *bare* wings are generally well confirmed, at least at subsonic speeds, so that the analyst can have confidence in them for many situations which do not permit exact treatment.

We might also mention that, in addition to its clear utility for cases of supersonic and hypersonic flow, the foregoing scheme is capable of being adapted to a uniform swept wing which behaves as a beam-rod with a rectilinear E.A. Aerodynamic strip theory can be introduced after the fashion of Sec. 4–4(e), but a minor complication arises because the differential equations replacing (7–308) will no longer contain only even-order derivatives of the dependent variables.

(b) Solution of the flutter problem by generalized coordinates

Flutter prediction by superposition of normal vibration modes or artificially defined generalized coordinates is the familiar procedure used regularly in the United States at least since the publication of the Smilg-Wasserman report (Ref. 4–56), and probably longer in certain European countries such as Great Britain (cf. Refs. 6–19, 7–68). Here we describe its application to computing the primary bending-torsion stability boundary for an essentially unswept ($|\Lambda| < 15°$) lifting surface with one-dimensional structural properties.

Techniques for setting up the equations of motion are reviewed in Sec. 3–5, whence we may take Eq. 3–118, for example, and specialize it to the case where all loads and deflections occur only in the z-direction. It is generally agreed that the "camber-bending" type of chordwise deformation (Ref. 7–69) has little influence on the flutter of moderate-to large-aspect-ratio straight wings; therefore, we deal with modes $\gamma_i(x, y)$ which are no more than linear in the streamwise variable x. As in foregoing sections, the entire deflection consists of a flexure $w(y, t)$ and twist $\theta(y, t)$. There are two common ways of defining the generalized coordinates: on a structure with a true elastic axis, it is aerodynamically convenient to adopt "uncoupled" modes of pure bending and torsion; whereas, for more complicated elastic configurations, the orthogonal normal modes are more suitable. Taking up the former first, we choose a total of n coordinates

$q_i(t)$, r of them associated with bending modes $f_{w_i}(y)$, the rest with torsion $f_{\theta_i}(y)$:

$$w(y, t) = \sum_{i=1}^{r} f_{w_i}(y)q_i(t) \qquad (7\text{--}324)$$

$$\theta(y, t) = \sum_{i=r+1}^{n} f_{\theta_i}(y)q_i(t) \qquad (7\text{--}325)$$

Since the f-functions are dimensionless, the q_i have dimensions of length for $1 \leq i \leq r$ but are angles for $r < i$. The instantaneous kinetic energy of one wing, lying between a root at $y = 0$ and tip at $y = l$, is easily shown to be

$$\begin{aligned}
T = &\frac{1}{2} \sum_{i=1}^{r} \sum_{j=1}^{r} \dot{q}_i \dot{q}_j \int_0^l m f_{w_i} f_{w_j} \, dy \\
&- \sum_{i=1}^{r} \sum_{j=r+1}^{n} \dot{q}_i \dot{q}_j \int_0^l S_\alpha f_{w_i} f_{\theta_j} \, dy \\
&+ \frac{1}{2} \sum_{i=r+1}^{n} \sum_{j=r+1}^{n} \dot{q}_i \dot{q}_j \int_0^l I_\alpha f_{\theta_i} f_{\theta_j} \, dy \qquad (7\text{--}326)
\end{aligned}$$

From Eq. 7–326 we can immediately deduce the form of each term appearing in Eq. 3–118. The running inertial properties m, S_α, I_α have the same definitions as in the preceding subsection, but they vary with y and may even contain delta functions to represent a fuselage, nacelles, or other concentrated masses. Since y is supposed to be an E.A., it follows that the potential energies of flexural and torsional deformation can be separately computed and added. Furthermore, it is customary to eliminate all elastic coupling by assuming a pseudo-orthogonality among the shapes $f_{w_1}, \cdots, f_{w_r}$ and $f_{\theta_{r+1}}, \cdots, f_{\theta_n}$, so that each of their contributions to U is independent. Then, as set forth on pp. 559–560 of Ref. 1–2, a series of natural frequencies ω_{w_i} and ω_{θ_i} can be invented by imagining a constrained free vibration in each mode. These considerations lead to

$$U = \frac{1}{2} \sum_{i=1}^{r} q_i^2 \omega_{w_i}^2 \int_0^l m f_{w_i}^2 \, dy + \frac{1}{2} \sum_{i=r+1}^{n} q_i^2 \omega_{\theta i}^2 \int_0^l I_\alpha f_{\theta i}^2 \, dy \quad (7\text{--}327)$$

Since the only external loads are lift $L(y, t)$ and moment $M_y(y, t)$, the generalized forces on the various degrees of freedom are

$$Q_i(t) = \begin{cases} \displaystyle\int_0^l L(y, t)f_{w_i}(y) \, dy, & (1 \leq i \leq r) \\[2ex] \displaystyle\int_0^l M_y(y, t)f_{\theta_i}(y) \, dy, & (r + 1 \leq i \leq n) \end{cases} \qquad (7\text{--}328)$$

Before writing down the equations of motion, we introduce structural friction in the customary fashion. Since this phenomenon may have a strong influence on the location of the stability boundary or even suppress the instability altogether in certain critical cases, its inclusion is much more important when treating flutter than when treating forced response. As discussed in Ref. 1–2, friction can be satisfactorily approximated with small forces, opposite in phase to the velocity but distributed in direct proportion to the elastic restoring force due to each mode of bending, and with a similarly distributed torque on each twisting mode. One thus represents a damping effect which is observed to destroy an amount of mechanical energy per cycle of vibration that is proportional to the square of the amplitude but independent of frequency. All this is accomplished by multiplying each frequency $\omega_{w_i}{}^2$ or $\omega_{\theta_i}{}^2$ in Eq. 7–327 with a complex constant $[1 + ig_{w_i}]$ or $[1 + ig_{\theta_i}]$. The g-factors are positive, with values in the range 0.005–0.05, depending on the structural material, design, and fabrication; they can often be estimated from free-vibration tests.

Given Eq. 7–326, Eqs. 7–328, and Eq. 7–327 with the added structural damping, one applies Lagrange's procedure or Eq. 3–118 to obtain the following:

$$\sum_{i=1}^{r} \ddot{q}_i \int_0^l m f_{w_i} f_{w_j}\,dy - \sum_{i=r+1}^{n} \ddot{q}_i \int_0^l S_\alpha f_{\theta_i} f_{w_j}\,dy$$

$$+ q_j \omega_{w_j}{}^2 [1 + ig_{w_j}] \int_0^l m f_{w_j}{}^2\,dy = \int_0^l L(y,t) f_{w_j}\,dy, \qquad (1 \le j \le r)$$

$$(7\text{--}329)$$

$$-\sum_{i=1}^{r} \ddot{q}_i \int_0^l S_\alpha f_{w_i} f_{\theta_j}\,dy + \sum_{i=r+1}^{n} \ddot{q}_i \int_0^l I_\alpha f_{\theta_i} f_{\theta_j}\,dy$$

$$+ q_j \omega_{\theta_j}{}^2 [1 + ig_{\theta_j}] \int_0^l I_\alpha f_{\theta_j}{}^2\,dy = \int_0^l M_y(y,t) f_{\theta_j}\,dy, \quad (r+1 \le j \le n)$$

$$(7\text{--}330)$$

As a rule, the integrals of $m f_{w_i} f_{w_j}$ and $I_\alpha f_{\theta_i} f_{\theta_j}$ for $i \ne j$ are made to vanish by means of the same pseudo-orthogonalization of the artificial modes that is involved in Eq. 7–327.

Since neutral stability is under investigation, we next assume each generalized coordinate to have the form $\bar{q}_j e^{i\omega t}$. It is then possible to arrive at formulas for the running lift and moment, based on any of the linear strip-theory aerodynamic operators from Sec. 4–5 or the three-dimensional operators from Sec. 4–7. Inasmuch as this sort of computation has been fully developed elsewhere, let us merely point out that, in the homogeneous case, each of the generalized forces can ultimately be reduced to the form

$$
Q_j = \begin{cases} 4\rho_\infty \omega^2 b_R^{\;3} l \left[\sum_{i=1}^{r} \bar{Q}_{ji}\dfrac{\bar{q}_i}{b_R} + \sum_{i=r+1}^{n} \bar{Q}_{ji}\bar{q}_i \right] e^{i\omega t}, \quad (1 \le j \le r) \\[2em] 4\rho_\infty \omega^2 b_R^{\;4} l \left[\sum_{i=1}^{r} \bar{Q}_{ji}\dfrac{\bar{q}_i}{b_R} + \sum_{i=r+1}^{n} \bar{Q}_{ji}\bar{q}_i \right] e^{i\omega t}, \quad (r+1 \le j \le n) \end{cases}
$$

$$(7\text{–}331)$$

Here b_R is the semichord at some reference station such as the wing root, and the $\bar{Q}_{ji}$ are dimensionless unit generalized forces, each depending on wing geometry, Mach number, reduced frequency $\omega b_R/U$, and the two mode shapes that are identified by its subscripts.*

When Eqs. 7–331 are introduced into Eqs. 7–329 and 7–330 and the common factor $e^{i\omega t}$ is cancelled, we can divide the bending equations by $4\rho_\infty \omega^2 b_R^{\;3} l$ and the torsional equations by $4\rho_\infty \omega^2 b_R^{\;4} l$ to obtain the following:

$$
\sum_{i=1}^{r} \frac{\bar{q}_i}{b_R} \left\{ \mu_R \int_0^1 \frac{m}{m_R} f_{w_i} f_{w_j}\, d\tilde{y} + \bar{Q}_{ji} \right\}
$$

$$
- \frac{\bar{q}_j}{b_R} \mu_R [1 + ig_{w_j}] \left(\frac{\omega_{w_j}}{\omega_R}\right)^2 \left(\frac{\omega_R}{\omega}\right)^2 \int_0^1 \frac{m}{m_R} f_{w_j}^{\;2}\, d\tilde{y}
$$

$$
+ \sum_{i=r+1}^{n} \bar{q}_i \left\{ -\mu_R x_{\alpha_R} \int_0^1 \frac{S_\alpha}{S_{\alpha_R}} f_{\theta_i} f_{w_j}\, d\tilde{y} + \bar{Q}_{ji} \right\} = 0, \quad (1 \le j \le r)
$$

$$(7\text{–}332)$$

$$
\sum_{i=1}^{r} \frac{\bar{q}_i}{b_R} \left\{ -\mu_R x_{\alpha_R} \int_0^1 \frac{S_\alpha}{S_{\alpha_R}} f_{w_i} f_{\theta_j}\, d\tilde{y} + \bar{Q}_{ji} \right\}
$$

$$
+ \sum_{i=r+1}^{n} \bar{q}_i \left\{ \mu_R r_{\alpha_R}^{\;2} \int_0^1 \frac{I_\alpha}{I_{\alpha_R}} f_{\theta_i} f_{\theta_j}\, d\tilde{y} + \bar{Q}_{ji} \right\}
$$

$$
- \bar{q}_j \mu_R r_{\alpha_R}^{\;2} [1 + ig_{\theta_j}] \left(\frac{\omega_{\theta_j}}{\omega_R}\right)^2 \left(\frac{\omega_R}{\omega}\right)^2 \int_0^1 \frac{I_\alpha}{I_{\alpha_R}} f_{\theta_j}^{\;2}\, d\tilde{y} = 0, \quad (r+1 \le j \le n)
$$

$$(7\text{–}333)$$

Properties of the reference station are marked with subscripts 'R'; thus,

$$
\mu_R \equiv \frac{m_R}{4\rho_\infty b_R^{\;2}}
$$

$$(7\text{–}334)$$

As before, x_α and r_α are the static unbalance and radius of gyration about the E.A., made dimensionless with respect to local semichord b.

* Note that $\bar{Q}_{ji} \neq \bar{Q}_{ij}$, as is easily proved by example.

The flutter determinant has order n and is formed from the coefficients of the $\bar{q}_i/b_R$ and $\bar{q}_i$. For a given configuration, ambient air density, and flight Mach number, this complex determinant amounts to two real equations for speed U_F and frequency ω_F. There are, at most, n pairs of physically significant roots, the one of greatest practical interest being that which yields the lowest speed.

For illustrative purposes, consider a cantilever with the calculation based on a single bending mode $f_w(y)$ and a single torsion mode $f_\theta(y)$. The generalized coordinates are q_1 and q_2, respectively. If we then neglect spanwise induction, it is an easy matter to express the generalized aerodynamic forces in supersonic notation as follows:

$$
\left.
\begin{aligned}
\bar{Q}_{11} &= -\int_0^1 \left(\frac{b}{b_R}\right)^2 [L_1 + iL_2] f_w^2 \, d\tilde{y} \\[6pt]
\bar{Q}_{12} &= \int_0^1 \left(\frac{b}{b_R}\right)^3 [L_3 + iL_4] f_w f_\theta \, d\tilde{y} \\[6pt]
\bar{Q}_{21} &= \int_0^1 \left(\frac{b}{b_R}\right)^3 [M_1 + iM_2] f_w f_\theta \, d\tilde{y} \\[6pt]
\bar{Q}_{22} &= -\int_0^1 \left(\frac{b}{b_R}\right)^4 [M_3 + iM_4] f_\theta^2 \, d\tilde{y}
\end{aligned}
\right\}
\tag{7-335}
$$

If desired, the dimensionless coefficients in Eqs. 7-335 may be multiplied by overall functions of y or otherwise manipulated to approximate three-dimensional aerodynamic effects. The determinantal flutter equation finally reads

$$
\begin{vmatrix}
\begin{aligned}
&\left\{\mu_R\left[1 - \left(\frac{\omega_w}{\omega_\theta}\right)^2\left(\frac{\omega_\theta}{\omega}\right)^2(1 + ig_w)\right]\int_0^1 \frac{m}{m_R} f_w^2 \, d\tilde{y}\right. \\
&\left. - \int_0^1 \left(\frac{b}{b_R}\right)^2 [L_1 + iL_2] f_w^2 \, d\tilde{y}\right\}
\end{aligned}
&
\begin{aligned}
&\left\{-\mu_R x_{\alpha_R}\int_0^1 \frac{S_\alpha}{S_{\alpha_R}} f_w f_\theta \, d\tilde{y}\right. \\
&\left. + \int_0^1 \left(\frac{b}{b_R}\right)^3 [L_3 + iL_4] f_w f_\theta \, d\tilde{y}\right\}
\end{aligned}
\\[20pt]
\begin{aligned}
&\left\{-\mu_R x_{\alpha_R}\int_0^1 \frac{S_\alpha}{S_{\alpha_R}} f_w f_\theta \, d\tilde{y}\right. \\
&\left. + \int_0^1 \left(\frac{b}{b_R}\right)^3 [M_1 + iM_2] f_w f_\theta \, d\tilde{y}\right\}
\end{aligned}
&
\begin{aligned}
&\left\{\mu_R r_{\alpha_R}^2\left[1 - \left(\frac{\omega_\theta}{\omega}\right)^2(1 + ig_\theta)\right]\int_0^1 \frac{I_\alpha}{I_{\alpha_R}} f_\theta^2 \, d\tilde{y}\right. \\
&\left. - \int_0^1 \left(\frac{b}{b_R}\right)^4 [M_3 + iM_4] f_\theta^2 \, d\tilde{y}\right\}
\end{aligned}
\end{vmatrix}
= 0 \tag{7-336}
$$

For flutter calculation based on true normal modes, $\varphi_i(x, y)$, the y-axis is simply a convenient structural reference line and need not play any special role in decoupling bending and torsion. No longer can the potential energy be cast in such a simple form as Eq. 7–327 except by taking advantage of the general orthogonality of the φ_i's. We write the normal displacement of the wing as

$$w(x, y, t) = \sum_{i=1} \varphi_i(x, y)\xi_i(t)$$
$$= e^{i\omega t} \sum_{i=1} \varphi_i(x, y)\bar{\xi}_i \qquad (7\text{–}337)$$

Here the φ_i are usually no more than linear in x. The flutter equations are

$$M_j[\omega_j^2 - \omega^2]\bar{\xi}_j = \bar{\Xi}_j{}^M, \qquad (j = 1, 2, 3, \cdots) \qquad (7\text{–}338)$$

where

$$M_j = \iint\limits_{\text{planform}} \rho(x, y)\varphi_j^2(x, y)\, dx\, dy \qquad (7\text{–}339)$$

and

$$\bar{\Xi}_j{}^M = \iint\limits_{\text{planform}} \Delta\bar{p}^M(x, y)\varphi_j(x, y)\, dx\, dy \qquad (7\text{–}340)$$

The only coupling among the degrees of freedom is aerodynamic and arises because the pressure difference $\Delta\bar{p}^M(x, y)e^{i\omega t}$ is linearly related to all of the $\bar{\xi}_j$. By analogy with Eq. 7–331, we can say

$$\bar{\Xi}_j{}^M = \sum_{i=1} \overline{Q_{ji}{}'}\bar{\xi}_i \qquad (7\text{–}341)$$

Evidently the flutter determinant is

$$\begin{vmatrix} \{M_1[\omega^2 - \omega_1^2] + \overline{Q_{11}{}'}\} & \overline{Q_{12}{}'} & \overline{Q_{13}{}'} & \cdots & \cdot \\ \overline{Q_{21}{}'} & \{M_2[\omega^2 - \omega_2^2] + \overline{Q_{22}{}'}\} & \overline{Q_{23}{}'} & & \cdot \\ \overline{Q_{31}{}'} & \overline{Q_{32}{}'} & \{M_3[\omega^2 - \omega_3^2] + \overline{Q_{33}{}'}\} & & \cdot \\ \cdot & & & & \cdot \\ \cdot & \cdot & & \cdots & \cdot \end{vmatrix} = 0$$

$$(7\text{–}342)$$

As in the case of Eq. 7–336, this can be regarded as the determinant of a sum of inertial, structural, and aerodynamic matrices. The order is equal to the number of modes needed to assure an accurate solution, which may vary from two on simple cantilevers to more than twenty on large, flexible airplanes with elastically mounted external stores. As a rule, a division can be made between modes which are symmetrical and antisymmetrical with respect to the vehicle's midplane, for no coupling exists between these.

Structural friction is readily introduced into Eq. 7–342 by replacing each ω_j^2 with $\omega_j^2[1 + ig_j]$. Also it is good practice to render the flutter equations dimensionless, thus reducing the amount of computation, whenever making parametric studies or extensive configuration changes. If strip theory or its equivalent is used for the aerodynamic operators, we find that the normal-mode scheme is the more laborious. This is because each shape function $\varphi_i(x, y)$ has both bending and torsion contributions, so that each of the $\overline{Q_{ij}}'$ contains all of the coefficients $L_1, L_2, \cdots, M_4$. By contrast, Eqs. 7–335 show how these coefficients appear only two at a time when artificially decoupled modes are employed. The wide availability of high-speed computers has tended lately to minimize this distinction as a factor in deciding the best way to proceed with a particular flutter analysis.

Another question which has lost some of its erstwhile significance is how to go about the mathematical determination of the eigenvalues of the flutter determinant. We have already reviewed this to some extent in Sec. 6–5. All that needs to be added is that the trial-and-error nature of the computation, which results even in the simplest cases from the transcendental dependence of the aerodynamic forces on reduced frequency and Mach number, is reinforced when modal methods and larger numbers of generalized coordinates are used.

Among the various techniques of solution, the U-g-method perhaps merits special mention because it replaces the problem of finding pairs of real roots of two simultaneous equations by the somewhat easier calcu-lation of complex roots of a single polynomial with complex coefficients. The latter factorization can be done in closed mathematical form for polynomials up to the fourth degree, which correspond to systems with four generalized coordinates. This advantage is gained only at the price of a certain artificiality: the structural damping coefficients of all the degrees of freedom are assumed equal to a single constant g, which is then adopted as one of the unknowns. Under these assumptions, one observes from Eq. 7–336 or any other dimensionless modal flutter determinant that

$$\Omega \equiv \left(\frac{\omega_R}{\omega}\right)^2 [1 + ig] \qquad (7\text{–}343)$$

occurs linearly in each term of the principal diagonal. (The reference frequency ω_R would be ω_θ in Eq. 7–336.) Furthermore, these are the only locations where ω and g appear explicitly in the equation. If the configuration, density, Mach number, and *reduced* frequency $\omega b_R/U$ are picked in advance, the determinant expands into a polynomial equation. Special procedures have been developed (e.g., Ref. 7–70) for getting the roots directly from the determinant when its order is too large to allow convenient expansion. The output of a U-g-calculation is generally displayed as a plot of artificial structural damping versus air-speed with the actual U_F occurring at the point where g equals the true

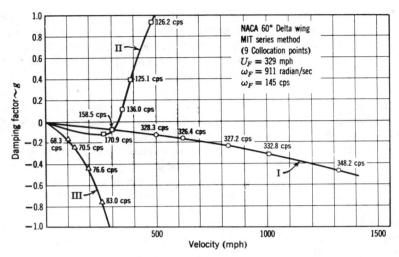

Fig. 7–18. Typical U-g curves from flutter analysis of a $60°$ delta wing.

friction. Figure 7–18 shows an example of such curves. It is often but not always a fact that the steepness of the slope at the flutter point is a qualitative measure of the violence of the onset of flutter with increasing flight speed. Zisfein and Frueh (Ref. 6–16) have published an illuminating comparison between values of g and the true decay ratios of aeroelastic modes at hypersonic speeds. They derive an interesting scheme for

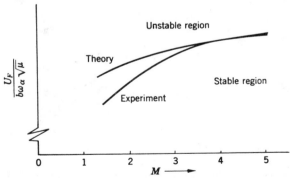

Fig. 7–19. Dimensionless flutter speed vs. Mach number M, obtained from wind-tunnel tests and piston-strip theory calculations on a series of wings like the one of Fig. 7–18. ω_α is the frequency of artificially uncoupled cantilever torsional vibration. (Note suppressed zero on ordinate scale. Adapted from Ref. 7–71; remaining details unavailable for reasons of security.)

interpreting values of g, which appears to be capable of generalization to other flight regimes and more complicated configurations.

Many examples of flutter stability boundaries will be found in the figures of Sec. 6–5 and in the parametric studies referenced there. Here we present Fig. 7–19 as a typical curve computed by modal methods (here by Eqs. 7–336). The lifting surfaces represented by these experimental and theoretical plots versus Mach number had aspect ratio two, rectangular planform, and other parameters characteristic of modern design practice on supersonic aircraft. This figure is especially interesting because of the good agreement with elementary piston-strip theory. This correlation can be improved below $M = 2.5$, where the piston operators are invalid, by using three-dimensional theory of the type discussed in Sec. 4–7(c) and adding an approximate correction for thickness effects.

Regarding the *convergence* of flutter computations based on generalized coordinates, it would appear that this is a much-neglected topic of research. One valuable study was given by Woolston and Runyan in Ref. 7–67. Additional light is thrown on this question by some data sent to the authors in a private communication from Dr. W. P. Rodden of Aerospace

TABLE 7–2

Speed eigenvalues of the flutter determinant, obtained using N normal vibration modes on a jet transport wing

$k_R = 0.06$

N	U_1 (knots)	U_2	U_3	U_4
2	650.00	934.00		
5	633.49	953.80	1450.06	3351.34
6	634.07	953.26	1452.90	3208.97
7	633.98	952.36	1450.10	2985.41
8	632.93	955.07	1401.94	2637.45
9	627.81	961.57	1332.53	2345.84
10	627.82	961.64	1332.54	2344.98

$k_R = 0.05$

N	U_1 (knots)	U_2	U_3	U_4
5	703.67	1095.43	1551.95	4210.90
6	704.42	1095.46	1552.94	4175.75
7	704.29	1093.07	1549.30	3179.79
8	702.70	1090.34	1501.40	2913.81
9	694.70	1090.61	1440.91	2503.20
10	694.71	1090.77	1440.91	2502.87

Corporation. Rodden's results were obtained by the U-g-method, using numbers N of normal modes varying from 5 to 10, on the jet-transport wing which forms the basis of Example 9–1, pp. 565–568, of Ref. 1–2. There is no need to reproduce here the detailed characteristics of this wing, except to say that it is straight, of moderate aspect ratio, and fitted with a large, concentrated nacelle mass about a third of the way out along the semispan. Using strip-theory aerodynamic operators, Rodden has computed the first four U-g-roots for two values of the reduced frequency k_R based on reference semichord $b_R = 5.468$ ft. Only the speeds are listed in Table 7–2, since convergence in any mode is clearly indicated by the way in which they vary with N. Neither of these k_R corresponds exactly to the flutter condition, but the minimum critical speed is quite close to U_2 at $k_R = 0.06$, this aeroelastic mode being the one which coincides with the first torsional mode of the structure when the airspeed is reduced to zero. We have added to Rodden's table some roots obtained in Ref. 1–2, using only two *artificially uncoupled* modes. This simplified approach is seen to be off by only a few per cent for estimating U_2; better agreement might be expected on an elementary structure without concentrated masses, but on many practical designs the convergence would be much poorer than that indicated here.

(c) Primary control surfaces

In the "primary" category we include those aerodynamic controls, such as ailerons, rudders, and leading- and trailing-edge flaps, which are intended to modify the lift and moment on the main surface in direct proportion to control displacement and which do not produce extensive separated flow. Thus we omit reference to the general category of spoilers and tabs, for which flutter prediction remains something of an art, unless full-scale tests are made. Experience shows that aeroelastic instability of controls which destroy circulation is less of a problem than on those which augment it. Such instability is usually susceptible of an *ad hoc* remedy. The authors are aware of at least one case where spoiler flutter was successfully predicted by leaving out the spoiler airloads altogether and merely allowing for the inertial and elastic coupling with the wing.

Any treatment of control-surface dynamics should start by calling the reader's attention to Templeton's book (Ref. 1–6), which is not only a valuable compendium of solution techniques but contains physical insights regarding instabilities and how they are modified by mass-balancing and similar changes in the system.

From the viewpoint of the present discussion, we can distinguish three types of primary flapped control: the all-movable stabilizer that rotates in pitch (or yaw) about an axis close to its root, the powerfully boosted control which is essentially irreversible, and the flexibly restrained or manually operated control with moderate to high reversibility. For flutter-prediction purposes, the all-movable surface is simply a bending-torsion structure in which the fundamental torsional mode has a much-diminished frequency and is dominated by the flexibility of the root restraint; when the deformations are one-dimensional, it can be treated as in Sec. 7-7(b), and we may gain some idea of the consequences of the lowered ω_α by examining such figures as 6–18, 6–25, and 6–28. The highly-irreversible surface can usually be visualized as a continuation of the main structure, particularly on low-aspect-ratio lifting surfaces. It results in larger amounts of chordwise or camber deformation in the overall normal modes, with slope discontinuities along the leading edge of the control, so that the flutter analysis involves plate-like, two-dimensional deflections. Although this problem falls within the scope of Chap. 8, it is here worth mentioning that both the subsonic and supersonic finite-span aerodynamic operators from Secs. 4–7(a) and (c) can be adapted to account for the discontinuities of chordwise slope (cf. Ref. 4–164).

Reversible and free-floating controls can often be analyzed by a slight extension of the one-dimensional modal approach of the preceding sub-section, wherein the nose-up rotation $\delta(t)$ relative to the chordline of the main surface is made an additional generalized coordinate. Consider, for example, a trailing-edge flap which extends from $y = l_1$ to $y = l_2$ and whose torsional rigidity is large enough that twisting about its own hinge-line may be neglected (both bending and torsion of the flap can be intro-duced by means of further artificial modes). Actually δ is an averaged angle over the flap span, since the vibrating wing undergoes some twist between l_1 and l_2.

Looking at the development of Eqs. 7–329 and 7–330, it is not difficult to work out corrections to the kinetic and potential energies and to the aerodynamic forces. Thus the potential energy must be augmented by

$$\Delta U = \tfrac{1}{2} K_\delta \delta^2 = \tfrac{1}{2} \omega_\delta^2 I_\delta \delta^2 \qquad (7\text{–}344)$$

where I_δ is the moment of inertia about the hingeline, K_δ is an equivalent torsion-spring constant of the rotational restraint, and

$$\omega_\delta \equiv \sqrt{K_\delta / I_\delta} \qquad (7\text{–}345)$$

denotes the natural frequency with which the flap would oscillate if the remaining structure were wholly constrained. In contrast to the lack of elastic coupling implied by Eq. 7–344, the control is inertially coupled with both wing bending and torsion through the running moment of inertia $i_\delta(y)$ about the hingeline and the static unbalance per unit span $S_\delta(y)$

(positive when the C.G. of the flap mass is behind the hingeline). The correction to Eq. 7-326 reads

$$\Delta\tau = \tfrac{1}{2}I_\delta \dot{\delta}^2 - \delta\sum_{i=1}^{r}\dot{q}_i\int_{l_1}^{l_2}S_\delta f_{w_i}\,dy + \delta\sum_{i=r+1}^{n}\dot{q}_i\int_{l_1}^{l_2}[i_\delta + b(e-a)S_\delta]f_{\theta_i}\,dy$$

(7-346)

Here b is local semichord, as before, and $(e-a)$ is a commonly used symbol for the dimensionless distance by which the hingeline lies aft of the wing E.A. Equation 7-346 reveals the interesting fact that complete mass-balancing, $S_\delta = 0$, can eliminate all coupling between flap rotation and pure bending vibration but not that with torsion.

The generalized force for the δ-degree of freedom is the total hinge moment on the flap,

$$Q_\delta = \int_{l_1}^{l_2}H(y,t)\,dy$$

(7-347)

As shown, for instance, by Eq. 4-131, H will receive contributions from all the linear and angular q_i's. Moreover, the lift and pitching moment in Eqs. 7-328 now contain terms linearly related to δ. Equations 4-129 and 4-130 demonstrate how these might involve the strip-theory aerodynamic coefficients L_5, L_6, M_5, and M_6, which are, in turn, functions of k, M, dimensionless flap chord, and hinge location. We are at length led to the following equations of motion to replace Eqs. 7-329 and 7-330:

$$\sum_{i=1}^{r}\ddot{q}_i\int_0^l mf_{w_i}f_{w_j}\,dy - \sum_{i=r+1}^{n}\ddot{q}_i\int_0^l S_\alpha f_{\theta_i}f_{w_j}\,dy - \ddot{\delta}\int_{l_1}^{l_2}S_\delta f_{w_j}\,dy$$

$$+ q_j\omega_{w_j}^2[1+ig_{w_j}]\int_0^l mf_{w_j}^2\,dy = \int_0^l L(y,t)f_{w_j}\,dy, \qquad (1 \le j \le r) \quad (7\text{-}348)$$

$$-\sum_{i=1}^{r}\ddot{q}_i\int_0^l S_\alpha f_{w_i}f_{\theta_j}\,dy + \sum_{i=r+1}^{n}\ddot{q}_i\int_0^l I_\alpha f_{\theta_i}f_{\theta_j}\,dy + \ddot{\delta}\int_{l_1}^{l_2}[i_\delta + b(e-a)S_\delta]f_{\theta_j}\,dy$$

$$+ q_j\omega_{\theta_j}^2[1+ig_{\theta_j}]\int_0^l I_\alpha f_{\theta_j}^2\,dy = \int_0^l M_y(y,t)f_{\theta_j}\,dy, \qquad (r+1 \le j \le n)$$

(7-349)

$$-\sum_{i=1}^{r}\ddot{q}_i\int_{l_1}^{l_2}S_\delta f_{w_i}\,dy + \sum_{i=r+1}^{n}\ddot{q}_i\int_{l_1}^{l_2}[i_\delta + b(e-a)S_\delta]f_{\theta_i}\,dy + I_\delta\ddot{\delta}$$

$$+ \delta\omega_\delta^2[1+ig_\delta]I_\delta = \int_{l_1}^{l_2}H(y,t)\,dy \quad (7\text{-}350)$$

The flutter determinant is readily constructed in particular cases. For instance, adding a control-surface degree of freedom to Eq. 7-336 while retaining the assumption of strip theory leads to Eq. 7-351 on page 391.

$$
\left\{ \mu_R \left[1 - \left(\frac{\omega_w}{\omega}\right)^2 \left(\frac{\omega_\theta}{\omega}\right)^2 (1+ig_w) \right] \int_0^1 \frac{m}{m_R} f_w^2 \, d\tilde{y} - \int_0^1 \left(\frac{b}{b_R}\right)^2 [L_1 + iL_2] f_w^2 \, d\tilde{y} \right\}
$$

$$
\left\{ -\mu_R x_{\alpha_R} \int_0^1 \frac{S_\alpha}{S_{\alpha_R}} f_w f_0 \, d\tilde{y} + \int_0^1 \left(\frac{b}{b_R}\right)^3 [M_1 + iM_2] f_w f_0 \, d\tilde{y} \right\}
$$

$$
\left\{ -\mu_R \,_{\delta_R} \int_{l_1}^{l_2} \frac{S_\delta}{S_{\delta_R}} f_w \, d\tilde{y} + \int_{l_1}^{l_2} \left(\frac{b}{b_R}\right)^3 [N_1 + iN_2] f_w \, d\tilde{y} \right\}
$$

$$
\left\{ -\mu_R x_{\alpha_R} \int_0^1 \frac{S_\alpha}{S_{\alpha_R}} f_w f_0 \, d\tilde{y} + \int_0^1 \left(\frac{b}{b_R}\right)^3 [L_3 + iL_4] f_w f_0 \, d\tilde{y} \right\}
$$

$$
\left\{ \mu_R r_{\alpha_R}^2 \left[1 - \left(\frac{\omega_0}{\omega}\right)^2 (1+ig_0) \right] \int_0^1 \frac{I_\alpha}{I_{\alpha_R}} f_0^2 \, d\tilde{y} - \int_0^1 \left(\frac{b}{b_R}\right)^4 \times [M_3 + iM_4] f_0^2 \, d\tilde{y} \right\}
$$

$$
\left\{ +\mu_R r'^2_{\delta_R} \int_{l_1}^{l_2} \frac{i_\delta}{i_{\delta_R}} f_0 \, d\tilde{y} + \mu_R x_{\delta_R} \int_{l_1}^{l_2} \left(\frac{b}{b_R}\right)(e-a) \frac{S_\delta}{S_{\delta_R}} f_0 \, d\tilde{y} - \int_{l_1}^{l_2} \left(\frac{b}{b_R}\right)^4 [N_3 + iN_4] f_0 \, d\tilde{y} \right\}
$$

$$
\left\{ -\mu_R x_{\delta_R} \int_{l_1}^{l_2} \frac{S_\delta}{S_{\delta_R}} f_w \, d\tilde{y} + \int_{l_1}^{l_2} \left(\frac{b}{b_R}\right)^3 [L_5 + iL_6] f_w \, d\tilde{y} \right\}
$$

$$
\left\{ \mu_R r'^2_{\delta_R} \int_{l_1}^{l_2} \frac{i_\delta}{i_{\delta_R}} f_0 \, d\tilde{y} + \mu_R x_{\delta_R} \int_{l_1}^{l_2} \frac{b}{b_R}(e-a) \frac{S_\delta}{S_{\delta_R}} f_0 \, d\tilde{y} - \int_{l_1}^{l_2} \left(\frac{b}{b_R}\right)^4 [M_5 + iM_6] f_0 \, d\tilde{y} \right\}
$$

$$
\left\{ \mu_R \frac{I_\delta}{m b_R^3} \left[1 - \left(\frac{\omega_\delta}{\omega}\right)^2 \left(\frac{\omega_0}{\omega}\right)^2 (1+ig_\delta) \right] - \int_{l_1}^{l_2} \left(\frac{b}{b_R}\right)^4 [N_5 + iN_6] \, d\tilde{y} \right\}
$$

$$
= 0
$$

$$(7\text{-}351)$$

Besides the dimensionless extremities $\tilde{l}_1$ and $\tilde{l}_2$ of the flap, a new static-unbalance parameter $x_{\delta_R} \equiv S_{\delta_R}/m_R b_R$ and a new radius of gyration $r_{\delta_R} \equiv \sqrt{i_{\delta_R}/m_R b_R^2}$ are required in Eq. 7-351.

Techniques for solving control-surface flutter determinants are identical with those employed on bending-torsion flutter, except for the presence of the additional parameters; x_{δ_R} and $\omega_\delta/\omega_\theta$ are often adjusted in an effort to optimize the performance. In many cases we find two significant aeroelastic modes, one involving primarily the bending and control degrees of freedom, the other coupling torsion and δ. There is then a range of speeds within which instability occurs, the lower end being at the U_F value for one mode, the higher at the U_F for the other. Many parametric studies of such flutter have been published. References 6–2 and 6–22 give curves based on incompressible aerodynamic operators, whereas the data of Refs. 6–33 and 6–34 are for the piston-theory range.

A type of instability peculiar to trailing-edge controls is the transonic single-degree-of-freedom phenomenon known as "aileron buzz." Until recently, this was generally dealt with in an empirical fashion by increasing irreversibility, adding viscous damping, changing the profile shape, or a combination of these measures. Since the appearance of studies by Coupry and Piazzoli (Ref. 4–73) and Eckhaus (Ref. 4–74), however, it now seems possible that buzz may be predictable on a rational basis. This is true especially on very thin airfoils and to the extent that the location of the boundary for the first onset of buzz instability is not affected by viscosity and shock-induced separation of the boundary layer.

(d) The influence of sweep

Several coordinated theoretical-experimental investigations have been made into the way flutter eigenvalues are affected by varying the sweep angle on a large-aspect-ratio wing; the papers by Molyneux (Ref. 7–72) and Barmby, Cunningham, and Garrick (Ref. 7–73) are excellent examples. So long as the structure remains one-dimensional, we may say that the general form of the equations of motion is unchanged from the sort of thing developed in subsection (b) above. An important new aerodynamic effect is introduced, however, by the incremental angle of attack due to the bending slope of the structural reference axis. Indeed, Cunningham (Ref. 7–74) has even demonstrated the possibility of a single-degree-of-freedom, pure bending flutter when the parameter $\bar{b}_R \tan \Lambda/l$ is large enough ($\bar{b}_R$ is a reference semichord measured normal to the swept span, and l is root-to-tip distance along the structural reference line).

As set forth in connection with static aeroelasticity in Secs. 7–2 and 7–3, we have the choice of describing deformations with respect either to cross sections taken parallel to the flight direction or to sections normal to the swept y-axis. The former representation is more compatible with the aero-dynamicist's viewpoint, especially if lifting-line operators are to be used. The latter tends to be more realistic structurally, since the ribs are usually oriented perpendicular to the span. Strip theory can readily be in-tegrated into either picture, as discussed in Sec. 4–4(e) and in Sec. 7–3 of Ref. 1–2.

We can conceive of varying the sweep of a given configuration in two ways: the lifting surface may be rotated about an axis normal to its plane at the root without changing its physical proportions; or it may be sheared so as to keep the streamwise chords $2b$ and tip-to-tip distance $2l$ unaltered. The former operation does not vary such structural properties as the natural frequencies, except for slight effects of clamping the root always along a streamwise line, but lowers the "aerodynamic efficiency." Accordingly, we expect the flutter speed to increase with increasing $|\Lambda|$. This is just what Ref. 7–73 finds for subsonic speeds. U_F rises at a rate somewhere between $1/\cos \Lambda$ and $1/\sqrt{\cos \Lambda}$ but closer to the latter. Molyneux (Ref. 7–72) concludes that "speed decreases slightly for small angles of sweepback and then increases rapidly as sweepback increases." A fruitful device for improving the flutter behavior of large-aspect-ratio swept wings has been to optimize the locations of one or more large concentrated masses, such as jet pods or external tanks. This process is best done on a wind-tunnel model and is reminiscent of the mass-balancing of controls.

The interpretation of results on sheared wings is somewhat more complicated. Clearly the structural aspect ratio is increased by shearing. In the experiments of Ref. 7–73, this caused both the fundamental fre-quencies of bending and torsion to diminish, the former faster than the latter; U_F then fell off slightly more slowly than $\sqrt{\cos \Lambda}$ (in the cantilever condition). If, however, we shear a wing while keeping GJ measured normal to the structural axis fixed, the length and torsional radius of gyration are proportional to $1/\cos \Lambda$ and $\cos \Lambda$, respectively. Theoretically there is no change in the "uncoupled" torsional frequency, and the slight loss of aerodynamic effectiveness might cause U_F to rise. This question has only minor practical interest, since the stability boundary can be predicted quite accurately on large-aspect-ratio swept wings of known elastic properties. One qualification to this statement is that strip theory may be excessively conservative in the high-subsonic and transonic regimes, so that suitable three-dimensional aerodynamic operators are recom-mended.

For lack of a better place, it is desirable to point out here that aero-thermoelastic effects can be incorporated into flutter computations by modal methods exactly as is done with any other system modification. The difficult task (cf. Sec. 2–3) is the structural one of finding how stiffnesses and vibration modes are affected because of thermal stresses and property deterioration. There is no novelty to the process of introducing these altered modes, corresponding to a series of points along a vehicle's flight trajectory, into the flutter determinant and recalculating eigenvalues. Often simple shifts in the natural frequencies will suffice. Of interest is the fact that stiffness increases due to rapid deceleration from a high-speed cruise or dash may be encountered as frequently as stiffness decreases. The former might perhaps be employed to *avoid* flutter, on a transient basis, in the low-altitude high-dynamic-pressure range.

(e) Very slender configurations; servo-coupled instability of guided missiles

It is no accident that the first topic taken up by Johnson in his listing of *Pitfalls in Missile Control* (Ref. 7–75) was titled "Aeroelasticity." All categories of guided and unguided missiles are, in varying degrees, victims of the same static and dynamic instabilities previously encountered on manned aircraft. Moreover, they add a few of their own.

Some features which can lead to difficulty are the following: missiles often have very slender bodies with low bending frequencies; their stabilization is accomplished by automatic means with limited adaptability; they may contain large masses of liquid fuel having coupled sloshing modes; they sometimes undergo enormous longitudinal acceleration with a force system that may reduce lateral bending stiffness significantly; and their engines are a high-powered source of acoustical and vibrational excitation. Connected with the first of the foregoing items, in the case of missiles launched vertically from a pad, is the large but sometimes-forgotten change in structural boundary conditions which accompanies the takeoff. Idealizing the missile with a uniform beam of length l, the fundamental natural frequency in bending can be written

$$\omega_1 = \frac{N\pi^2}{l^2}\sqrt{\frac{EI}{m}} \tag{7-352}$$

where m is the mass per unit length and EI the flexural rigidity. On the pad N may be anywhere from 0.356 (cantilever support) to 1.562 (free simple support). Once thrust-borne,* it rises to $N = 2.28$. Thus there could conceivably occur a sudden, sixfold jump in the most significant structural frequencies. It is reported that the control systems of some

---

* Any effect of the operation of the rocket engines on natural vibration is overlooked. It is recognized that zero-frequency, rigid-body modes are present after launching.

ballistic missiles are not activated until after clearing the pad, because they would be unstable on the ground.

These lower pre-launch frequencies can also make the structure liable to resonate with the periodic air forces due to its own vortex wake in a wind (Fung, Ref. 7–78), as tall smokestacks are prone to do. Special measures, such as spoilers or protective fairings, have had to be used in certain cases. After launching, the elastic missile is subject to a kind of gust excitation that differs from conventional atmospheric turbulence in horizontal flight. Even on essentially turbulence-free days, the variation of wind speed with height produces a time-dependent crosswind velocity. More experimental evidence is needed, especially at the shorter wavelengths, on statistical properties both of this wind shear and of the horizontal component of conventional turbulence. The assumption that such turbulence is isotropic in all three of its components does not always seem justified, particularly near the ground and in stable layers of the higher atmosphere.

Turning to dynamic instability of the vehicle in flight, we note that this problem is related to that of "feedback coupling," which has often been discussed in connection with flutter of airplanes (see, for instance, Ref. 7–76). Although other papers undoubtedly exist, the authors have been able to find only one unclassified report (Edelen, Ref. 7–77) which contains a full analytical treatment. Motion of the missile relative to inertial space is usually sensed by a system of linear accelerometers, plus gyroscopes arranged to measure angular velocity about each of three axes fixed to the structure. The possibility of coupling exists because bending vibrations can produce signals in these instruments. If these signals come through at frequencies below the cut-off frequencies of the various electronic and mechanical elements of the control systems, they will cause spurious corrective forces and moments to be applied to the vehicle. The structural feedback loops thus established may produce undesirable waste of power or even a vibrational instability of the missile, which is akin to flutter but is probably better described as "dynamic aeroelastic instability."

The various loops are illustrated schematically by the block diagram in Fig. 7–20. This refers to lateral and angular motion in one plane by a missile which is assumed to be effectively roll-stabilized. Since each system differs in many details from every other one, the situation can only be represented by typical examples. A rather general "fix" which is often suggested, however, consists of placing the accelerometers at the nodes and the rate gyros at the loops of the vibration mode that is expected to be most strongly coupled. The difficulty with this scheme, which seems to invalidate it for many practical applications, is that often more than one mode is involved, and the node and loop positions of different mode shapes

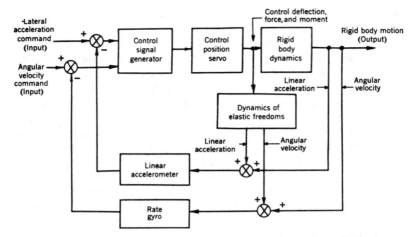

Fig. 7–20. Block diagram of missile control system with aeroelastic feedback.

never coincide. It would appear better in most cases to adopt the more sophisticated approach of including important elastic degrees of freedom in the analysis by which the control system is designed. Unfortunately, this may then create requirements for more elaborate computation in flight and the use of more than one sensing element in different positions along the length of the missile.

The foregoing discussion applies regardless of whether the control actions are produced by aerodynamic surfaces, auxiliary vernier rockets, or swiveling of the principal rockets. (The transfer function relating time-dependent control motion to the forces and moments generated thereby is more difficult to determine in the first case.) As a simple example of how the equations of motion are set up, we consider the dynamics in a single plane of a roll-stabilized missile with a single, gimballed rocket engine whose angular position is adjusted in response to signals generated by one rate gyro located at x_g on the axis (Fig. 7–21). This represents a considerable idealization of the control system of the Vanguard. The rocket thrust-line is rotated through an angle δ_r relative to the lowest slope of the deflected centerline. This δ_r is related to the local absolute slope θ_g of the centerline at the gyro station by the transfer function of the control system and actuator dynamics

$$\delta_r = A\{\theta_g\} \tag{7–353}$$

A is a time-dependent control system operator, whose Laplace transform can usually be expressed as the ratio of two finite polynomials in the transform variables.

For convenience, all structure attaching the engine and its actuator to the missile is assumed infinitely stiff, and inertia forces connected with the angular acceleration $\ddot{\delta}_r$ are ignored. The absolute displacement $w(x, t)$ of the centerline is conveniently represented by the summation

$$w(x, t) = w_R(t) + \theta_R(t)x + \sum_{i=1} \varphi_i(x)\xi_i(t) \qquad (7\text{-}354)$$

of two rigid-body modes plus the normal modes $\varphi_i(x)$ of vibration of the elastic structure. The external force system in the z-direction consists of an aerodynamic force $L(x, t)$ per unit length plus the concentrated force (small angles assumed)

$$T\delta_r = T\left[A\{\theta_g\} + \theta_R(t) + \sum_{i=1} \xi_i(t) \frac{d\varphi_i(x_r)}{dx} \right] \qquad (7\text{-}355)$$

Finally,

$$\theta_g(t) = \theta_R(t) + \sum_{i=1} \xi_i(t) \frac{d\varphi_i(x_g)}{dx} \qquad (7\text{-}356)$$

It is easily shown that the equations of motion of the system are as follows:

(i) Lateral translation:

$$M_{\text{tot}}\ddot{w}_R = T\delta_r + \int_{\text{body}} L(x, t)\, dx \qquad (7\text{-}357)$$

(ii) Pitching (x_r is negative):

$$I_{\text{tot}}\ddot{\theta}_R = x_r T\delta_r + \int_{\text{body}} xL(x, t)\, dx \qquad (7\text{-}358)$$

(iii) Elastic degree of freedom i (generalized mass M_i, frequency ω_i):

$$M_i\ddot{\xi}_i + \omega_i^2 M_i\xi_i = \Xi_i(t) \equiv \varphi_i(x_r)T\delta_r + \int_{\text{body}} L(x, t)\varphi_i(x)\, dx$$
$$(7\text{-}359)$$

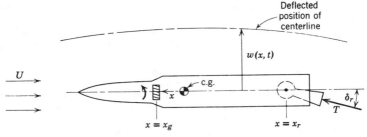

Fig. 7–21. Simplified elastic missile controlled by a gimbal-mounted rocket engine.

The distributed airload $L(x, t)$ includes contributions from all the degrees of freedom and from any external disturbances such as gusts. It could be computed by slender-body theory or by any other appropriate scheme for predicting unsteady aerodynamic forces. Stability of the system is determined from the homogeneous system of equations obtained by dropping external disturbances and substituting for θ_g through Eq. 7–356. Because of the form of the operator A, the Laplace transformation would probably be applied. Such techniques as the Nyquist criterion and roots locus method are convenient for designing A so as to furnish satisfactory performance; analog computers are, of course, frequently employed in these studies. The same equations are suitable for analyzing stability on the launching pad or in the absence of appreciable airloads ($L = 0$). In the former case, the rigid body freedoms would have to be dropped, and modes $\varphi_i(x)$ appropriate to the restrained position on the ground would be introduced.

It is recognized that the foregoing illustration has been excessively simplified. Factors which might have to be accounted for in a more complete analysis are the following: (*i*) inertia of the swiveling rocket (see Ref. 7–77); (*ii*) elasticity of the back-up structure and differences between the elastic slopes at the gimbal and actuator attachment points (see Ref. 7–77); (*iii*) dynamics of liquid fuel; (*iv*) dynamics of the flow of hot gases within the rocket chamber and exhaust jet; (*v*) effects of rolling motion and "cross-talk"; and (*vi*) thrust variations. The knowledgeable reader can add other items to this list.

REFERENCES

7–1. Traill-Nash, R. W., and A. R. Collar, "Effects of Shear Flexibility and Rotary Inertia on the Bending Vibrations of Beams," *Quart. J. Mechanics and Applied Math.*, Vol. 6, No. 2, June 1953, pp. 186–222.

7–2. Leonard, R. W., *On Solutions for the Transient Response of Beams*, NASA Technical Report R-21, February 1958.

7–3. Young, D., and R. P. Felgar, Jr., *Tables of Characteristic Functions Representing Normal Modes of Vibration of a Beam*, The University of Texas Publication 4913, Engineering Research Series 44, 1949.

7–4. Felgar, R. P., Jr., *Formulas for Integrals Containing Characteristic Functions of a Vibrating Beam*, Circular 14, Bureau of Engineering Research, The University of Texas, 1950.

7–5. Morse, P. M., and H. Feshbach, *Methods of Theoretical Physics*, Vols. I and II, McGraw-Hill Book Company, New York, 1953.

7–6. Hildebrand, F. B., and E. Reissner, *The Influence of the Aerodynamic Span Effect on the Magnitude of the Torsional Divergence Velocity and on the Shape of the Corresponding Deflection Mode*, NACA Technical Note 926, 1944.

7–7. von Kármán, T., and M. A. Biot, *Mathematical Methods in Engineering*, McGraw-Hill Book Company, New York, 1940.

7–8. Budiansky, B., and J. Mayers, "Influence of Aerodynamic Heating on the Effective Torsional Stiffness of Thin Wings," *J. Aero. Sciences*, Vol. 23, No. 12, December 1956, pp. 1081–1093, 1108.

7–9. Vosteen, L. F., *Effect of Temperature on Dynamic Modulus of Elasticity of Some Structural Alloys*, NACA Technical Note 4348, 1958.

7–10. Bisplinghoff, R. L., *The Finite Twisting and Bending of Heated Elastic Lifting Surfaces*, Mitteilung aus dem *Institut für Flugzeugstatik und Leichtbau*, E.T.H., Zürich, Nr. 4, 1958.

7–11. Dryden, H. L., and J. Duberg, *Aeroelastic Effects of Aerodynamic Heating*, paper presented to Fifth General Assembly of AGARD, Ottawa, June 1955.

7–12. Bisplinghoff, R. L., and T. H. H. Pian, "On the Vibrations of Thermally Buckled Bars and Plates," *Proc.* Ninth Internat. Congress of Applied Mechanics, Brussels, 1956.

7–13. Singer, Joseph, and N. J. Hoff, "Effect of the Change in Thermal Stresses Due to Large Deflections on the Torsional Rigidity of Wings," *J. Aero. Sciences*, Vol. 24, No. 4, April 1957, p. 310.

7–14. Johnson, H. C., and G. Foteio, *Rolling Effectiveness and Aileron Reversal Characteristics of Straight and Swept-Back Wings*, Air Force Technical Report 6198, 1951.

7–15. Hedgepeth, J. M., and R. J. Kell, *Comparison Between Theoretical and Experimental Rates of Roll of Two Models with Flexible Rectangular Wings at Supersonic Speeds*, NACA Research Memorandum L54F23, 1954.

7–16. Hildebrand, F. B., *Introduction to Numerical Analysis*, McGraw-Hill Book Company, New York, 1956.

7–17. Michal, A. D., *Matrix and Tensor Calculus*, John Wiley and Sons, New York, 1947 (see Chapter 6).

7–18. Pearson, H. A., and W. S. Aiken, Jr., *Charts for Determination of Wing Torsional Stiffness Required for Specified Rolling Characteristics or Aileron Reversal Speed*, NACA Report 799, 1944.

7–19. See, for example, *Flight*, Vol. LXIV, No. 2339, November 20, 1953, pp. 680–681.

7–20. Diederich, F. W., and B. Budiansky, *Divergence of Swept Wings*, NACA Technical Note 1680, 1948.

7–21. Diederich, F. W., *Calculation of the Aerodynamic Loading of Swept and Unswept Flexible Wings of Arbitrary Stiffness*, NACA Report 1000, 1950.

7–22. Diederich, F. W., and K. A. Foss, *Charts and Approximate Formulas for the Estimation of Aeroelastic Effects on the Loading of Swept and Unswept Wings*, NACA Report 1140, 1953.

7–23. Pai, S. I., and W. R. Sears, "Some Aeroelastic Properties of Swept Wings," *J. Aero. Sciences*, Vol. 16, No. 2, February 1949, pp. 105–115, 119.

7–24. Pian, T. H. H., and H. Lin, *Effect of Structural Flexibility on Aircraft Loading, Part II—Spanwise Airload Distribution*, Air Force Technical Report 6358, Part II, 1951.

7–25. Flax, A. H., "Aeroelastic Problems at Supersonic Speeds," *Proc.* Second Internat. Aero. Congress, New York, 1949.

7–26. Seifert, George, "A Third Order Boundary Value Problem Arising in Aeroelastic Wing Theory," *Quart. Applied Math.*, Vol. IX, No. 2, 1951.

7–27. Diederich, F. W., and K. A. Foss, *Static Aeroelastic Phenomena of M-, W-, and Λ-Wings*, NACA RM L52J21, February 1953.

7–28. Foss, K. A., *Charts and Approximate Formulas for the Estimation of Aeroelastic Effects on the Lateral Control of Swept and Unswept Wings*, NACA Report 1139, 1953.

7-29. Brown, S. C., *Predicted Static Aeroelastic Effects on Wings With Supersonic .Leading Edges and Streamwise Tips*, NASA Memorandum 4-18-59A, April 1959.

7-30. Lighthill, M. J., "Higher Approximations," Article E, Volume VI, of *High Speed Aerodynamics and Jet Propulsion*, W. R. Sears, Ed., Princeton University Press, Princeton, N.J., 1954.

7-31. Biot, M. A., *Aeroelastic Stability of Supersonic Wings*, Report 1: "Chordwise Divergence—The Two-Dimensional Case," Cornell Aeronautical Laboratory, Report CAL/CM-427, CAL-1-E-1, December 1947.

7-32. Biot, M. A., "Divergence of Supersonic Wings Including Chordwise Bending," *J. Aero. Sciences*, Vol. 23, No. 3, March 1956, pp. 237–251, 271.

7-33. Biot, M. A., *Aeroelastic Stability of Supersonic Wings*, Report 3: "General Method for the Two-Dimensional Case and Its Application to the Chordwise Divergence of a Biconvex Section," Cornell Aeronautical Laboratory Report CAL/CM-506, CAL-1-E-1, September 1948.

7-34. Dugundji, J., and J. D. C. Crisp, *On the Aeroelastic Characteristics of Low-Aspect-Ratio Wings with Chordwise Deformations*, USAF Office of Scientific Research Technical Note 59–787, July 1959.

7-35. Martin, D. J., and C. E. Watkins, "Transonic and Supersonic Divergence Characteristics of Low-Aspect-Ratio Wings and Controls," IAS Report No. 59-58.

7-36. Hedgepeth, J. M., and P. G. Waner, *Analysis of Static Aeroelastic Behavior of Low-Aspect-Ratio Rectangular Wings*, NACA Technical Note 3958, April 1957.

7-37. Hancock, G. J., "Divergence of Plate Airfoils of Low Aspect Ratio at Supersonic Speeds," *J. Aero/Space Sciences*, Vol. 26, No. 8, August 1959, pp. 495–507, 517.

7-38. Williams, D., *Displacements of a Linear Elastic System Under a Given Transient Load*, British R.A.E. Report S.M.E. C/7219/DW/19, 1946 (see also Reports S.M.E. 3309 and 3316, 1945).

7-39. Mazet, R., *Some Aspects of Ground and Flight Vibration Tests*, AGARD Report 40-T, April 1956.

7-40. Lewis, R. C., and D. L. Wrisley, "A System for the Excitation of Pure Natural Modes of Complex Structures," *J. Aero. Sciences*, Vol. 17, No. 11, November 1950, pp. 705–722, 735.

7-41. Mar, J. W., T. H. H. Pian, and J. M. Calligeros, "A Note on Methods for the Determination of Transient Stresses," *J. Aero.Sciences*, Readers' Forum, Vol. 23, No. 1, January 1956, pp. 94–95.

7-42. Houbolt, J. C., *A Recurrence Matrix Solution for the Dynamic Response of Aircraft in Gusts*, NACA Report 1010, 1951.

7-43. Ramberg, W., and A. E. McPherson, "Experimental Verification of Theory of Landing Impact," *Proc.* Sixth Internat. Congress for Applied Mechanics, Paris, 1946 (also available as a National Bureau of Standards Report).

7-44. Pian, T. H. H., and H. I. Flomenhoft, "Analytical and Experimental Studies on Dynamic Loads in Airplane Structures During Landing," *J. Aero. Sciences*, Vol. 17, No. 12, December 1950, pp. 765–774, 786.

7-45. Fung, Y. C., "The Analysis of Dynamic Stresses in Aircraft Structures During Landing as Nonstationary Random Processes," *J. Applied Mechanics*, Vol. 22, No. 4, December 1955, pp. 449–457.

7-46. Biot, M. A., and R. L. Bisplinghoff, *Dynamic Loads on Airplane Structures During Landing*, NACA Wartime Report No. W-92, 1944.

7-47. Eisenman, R. L., and E. H. Kramer, *A Method for Predicting Dynamic Landing Loads*, USAF Wright Air Development Center Technical Report 54-28, 1954.

7-48. Zahorski, A. H., "Remarks on Dynamic Loads in Landing," *J. Aero. Sciences*, Vol. 19, No. 4, April 1952, pp. 258–264.

7-49. Cook, F. E., and B. Milwitzky, *Effect of Interaction on Landing Gear Behavior and Dynamic Loads in a Flexible Airplane Structure*, NACA Report 1278, 1956.

7-50. Locke, F. W. S., *A Statistical Study of Maximum Vertical Acceleration Encountered in Flying Boats in Rough Water Landings*, U.S.N. Bureau of Aeronautics, Research Division Report 1184, 1952.

7-51. McBrearty, J. F., *A Review of Landing Gear and Ground Loads Problems*, AGARD Report 118, May 1957.

7-52. Many authors, *Proceedings of Flight Flutter Testing Symposium*, sponsored by Aircraft Industries Association and USAF Office of Scientific Research, Washington, D.C., May 1958.

7-53. Bisplinghoff, R. L., and E. A. Witmer, "Blast Loading of Aircraft Structures," *Proc.* Sixth Anglo-Amer. Aero. Conf., Folkestone, 1957, pp. 205–234.

7-54. Codik, A., H. Lin, and T. H. H. Pian, *Effect of Structural Flexibility on Aircraft Loading*, Part XII: "The Gust Response of a Sweptback Tapered Wing Including Bending Flexibility," Air Force Technical Report 6358, Part XII, 1953.

7-55. Carta, F. O., T. H. H. Pian, and H. Lin, *Effect of Structural Flexibility on Aircraft Loading*, Part IX: "Lift and Moment Growths on a Swept, Tapered, Rigid Wing upon Entering a Gust," Air Force Technical Report 6358, Part IX, 1953.

7-56. Foss, K. A., D. Sternlight, and T. H. H. Pian, *Effect of Structural Flexibility on Aircraft Loading*, Part XIX: "A Parametric Study of the Gust Response of Swept Wing Airplanes Including a Wing-Bending Degree of Freedom," Air Force Technical Report 6358, Part XIX, 1954.

7-57. Liepmann, H. W., "Extension of the Statistical Approach to Buffeting and Gust Response of Wings of Finite Span," *J. Aero. Sciences*, Vol. 22, No. 3, March 1955, pp. 197–200.

7-58. Diederich, F. W., "The Dynamic Response of a Large Airplane to Continuous Random Atmospheric Disturbances," *J. Aero. Sciences*, Vol. 23, No. 10, October 1956, pp. 917–930.

7-59. Ribner, H. S., "Spectral Theory of Buffeting and Gust Response: Unification and Extension," *J. Aero. Sciences*, Vol. 23, No. 12, December 1956, pp. 1075–1077, 1118.

7-60. Foss, K. A., and W. L. McCabe, *Gust Loading of Rigid and Flexible Aircraft in Continuous Atmospheric Turbulence*, USAF Wright Air Development Center Technical Report 57-704, January 1958.

7-61. Bisplinghoff, R. L., T. H. H. Pian, and K. A. Foss, *Response of Elastic Aircraft to Continuous Turbulence*, AGARD Report 117, April–May 1957.

7-62. Press, H., M. T. Meadows, and I. Hadlock, *Estimates of Probability Distribution of Root-Mean-Square Gust Velocity of Atmospheric Turbulence from Operational Gust-Load Data by Random-Process Theory*, NACA Technical Note 3362, 1955.

7-63. Coleman, T. L., H. Press, and M. T. Meadows, *An Evaluation of Effects of Flexibility on Wing Strains in Rough Air for a Large Swept-Wing Airplane by Means of Experimentally Determined Frequency-Response Functions with an Assessment of Random-Process Techniques Employed*, NACA Technical Note 4291, 1958.

7-64. Goland, M., "The Flutter of a Uniform Cantilever Wing," *Jour. Applied Mechanics*, Vol. 12, No. 4, December 1945, pp. A-197-A-208.

7-65. Goland, M., and Y. L. Luke, "The Flutter of a Uniform Wing with Tip Weights," *Jour. Applied Mechanics*, Vol. 15, No. 1, March 1948, pp. 13–20.

7–66. Runyan, H. L., and C. E. Watkins, *Flutter of a Uniform Wing with an Arbitrarily Placed Mass According to a Differential-Equation Analysis and a Comparison with Experiment*, NACA Report 966, 1950.

7–67. Woolston, D. S., and H. L. Runyan, *Appraisal of a Method of Flutter Analysis Based on Chosen Modes by Comparison with Experiment for Cases of Large Mass Coupling*, NACA Technical Note 1902, 1949.

7–68. Duncan, W. J., and C. L. T. Griffith, *The Influence of Wing Taper on the Flutter of Cantilever Wings*, British ARC Reports and Memoranda 1869, 1939.

7–69. Spielberg, I. N., *The Two-Dimensional Incompressible Aerodynamic Coefficients for Oscillatory Changes in Airfoil Camber*, U.S. Air Force Technical Note WCNS 52-7, 1952.

7–70. Leppert, E. L., Jr., "An Application of IBM Machines to the Solution of the Flutter Determinant," *J. Aero. Sciences*, Vol. 14, No. 3, March 1947, pp. 171–174.

7–71. Ashley, H., W. J. Mykytow, and J. R. Martuccelli, "Prediction of Lifting Surface Flutter at Supersonic Speeds," *Proc.* Second Internat. Congress of the Aero. Sciences, Zürich, September 1960.

7–72. Molyneux, W. G., *The Flutter of Swept and Unswept Wings with Fixed-Root Conditions*, British A.R.C. Reports and Memoranda 2796, 1950.

7–73. Barmby, J. G., H. J. Cunningham, and I. E. Garrick, *Study of Effects of Sweep on the Flutter of Cantilever Wings*, NACA Report 1014, 1951.

7–74. Cunningham, H. J., *Analysis of Pure-Bending Flutter of a Cantilever Swept Wing and Its Relation to Bending-Torsion Flutter*, NACA Technical Note 2461, 1951.

7–75. Johnson, R. L., "Pitfalls in Missile Control," *Proc.* the AGARD Second Guided Missiles Seminar—Guidance and Control, AGARDOGRAPH 21, September 1956, pp. 115–138. (See also the paper by F. E. Perry in the same volume.)

7–76. McRuer, D. T., D. Benum, and G. E. Click, *The Influence of Servomechanisms on the Flutter of Servo-Controlled Aircraft*, Air Force Technical Report No. 6287, October 1953.

7–77. Edelen, D. G. B., *The Problem of Structural Feedback for a Rocket Structure*, unnumbered report of the Martin Company, Baltimore, 1957.

7–78. Fung, Y. C., "Fluctuating Lift and Drag Acting on a Cylinder in a Flow at Supercritical Reynolds Numbers," *J. Aero. Sciences*, Vol. 27, No. 11, November 1960, pp. 801–814.

8

TWO-DIMENSIONAL STRUCTURES

8-1 INTRODUCTION

In Chap. 7, we considered in some detail one-dimensional structures, that is, structures whose state of deformation can be adequately described by a set of functions of a single space coordinate. In the present chapter, we extend these concepts to structures where two space coordinates are required, and we define such structures as two-dimensional. In the case of one-dimensional lifting surfaces such as slender wings, we assume that camber bending in the streamwise direction can be neglected. A sufficient degree of slenderness is implied so that it is possible to apply aerodynamic and structural theories that are reasonably well established. This fortuitous state of affairs may be said to be due to the simplifying assumptions that spanwise flow components are small compared with chordwise components and that spanwise normal stresses are large compared with chordwise values. The aeroelastic problem is thus characterized by a single independent variable along a spanwise axis. In the case of two-dimensional structures, camber bending in the streamwise direction can no longer be ignored because the lifting surface may have spanwise and chordwise dimensions of the same order of magnitude. Here we are confronted with problems in which spanwise and chordwise components of flow and normal stresses assume nearly equal proportions. The complexities are thus multiplied and the techniques available for obtaining practical solutions are fewer. We may add, however, that aeroelastic phenomena associated with two-dimensional lifting surfaces are often less severe than those with the more slender one-dimensional structures. Since low-aspect-ratio

403

lifting surfaces are generally found on very high-speed vehicles, we may say also that the supersonic speed range is of most interest to us in this connection.

8–2 TWO-DIMENSIONAL LIFTING SURFACES IN GENERAL

Prior to discussing some particular types of two-dimensional lifting surfaces, we shall consider in general terms the form of the aeroelastic equations which apply to all such systems.

(a) Aeroelastic equations

The behavior of a two-dimensional lifting surface can be described in terms of the operator equation (3–79),

$$(\mathscr{S} - \mathscr{A} - \mathscr{I})(w) = Z_D \qquad (3\text{–}79)$$

In Sec. 3–5, several methods of solution of Eq. 3–79 were described. These methods are largely approximate in nature since the geometrical and stiffness properties of actual structures are usually known only in a numerical sense. It was pointed out that approximate methods of solution can be broken down into two steps. In the first, the space configuration of the deformed structure is approximated by an equivalent system with finite degrees of freedom. This reduces the mathematics of the continuous structure to a system of simultaneous equations. The second step is that of finding solutions to these simultaneous equations.

In the present section, one of the methods of Sec. 3–5, namely Galerkin's method [cf. Sec. 3–5(c)], is selected as a basis for approximate analyses. The deformation of the lifting surface is represented in this method by a superposition of n assumed modes, $\gamma_j(x, y)$, as stated by Eq. 3–103. By means of this representation, the n simultaneous equations of the system are those given by Eq. 3–105.

$$\sum_{j=1}^{n} \iint_S (\mathscr{S} - \mathscr{A} - \mathscr{I})(\gamma_j q_j)\gamma_i \, dx \, dy$$
$$= \iint_S Z_D \gamma_i \, dx \, dy, \qquad (i = 1, 2, \cdots, n) \quad (3\text{–}105)$$

We may evaluate explicitly the integrals on the left-hand side of Eq. 3–106 for the case of a plate-like lifting surface.

Let us consider first the term involving the inertial operator $\mathcal{I}$. By introducing $\mathcal{I}(\quad) = -m\dfrac{\partial^2}{\partial t^2}(\quad)$, where m is the mass per unit area of lifting surface,

$$\sum_{j=1}^{n}\iint_{S}\mathcal{I}(\gamma_j q_j)\gamma_i\,dx\,dy = -\sum_{j=1}^{n}m_{ij}\ddot{q}_j, \quad (i=1,2,\cdots,n) \quad (8\text{-}1)$$

where m_{ij} is a generalized mass defined by

$$m_{ij} = \iint_{S}\gamma_i\gamma_j m\,dx\,dy, \quad (i,j=1,2,\cdots,n)$$

which has the property that $m_{ij} = m_{ji}$.

The term involving the structural operator $\mathcal{S}$ may be simplified in the following way:

$$\sum_{j=1}^{n}\iint_{S}\mathcal{S}(\gamma_j q_j)\gamma_i\,dx\,dy = \sum_{j=1}^{n}k_{ij}q_j, \quad (i=1,2,\cdots,n) \quad (8\text{-}2)$$

where

$$k_{ij} = \iint_{S}\mathcal{S}(\gamma_j)\gamma_i\,dx\,dy, \quad (i,j=1,2,\cdots,n)$$

with the property, in general, that $k_{ij} = k_{ji}$.

The structural operator, $\mathcal{S}$, may, for example, take the form

$$\mathcal{S}(\quad) = \left[\frac{\partial^2}{\partial x^2}\left(D_{xx}\frac{\partial^2}{\partial x^2} + D_{xy}\frac{\partial^2}{\partial y^2}\right) + 4\frac{\partial^2}{\partial x\,\partial y}\left(D_u\frac{\partial^2}{\partial x\,\partial y}\right)\right.$$
$$\left. + \frac{\partial^2}{\partial y^2}\left(D_{yx}\frac{\partial^2}{\partial x^2} + D_{yy}\frac{\partial^2}{\partial y^2}\right)\right](\quad)$$

which applies to small deformations of plates of variable thickness.

In evaluating the term involving the aerodynamic operator, $\mathcal{A}$, we shall illustrate the method by assuming that $\delta \gg 1/M^3$, $M^2 \gg 1$, and $M\delta \ll 1$ and take advantage of the relatively simple point function properties of piston theory discussed in Secs. 4–5(d) and 4–6(d). Under these assumptions, we have

$$\sum_{j=1}^{N}\iint_{S}\mathcal{A}(\gamma_j q_j)\gamma_i\,dx\,dy = -\sum_{j=1}^{N}a_{ij}\dot{q}_j - \sum_{j=1}^{N}b_{ij}q_j \quad (i=1,2,\cdots,N) \quad (8\text{-}3)$$

where

$$\mathcal{A}(\quad) = -\frac{4q}{M}\left[1 + \frac{(\gamma+1)}{2}M\frac{dz_t}{dx}\right]\left(\frac{\partial}{\partial x} + \frac{1}{U}\frac{\partial}{\partial t}\right)(\quad)$$

and

$$a_{ij} = \frac{2\rho U}{M} \iint_S \left[1 + \frac{(\gamma + 1)}{2} M \frac{dz_t}{dx}\right] \gamma_i \gamma_j \, dx \, dy$$

$$b_{ij} = \frac{2\rho U^2}{M} \iint_S \left[1 + \frac{(\gamma + 1)}{2} M \frac{dz_t}{dx}\right] \frac{d\gamma_j}{dx} \gamma_i \, dx \, dy$$

Finally, since Z_D is a function only of time, the term on the right-hand side of Eq. 3–105 is an explicit function of time; and we write that

$$F_i{}^D(t) = \iint_S Z_D \gamma_i \, dx \, dy \qquad (i = 1, 2, \cdots, n) \tag{8–4}$$

Equation 3–105 reduces then, by means of Eqs. 8–1, 8–2, 8–3, and 8–4, to the following set of ordinary simultaneous differential equations:

$$\sum_{j=1}^{n} [m_{ij}\ddot{q}_j + a_{ij}\dot{q}_j + (k_{ij} + b_{ij})q_j] = F_i{}^D(t) \qquad (i = 1, 2, \cdots, n) \tag{8–5}$$

Equations 8–5 contain complete coupling among all the coordinates, q_j, due to inertial, elastic, and aerodynamic effects. A simplification in these equations can be obtained by eliminating the inertial and elastic coupling. This is accomplished by expressing the assumed mode shapes γ_j in terms of the normal vibration modes, ϕ_j of the free system in a vacuum, or in mathematical terms, the eigenfunctions of the homogeneous equation

$$(\mathscr{S} - \mathscr{I})(w) = 0 \qquad \text{or} \qquad m \frac{\partial^2 w}{\partial t^2} + \mathscr{S}(w) = 0 \tag{8–6}$$

If ϕ_j and ω_j are eigenfunctions and eigenvalues of Eq. 8–6, then we may put $w = \phi_j e^{i\omega t}$, and we obtain

$$m\omega_j{}^2 \phi_j = \mathscr{S}(\phi_j) \tag{8–7}$$

In addition,

$$\iint_S \mathscr{S}(\phi_j)\phi_i \, dx \, dy = \omega_j{}^2 \iint_S \phi_j \phi_i m \, dx \, dy \tag{8–8}$$

Because of the orthogonality of the eigenfunctions, we can reduce Eq. 8–8 to

$$\iint_S \mathscr{S}(\phi_j)\phi_i \, dx \, dy = M_j \omega_j{}^2 \tag{8–9}$$

where

$$M_j = \iint_S \phi_j^2 m \, dx \, dy \qquad (8\text{--}10)$$

is now the generalized mass. When we take as a solution

$$w = \sum_{j=1}^{n} \phi_j(x, y)\xi_j \qquad (8\text{--}11)$$

and bring to bear Eqs. 8–9 and 8–10, we obtain the simpler set of ordinary simultaneous differential equations in which the elastic and inertial coupling has been eliminated as follows:

$$M_i \ddot{\xi}_i + \sum_{j=1}^{n} a_{ij} \dot{\xi}_j + M_i \omega_i^2 \xi_i + \sum_{j=1}^{n} b_{ij} \xi_j = \Xi_i(t) \qquad (i = 1, 2, \cdots, n) \qquad (8\text{--}12)$$

where

$$M_i = \iint_S \phi_i^2 m \, dx \, dy$$

$$a_{ij} = \frac{2\rho U}{M} \iint_S \left[1 + \left(\frac{\gamma + 1}{2} \right) M \frac{dz_t}{dx} \right] \phi_i \phi_j \, dx \, dy$$

$$b_{ij} = \frac{2\rho U^2}{M} \iint_S \left[1 + \left(\frac{\gamma + 1}{2} \right) M \frac{dz_t}{dx} \right] \frac{\partial \phi_j}{\partial x} \phi_i \, dx \, dy$$

$$\Xi_i^D(t) = \iint_S Z_D \phi_i \, dx \, dy$$

It will be convenient to use the notation of Dugundji and Crisp (Ref. 8–1) and rewrite Eqs. 8–12 in nondimensional form. We select the nondimensional space coordinates

$$\xi = \frac{x}{2b}, \qquad \eta = \frac{y}{l} \qquad (8\text{--}13)$$

where b is a reference semichord taken at the wing root and l is the semispan. A nondimensional time variable is taken as $s = Ut/b$. If we assume nondimensional deformation mode shapes $\bar{\phi}_j(\xi, \eta)$ such that

$$\phi_j = 2b\bar{\phi}_j(\xi, \eta)$$

then Eq. 8–12 reduces to the dimensionless form

$$2\bar{\mu} M \bar{M}_i \frac{\partial^2 \xi_i}{\partial s^2} + \sum_{j=1}^{n} \bar{a}_{ij} \frac{\partial \xi_j}{\partial s} + 2\bar{\mu} M \bar{M}_i k_i^2 \xi_i + \sum_{j=1}^{n} \bar{b}_{ij} \xi_j = \frac{M}{2\rho U^2} \bar{\Xi}_i^D(s)$$

$$(i = 1, 2, \cdots, n) \qquad (8\text{--}14)$$

where $\bar{\mu} = \dfrac{m_0}{2\rho b} = $ mass density ratio

$m_0 = $ maximum value of m

$$\bar{M}_i = \iint_S \frac{m}{m_0} \bar{\phi}_i^2 \, d\xi \, d\eta$$

$$\bar{a}_{ij} = 2 \iint_S (1 + \Gamma)\bar{\phi}_i\bar{\phi}_j \, d\xi \, d\eta$$

$$\bar{b}_{ij} = \iint_S (1 + \Gamma)\bar{\phi}_i \frac{\partial \bar{\phi}_j}{\partial \xi} \, d\xi \, d\eta$$

$$\Gamma = \frac{(\gamma + 1)}{4} M\left(\frac{h_{\max}}{2b}\right) \frac{d}{d\xi}\left[\frac{h(\xi, \eta)}{h_{\max}}\right]$$

where $h = $ wing thickness

$h_{\max} = $ maximum wing thickness

$$k_j = \frac{\omega_j b}{U} = \text{reduced frequency parameter}$$

$$\Xi_i^D = \iint_S Z_D \bar{\phi}_i \, d\xi \, d\eta$$

(b) Effects of aerodynamic heating

In assessing the influence of aerodynamic heating on aeroelastic phenomena, we find two principal effects. The first is simply a reduction in Young's modulus, E, with elevated temperature. All heated structural components are affected, and their resulting loss in stiffness and frequency can be predicted by merely reducing the value of E in an appropriate manner. A second effect is a reduction in stiffness due to thermal stresses. Since thermal stresses arise from unequal thermal expansion of the structure resulting from transient nonuniform temperature changes, they die out when the entire structure reaches a uniform temperature level.

In discussing aerodynamic heating effects, we shall assume at first that the two-dimensional structure is plate-like in character and that temperature distributions over the surface and throughout the thickness are given explicitly. The two-dimensional equilibrium equation of elasticity, assuming a temperature distribution throughout the thickness of a plate-like surface, is

$$(\mathscr{S} - \mathscr{A} - \mathscr{I})(w) = Z_D - \frac{\partial^2 \bar{\bar{T}}_x}{\partial x^2} - \frac{\partial^2 \bar{\bar{T}}_y}{\partial y^2} \tag{8--15}$$

where the definitions of $\bar{\bar{T}}_x$ and $\bar{\bar{T}}_y$ are indicated by Eqs. 5–40 and where the structural operator is defined by (cf. Eq. 5–62)

$$\mathscr{S}(\quad) = \left[\frac{\partial^2}{\partial x^2}\left(D_{xx}\frac{\partial^2}{\partial x^2} + D_{xy}\frac{\partial^2}{\partial y^2}\right) + 4\frac{\partial^2}{\partial x\,\partial y}\left(D_u\frac{\partial^2}{\partial x\,\partial y}\right)\right.$$

$$+ \frac{\partial^2}{\partial y^2}\left(D_{yx}\frac{\partial^2}{\partial x^2} + D_{yy}\frac{\partial^2}{\partial y^2}\right)$$

$$\left.- \left(\frac{\partial^2 F}{\partial y^2}\frac{\partial^2}{\partial x^2} - 2\frac{\partial^2 F}{\partial x\,\partial y}\frac{\partial^2}{\partial x\,\partial y} + \frac{\partial^2 F}{\partial x^2}\frac{\partial^2}{\partial y^2}\right)\right](\quad)$$

Since two unknowns, w, the lateral displacement, and F, the midplane stress function, are involved in Eq. 8–15, an additional equation must be supplied. This is the compatibility equation (5–63), which involves the same two dependent variables w and F. A simultaneous solution of Eqs. 8–15 and 5–63 for the prevailing boundary conditions is thus required; such a solution is appropriate for very strong temperature gradients and for large lateral displacements.

A simplification of the system of equations (8–15 and 5–63) may, however, be obtained if it is assumed that the lateral displacements are sufficiently small so that the midplane stresses are unaffected by them. This assumption permits Eqs. 8–15 and 5–63 to be uncoupled. For example, when the higher order terms in w are discarded, we obtain from Eq. 5–63

$$\frac{\partial^2}{\partial x^2}\left(R_{yy}\frac{\partial^2 F}{\partial x^2} + R_{yx}\frac{\partial^2 F}{\partial y^2}\right) + \frac{\partial^2}{\partial x\,\partial y}\left(\frac{1}{K_u}\frac{\partial^2 F}{\partial x\,\partial y}\right) + \frac{\partial^2}{\partial y^2}\left(R_{xy}\frac{\partial^2 F}{\partial x^2} + R_{xx}\frac{\partial^2 F}{\partial y^2}\right)$$

$$= -\frac{\partial^2}{\partial x^2}(R_{yx}\bar{T}_x + R_{yy}\bar{T}_y) - \frac{\partial^2}{\partial y^2}(R_{xx}\bar{T}_x + R_{xy}\bar{T}_y) \quad (8\text{–}16)$$

When Eq. 8–16 is solved with appropriate boundary conditions, the stress function F is obtained. The midplane stress resultants N_{xx}, N_{xy}, and N_{yy} can be computed from this stress function by means of Eqs. 5–54. When this step is accomplished, Eq. 8–15 can be solved by Galerkin's method; and we obtain thereby, for the case of an aerodynamically heated plate-like surface, the following simultaneous equations:

$$2\bar{\mu}M\bar{M}_i\frac{\partial^2\xi_i}{\partial s^2} + \sum_{j=1}^{n}\bar{a}_{ij}\frac{\partial\xi_j}{\partial s} + 2\bar{\mu}M\bar{M}_i k_i^2 + \sum_{j=1}^{n}(\bar{b}_{ij} - \bar{C}_{ij}^{\,T})\xi_j$$

$$= \frac{M}{2\rho U^2}\bar{\Xi}_i^{\,D}(s) - \bar{C}_i^{\,\bar{\bar{T}}}, \quad (i = 1, 2, \cdots, n) \quad (8\text{–}17)$$

where

$$\bar{C}_{ij}{}^T = \frac{M}{4\rho U^2 b} \iint\limits_S \left[N_{xx}^{(0)} \frac{\partial^2 \bar{\phi}_j}{\partial \xi^2} + \left(\frac{2b}{l}\right) 2 N_{xy}^{(0)} \frac{\partial^2 \bar{\phi}_j}{\partial \xi\, \partial \eta} \right.$$

$$\left. + \left(\frac{2b}{l}\right)^2 N_{yy}^{(0)} \frac{\partial^2 \bar{\phi}_j}{\partial \eta^2} \right] \bar{\phi}_i\, d\xi\, d\eta, \qquad (i, j = 1, 2, \cdots, n)$$

are correction terms due to the explicitly defined midplane thermal stresses $N_{xx}^{(0)}$, $N_{xy}^{(0)}$, and $N_{yy}^{(0)}$, and

$$\bar{C}_i{}^{\bar{\bar{T}}} = \frac{M}{8\rho U^2 b^2} \iint\limits_S \left[\frac{\partial^2 \bar{\bar{T}}_\xi}{\partial \xi^2} + \left(\frac{2b}{l}\right)^2 \frac{\partial^2 \bar{\bar{T}}_\eta}{\partial \eta^2} \right] \bar{\phi}_i\, d\xi\, d\eta$$

are correction terms due to temperature gradients across the plate thickness.

8-3 LOW-ASPECT-RATIO CANTILEVER WINGS

Differences in aeroelastic effects between high- and low-aspect-ratio wings are largely due to the existence of chordwise deformation modes in the latter. In the present section, we shall illustrate some of these effects on flutter and divergence properties by means of a cantilever wing of rectangular planform and very low aspect ratio.

(a) Aeroelastic equations

The rectangular wing chosen for illustration was discussed by Dugundji and Crisp in Ref. 8-1 and was designated MW-2 in the NACA tests performed by Heldenfels and Rosencrans, described by Ref. 8-2. The wing has an aspect ratio of 1, uniform spanwise properties, and a hollow built-up biconvex section of 5% thickness ratio. By employing piston theory in the analysis of this wing, aerodynamic profile thickness effects are introduced; but tip Mach-cone effects are neglected. In this case, the former effects should be considerably more significant than the latter. Three mode shapes, first bending, first torsion, and first chordwise, as illustrated by Fig. 8-1, are employed as degrees of freedom. They may be represented analytically by the following orthogonal functions:

$$\phi_1 = [\cosh \beta_1 \eta - \cos \beta_1 \eta - \alpha_1(\sinh \beta_1 \eta - \sin \beta_1 \eta)] \qquad \text{(Bending)}$$

$$\phi_2 = 2(1 - 2\xi) \sin (\pi \eta / 2) \qquad\qquad\qquad\quad \text{(Torsion)} \quad (8\text{-}18)$$

$$\phi_3 = [\cosh \beta_3 \xi + \cos \beta_3 \xi - \alpha_3(\sinh \beta_3 \xi + \sin \beta_3 \xi)]\eta^2 \quad \text{(Chordwise)}$$

where β_1, β_3, α_1, and α_3 are given by

$$\beta_1 = 1.8751 \qquad \alpha_1 = 0.7341$$
$$\beta_3 = 4.7300 \qquad \alpha_3 = 0.9825$$

The natural frequencies corresponding to these modes are the experimentally determined frequencies of the MW–2 wing given in Ref. 8–2 as

$$\omega_1 = 67 \text{ cps}, \qquad \omega_2 = 143 \text{ cps}, \qquad \omega_3 = 265 \text{ cps} \qquad (8\text{–}19)$$

The mass distribution is assumed uniform, an assumption which is nearly satisfied by the MW–2 wing. The biconvex profile cross section is taken according to the form

$$\frac{h}{h_{\max}} = 4\xi(1 - \xi) \qquad (8\text{–}20)$$

and then

$$\Gamma = (\gamma + 1)M\left(\frac{h_{\max}}{2b}\right)(1 - 2\xi) \qquad (8\text{–}21)$$

Making use of the data of Eqs. 8–18 through 8–21, the $\bar{M}_i$, $\bar{a}_{ij}$, and $\bar{b}_{ij}$

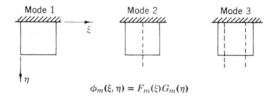

$$\phi_m(\xi, \eta) = F_m(\xi)G_m(\eta)$$

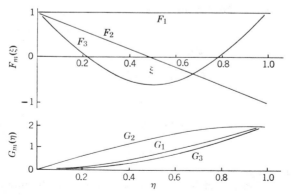

Fig. 8–1. Mode shapes employed in rectangular wing analysis.

coefficients of Eqs. 8–14 may be evaluated and expressed in the following matrix forms applicable to a three-degree-of-freedom system:

$$[\bar{M}_i] = \begin{bmatrix} 1 & 0 & 0 \\ 0 & 0.6667 & 0 \\ 0 & 0 & 0.2000 \end{bmatrix}$$

$$[\bar{a}_{ij}] = \begin{bmatrix} 2 & 0 & 0 \\ 0 & 1.333 & 0 \\ 0 & 0 & 0.400 \end{bmatrix} + (\gamma + 1)M\left(\frac{h_{max}}{2b}\right)\begin{bmatrix} 0 & 0.905 & 0 \\ 0.905 & 0 & 0.351 \\ 0 & 0.351 & 0 \end{bmatrix}$$

$$[\bar{b}_{ij}] = \begin{bmatrix} 0 & -2.720 & 0 \\ 0 & 0 & -2.354 \\ 0 & 0 & 0 \end{bmatrix}$$

$$+ (\gamma + 1)M\left(\frac{h_{max}}{2b}\right)\begin{bmatrix} 0 & 0 & -1.783 \\ 0 & -1.333 & 0 \\ 0 & 0 & -0.600 \end{bmatrix} \qquad (8\text{--}22)$$

(b) Flutter

When we specialize Eqs. 8–14 to the case of flutter by putting the right-hand side equal to zero and setting $\xi_j(s) = \bar{\xi}_j e^{iks}$, the following matrix equation is obtained:

$$[\bar{C}_{ij}]\{\bar{\xi}_j\} = 0 \qquad (8\text{--}23)$$

where the elements of the square matrix are complex quantities with the form

$$\bar{C}_{ij} = 2\bar{\mu}Mk^2\bar{M}_i(\Lambda_i^2\Omega - 1)\delta_{ij} + ik\bar{a}_{ij} + \bar{b}_{ij}$$

and where

$$\Lambda_i = \frac{\omega_i}{\omega_R}$$

$$\Omega = (\omega_R/\omega)^2$$

$$\delta_{ij} = \text{Kronecker delta}$$

The values of k and ω which make the determinant of Eq. 8–23 equal to zero,

$$|\bar{C}_{ij}| = 0 \qquad (8\text{--}24)$$

define the flutter condition.

The MW-2 wing of Ref. 8-2 is a square cantilever aluminum wing of 20-inch chord and 20-inch span with a 5% thickness ratio. Under sea level conditions at $M = 2$, we have the parameters $\bar{\mu}M = 32$, $M(h_{max}/2b)$ $= 0.1$, and the frequency ratios $\Lambda_1 = 0.466$, $\Lambda_2 = 1.00$, and $\Lambda_3 = 1.85$, where ω_2 is chosen as the reference frequency, ω_R.

The U-g-method of flutter analysis is employed [cf. Sec. 7–7(b)] in which flutter frequency parameter Ω is interpreted as $\Omega = (\omega_R/\omega)^2(1 + ig)$, where g is the structural damping. Figures 8–2(a) through (e), taken from Ref. 8–3, are velocity versus damping $\left(\dfrac{U}{b\omega_2} \text{ vs. } g\right)$ plots obtained by solving Eq. 8–24 for the three-degree-of-freedom case outlined. Each of the five figures corresponds to a different value of the frequency ratio, Λ_3, the latter numbers being chosen as 1.0, 1.333, 1.833, 2.1, and ∞. The role of the chordwise mode is interestingly revealed by the curves of Fig. 8–2. It is evident that as the frequency ratio Λ_3 is reduced, the lowest unstable branch changes from the torsion to the chordwise branch. The numbers on the figures indicate the flutter frequency ratio, ω/ω_2, and they serve to identify the unstable branches. Two distinct types of flutter modes are indicated by this behavior. One is associated primarily with the torsion mode and the other with the chordwise mode. It is of interest to observe that the slope of the curve at flutter, that is, $\left(\dfrac{\partial g}{\partial(U/b\omega_2)}\right)_{g=0}$, is of the same order of magnitude for both types of flutter. Both can be said to be of the "strong flutter" type. Figure 8–3 illustrates a summary of the critical flutter boundaries of Fig. 8–2 in terms of a plot of the variation of $U/b\omega_2$ with Λ_3.

In order to reveal more clearly the nature of the two types of flutter encountered with this system, Dugundji and Crisp (Ref. 8–1) conducted separate two-degree-of-freedom analyses, using the constituent binary systems 1–2, 1–3, and 2–3. It is evident *a priori* from the form of the matrices of Eqs. 9–22 that the bending-chordwise binary system (1–3) is stable since the ξ_1 and ξ_3 coordinates are uncoupled and since both single-degree-of-freedom systems are stable. The 1–2 binary system is the common bending-torsion system which is, in fact, illustrated already by Fig. 8–2(e) as the degenerate three-degree-of-freedom case with $\Lambda_3 = \infty$. The 2–3 binary system couples the torsion with the chordwise modes, and this combination leads to a flutter speed, the value of which obviously depends on the parameter Λ_3. This results obtained by Dugundji and Crisp for the 1–2 and 2–3 binary systems are plotted also in Fig. 8–3. It is evident that these binary systems provide a rough quantitative estimate and a qualitative delineation of the behavior of the more general ternary system.

It would seem evident from studying Fig. 8–3 that when the chordwise mode approaches the torsion mode frequency ($\Lambda_3 < 1.4$) there can be a marked reduction in flutter speed below that indicated by the

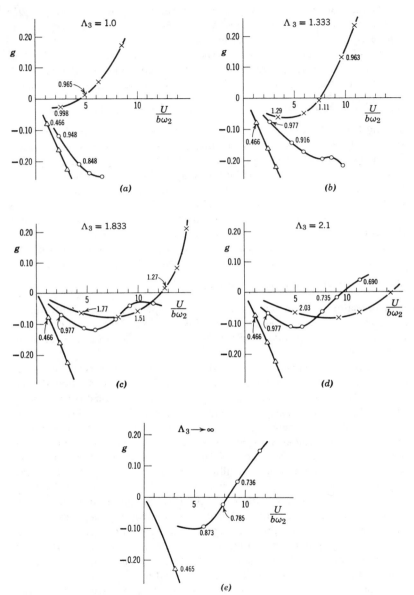

Fig. 8–2. *U-g* plot for hollow built-up rectangular wing.

bending torsion system. This range of action involves essentially then a coupled torsion-chordwise flutter mode with very little of the bending mode present. Above $\Lambda_3 = 1.4$, up to values of the order of $\Lambda_3 = 3$, the chordwise mode appears to increase the flutter speed above that predicted by the bending-torsion system.

(c) Divergence

The divergence behavior of a low-aspect-ratio wing may be extracted also from Eq. 8–14 by putting

$$\frac{\partial^2 \xi_i}{\partial s^2} = \frac{\partial \xi_j}{\partial s} = \bar{\Xi}_D(s) = 0$$

This yields the matrix equation

$$[D_{ij}]\{\xi_j\} = 0 \tag{8–25}$$

where

$$D_{ij} = 2\bar{\mu}M\bar{M}_i k_i^2 \delta_{ij} + \bar{b}_{ij} \tag{8–26}$$

and where

$$\delta_{ij} = \text{Kronecker delta}$$

Introducing the notational changes, $k_i = \omega_i b/U$ and $\Lambda_i = \omega_i/\omega_R$, we require that for static instability the determinant of the matrix be put to zero, and we obtain therefrom as the divergence condition

$$|2\bar{\mu}M\bar{M}_i k_R^2 \Lambda_i^2 \delta_{ij} + \bar{b}_{ij}| = 0 \tag{8–27}$$

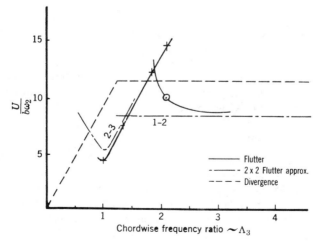

Fig. 8–3. Flutter and divergence of hollow built-up rectangular wing.

In the case of the NACA MW-2 wing, the divergence determinant (8-27) reduces to

$$\begin{vmatrix} 0.434\bar{\mu}Mk_2{}^2 & -2.720 & -4.28M\left(\frac{h_{\max}}{2b}\right) \\ 0 & -3.2M\left(\frac{h_{\max}}{2b}\right) + 1.333\bar{\mu}Mk_2{}^2 & -2.354 \\ 0 & 0 & -1.44M\left(\frac{h_{\max}}{2b}\right) + 1.40\Lambda_3{}^2\bar{\mu}Mk_2{}^2 \end{vmatrix} = 0$$

(8-28)

where $k_R = k_2$, and $\gamma = 1.4$.

When the determinant is expanded, there are obtained the three roots

$$\frac{U}{b\omega_2}\sqrt{\frac{1}{\bar{\mu}}\left(\frac{h_{\max}}{2b}\right)} = \infty,\ 0.646,\ 0.527\Lambda_3 \qquad (8\text{-}29)$$

Since the determinant is of triangular form, the divergence speeds are evidently those of the individual bending, torsion, and chordwise modes. The boundaries represented by these divergence speeds are superimposed upon the flutter boundaries in Fig. 8-3. It is evident that the divergence boundary occurs above the flutter boundary over most of the practical range except for a narrow region in the vicinity of $\Lambda_3 = 1.8$.

8-4 SURFACE SKIN PANELS

(a) Introduction

Other two-dimensional aeroelastic problems of interest are those concerned with the behavior of surface skin panels that have one side exposed to an airstream and the other side to still air. Dynamic aeroelastic instability, or panel flutter, is the principal problem of this group, although static aeroelastic instability and dynamic response to aerodynamic excitation are matters of some interest.

(b) Aeroelastic equations

Let us consider a thin panel which is located on the surface of a flight vehicle and has one side exposed to a supersonic airstream and the other side to still air, as illustrated by Fig. 8-4. The initial geometry of the surface may be arbitrary, and the panel may be subjected to midplane stresses arising from internal pressurization or from thermal effects. We shall assume that the small lateral deflection, w, constitutes the sole dependent variable of the problem.

Based upon shallow shell theory [cf. Sec. 5–2(g)], the following operator equation governs the phenomenon of skin panel aeroelasticity when the surface has an initial midplane shape specified by $z = z(x, y)$ and when the panel is homogeneous, isotropic, and of constant thickness, and has a temperature distribution $T(x, y)$ constant throughout its thickness:

$$(\mathscr{S} - \mathscr{A} - \mathscr{I})(w) = Z_D + \left(\frac{\partial^2 F}{\partial y^2}\frac{\partial^2}{\partial x^2} - 2\frac{\partial^2 F}{\partial x\,\partial y}\frac{\partial^2}{\partial x\,\partial y} + \frac{\partial^2 F}{\partial x^2}\frac{\partial^2}{\partial y^2}\right)(z)$$

(8–30)

where the stress function F is governed by the compatibility equation

$$\frac{1}{Eh}\nabla^4 F = -\alpha\nabla^2 T + \left(\frac{\partial^2 w}{\partial x\,\partial y}\right)^2 - \frac{\partial^2 w}{\partial x^2}\frac{\partial^2 w}{\partial y^2} - \frac{\partial^2 z}{\partial x^2}\frac{\partial^2 w}{\partial y^2}$$

$$- \frac{\partial^2 z}{\partial y^2}\frac{\partial^2 w}{\partial x^2} + 2\frac{\partial^2 z}{\partial x\,\partial y}\frac{\partial^2 w}{\partial x\,\partial y}$$

The structural and inertial operators are

$$\mathscr{S}(\quad) = \left[D\nabla^4 - \frac{\partial^2 F}{\partial y^2}\frac{\partial^2}{\partial x^2} + 2\frac{\partial^2 F}{\partial x\,\partial y}\cdot\frac{\partial^2}{\partial x\,\partial y} - \frac{\partial^2 F}{\partial x^2}\frac{\partial^2}{\partial y^2}\right](\quad)\qquad(8\text{–}31)$$

$$\mathscr{I}(\quad) = -\left[m_0\frac{\partial^2}{\partial t^2}\right](\quad)\qquad(8\text{–}32)$$

We shall employ the aerodynamic operator

$$\mathscr{A}(\quad) = -\frac{2q}{U\sqrt{M^2 - 1}}\left[U\frac{\partial}{\partial x} + \frac{M^2 - 2}{M^2 - 1}\frac{\partial}{\partial t}\right](\quad)\qquad(8\text{–}33)$$

which reduces to the case of piston theory when $\sqrt{M^2 - 1} \rightarrow M$ and $[(M^2 - 2)/(M^2 - 1)] \rightarrow 1$. Equation 8–33 represents a first-order approximation to the aerodynamic theory in which we neglect the influence of three-dimensional aerodynamic effects and limit our results to Mach numbers beyond approximately 1.6.

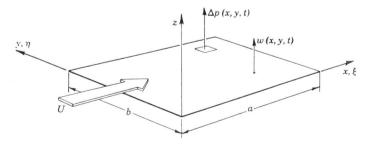

Fig. 8–4. Surface skin panel.

Combining Eqs. 8–30, 8–31, 8–32, and 8–33, we obtain the following equilibrium aeroelastic equation of skin panels:

$$
\begin{aligned}
\Bigg[D\nabla^4 &- \frac{\partial^2 F}{\partial y^2}\frac{\partial^2}{\partial x^2} + 2\frac{\partial^2 F}{\partial x\,\partial y}\frac{\partial^2}{\partial x\,\partial y} - \frac{\partial^2 F}{\partial x^2}\frac{\partial^2}{\partial y^2} \\
&+ \frac{2q}{\sqrt{M^2-1}}\left(\frac{\partial}{\partial x} + \frac{1}{U}\frac{M^2-2}{M^2-1}\frac{\partial}{\partial t}\right) + m_0\frac{\partial^2}{\partial t^2}\Bigg](w) \\
&= Z_D + \left(\frac{\partial^2 F}{\partial y^2}\frac{\partial^2}{\partial x^2} - 2\frac{\partial^2 F}{\partial x\,\partial y}\frac{\partial^2}{\partial x\,\partial y} + \frac{\partial^2 F}{\partial x^2}\frac{\partial^2}{\partial y^2}\right)(z) \quad (8\text{--}34)
\end{aligned}
$$

where the compatibility equation in the stress function F is the same as that of Eq. 8–30.

8–5 FLAT PANELS

We treat first some of those aeroelastic problems which concern flat panels. Equation 8–34 is rewritten by assuming that $w(x, y)$ represents small perturbations from an initially flat and initially stressed condition characterized by the stress resultants $N_{xx}^{(0)}$, $N_{yy}^{(0)}$, and $N_{xy}^{(0)}$. This results in the following differential equation:

$$
\begin{aligned}
\frac{\partial^4 W}{\partial \xi^4} &+ 2\left(\frac{a}{b}\right)^2\frac{\partial^4 W}{\partial \xi^2\,\partial \eta^2} + \left(\frac{a}{b}\right)^4\frac{\partial^4 W}{\partial \eta^4} + R_{xx}\frac{\partial^2 W}{\partial \xi^2} + 2R_{xy}\frac{\partial^2 W}{\partial \xi\,\partial \eta} \\
&+ R_{yy}\frac{\partial^2 W}{\partial \eta^2} + \lambda\frac{\partial W}{\partial \xi} + \lambda\frac{M^2-2}{M^2-1}\frac{a}{U}\frac{\partial W}{\partial t} + \frac{m_0 a^4}{D}\frac{\partial^2 W}{\partial t^2} = \frac{a^3}{D}Z_D \quad (8\text{--}35)
\end{aligned}
$$

where we have made use of Eqs. 5–54 and introduced the nondimensional parameters

$$
\left.
\begin{aligned}
R_{xx} &= -\frac{N_{xx}^{(0)}a^2}{D} & R_{yy} &= -\frac{N_{yy}^{(0)}a^4}{b^2 D} \\[2mm]
R_{xy} &= -\frac{N_{xy}^{(0)}a^3}{b\,D} & \lambda &= \frac{2q a^3}{D\sqrt{M^2-1}} \\[2mm]
W &= \frac{w}{a} & \xi &= \frac{x}{a} \\[2mm]
& & \eta &= \frac{y}{b}
\end{aligned}
\right\}
\quad (8\text{--}36)
$$

and where R_{xx}, R_{xy}, and R_{yy} are regarded as constants. It will be observed that R_{xx} and R_{yy} represent compressive stress resultants.

(a) Instability of panels.   Panel flutter

At first, we shall focus our attention on the problems of panel instability. Experiments have shown conclusively that flutter of surface skin panels can exist (Refs. 8–4 and 8–5), and Jordan (Ref. 8–6) has suggested that such flutter contributed to early German V-2 rocket failures during World War II. Panel flutter differs from the more conventional lifting surface flutter in at least two important respects: first, it is entirely a supersonic phenomenon; and second, structural nonlinearities associated with the lateral deformations of panels tend strongly to limit the flutter amplitudes. The latter limitation often causes the modes of structural failure to be those peculiar to fatigue rather than explosive fracture of the skin surface.

Theoretical attacks on the panel flutter problem have been diverse and have led on occasions to different conclusions. The widest attention has been given to the relatively simple problem of a flat, simply supported plate in a supersonic airstream. Investigators who have made significant contributions to this problem include Miles (Ref. 8–7), Shen (Ref. 8–8), Nelson and Cunningham (Ref. 8–9), Goland and Luke (Ref. 8–10), Hedgepeth (Ref. 8–11), Eisley (Ref. 8–12), Movchan (Ref. 8–13), Houbolt (Ref. 8–14), Dugundji (Ref. 8–15), and Shulman (Ref. 8–16). These authors have considered many aspects of the flutter of flat panels, such as the applicability of the Rayleigh-Ritz method, the employment of traveling wave solutions, and a wide variety of aerodynamic theories. A reasonably clear picture of the physical nature of panel flutter and the tools which are required for obtaining theoretical solutions for such problems has emerged from their work.

The flutter of buckled or slightly curved panels has also received attention, principally by Fung (Refs. 8–17 and 8–18), Houbolt (Ref. 8–14), and Eisley (Ref. 8–19). The influence of initial imperfections and the implication that large-deflection structural operators are required are features which distinguish this work from the earlier work on flat panels.

(*i*) *Instability of flat panels of infinite aspect ratio.*   If we suppose that the panel is free of initial stresses and is long in the *y*-direction, we may assume that the nondimensional deformation W is independent of y and rewrite Eq. 8–35 as

$$\frac{\partial^4 W}{\partial \xi^4} + R_{xx} \frac{\partial^2 W}{\partial \xi^2} + \lambda \frac{\partial W}{\partial \xi} + \lambda \frac{M^2 - 2}{M^2 - 1} \frac{a}{U} \frac{\partial W}{\partial t} + \frac{m_0 a^4}{D} \frac{\partial^2 W}{\partial t^2} = 0 \quad (8\text{–}37)$$

with boundary conditions

$$W(0) = W(1) = \frac{\partial^2 W(0)}{\partial \xi^2} = \frac{\partial^2 W(1)}{\partial \xi^2} = 0 \quad (8\text{–}38)$$

for a simply supported panel and

$$W(0) = W(1) = \frac{\partial W(0)}{\partial \xi} = \frac{\partial W(1)}{\partial \xi} = 0 \qquad (8\text{-}39)$$

for a clamped panel.

EXACT SOLUTION FOR A SEMI-INFINITE FLAT PANEL.   An exact solution of Eq. 8-37 may be found by introducing

$$W = \overline{W}e^{\alpha t} \qquad (8\text{-}40)$$

where, in general, α is a complex number, $\alpha = \beta + i\omega$. The problem of defining the borderline between stability and instability is one of determining the value of M which will cause β to vanish. The procedure followed here for obtaining an exact solution is that used by Houbolt in Ref. 8-14. When Eq. 8-40 is introduced into Eq. 8-37, we obtain

$$\frac{d^4 \overline{W}}{d\xi^4} + R_{xx}\frac{d^2 \overline{W}}{d\xi^2} + \lambda \frac{d\overline{W}}{d\xi} - k\overline{W} = 0 \qquad (8\text{-}41)$$

where k is an eigenvalue of the form

$$k = -\lambda \frac{M^2 - 2}{M^2 - 1}\frac{a}{U}\alpha - \frac{m_0 a^4}{D}\alpha^2 = -\pi^4 g_\alpha \frac{\alpha}{\omega_r} - \pi^4 \frac{\alpha^2}{\omega_r^2}$$

$$\omega_r = \pi^2 \sqrt{\frac{D}{m_0 a^4}} = \begin{array}{l}\text{First natural frequency for a semi-}\\ \text{infinite simply supported flat panel.}\end{array}$$

$$g_\alpha = \frac{M^2 - 2}{(M^2 - 1)^{3/2}}\frac{\rho U}{m_0 \omega_r} = \text{Damping coefficient based on } \omega_r.$$

If we put $\overline{W} = e^{p\xi}$, we obtain the characteristic equation

$$p^4 + R_{xx}p^2 + \lambda p - k = 0 \qquad (8\text{-}42)$$

with the four roots

$$\begin{aligned} p_1 &= -\epsilon + b \\ p_2 &= -\epsilon - b \\ p_3 &= \epsilon + ic \\ p_4 &= \epsilon - ic \end{aligned} \qquad (8\text{-}43)$$

where the coefficients of the equation are related to the roots by

$$b^2 = \frac{\lambda}{4\epsilon} - \left(\epsilon^2 + \frac{R_{xx}}{2}\right)$$

$$c^2 = \frac{\lambda}{4\epsilon} + \left(\epsilon^2 + \frac{R_{xx}}{2}\right) \qquad (8\text{-}44)$$

$$k = -\left(2\epsilon^2 + \frac{R_{xx}}{2}\right)^2 + \frac{\lambda^2}{16\epsilon^2}$$

In terms of the four roots of Eq. 8-43, the amplitude $\overline{W}$ is

$$\overline{W} = A_1 e^{p_1 \xi} + A_2 e^{p_2 \xi} + A_3 e^{p_3 \xi} + A_4 e^{p_4 \xi} \qquad (8\text{-}45)$$

where the A_i's are obtained by applying the boundary conditions on the plate. When these boundary conditions are satisfied, a stability condition for flutter is obtained.

For the case of a plate with simply supported edges, we have

$$\overline{W}(0) = \overline{W}''(0) = \overline{W}(1) = \overline{W}''(1) = 0$$

and we obtain, by making use of Eq. 8-45, the following stability determinant:

$$\begin{vmatrix} 1 & 1 & 1 & 1 \\ p_1^{\,2} & p_2^{\,2} & p_3^{\,2} & p_4^{\,2} \\ e^{p_1} & e^{p_2} & e^{p_3} & e^{p_4} \\ p_1^{\,2} e^{p_1} & p_2^{\,2} e^{p_2} & p_3^{\,2} e^{p_3} & p_4^{\,2} e^{p_4} \end{vmatrix} = 0 \qquad (8\text{-}46)$$

Expanding Eq. 8-46, we find after some reduction the condition

$$16 i \epsilon^2 bc \cosh 2\epsilon - (b + ic)^2 [4\epsilon^2 - (b - ic)^2] \cosh (b + ic)$$
$$+ (b - ic)^2 [4\epsilon^2 - (b + ic)^2] \cosh (b - ic) = 0 \qquad (8\text{-}47)$$

For clamped edges, where $\overline{W}(0) = \dot{\overline{W}}'(0) = \overline{W}(1) = \overline{W}'(1) = 0$, there is a somewhat similar condition:

$$4 ibc \cosh 2\epsilon + [-4\epsilon^2 + (b - ic)^2] \cosh (b + ic)$$
$$- [-4\epsilon^2 + (b + ic)^2] \cosh (b - ic) = 0 \qquad (8\text{-}48)$$

In order to compute the flutter boundaries, Houbolt used the following procedure: for a given R_{xx}, a value of λ is chosen; then $\epsilon = \epsilon_1 + i\epsilon_2$ is varied until the stability equation is satisfied. Having found ϵ, the last of Eqs. 8-44 is employed to compute $k = k_1 + ik_2$; and then a solution for α is obtained in terms of this k (cf. Eq. 8-41) by means of the following formula:

$$\alpha = \beta + i\omega = -\frac{\omega_r g_\alpha}{2} \pm \sqrt{\left(\frac{\omega_r g_\alpha}{2}\right)^2 - \frac{\omega_r^{\,2}}{\pi^4}(k_1 + ik_2)} \qquad (8\text{-}49)$$

Finally, this result is examined to obtain the conditions on g_α which are required in order to make β pass from a negative to a positive value. It can be shown that for flutter—that is, for $\beta \geq 0$—the following condition on g_α must be satisfied:

$$\pi^4 g_\alpha^{\,2} < \frac{(k_2)^2}{k_1} \qquad (8\text{-}50)$$

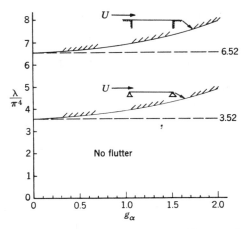

Fig. 8–5. Flutter boundaries for semi-infinite plate with $R_{xx} = 0$. (Adapted from Ref. 8–14.)

The corresponding flutter frequency is obtained from

$$\omega_f{}^2 = \frac{k_1\omega_r{}^2}{\pi^4} \qquad (8\text{–}51a)$$

Some results due to Houbolt's work (Ref. 8–14), using the above solution, are shown in Fig. 8–5. The solid lines represent flutter boundaries obtained for $R_{xx} = 0$. The horizontal dotted lines corresponding to $\lambda/\pi^4 = 3.52$ and 6.52, for simply supported and clamped edges, respectively, are the results that would be obtained if the damping term proportional to $\partial W/\partial t$ in Eq. 8–37 were omitted. The value for pin-ends corresponds to the critical value obtained by Hedgepeth (Ref. 8–20) who also presents an exact analysis in which the damping term is neglected. Values of g_α usually found in practice are for the most part less than approximately 0.5, where there is very little difference between the actual flutter boundary and the horizontal dotted line. Thus, for most practical purposes, the damping term may be neglected in panel flutter analyses. It would appear that a necessary requirement for the validity of panel flutter analyses with zero damping is the presence of mass-density parameters, $\mu = m_0/\rho a$, of ten or greater. It has been observed on other occasions (cf. Ref. 8–1) that a large mass-density parameter is a necessary requirement for the success of zero-damping analyses of wing-like surfaces.

The analysis outlined above taking account of the damping term is, of course, simplified considerably when this term is neglected. Regardless of whether this term is included or not, the analysis in both cases is the same through Eq. 8–48. Beyond this point, when the damping term is neglected, we can proceed more simply by first choosing values of λ and R_{xx}, and then varying ϵ until Eqs. 8–47 or 8–48 are satisfied. Then with ϵ, λ, and R_{xx} known, the eigenvalue, k, may be evaluated by the last of Eqs. 8–44. Finally, when k is known, the flutter frequency is obtained directly from

$$\omega_f{}^2 = \frac{k\omega_R{}^2}{\pi^4} \qquad (8\text{–}51b)$$

Figure 8–6, based upon the work of Hedgepeth (Ref. 8-11), illustrates a plot of the relationship between the parameters λ and k for the case of $R_{xx} = 0$. It is apparent that for $\lambda = 0$ there exists an infinite number of discrete values of k which correspond to the natural frequencies of the plate. When the dynamic pressure parameter, λ, is increased, the values of k— that is, the values of the natural frequencies—change slowly until two of the natural frequencies coalesce at λ_{cr}, which corresponds to the flutter speed. More specifically, when $\lambda < \lambda_{cr}$, the values of the eigenvalue k remain real; when $\lambda = \lambda_{cr}$, two of the values of k become equal; and when $\lambda > \lambda_{cr}$, these two values become complex. As λ is increased further, other natural frequencies of the panel would coalesce, although the speed at which this occurs would have only academic interest. Methods of flutter analysis based upon the frequency coalescence behavior of conservative undamped systems appear to be useful only when the damping coupling is small compared to stiffness or inertial coupling.

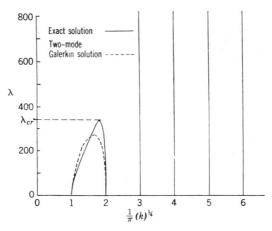

Fig. 8–6. Interaction between λ and k for semi-infinite simply supported panels. (Reproduced from Ref. 8–11.)

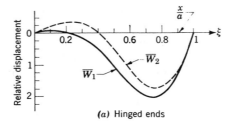

(a) Hinged ends

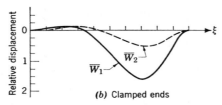

(b) Clamped ends

Fig. 8–7. Flutter mode shapes for semi-infinite panels. ($\overline{W} = \overline{W}_1 + i\overline{W}_2$.) (Reproduced from Ref. 8–14.)

Typical mode shapes for semi-infinite panels under flutter conditions have been computed by Houbolt (Ref. 8–14) and are as illustrated by Fig. 8–7(a) and (b) for pinned and clamped edges, respectively. Both figures illustrate the interesting features that the lateral deflection is concentrated near the rear of the panel and that there are appreciable second mode components in the flutter modes.

GALERKIN SOLUTION FOR A SEMI-INFINITE FLAT PANEL. Examination of Fig. 8–7 would suggest that a Rayleigh-Ritz or Galerkin solution using three or four modes may provide an adequate approximate solution. Let us suppose that we take as a solution to Eq. 8–37 the following series representation which applies to a simply supported panel:

$$W(\xi, t) = \sum_{n=1}^{N} \sin(n\pi\xi) \cdot q_n(t) \tag{8–52}$$

Introducing this series into Eq. 8–37 and applying Galerkin's method [cf. Sec. 3–5(c)], we obtain

$$m^2\left(m^2 - \frac{R_{xx}}{\pi^2}\right)q_m + \frac{\lambda}{\pi^4}\left[\frac{(M^2 - 2)}{(M^2 - 1)}\frac{dq_m}{ds} + \right.$$
$$\left. \sqrt{M^2 - 1}\,\mu\,\frac{d^2 q_m}{ds^2} + \sum_{(m+n)=1,3,5,\cdots}^{N} \frac{4mn}{m^2 - n^2}q_n\right] = 0$$
$$(m = 1, 2, \cdots, N) \quad (8–53)$$

where $\mu = m_0/\rho a$ is a mass density parameter and $s = Ut/a$ is a non-dimensional time. It should be noted that the last summation in Eq. 8–53 is only over those values of n where $(n + m)$ is odd. The stability of Eqs. 8–53 may be examined by Routh's criteria, or more directly by putting as the condition for flutter or neutral stability $q_m(s) = \bar{q}_m e^{i\omega as/U}$. This yields

$$\left[m^2 \left(m^2 - \frac{R_{xx}}{\pi^2} \right) + \frac{i\lambda}{\pi^4} \frac{(M^2 - 2)}{(M^2 - 1)} \frac{\omega a}{U} - \sqrt{M^2 - 1} \frac{\mu\lambda}{\pi^4} \left(\frac{\omega a}{U} \right)^2 \right] \bar{q}_m$$

$$+ \frac{\lambda}{\pi^4} \sum_{(m+n)=1,3,5,\cdots}^{N} \frac{4mn}{m^2 - n^2} \bar{q}_n = 0 \quad (8\text{–}54)$$

When the determinant of Eqs. 8–54 is set equal to zero, the eigenvalues of the system may be computed. This coupled determinant can be solved by putting its real and imaginary parts equal to zero, as in conventional flutter analyses. Since

$$\lambda = \frac{m_0 a^4}{\mu D \sqrt{M^2 - 1}} \frac{\omega^2}{\left(\dfrac{\omega a}{U} \right)^2}$$

the complex determinant, after dividing through the original equation by ω^2, can be solved for ω and $\omega a/U$. The Galerkin method of analysis can easily be extended to include more exact supersonic air forces [cf. Sec. 4–7(c)].

If the damping term is neglected, Galerkin's method provides an exceptionally simple and direct approach. When two terms are taken, designated by the integers m and n, we obtain by neglecting the damping term and setting the determinant of Eqs. 8–54 equal to zero

$$\lambda = \frac{\pi^4(n^2 - m^2)}{4mn} \sqrt{\left(-m^4 + \frac{R_{xx}}{\pi^2} m^2 + \frac{k}{\pi^4} \right) \left(n^4 - \frac{R_{xx}}{\pi^2} n^2 - \frac{k}{\pi^4} \right)}$$

$$(m + n \text{ odd}) \quad (8\text{–}55)$$

The dashed curve in Fig. 8–6 illustrates an example of a plot of λ versus $(1/\pi)(k)^{1/4}$ as given by Eq. 8–55 for $R_{xx} = 0$, $m = 1$, and $n = 2$. The peak value of the λ versus k curve, obtained by putting $d\lambda/dk = 0$, defines the critical values of the parameters λ and k, as follows:

$$\lambda_{cr} = \frac{\pi^4}{8mn} (n^2 - m^2)^2 \left[(m^2 + n^2) - \frac{R_{xx}}{\pi^2} \right], \quad (m + n \text{ odd}) \quad (8\text{–}56a)$$

$$k_{cr} = \frac{\pi^4}{2} \left[m^4 + n^4 - \frac{R_{xx}}{\pi^2} (m^2 + n^2) \right], \quad (m + n \text{ odd}) \quad (8\text{–}56b)$$

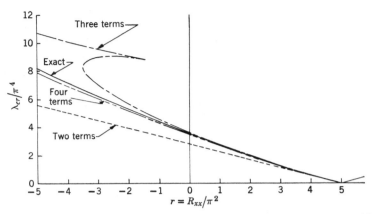

Fig. 8–8. Variation of λ_{cr}/π^4 with R_{xx}/π^2 for two-dimensional panels. (Reproduced from Ref. 8–11.)

Hedgepeth's results (Ref. 8–20), based on an exact solution in which the damping term is neglected, are compared to Galerkin's solutions involving two, three, .and four assumed modes in Fig. 8–8. The two-term approximation shown in Fig. 8–8 corresponds to Eq. 8–56a with $m = 1$, $n = 2$. The two-mode solution is seen to be conservative. The addition of another mode tends to stabilize the two-mode solution and produce a slightly nonconservative result. Finally, a fourth mode destabilizes the three-mode solution and partly offsets the influence of the third mode. It would appear that for a system well-ordered in both mode number and frequency, an even number of modes should be included in order to obtain a conservative answer. It is also apparent from Fig. 8–8 that two- and three-mode Galerkin solutions give reasonable results for panels in compression; but for panels in tension, that is, for negative values of R_{xx}, more modes are required. Indeed, for the limiting case of a membrane when R_{xx} approaches minus infinity, the flutter speed is infinite and an infinite number of modes is required.

REMARKS ON THE FLUTTER OF SEMI-INFINITE FLAT PANELS. Figure 8–9 shows a rather interesting summary of the complete stability behavior of flat panels having simply supported and clamped edges. In this diagram, we have included the possibility of a static instability when the midplane compressive stress resultants, $-N_{xx}$, are increased in magnitude. The conditions required to produce static buckling are examined by putting the time-dependent terms in Eqs. 8–53 equal to zero. With regard to static stability, we observe that the pressure of the airstream tends to stabilize the panel and permits a midplane stress resultant somewhat greater than the conventional buckling value. In fact, when the speed is such that $\lambda/\pi^4 = 9/8$ for a simply supported panel, the compressive load required for buckling is 2.5 times as great as the conventional Euler load. The stabilizing influence of tension and the destabilizing influence of compression on panel flutter are evident from Figs. 8–8 and 8–9.

It can be concluded at the time of this writing that, although the physical understanding of panel flutter is in an embryonic state, some reasonably clear patterns of analyses are emerging. For Mach numbers beyond 1.6, it appears that a first-order aerodynamic theory such as piston theory is adequate. For μ greater than approximately ten, the "static" aerodynamic forces employed by Hedgepeth are probably admissible. The type of panel flutter predicted by these analyses is due primarily to a coupling of the two lowest vibration modes of the panel, much as in the case of classical bending-torsion flutter (cf. Fig. 8–6). The flutter frequency is slightly below that of the second natural vibration mode, and the flutter speeds are as defined by the parameter λ_{cr}. Below $M = 1.6$, a first-order aerodynamic theory is inadmissible, and it is necessary to employ more exact air forces, such as those described in Sec. 4–7(c). It has been shown by Nelson and Cunningham (Ref. 8–9) that the flutter speed drops sharply near $M = \sqrt{2}$ due to losses in aerodynamic damping. Under these

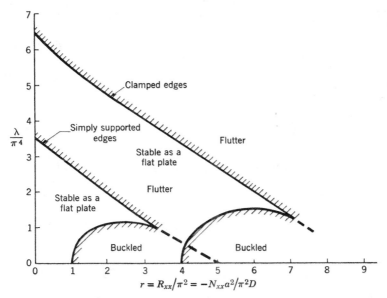

Fig. 8–9. Stability boundaries of two-dimensional flat panels.

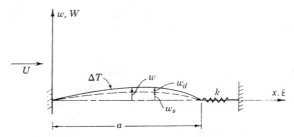

Fig. 8–10. Buckled two-dimensional panel. (Adapted from Ref. 8–14.)

circumstances, the flutter frequencies are near that of the first vibration mode, and the process resembles that of a soft single-degree-of-freedom flutter.

(*ii*) *Instability of semi-infinite buckled flat panels.* In the previous section, it was shown that the nondimensional stress resultant, R_{xx}, has a profound influence on the flutter speed. We have implicitly assumed, however, that the values of R_{xx} which are allowed are insufficient to produce plate buckling. We consider now the consequences of permitting the panel to buckle. We assume that the buckling is produced by allowing the temperature of the panel to be uniformly raised by an amount ΔT above that of its surroundings, as illustrated by Fig. 8–10. This produces compressive thermal stresses in the plate and results finally in buckling when $N_{xx}^{(0)} = -\pi^2 D/a^2$. After buckling has taken place, the midplane stress resultant is no longer constant; but it has the nondimensional form

$$R_{xx} = -\frac{N_{xx}^{(0)}a^2}{D} = \frac{Eha^2\beta_k}{D}\left[\alpha\Delta T - \frac{1}{2}\int_0^1\left(\frac{\partial W}{\partial \xi}\right)^2 d\xi\right] \qquad (8-57)$$

where β_k is a support factor which accounts for the effective stiffness of the supporting structure defined by

$$\beta_k = \frac{1}{1 + (Eh/ak)}$$

and where k is a spring constant per unit spanwise length of panel (cf. Fig. 8-10). The aeroelastic equation of the buckled, initially flat, infinite-aspect-ratio panel is obtained by inserting the value of R_{xx} given by Eq. 8–57 into Eq. 8–37.

$$\frac{\partial^4 W}{\partial \xi^4} + \frac{Eha^2\beta_k}{D}\left[\alpha\Delta T - \frac{1}{2}\int_0^1\left(\frac{\partial W}{\partial \xi}\right)^2 d\xi\right]\frac{\partial^2 W}{\partial \xi^2}$$
$$+ \lambda\frac{\partial W}{\partial \xi} + \lambda\frac{M^2 - 2}{M^2 - 1}\frac{a}{U}\frac{\partial W}{\partial t} + \frac{m_0 a^4}{D}\frac{\partial^2 W}{\partial t^2} = 0 \qquad (8-58)$$

Let us put $W = W_s + W_d$, where W_s represents the nondimensional static, and W_d the nondimensional dynamic, deflections of the panel. After making this substitution, we may separate Eq. 8–58 into two equations as follows:

$$\frac{d^4 W_s}{d\xi^4} + \frac{Eh\,a^2\beta_k}{D}\left[\alpha\Delta T - \frac{1}{2}\int_0^1\left(\frac{dW_s}{d\xi}\right)^2 d\xi\right]\frac{d^2 W_s}{d\xi^2} + \lambda\frac{dW_s}{d\xi} = 0 \quad (8\text{–}59)$$

$$\frac{\partial^4 W_d}{\partial\xi^4} + \frac{Eh\,a^2\beta_k}{D}\left[\alpha\Delta T - \frac{1}{2}\int_0^1\left(\frac{dW_s}{d\xi} + \frac{\partial W_d}{\partial\xi}\right)^2 d\xi\right]\frac{\partial^2 W_d}{\partial\xi^2} + \lambda\frac{\partial W_d}{\partial\xi}$$

$$- \frac{Eha^2\beta_k}{2D}\left[\int_0^1\frac{\partial W_d}{\partial\xi}\left(2\frac{dW_s}{d\xi} + \frac{\partial W_d}{\partial\xi}\right) d\xi\right]\frac{d^2 W_s}{d\xi^2}$$

$$+ \lambda\frac{M^2 - 2}{M^2 - 1}\frac{a}{U}\frac{\partial W_d}{\partial t} + \frac{m_0 a^4}{D}\frac{\partial^2 W_d}{\partial t^2} = 0 \quad (8\text{–}60)$$

The first of these nonlinear equations may be employed to compute the static buckled deformation shape, W_s; and once this result is obtained, the second may be used to compute the additional dynamic deformation, W_d.

We consider first a Galerkin type solution of Eq. 8–59 in which we put

$$W_s = \sum_{i=1}^{n} \phi_i q_i \quad (8\text{–}61)$$

where the ϕ_i are modal functions which satisfy the boundary conditions on the panel. Substituting Eq. 8–61 into Eq. 8–59, and applying Galerkin's method, yields the equations

$$\sum_{i=1}^{n}\left\{A_{ki}{}^\phi + \frac{Eha^2\beta_k}{D}\alpha\Delta T B_{ki}{}^\phi + \lambda C_{ki}{}^\phi - \frac{B_{ki}{}^\phi Eha^2\beta_k}{2D}\left[\sum_l^n\sum_m^n F_{lm}{}^\phi q_l q_m\right]\right\}q_i = 0$$
$$(8\text{–}62)$$

where

$$A_{ki}{}^\phi = \int_0^1\frac{d^2\phi_k}{d\xi^2}\frac{d^2\phi_i}{d\xi^2}\,d\xi \qquad C_{ki}{}^\phi = \int_0^1\phi_k\frac{d^2\phi_i}{d\xi^2}\,d\xi$$

$$B_{ki}{}^\phi = -\int_0^1\frac{d\phi_k}{d\xi}\frac{d\phi_i}{d\xi}\,d\xi \qquad F_{lm}{}^\phi = \int_0^1\frac{d\phi_l}{d\xi}\frac{d\phi_m}{d\xi}\,d\xi$$

If we take, for example, the case of simply supported edges, we put

$$\phi_i = \sin(i\pi\xi) \quad (8\text{–}63)$$

and when two modes are taken, two algebraic equations are obtained:

$$\left[1 - \frac{\Delta T}{\Delta T_{cr}} + \frac{E\beta_k h a^2}{4D}(q_1{}^2 + 4q_2{}^2)\right]q_1 - \frac{8}{3}\frac{\lambda}{\pi^4}q_2 = 0$$

$$\frac{8}{3}\frac{\lambda}{\pi^4}q_1 + \left[16 - 4\frac{\Delta T}{\Delta T_{cr}} + \frac{E\beta_k h a^2}{D}(q_1{}^2 + 4q_2{}^2)\right]q_2 = 0$$
$$(8\text{–}64)$$

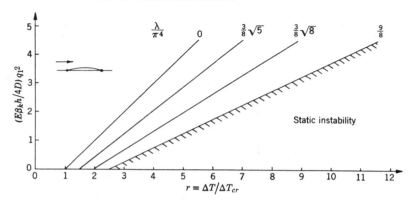

Fig. 8–11. Relations among first mode component, velocity, and temperature for buckled panel. (Reproduced from Ref. 8–14.)

where

$$\Delta T_{cr} = \frac{\pi^2 D}{E\beta_k \alpha h a^2} \qquad (8\text{–}65)$$

is the rise in temperature that is required to buckle the panel.

The solution of Eqs. 8–64 leads to a quadratic equation in the non-dimensional generalized coordinate q_1, which can be solved explicitly to give

$$q_1 = \sqrt{\left\{\frac{4D}{E\beta_k h a^2}\left[\frac{1}{2}\left(\frac{\Delta T}{\Delta T_{cr}} - 1\right)\left(1 + \sqrt{1 - \frac{8}{9}\left(\frac{\lambda}{\pi^4}\right)^2}\right) - \frac{16}{27}\left(\frac{\lambda}{\pi^4}\right)^2\right]\right\}} \qquad (8\text{–}66)$$

The physical nature of the solution given by Eq. 8–66 may be portrayed by Fig. 8–11, which is due to Houbolt (Ref. 8–14). It is apparent that for $\Delta T/\Delta T_{cr} < 1$, only the trivial solution $q_1 = q_2 = 0$ is obtained. Thus, in Fig. 8–11, we observe that $q_1 = 0$ in the region where $\Delta T/\Delta T_{cr} < 1.0$. In the region where $1 < \Delta T/\Delta T_{cr} < 2.5$, we observe that an increase in λ, that is, an increase in velocity, will cause the buckled panel to be blown flat. However, for $\Delta T/\Delta T_{cr} > 2.5$, an increase in velocity will cause the buckle depth to decrease until a static instability is reached. The velocity at which the buckled panel is statically unstable is independent of buckle depth and temperature. In Fig. 8-12, plots of the boundaries of static instability for simply supported and clamped plates of infinite aspect ratio $(b/a = \infty)$ are illustrated.

Let us turn our attention next to the conditions for dynamic instability of buckled panels. This may be accomplished by applying a Galerkin type solution to Eq. 8–60. We observe that Eq. 8–60 is nonlinear. We shall be

content, however, with a solution in which W_d represents a small perturbation in W and hence is small in comparison with W_s. Neglecting squares
and products of terms involving W_d in Eq. 8–60 results in a linearized
version of Eq. 8–60, as follows:

$$\frac{\partial^4 W_d}{\partial \xi^4} + \frac{Eha^2\beta_k}{D} \left[\alpha\Delta T - \frac{1}{2} \int_0^1 \left(\frac{dW_s}{d\xi}\right)^2 d\xi \right] \frac{\partial^2 W_d}{\partial \xi^2}$$

$$- \frac{a^2 Eh\beta_k}{D} \left[\int_0^1 \frac{\partial W_d}{\partial \xi} \cdot \frac{dW_s}{d\xi} d\xi \right] \frac{d^2 W_s}{d\xi^2} + \lambda \frac{\partial W_d}{d\xi}$$

$$+ \lambda \left(\frac{M^2 - 2}{M^2 - 1}\right) \frac{a}{U} \frac{\partial W_d}{\partial t} + \frac{m_0 a^4}{D} \frac{\partial^2 W_d}{\partial t^2} = 0 \quad (8\text{–}67)$$

Substituting into this equation the series

$$W_d = \sum_{i=1}^N \psi_i \zeta_i \quad (8\text{–}68)$$

where ψ_i are modal functions which satisfy the boundary conditions on the
panel and ζ_i are generalized coordinates, we obtain by applying Galerkin's
method the following differential equations:

$$\sum_{i=1}^N \left\{ \mu\lambda\sqrt{M^2 - 1} \, H_{ki}{}^\psi \frac{d^2}{ds^2} + \lambda \frac{M^2 - 2}{M^2 - 1} H_{ki}{}^\psi \frac{d}{ds} + \lambda C_{ki}{}^\psi \right.$$

$$- \frac{Eha^2\beta_k}{2D} B_{ki}{}^\psi \sum_l^N \sum_m^N F_{lm}{}^\phi q_l q_m - \frac{Eha^2\beta_k}{D}$$

$$\times J_{ki}{}^{\phi\psi} \sum_l^N \sum_m^N F_{lm}{}^{\phi\psi} q_l \zeta_m + A_{ki}{}^\psi + \frac{Eha^2\beta_k}{D} \alpha\Delta T B_{ki}{}^\psi \left. \right\} \zeta_i = 0 \quad (8\text{–}69)$$

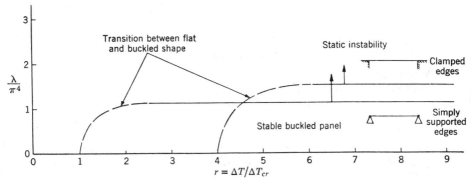

Fig. 8–12. Static stability boundaries of buckled panels. (Reproduced from Ref. 8–14.)

where

$$A_{ki}{}^{\psi} = \int_0^1 \frac{\partial^2 \psi_k}{\partial \xi^2} \frac{\partial^2 \psi_i}{\partial \xi^2} d\xi \qquad F_{lm}{}^{\phi} = \int_0^1 \frac{\partial \phi_l}{\partial \xi} \frac{\partial \phi_m}{\partial \xi} d\xi$$

$$B_{ki}{}^{\psi} = -\int_0^1 \frac{\partial \psi_k}{\partial \xi} \frac{\partial \psi_i}{\partial \xi} d\xi \qquad F_{lm}{}^{\phi\psi} = \int_0^1 \frac{\partial \phi_l}{\partial \xi} \cdot \frac{\partial \psi_m}{\partial \xi} d\xi$$

$$C_{ki}{}^{\psi} = \int_0^1 \psi_k \frac{\partial \psi_i}{\partial \xi} d\xi \qquad H_{ki}{}^{\psi} = \int_0^1 \psi_k \psi_i \, d\xi$$

$$J_{ki}{}^{\phi\psi} = \int_0^1 \frac{\partial^2 \phi_i}{\partial \xi^2} \cdot \frac{\partial \psi_k}{\partial \xi} d\xi$$

Selecting again the simply supported panel for illustrative purposes, we take ϕ_i and ψ_i both according to Eq. 8–63; and we obtain the following equations for two modes:

$$\left[\mu\sqrt{M^2 - 1} \, \frac{\lambda}{\pi^4} \frac{d^2}{ds^2} + \frac{\lambda}{\pi^4} \frac{d}{ds} + 1 - \frac{\Delta T}{\Delta T_{cr}} + \frac{hE\beta_k a^2}{D}(\tfrac{3}{4}q_1{}^2 + q_2{}^2) \right] \zeta_1$$

$$+ \left[\frac{hE\beta_k a^2}{D} 2q_1 q_2 - \frac{8}{3} \frac{\lambda}{\pi^4} \right] \zeta_2 = 0$$

$$\left[\frac{hE\beta_k a^2}{D} 2q_1 q_2 + \frac{8}{3} \frac{\lambda}{\pi^4} \right] \zeta_1 + \left[\mu\sqrt{M^2 - 1} \, \frac{\lambda}{\pi^4} \frac{d^2}{ds^2} \right. \tag{8-70}$$

$$\left. + 16 - 4\frac{\Delta T}{\Delta T_{cr}} + \frac{hE\beta_k a^2}{D}(q_1{}^2 + 12q_2{}^2) \right] \zeta_2 = 0$$

Equations 8–70 lead to a fourth-order characteristic equation in r when solutions of the form e^{rs} are assumed for ζ_1 and ζ_2. The values of q_1 and q_2 as obtained from Eq. 8–64 are inserted into the coefficients, and the resulting characteristic equation is examined for static and dynamic stability by means of Routh's criteria. When the characteristic equation has the form

$$a_0 r^4 + a_1 r^3 + a_2 r^2 + a_3 r + a_4 = 0 \tag{8-71}$$

we have, from Routh's criteria,

(a) Static stability when all coefficients > 0.

(b) Dynamic stability when $a_1 a_2 a_3 - a_0 a_3{}^2 - a_4 a_1{}^2 > 0$.

When these results are investigated, we find that no dynamic instability condition is possible in the stable buckled range. For dynamic instability to occur for $\Delta T / \Delta T_{cr}$ less than 2.5 and 6.06 for the simply supported and clamped panels, respectively, the panel must be unbuckled. Thus the complete stability boundaries of the panel may be obtained by combining Figs. 8–9 and 8–12 as illustrated by Fig. 8–13. It will be observed that the value of R_{xx}/π^2 in Fig. 8–9 has the same significance as $\Delta T / \Delta T_{cr}$ in Fig. 8–12.

(*iii*) *Instability of rectangular flat panels of finite aspect ratio.* In the sections directly above, we have concentrated attention on semi-infinite panels with the air blowing in the direction of the finite dimension. We shall discuss in the present section the flutter of finite-aspect-ratio panels. It will be shown that the solutions presented above for semi-infinite panels apply also to finite-aspect-ratio panels, with changes in the definitions of some of the parameters. Let us put as a solution of Eq. 8–35 the product

$$W = \bar{F}(\xi)\bar{G}(\eta)e^{\alpha t} \tag{8–72}$$

Substituting Eq. 8–72 into Eq. 8–35, multiplying through by $\bar{G}(\eta)$, and integrating on η from 0 to b, the following result is obtained:

$$\frac{\partial^4 \bar{F}}{\partial \xi^4} + R \frac{\partial^2 \bar{F}}{\partial \xi^2} + \lambda \frac{\partial \bar{F}}{\partial \xi} - k'\bar{F} = 0 \tag{8–73}$$

where

$$R = R_{xx} - 2\frac{a^2}{b^2}\frac{\displaystyle\int_0^1 \left(\frac{\partial \bar{G}}{\partial \eta}\right)^2 d\eta}{\displaystyle\int_0^1 \bar{G}^2 \, d\eta}$$

$$k' = -\frac{m_0 a^4}{D}\alpha^2 + R_{yy}\frac{\displaystyle\int_0^1 \left(\frac{\partial \bar{G}}{\partial \eta}\right)^2 d\eta}{\displaystyle\int_0^1 \bar{G}^2 \, d\eta} - \left(\frac{a}{b}\right)^4 \frac{\displaystyle\int_0^1 \left(\frac{\partial^2 \bar{G}}{\partial \eta^2}\right)^2 d\eta}{\displaystyle\int_0^1 \bar{G}^2 \, d\eta}$$

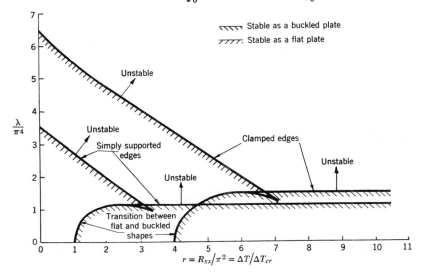

Fig. 8–13. Complete stability boundaries of flat and buckled panels.

In forming Eq. 8–73, the damping term proportional to $\partial W/\partial t$ in Eq. 8–35 has been omitted, and the possibility of midplane stress resultants N_{xx} and N_{yy} has been allowed for. The quantity k' now takes on the role of the eigenvalue.

We observe that if the nondimensional deformation shape, $\bar{G}(\eta)$, is specified, the form of Eq. 8–73 is identical to that of Eq. 8–41. The solutions of the latter equation, given in the previous sections for semi-infinite panels, may, therefore, be adapted to the present problem. For example, Figs. 8–8, 8–9, 8–11, 8–12, and 8–13, defining the static and dynamic instability boundaries, may be applied to finite-aspect-ratio panels by using the same interpretation of λ (cf. Eq. 8–36), but now defining r according to

$$r = \frac{R}{\pi^2} = \frac{R_{xx}}{\pi^2} - \frac{2a^2}{\pi^2 b^2} \frac{\displaystyle\int_0^1 \left(\frac{\partial \bar{G}}{\partial \eta}\right)^2 d\eta}{\displaystyle\int_0^1 \bar{G}^2\, d\eta} \tag{8–74}$$

In all of the figures just mentioned, r is effectively shifted to the left when finite-aspect-ratio effects are included.

Interpretation of the flutter frequency requires, however, consideration of the difference between the quantities k and k'. Omitting the damping term and putting $\alpha = \omega_f i$, we have for the semi-infinite panel from Eq. 8–41

$$k = \frac{\pi^4 \omega_f{}^2}{\omega_r{}^2} \tag{8–75}$$

Correspondingly, we have for the finite-aspect-ratio panel, from Eq. 8–73

$$k' = \frac{\pi^4 \omega_f'^2}{\omega_r{}^2} + R_{yy} \frac{\displaystyle\int_0^1 \left(\frac{\partial \bar{G}}{\partial \eta}\right)^2 d\eta}{\displaystyle\int_0^1 \bar{G}^2\, d\eta} - \left(\frac{a}{b}\right)^4 \frac{\displaystyle\int_0^1 \left(\frac{\partial^2 \bar{G}}{\partial \eta^2}\right)^2 d\eta}{\displaystyle\int_0^1 \bar{G}^2\, d\eta} \tag{8–76}$$

Since k and k' are eigenvalues of the same differential equation, they may be equated to each other; thus, we obtain

$$\frac{\omega_f'^2}{\omega_r{}^2} = \frac{\omega_f{}^2}{\omega_r{}^2} - \frac{R_{yy}}{\pi^4} \frac{\displaystyle\int_0^1 \left(\frac{\partial \bar{G}}{\partial \eta}\right)^2 d\eta}{\displaystyle\int_0^1 \bar{G}^2\, d\eta} + \left(\frac{a}{\pi b}\right)^4 \frac{\displaystyle\int_0^1 \left(\frac{\partial^2 \bar{G}}{\partial \eta^2}\right)^2 d\eta}{\displaystyle\int_0^1 \bar{G}^2\, d\eta} \tag{8–77}$$

Equation 8–77 may be employed to compute the flutter frequency, ω_f', of a rectangular panel of dimensions $x = a$ and $y = b$ when the flutter frequency, ω_f, is known for a panel of dimensions $x = a$ and $y = \infty$.

As a simple illustration, it can be easily shown that for a simply supported panel where $\bar{G} = \sin(\pi\eta)$, we have

$$r = \frac{R_{xx}}{\pi^2} - 2\left(\frac{a}{b}\right)^2$$

$$\frac{{\omega_f'}^2}{{\omega_r}^2} = \frac{{\omega_f}^2}{{\omega_r}^2} - \frac{R_{yy}}{\pi^2} + \frac{a^4}{b^4}$$

(8–78)

The variation of thickness ratio, h/a, required to prevent flutter of an uncompressed aluminum alloy panel at 50,000 ft, is shown by the diagram of Fig. 8–14. This diagram, taken from Ref. 8–11, illustrates the fact that the low supersonic range is the most critical and shows that the effect of reducing aspect ratio is a reduction of the required h/a. The dashed lines of the figure are results obtained at the low supersonic Mach numbers by Luke and St. John (Ref. 8–21). Although their results were obtained from a two-mode solution, and the solid lines were obtained from a four-mode solution, the merging of the lines is remarkably close. It is evident also from Fig. 8–14 that, for $a/b = \infty$, the most critical design condition occurs in the vicinity of $M = 1.3$, with the thickness decreasing rapidly between $M = 1.3$ and $M = 1.5$ and then increasing slowly. On the other hand, for $a/b = 1$, the required thickness ratio to prevent flutter increases slowly with Mach number. Although the critical thickness ratios indicated by Fig. 8–14 would correspond to "minimum gages" in most designs, we

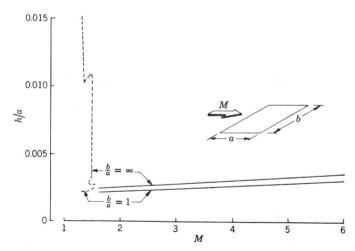

Fig. 8–14. Variation of thickness ratio with Mach number for unstressed aluminum panels at 50,000 ft. (Reproduced from Ref. 8–11.)

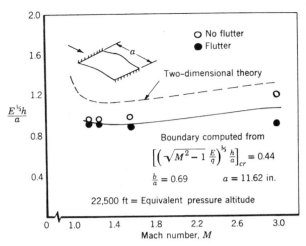

Fig. 8–15. Comparison between theory and experiment for panel flutter of buckled panels. (Reproduced from Ref. 8–25.)

cannot conclude that this is always the case since, for example, low-aspect-ratio panels compressively loaded in the chordwise, or x, direction may require appreciably higher thickness ratios in order to prevent flutter. It is interesting to observe that the flutter speed is independent of N_{yy} although the flutter frequency is affected as shown by Eq. 8–77. This result is of interest since the stress resultant, N_{yy}, may be appreciable in wings or pressurized fuselage and missile bodies.

(iv) *Experimental confirmation of panel flutter theory.* In order finally to assess the validity of the panel flutter theory, careful comparisons to experimental evidence must eventually be made. At the time of this writing, such comparisons are not possible because of a lack of experimental data, although the limited amount of such data available tends to indicate that these panel flutter theories may be slightly conservative (Refs. 8–24 and 8–25). For example, Fig. 8–15 indicates a comparison between theory and experiment taken from Ref. 8–25. Here, the effect of Mach number is shown on the flutter of buckled panels clamped at the front and rear edges and at an equivalent altitude of 22,500 ft. The structural stiffness parameter $E^{1/3}h/a$ is plotted against Mach number in this figure. The dotted theoretical curve is for a two-dimensional panel; whereas the actual panel is of finite width, clamped only at front and rear with the sides free, and of aspect ratio $b/a = 0.69$. The plotted points denote experimental data, and the solid line represents an empirically derived flutter boundary. The conservatism of the theory in this instance is evident from the figure.

Panel flutter experiments have been especially illuminating in indicating the destructive nature of the phenomenon. In general, panel flutter appears to be not immediately destructive, and designers will probably be concerned largely with its fatigue implications.

8-6 THIN CYLINDRICAL PANELS

In Secs. 8-4 and 8-5, we considered the aeroelastic equations and certain of their instability solutions for flat and buckled panels. In practical design applications, however, we are frequently confronted with curved panels as illustrated by Fig. 8-16, or complete structures, such as cylindrical shells. We shall introduce this subject by retaining the shallow shell terms in Eqs. 8-34 as suggested by Shulman (Ref. 8-16) and Voss (Ref. 8-26). Let us rewrite these equations by assuming that the deflections $w(x, y)$ represent small perturbations from a prestressed condition defined by specified values of $N_{xx}^{(0)}$, $N_{xy}^{(0)}$, and $N_{yy}^{(0)}$. If we linearize Eqs. 8-34 with respect to deflections and midplane stress resultants, we obtain

$$\left[D\nabla^4 - N_{xx}^{(0)} \frac{\partial^2}{\partial x^2} - 2N_{xy}^{(0)} \frac{\partial^2}{\partial x\, \partial y} - N_{yy}^{(0)} \frac{\partial^2}{\partial y^2} \right.$$

$$\left. + \frac{2q}{\sqrt{M^2 - 1}} \left(\frac{\partial}{\partial x} + \frac{1}{U} \frac{M^2 - 2}{M^2 - 1} \frac{\partial}{\partial t} \right) + m_0 \frac{\partial^2}{\partial t^2} \right](w)$$

$$- \left(\frac{\partial^2 z}{\partial x^2} \frac{\partial^2}{\partial y^2} + \frac{\partial^2 z}{\partial y^2} \frac{\partial^2}{\partial x^2} - 2 \frac{\partial^2 z}{\partial x\, \partial y} \frac{\partial^2}{\partial x\, \partial y} \right)(F) = Z_D \quad (8\text{-}79)$$

$$\nabla^4(F) + Eh \left(\frac{\partial^2 z}{\partial x^2} \frac{\partial^2}{\partial y^2} + \frac{\partial^2 z}{\partial y^2} \frac{\partial^2}{\partial x^2} - 2 \frac{\partial^2 z}{\partial x\, \partial y} \frac{\partial^2}{\partial x\, \partial y} \right)(w) = 0 \quad (8\text{-}80)$$

where we have omitted the temperature terms for brevity. Specializing

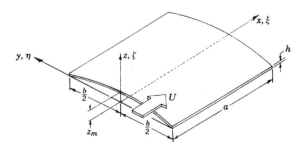

Fig. 8-16. Shallow cylindrical panel.

Eqs. 8–79 and 8–80 to the case of a shallow cylindrical shell of radius R, we have*

$$\left[D\nabla^4 - N_{xx}^{(0)} \frac{\partial^2}{\partial x^2} - 2N_{xy}^{(0)} \frac{\partial^2}{\partial x\, \partial y} - N_{yy}^{(0)} \frac{\partial^2}{\partial y^2} + \frac{2q}{\sqrt{M^2 - 1}} \right.$$

$$\left. \times \left(\frac{\partial}{\partial x} + \frac{1}{U} \frac{M^2 - 2}{M^2 - 1} \frac{\partial}{\partial t} \right) + m_0 \frac{\partial^2}{\partial t^2} \right] (w) + \frac{1}{R} \frac{\partial^2 F}{\partial x^2} = Z_D \quad (8\text{–}81)$$

$$\nabla^4(F) - \frac{Eh}{R} \frac{\partial^2 w}{\partial x^2} = 0 \quad (8\text{–}82)$$

We shall consider in the following paragraphs solutions of Eqs. 8–81 and 8–82 for the instability of a shallow, simply supported rectangular panel of dimensions a and b as shown by Fig. 8–16. By introducing the nondimensional parameters of Eq. 8–36, we obtain the following forms of Eqs. 8–81 and 8–82:

$$\frac{\partial^4 W}{\partial \xi^4} + 2\left(\frac{a}{b}\right)^2 \frac{\partial^4 W}{\partial \xi^2\, \partial \eta^2} + \left(\frac{a}{b}\right)^4 \frac{\partial^4 W}{\partial \eta^4} + R_{xx} \frac{\partial^2 W}{\partial \xi^2} + 2R_{xy} \frac{\partial^2 W}{\partial \xi\, \partial \eta} + R_{yy} \frac{\partial^2 W}{\partial \eta^2}$$

$$+ \lambda \frac{\partial W}{\partial \xi} + \lambda \frac{M^2 - 2}{M^2 - 1} \frac{a}{U} \frac{\partial W}{\partial t} + \frac{m_0 a^4}{D} \frac{\partial^2 W}{\partial t^2} + \frac{1}{R'} \frac{\partial^2 F'}{\partial \xi^2} = \frac{a^3}{D} Z_D \quad (8\text{–}83)$$

$$\frac{\partial^4 F'}{\partial \xi^4} + 2\left(\frac{a}{b}\right)^2 \frac{\partial^4 F'}{\partial \xi^2\, \partial \eta^2} + \left(\frac{a}{b}\right)^4 \frac{\partial^4 F'}{\partial \eta^4} - \gamma' \frac{\partial^2 W}{\partial \xi^2} = 0 \quad (8\text{–}84)$$

where the additional parameters $F' = F/D$, $R' = R/a$, and $\gamma' = Eha^3/DR$ have been introduced. Equation 8–83 can be compared with Eq. 8–35 for the flat panel, and the former reduces to the latter when $R' \to \infty$.

Neglecting the damping term proportional to $\partial W/\partial t$ and the shear term proportional to N_{xy}, solutions are obtained by putting

$$W(\xi, \eta, t) = \overline{W}_n(\xi) \cos n\pi\eta\, e^{i\omega t}$$
$$F'(\xi, \eta, t) = \overline{F}_n{}'(\xi) \cos n\pi\eta\, e^{i\omega t} \quad (8\text{–}85)$$

where we introduce the assumption of simply supported edges. When these quantities are substituted into Eqs. 8–83 and 8–84, there is derived for

---

* Here we represent the shape of the cylindrical shell by the formula

$$z = z_m \left[1 - \left(\frac{y}{b/2} \right)^2 \right]$$

The maximum camber is designated as z_m (cf. Fig. 8–16); and this quantity relates to the radius by $z_m = b^2/8R$.

the homogeneous system

$$\frac{d^4\overline{W}_n}{d\xi^4} - 2n^2\pi^2\left(\frac{a}{b}\right)^2 \frac{d^2\overline{W}_n}{d\xi^2} + n^4\pi^4\left(\frac{a}{b}\right)^4 \overline{W}_n + R_{xx}\frac{d^2\overline{W}_n}{d\xi^2}$$

$$- R_{yy}n^2\pi^2\overline{W}_n + \lambda\frac{d\overline{W}_n}{d\xi} - m_0\frac{a^4\omega^2}{D}\overline{W}_n + \frac{1}{R'}\frac{d^2F}{d\xi^2} = 0 \quad (8\text{-}86)$$

$$\frac{d^4F_n{}'}{d\xi^4} - 2n^2\pi^2\left(\frac{a}{b}\right)^2 \frac{d^2F_n{}'}{d\xi^2} + n^4\pi^4\left(\frac{a}{b}\right)^4 F_n{}' - \gamma'\frac{d^2\overline{W}_n}{d\xi^2} = 0 \quad (8\text{-}87)$$

In obtaining solutions to Eqs. 8–86 and 8–87, we follow Voss (Ref. 8–26) and apply Galerkin's method by substituting

$$\overline{W}_n(\xi) = \sum_m a_{mn}\sin m\pi\xi$$

$$F_n(\xi) = \sum_m b_{mn}\sin m\pi\xi$$

$$(8\text{-}88)$$

From this substitution, the following equations are derived:

$$\left\{m^4 - \left[\frac{R_{xx}}{\pi^2} - 2n^2\left(\frac{a}{b}\right)^2\right]m^2 - \left[\frac{R_{yy}}{\pi^2}n^2\left(\frac{a}{b}\right)^2 - n^4\left(\frac{a}{b}\right)^4 + \frac{k}{\pi^4}\right] \right. \quad (8\text{-}89)$$

$$\left. + \left[\frac{1}{R'}\frac{\gamma'}{a^4}\frac{m^4}{\left[m^2 + n^2\frac{a^2}{b^2}\right]^2}\right]\right\} a_{mn} + \frac{\lambda}{\pi^4}\sum_{(m+q)=1,3,5,\cdots}^{N}\frac{4mq}{(m^2-q^2)}a_{qn} = 0$$

where $k = (m_0a^4/D)\omega^2$. Equations 8–89 reduce to Eqs. 8–54 for the case of a semi-infinite flat panel. They also reduce, in the absence of aerodynamic forces ($\lambda = 0$), to Reissner's shallow shell frequency equation

$$\omega^2 = \frac{\pi^4 D}{m_0a^4}\left\{\left[m^2 + n^2\left(\frac{a}{b}\right)^2\right]^2 + \frac{\gamma'}{R'a^4}\frac{m^4}{\left[m^2 + n^2\left(\frac{a}{b}\right)^2\right]^2}\right.$$

$$\left. - \frac{R_{xx}}{\pi^2}m^2 - \frac{R_{yy}}{\pi^2}n^2\left(\frac{a}{b}\right)^2\right\} \quad (8\text{-}89a)$$

The first term of Eq. 8–89a clearly gives the influence of bending of the shell, the second term the influence of stretching, and the last two terms the effects of initial stressing. We illustrate a solution of Eqs. 8–89 by assuming first a single value of n and then selecting single values of m and

q such that $(m + q)$ is odd. This leads to the following result for the parameter λ:

$$\lambda = \frac{\pi^4(q^2 - m^2)}{4mq} \sqrt{\left\{ -m^4 + \left[\frac{R_{xx}}{\pi^2} - 2n^2\left(\frac{a}{b}\right)^2\right]m^2 \right.}$$

$$+ \left[\frac{R_{yy}}{\pi^2} n^2\left(\frac{a}{b}\right)^2 - n^4\left(\frac{a}{b}\right)^4 + \frac{k}{\pi^4}\right] - \left[\frac{1}{R'}\frac{\gamma'}{a^4}\frac{m^4}{\left[m^2 + n^2\left(\frac{a}{b}\right)^2\right]^2}\right]\right\}$$

$$\times \left\{ q^4 - \left[\frac{R_{xx}}{\pi^2} - 2n^2\left(\frac{a}{b}\right)^2\right]q^2 - \left[\frac{R_{yy}}{\pi^2} n^2\left(\frac{a}{b}\right)^2 - n^4\left(\frac{a}{b}\right)^4 + \frac{k}{\pi^4}\right] \right.$$

$$\left. \left. + \left[\frac{1}{R'}\frac{\gamma'}{a^4}\frac{q^4}{\left[q^2 + n^2\left(\frac{a}{b}\right)^2\right]^2}\right]\right\}, \quad (m+q \text{ odd}) \right. \tag{8-90}$$

Since the peak value computed by putting $d\lambda/dk = 0$ defines the flutter speed, we obtain

$$\lambda_{cr} = \frac{\pi^4(q^2 - m^2)^2}{8mq}\left\{ q^2 + m^2 - \left[\frac{R_{xx}}{\pi^2} - 2n^2\left(\frac{a}{b}\right)^2\right] \right.$$

$$+ \frac{\frac{1}{R'}\frac{\gamma'}{a^4}\left[\frac{q^4}{\left[q^2 + n^2\left(\frac{a}{b}\right)^2\right]^2} - \frac{m^4}{\left[m^2 + n^2\left(\frac{a}{b}\right)^2\right]^2}\right]}{q^2 - m^2} \right\}, (m+q \text{ odd}) \tag{8-91a}$$

$$k_{cr} = \frac{\pi^4}{2}\left\{ m^4 + q^4 - \left[\frac{R_{xx}}{\pi^2} - 2n^2\left(\frac{a}{b}\right)^2\right](m^2 + q^2) \right.$$

$$\left. + \frac{1}{R'}\frac{\gamma'}{a^4}\left[\frac{m^4}{\left[m^2 + n^2\left(\frac{a}{b}\right)^2\right]^2} + \frac{q^4}{\left[q^2 + n^2\left(\frac{a}{b}\right)^2\right]^2}\right]\right\}, \quad (m+q \text{ odd}) \tag{8-91b}$$

It is easy to verify also that Eqs. 8–90, 8–91a, and 8–91b reduce to the semi-infinite flat plate equations (8–55, 8–56a, and 8–56b, respectively) when $R' \to \infty$ and $b \to \infty$.

It is instructive to observe the corrections to the flat panel results which arise due to curvature. Making use of Eq. 8–91a, let us put $m = 1$ and $q = 2$. This yields for the case of $R_{xx} = 0$,

$$\lambda_{cr} = \frac{9\pi^4}{16}\left[5 + 2n^2\left(\frac{a}{b}\right)^2\right] + \Delta\lambda_{cr} \tag{8-92}$$

where the first term is the flat panel result and where

$$\Delta \lambda_{cr} = \frac{9\pi^4}{16} \frac{1}{R'} \frac{\gamma'}{a^4} \frac{n^2 \left(\frac{a}{b}\right)^2 \left[8 + 5n^2 \left(\frac{a}{b}\right)^2\right]}{\left[4 + n^2 \left(\frac{a}{b}\right)^2\right]^2 \left[1 + n^2 \left(\frac{a}{b}\right)^2\right]^2}$$

is the correction due to curvature.

When we examine Eq. 8–92, using geometric data corresponding to practical cases, it becomes apparent that the corrective term may be several times larger than the flat panel term. Furthermore, there may be several modal combinations with nearly the same values of λ_{cr}; and the critical value does not necessarily occur, as it does with flat plates, with a combination of minimum mode numbers m and q. It would appear, however, that if attention is restricted to the lowest values of m and q, an analysis which requires a not unreasonable number of modes will give a good approximation to the true flutter behavior. With regard to the transverse modes and the integer n, we find a behavior unlike that of the flat plate where the lowest value of n produced a critical flutter speed. Here we find by a trial and error process that there can be a minimum flutter speed for a reasonably large value of n. This indicates then that the flutter speed decreases with increasing n, reaches a minimum, and then increases again. In each analysis, the analyst is obliged to find the critical value of n required for a minimum flutter speed.

The preceding analysis, based upon the shallow shell theory, is not only applicable to segments of circular cylinders but also provides a possible flutter condition for complete circular cylinders in which the number of circumferential waves in the flutter mode is large. This point is discussed at greater length in Sec. 8–7 which follows.

8–7   CIRCULAR CYLINDRICAL SHELLS

In Sec. 8–6, we discussed the aeroelastic equations and a theory of flutter of cylindrical panels in terms of the shallow shell theory. In the present section, we extend these notions to the case of a complete circular cylinder, as illustrated by Fig. 8–17. The selection of appropriate sets of equations governing the shell behavior is one of the most crucial points in the success of such an extension. We shall find it convenient to discuss the flutter behavior of the complete cylinder in terms of two regimes. The first is that in which the number of circumferential waves in the flutter mode is large, and the second is that in which the number of circumferential waves is small, approaching zero.

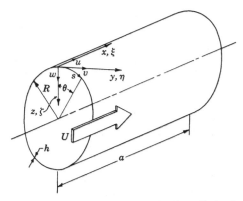

Fig. 8–17. Notation employed for circular cylinder flutter.

(a) Circular cylinder flutter involving a large number of circumferential waves

When the number of circumferential waves is sufficiently large, experience has shown that Donnell's theory provides a simple and reasonably accurate basis for the shell theory. If we assume, for example, that $w(\theta, x)$ is proportional to $\cos n\theta$, then we may say that Donnell's theory may be employed when n exceeds approximately 2 to 3. Making use of the notation of Sec. 5–2(j), applying Donnell's theory (Eq. 5–84) and the supersonic aerodynamic operator of Eq. 8–33, the aeroelastic equation of the complete cylinder reads

$$\left\{\frac{D}{R^8}\nabla^8 + \frac{Eh}{R^2}\frac{\partial^4}{\partial x^4} + \frac{1}{R^4}\nabla^4\left[\left(-N_{xx}^{(0)}\frac{\partial^2}{\partial x^2} - 2\frac{N_{x\theta}^{(0)}}{R}\frac{\partial^2}{\partial x\,\partial\theta} - \frac{N_{\theta\theta}^{(0)}}{R^2}\frac{\partial^2}{\partial\theta^2}\right)\right.\right.$$
$$\left.\left. + \frac{2q}{\sqrt{M^2-1}}\left(\frac{\partial}{\partial x} + \frac{1}{U}\frac{M^2-2}{M^2-1}\frac{\partial}{\partial t}\right) + m_0\frac{\partial^2}{\partial t^2}\right]\right\}(w) = \frac{1}{R^4}\nabla^4(Z_D) \quad (8\text{–}93)^*$$

where we have included (cf. Eq. 5–82) the influence of prescribed stress resultants $N_{xx}^{(0)}$, $N_{x\theta}^{(0)}$, and $N_{\theta\theta}^{(0)}$, in addition to the aerodynamic, inertial, and disturbance forces. When the combination of Eqs. 8–81 and 8–82 is compared with Eq. 8–93, we see that after the appropriate notational changes are made, the two are identical. That is, the shallow shell theory applied to a circular cylindrical segment is identical to Donnell's theory. We may conclude, therefore, that the methods of flutter analysis derived in Sec. 8-6 for circular cylindrical panels, and based on shallow shell theory, may be applied also to the flutter of complete cylinders, provided that the number of circumferential waves is sufficiently large.

* In Eqs. 8–93, the operator ∇^4 is defined as indicated in Eq. 5–79.

(b) Circular cylinder flutter, including the possibility of a small number of circumferential waves

It is logical to inquire next into the nature of a theory which includes the flutter of complete circular cylinders when the number of circumferential waves, n, may be small, approaching zero. Under these conditions we can expect the participation of vibration modes in the flutter mechanism which have considerable membrane as well as bending action. Such a theory must, of course, take account of longitudinal and tangential, as well as radial, inertial forces. Since the first-order aerodynamic theory which is available to us, for example, the operator form of Eq. 8–33, involves only forces normal to the surface of the cylinder, we must be content at the present time to include only those modes which have some radial component of displacement.

(*i*) *The equations of motion in terms of normal coordinates.* It will be convenient to express the aeroelastic equations of motion of the cylinder in the present application in terms of displacements of its natural modes of vibration. The form by which this may be accomplished is borrowed from Eq. 3–118, as follows:

$$M_{mn}\ddot{\xi}_{mn} + M_{mn}\omega_{mn}{}^2\xi_{mn} = \Xi_{mn} \qquad (8\text{--}94)$$

where

$$M_{mn} = \iint\limits_{S} (u_{mn}{}^2 + v_{mn}{}^2 + w_{mn}{}^2)m_0\,dS$$

is the generalized mass of the mode and ω_{mn} is the frequency of the mode identified by (m, n). ξ_{mn} are the normal coordinates and Ξ_{mn} are the generalized forces defined by

$$\Xi_{mn} = \iint\limits_{S} \mathscr{A}(w)w_{mn}\,dS + \iint\limits_{S} Z_D w_{mn}\,dS \qquad (8\text{--}95)$$

where the existence of a disturbance force Z_D is postulated. The modal functions u_{mn}, v_{mn}, and w_{mn} are defined for the cylinder according to

$$u_{mn} = \bar{u}_{mn} \cos\frac{m\pi x}{a} \cos n\theta$$

$$v_{mn} = \bar{v}_{mn} \sin\frac{m\pi x}{a} \sin n\theta \qquad (8\text{--}96)$$

$$w_{mn} = \bar{w}_{mn} \sin\frac{m\pi x}{a} \cos n\theta$$

where $\bar{u}_{mn}$, $\bar{v}_{mn}$, and $\bar{w}_{mn}$ are modal amplitudes corresponding to the frequency ω_{mn}, and derived from a free vibration analysis of the cylinders.

We proceed by introducing the aerodynamic operator $\mathscr{A}(w)$ given by Eq. 8–33 and by putting

$$w = \sum_{r}^{N} w_{rn}\xi_{rn} \tag{8-97}$$

where we assume that a particular value of n is selected for investigation. This yields as the equations of motion of the cylinder

$$T_{mn}\left\{\ddot{\xi}_{mn} + \omega_{mn}{}^{2}\xi_{mn} + \frac{2q}{\sqrt{M^2-1}}\frac{1}{U}\frac{M^2-2}{M^2-1}\frac{1}{m_0}\frac{1}{T_{mn}}\dot{\xi}_{mn}\right\}\bar{w}_{mn}$$

$$+ \frac{2q}{am_0\sqrt{M^2-1}}\sum_{\substack{r \\ (m+r)=1,3,5,\cdots\text{ odd}}}^{N}\frac{4mr}{(m^2-r^2)}\bar{w}_{rn}\xi_{rn}$$

$$= \frac{2}{Ra\pi m_0\bar{w}_{mn}}\iint_{S}Z_D w_{mn}\,dS \tag{8-98}$$

where

$$T_{mn} = \left[\left(\frac{\bar{u}_{mn}}{\bar{w}_{mn}}\right)^{2} + \left(\frac{\bar{v}_{mn}}{\bar{w}_{mn}}\right)^{2} + 1\right]$$

Let us concentrate our attention on the homogeneous form of Eq. 8–98 and introduce as a solution

$$\xi_{mn} = \bar{\xi}_{mn}e^{\alpha t} \tag{8-99}$$

where $\alpha = \beta + i\omega$. The following result is obtained:

$$T_{mn}\left\{\omega_{mn}{}^{2} - \omega^{2}(1+i\delta)^{2} + i\omega\frac{2q}{\sqrt{M^2-1}}\frac{1}{U}\frac{M^2-2}{M^2-1}\frac{1}{m_0}\frac{1}{T_{mn}}(1+i\delta)\right\}$$

$$\times \bar{w}_{mn}\bar{\xi}_{mn} + \frac{2q}{am_0\sqrt{M^2-1}}\sum_{\substack{r \\ (m+r)=1,3,5,\cdots}}^{N}\frac{4mr}{(m^2-r^2)}(\bar{w}_{rn}\bar{\xi}_{rn}) = 0 \tag{8-100}$$

where $\delta = -\beta/\sqrt{\beta^2-\omega^2} \approx -\dfrac{\beta}{\omega}$ is the damping ratio. The effect of structural damping may also be introduced into Eq. 8–100 by multiplying $\omega_{mn}{}^{2}$ by the factor $(1+ig_{mn})$, where g_{mn} is the structural damping coefficient of the mode (m, n).

By putting the determinant of the coefficients of $\bar{\xi}_{mn}\bar{w}_{mn}$ in Eq. 8–100 equal to zero, there may be computed a set of frequencies equal in number to the number of modes assumed. Corresponding to each frequency there is a damping ratio, the vanishing of which defines a flutter stability boundary.

(*ii*) *Natural modes of vibration of circular cylinders and their employment in flutter analyses.* In the previous section, the problem of cylinder flutter is formulated in terms of the natural modes of vibration. In particular, it was the object of this formulation to develop a theory which is not limited *a priori* to a particular range of the parameter k. It will be necessary, therefore, to derive the natural modes in the most general manner, including the effects of tangential as well as radial inertial forces. It is also desirable to include the influence of initial stressing in such calculations in order to provide for the effects of internal pressure or steady loading.

We use as a basis for this derivation the nonlinear equations for a circular cylinder (5–78), and express the shell stresses and displacements in terms of departures from an initially stressed configuration. These departures are small, and we may linearize the equations with respect to them. Let us suppose that the initially stressed configuration is characterized by the stress resultants $N_{xx}^{(0)}$ and $N_{\theta\theta}^{(0)}$ and that the externally applied loads per unit of shell area are $X^{(0)}$, $Y^{(0)}$, and $Z^{(0)}$. We superimpose in Eqs. 5–78 the initial stress resultants $N_{xx}^{(0)}$ and $N_{\theta\theta}^{(0)}$ upon each N_{xx} and $N_{\theta\theta}$ stress resultant which appears. That is, we introduce in place of N_{xx} the quantity $N_{xx}^{(0)} + N_{xx}$; and in place of $N_{\theta\theta}$, the quantity $N_{\theta\theta}^{(0)} + N_{\theta\theta}$. All stress resultants without the superscript (0), as well as all the stress couples and the displacements, are regarded as small quantities; and their squares and products are neglected. When this process is carried out, and when we introduce the stress-strain relations (Eqs. 5–38 and 5–39), and the inertial forces per unit of shell area in the form

$$X = X^{(0)} - m_0 \frac{\partial^2 u}{\partial t^2}, \quad Y = Y^{(0)} - m_0 \frac{\partial^2 v}{\partial t^2}, \quad Z = Z^{(0)} - m_0 \frac{\partial^2 w}{\partial t^2} \quad (8\text{–}101)$$

we obtain the following equilibrium equations of free vibration about an initially stressed configuration:

$$\frac{\partial^2 u}{\partial x^2} + \frac{1-\nu}{2R^2}\frac{\partial^2 u}{\partial \theta^2} + \frac{1+\nu}{2R}\frac{\partial^2 v}{\partial x\,\partial\theta} - \frac{\nu}{R}\frac{\partial w}{\partial x} + \frac{h^2}{12R^2}F_1(u,\,v,\,w)$$
$$+ \frac{1-\nu^2}{Eh}\left[G_1(N_{xx}^{(0)},\,N_{\theta\theta}^{(0)},\,u,\,v,\,w) - m_0\frac{\partial^2 u}{\partial t^2}\right] = 0$$

$$\frac{1+\nu}{2R}\frac{\partial^2 u}{\partial x\,\partial\theta} + \frac{1}{R^2}\frac{\partial^2 v}{\partial \theta^2} + \frac{1-\nu}{2}\frac{\partial^2 v}{\partial x^2} - \frac{1}{R^2}\frac{\partial w}{\partial \theta} + \frac{h^2}{12R^2}F_2(u,\,v,\,w)$$
$$\qquad\qquad (8\text{–}102)$$
$$+ \frac{1-\nu^2}{Eh}\left[G_2(N_{xx}^{(0)},\,N_{\theta\theta}^{(0)},\,u,\,v,\,w) - m_0\frac{\partial^2 v}{\partial t^2}\right] = 0$$

$$\frac{\nu}{R}\frac{\partial u}{\partial x} + \frac{1}{R^2}\frac{\partial v}{\partial \theta} - \frac{w}{R^2} - \frac{h^2}{12R^4}\nabla^4 w + \frac{h^2}{12R^2}F_3(u,\,v,\,w)$$
$$+ \frac{1-\nu^2}{Eh}\left[G_3(N_{xx}^{(0)},\,N_{\theta\theta}^{(0)},\,u,\,v,\,w) - m_0\frac{\partial^2 w}{\partial t^2}\right] = 0$$

In reaching the final form of Eqs. 8–102, we must, of course, make use of the fact that the equilibrium conditions among the initial membrane stresses are

$$\frac{\partial N_{xx}^{(0)}}{\partial x} + X^{(0)} = 0$$

$$\frac{1}{R}\frac{\partial N_{\theta\theta}^{(0)}}{\partial \theta} + Y^{(0)} = 0 \qquad \frac{1}{R}N_{\theta\theta}^{(0)} + Z^{(0)} = 0 \qquad (8\text{–}103)$$

In Eqs. 8–102, $F_1(u, v, w)$, $F_2(u, v, w)$, and $F_3(u, v, w)$ are as shown by Table 5–1. Various forms of these functions may be substituted into Eqs. 8–102, although the explicit results obtained from Eqs. 5–78 are those of Washizu as given by row 5 of Table 5–1. The functions $G_1(N_{xx}^{(0)}, N_{\theta\theta}^{(0)}, u, v, w)$, $G_2(N_{xx}^{(0)}, N_{\theta\theta}^{(0)}, u, v, w)$, and $G_3(N_{xx}^{(0)}, N_{\theta\theta}^{(0)}, u, v, w)$ introduce the effects of initial stressing into the vibration equations. Based upon Eqs. 5–78, these functions are derived as

$$G_1(N_{xx}^{(0)}, N_{\theta\theta}^{(0)}, u, v, w) = N_{xx}^{(0)}\frac{\partial^2 u}{\partial x^2} + \frac{1}{R^2}N_{\theta\theta}^{(0)}\frac{\partial^2 u}{\partial \theta^2}$$

$$G_2(N_{xx}^{(0)}, N_{\theta\theta}^{(0)}, u, v, w) = N_{xx}^{(0)}\frac{\partial^2 v}{\partial x^2} - \frac{1}{R^2}N_{\theta\theta}^{(0)}\left(v + \frac{\partial w}{\partial \theta}\right)$$

$$+ \frac{1}{R^2}N_{\theta\theta}^{(0)}\left(\frac{\partial^2 v}{\partial \theta^2} - \frac{\partial w}{\partial \theta}\right) \qquad (8\text{–}104a)$$

$$G_3(N_{xx}^{(0)}, N_{\theta\theta}^{(0)}, u, v, w) = N_{xx}^{(0)}\frac{\partial^2 w}{\partial x^2} + \frac{1}{R^2}N_{\theta\theta}^{(0)}\left(\frac{\partial v}{\partial \theta} + \frac{\partial^2 w}{\partial \theta^2}\right) + \frac{1}{R}\left(\frac{\partial v}{\partial \theta} - w\right)$$

When Timoshenko's nonlinear circular cylinder equations are employed (Eqs. 298 and 299 of Ref. 5–1) as a basis for deriving Eqs. 8–102, we obtain for $F_1(u, v, w)$, $F_2(u, v, w)$, and $F_3(u, v, w)$ the results shown by row 2 of Table 5–1, with the following initial stressing functions:

$$G_1(N_{xx}^{(0)}, N_{\theta\theta}^{(0)}, u, v, w) = -\frac{N_{\theta\theta}^{(0)}}{R}\left(\frac{\partial^2 v}{\partial x\,\partial \theta} - \frac{\partial w}{\partial x}\right)$$

$$G_2(N_{xx}^{(0)}, N_{\theta\theta}^{(0)}, u, v, w) = N_{xx}^{(0)}\frac{\partial^2 v}{\partial x^2} \qquad (8\text{–}104b)*$$

$$G_3(N_{xx}^{(0)}, N_{\theta\theta}^{(0)}, u, v, w) = N_{xx}^{(0)}\frac{\partial^2 w}{\partial x^2} + \frac{1}{R^2}N_{\theta\theta}^{(0)}\left(\frac{\partial v}{\partial \theta} + \frac{\partial^2 w}{\partial \theta^2}\right)$$

* Equations 8–104b, from Timoshenko, exhibit a lack of symmetry as compared to those of Washizu (8–104a) obtained by a variational process. This lack of symmetry of the Love-Timoshenko results has been criticized by other authors, e.g. Vlasov, Ref. 5–12.

We may take as one of the possible free vibration solutions to Eqs. 8–102 the following:

$$u = u_{mn} \cos \left(\frac{m\pi x}{a}\right) \cos n\theta e^{i\omega t}$$

$$v = v_{mn} \sin \left(\frac{m\pi x}{a}\right) \sin n\theta e^{i\omega t} \qquad (8\text{–}105)$$

$$w = w_{mn} \sin \left(\frac{m\pi x}{a}\right) \cos n\theta e^{i\omega t}$$

where m and n are integers and ω is the natural frequency of vibration. When Eqs. 8–105 are substituted into Eqs. 8–102, a set of linear homogeneous equations in u_{mn}, v_{mn}, and w_{mn} are obtained; and for a nontrivial solution of this set, the determinant of the coefficients must vanish. This provides the cubic frequency relation

$$\kappa^3 - \kappa_2'\kappa^2 + \kappa_1'\kappa - \kappa_0' = 0 \qquad (8\text{–}106)$$

where

$$\kappa = \frac{m_0 R^2 \omega^2}{Eh}$$

is a frequency parameter. For every assigned pair of values m and n, there are unique values of κ_0', κ_1', and κ_2', and there are three natural frequencies. Corresponding to each frequency, the modal ratios u_{mn}/w_{mn} and v_{mn}/w_{mn} may be computed, and the values of these ratios indicate whether the mode is predominantly longitudinal, circumferential, or radial. Space does not permit a comprehensive statement of the character of the solutions to Eqs. 8–102; the reader is referred to the rather extensive literature on the subject, of which Refs. 8–27 through 8–32 are examples. There have been, in particular, numerous studies made of freely vibrating cylinders based upon the Timoshenko formulation of the vibration equations, where Eqs. 8–102 are employed with F_1, F_2, and F_3 given by row 2 of Table 5–1 and G_1, G_2, and G_3 by Eqs. 8–104. Refs. 8–28 and 8–20 are examples of these applications.

The flutter analysis of a cylinder proceeds in a straightforward manner by the application of Eq. 8–100, once the natural modes and frequencies of free vibration of the cylinder have been computed. General statements concerning the type and number of modes which are required in cylinder flutter analysis are difficult to make, and the reader is referred to the work of Voss (Ref. 8–26) for an indication of these requirements. Voss' work, in which a single primarily radial mode was selected to represent each mode number combination, showed that there are two ranges of critical flutter modes for the cylinder. The first, which corresponds to low values

---

448 PRINCIPLES OF AEROELASTICITY

of the circumferential mode number n but high values of the longitudinal mode number m, is classified as membrane-type flutter. The second, which corresponds to high n and low m, is similar to the plate-type instability. According to Voss, the membrane-type of instability appears to be critical for $n = 0$. Since the frequency spectrum for $n = 0$ is very closely spaced, a modal solution may require a large number of longitudinal modes to obtain satisfactory precision. For this reason, the modal approach may not represent the most satisfactory procedure for the membrane-type flutter range. The theory of Sec. 8–6 may be said to provide a satisfactory approach for the high n and low m plate-type instability range. In general, however, both ranges appear to require investigation in a given problem.

REFERENCES

8–1. Dugundji, John, and J. D. C. Crisp, *On the Aeroelastic Characteristics of Low Aspect Ratio Wings With Chordwise Deformation*, OSR T.N. 59-787, July 1959.

8–2. Heldenfels, R. R., and R. Rosencrans, *Preliminary Results of Supersonic—Jet Tests of Simplified Wing Structure*, NACA RM L53E26a, July 1953.

8–3. Cooper, R. E., *The Effects of Chordwise Bending on Flutter*, M.I.T., S.B. Thesis, June 1958.

8–4. Sylvester, M. A., and J. E. Baker, *Some Experimental Studies of Panel Flutter at Mach Number 1.3*, NACA T.N. 3914, 1957.

8–5. Eisley, J. G., *The Flutter of a Two-Dimensional Buckled Plate With Clamped Edges in a Supersonic Flow*, AFOSR T.N. 56-296, 1956.

8–6. Jordan, P. F., "The Physical Nature of Panel Flutter," *Aero Digest*, pp. 34–38, February 1956.

8–7. Miles, J. W., *Dynamic Chordwise Stability at Supersonic Speeds*, North American Aviation Report AL-1140, 1950.

8–8. Shen, S. F., *Flutter of a Two-Dimensional Simply Supported Uniform Panel in a Supersonic Stream*, M.I.T. Aeroelastic and Structures Research Laboratory Report 25.10, 1952.

8–9. Nelson, H. C., and H. J. Cunningham, *Theoretical Investigation of Flutter of Two-Dimensional Flat Panels With One Surface Exposed to Supersonic Potential Flow*, NACA Report 1280, 1956.

8–10. Goland, M., and Y. L. Luke, "An Exact Solution for Two-Dimensional Linear Panel Flutter at Supersonic Speeds," *J. Aero. Sciences*, Vol. 21, No. 4, pp. 275–276, 1954.

8–11. Hedgepeth, J. M., "Flutter of Rectangular Simply Supported Panels at High Supersonic Speeds," *J. Aero. Sciences*, Vol. 24, No. 8, pp. 563–573, 1957.

8–12. Eisley, J. G., *The Flutter of Simply Supported Rectangular Plates in Supersonic Flow*, AFOSR T.N. 55-236, 1955.

8–13. Movchan, A. A., "Ob ustoichivosti paneli, dvizhushcheisia v gaze (On the Stability of a Panel Moving in a Gas)," *Prikladnaia Matematika i Mekhanika*, Vol. 21, No. 2, 1957. (Translated and Issued as NASA RE 11-21-58 W.)

8–14. Houbolt, J. C., *A Study of Several Aerothermoelastic Problems of Aircraft Structures*, Mitteilung aus dem Institut für Flugzeugstatik und Leichtbau, Nr. 5, E.T.H., Zürich, 1958.

8-15. Dugundji, John, *Panel Flutter*, Chapter 6, Vol. 2, M.I.T. Summer Session Notes on Aeroelasticity, 1958.

8-16. Shulman, Y., *Some Dynamic and Aeroelastic Problems of Plate and Shell Structures*, M.I.T., Sc.D. Thesis, 1959.

8-17. Fung, Y. C., "On Two-Dimensional Panel Flutter," *J. Aero. Sciences*, Vol. 25, No. 3, pp. 145–160, March 1958.

8-18. Fung, Y. C., *The Flutter of a Buckled Plate in a Supersonic Flow*, AFOSR T.N. 55-237, 1955.

8-19. Eisley, J. G., *The Flutter of a Two-Dimensional Buckled Plate With Clamped Edges in a Supersonic Flow*, AFOSR T.N. 56-296, 1956.

8-20. Hedgepeth, J. M., "On the Flutter of Panels at High Mach Numbers," *J. Aero. Sciences*, Readers' Forum, Vol. 23, No. 6, p. 609, 1956.

8-21. Luke, Y. L., and A. D. St. John, *Panel Flutter at Supersonic Speeds*, Midwest Research Institute Fifth Quarterly Progress Report, Contact No. AF 33(616)-2987, October 31, 1956.

8-22. Leonard, R. W., and J. M. Hedgepeth, "On the Flutter of Infinitely Long Panels on Many Supports," *J. Aero. Sciences*, Readers' Forum, Vol. 24, No. 4, pp. 381–383, May 1957.

8-23. Isaacs, R. P., *Transtability Flutter of Supersonic Aircraft Panels*, Report P-101, The Rand Corporation, Santa Monica, California, July 1949.

8-24. Sylvester, M. A., H. C. Nelson, and H. J. Cunningham, *Experimental and Theoretical Studies of Panel Flutter at Mach Numbers 1.2 to 3.0*, NACA RM L55 E186, July 1955.

8-25. Sylvester, M. A., *Experimental Studies of Flutter of Buckled Rectangular Panels at Mach Numbers From 1.2 to 3.0 Including Effects of Pressure Differential and Panel Width-Length Ratio*, NACA RM L55I30, December 1955.

8-26. Voss, H. M., *The Effect of an External Supersonic Flow on the Vibration Characteristics of Thin Cylindrical Shells*, Presented at the I.A.S. 28th Annual Meeting, New York, New York, January 1960, IAS Paper 60-45.

8-27. Reissner, E., "On Transverse Vibrations of Thin Shallow Elastic Shells," *Quart. Applied Math.*, Vol. 13, No. 2, pp. 169–176, July 1955.

8-28. Fung, Y. C., E. E. Sechler, and A. Kaplan, "On the Vibration of Thin Cylindrical Shells Under Internal Pressure," *J. Aero. Sciences*, Vol. 24, No. 9, September 1957.

8-29. Arnold, R. N., and G. B. Warburton, "Flexural Vibrations of the Walls of Thin Cylindrical Shells Having Freely Supported Ends," *Proc. Royal Soc.*, London, Series A, Vol. 197, p. 238, 1949.

8-30. Reissner, E., *Non-Linear Effects in Vibrations of Cylindrical Shells*, Aeromechanics Report No. AM5-6, Ramo-Wooldridge Corp., August 1955.

8-31. Reissner, E., *Notes on Vibrations of Thin, Pressurized Cylindrical Shells*, Aeromechanics Report AM5-4, Ramo-Wooldridge Corp., November 1954.

8-32. Baron, M. L., and H. H. Bleich, "Tables for Frequencies and Modes of Free Vibration of Infinitely Long Thin Cylindrical Shells," Transactions of the A.S.M.E., Vol. 21, No. 2, *J. Applied Mech.*, pp. 178–184, June 1954.

9

THE UNRESTRAINED VEHICLE

9-1 INTRODUCTION

In Chaps. 7 and 8, we discussed one- and two-dimensional dynamic aeroelastic problems in which the lifting surface is restrained against free motion. In dealing with the dynamic and aeroelastic behavior of complete vehicles in flight, we must release all of the external constraints and treat the vehicle as an object which is free in space. A completely general formulation of the aeroelastic equations of motion of a free vehicle in flight involves consideration of a three-dimensional body with six translational and rotational degrees of freedom.

Let us refer to Fig. 9–1 to indicate the nature of the body axes which are employed, together with the notation for the linear and angular velocities along and about these axes.* We take as a basis for the mathematical treatment of the unrestrained vehicle, the development at the beginning of Chap. 2, represented by Eqs. 2–1 through 2–26, inclusive. Let us suppose that the origin "0" of the x-y-z-system of body axes in Fig. 9–1 remains always at the center of gravity of the deformed structure and that the angular movements of these axes represent rigid body rotations. We shall assume also that the body axes may move linearly and angularly through large displacements, but that the elastic deformations of the vehicle with respect to these axes are small.

The arbitrary motion of an elastic three-dimensional body under

* Since we are concerned in Chap. 9 with an unrestrained vehicle, the axis system has been chosen to correspond with the standard system adopted for airplane stability analyses.

aerodynamic forces is defined by three vector equations of equilibrium. We assume that two types of forces are acting on the surface of the vehicle. The first is an explicitly defined force per unit area, $\mathbf{F}^D$, which is applied by a reactive source or an atmospheric disturbance and has arbitrary spatial and time dependence. The second is an aerodynamic force per unit area, $\mathbf{F}^M$, which arises from the vehicle's motion. The latter may depend not only on the instantaneous motion of the system, but also on all previous conditions of motion. The total surface force per unit area is thus represented by

$$\mathbf{F} = \mathbf{F}^D + \mathbf{F}^M \tag{9-1}$$

From Eqs. 2–1 and 2–5, we have for the first vector equation of equilibrium

$$\frac{d}{dt} \iiint_V \frac{d\mathbf{r}'}{dt} \rho \, dV = M \frac{d^2 \mathbf{r}_0'}{dt^2} = \iint_S (\mathbf{F}^D + \mathbf{F}^M) \, dS \tag{9-2}$$

The second vector equation of equilibrium, derived from Eq. 2–6, is the following:

$$\frac{d}{dt} \iiint_V \mathbf{r} \times \frac{d\mathbf{r}}{dt} \rho \, dV = \iint_S \mathbf{r} \times (\mathbf{F}^D + \mathbf{F}^M) \, dS \tag{9-3}$$

The third vector equation of equilibrium may be expressed in terms of the differential operator of Eq. 5–91 or, alternatively, in terms of the integral

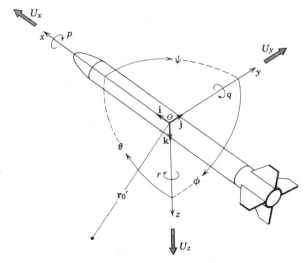

Fig. 9–1. Axis system for unrestrained vehicle.

operator of Eqs. 5–93 and 5–94. In the former case, we have the vector differential equation of equilibrium

$$\tilde{\mathscr{S}}(\mathbf{q}) + \rho \mathbf{a} = 0 \qquad (9\text{–}4a)$$

where $\mathbf{a}$ is the acceleration of an arbitrary particle according to

$$\mathbf{a} = \frac{d}{dt}\left(\frac{d\mathbf{r}'}{dt}\right) = \frac{d}{dt}\left(\frac{d\mathbf{r}_0'}{dt} + \frac{d\tilde{\mathbf{r}}}{dt} + \frac{d\mathbf{q}}{dt}\right)$$

and where the boundary conditions are those already stated by Eqs. 5–89. The vector $\tilde{\mathbf{r}}$ is a position vector from the center of gravity to a particle in the unstrained vehicle, and the vector $\mathbf{q}$ is an elastic deformation vector. These vectors relate to the position vector $\mathbf{r}$ through the relation $\mathbf{r} = \tilde{\mathbf{r}} + \mathbf{q}$. In making use of the third vector equation of equilibrium, it is important to recognize that, in the majority of cases, the vehicle with which we are dealing cannot be considered a simple homogeneous, isotropic body. It will be desirable, therefore, to construct this equation in the most general terms. Equations 5–93 and 5–94 provide a basis for a more general approach in which the elastic influence coefficients can be left in unspecified terms for later definition. These equations yield a vector integral equation, as follows:

$$\mathbf{q} - \mathbf{q}_0 - \tfrac{1}{2}\left[(\nabla \times \mathbf{q})_0\right] \times \tilde{\mathbf{r}} = -\iiint_V \Gamma \cdot \mathbf{a}\rho \, dV$$

$$+ \iiint_V \Gamma \cdot (\mathbf{F}^D + \mathbf{F}^M)\, \delta(\mathbf{r} - \mathbf{r}_s)\, dV \qquad (9\text{–}4b)$$

Equations 9–4a and 9–4b provide alternative forms of the required third vector equation of equilibrium. Initial conditions must be adjoined to both Eqs. 9–4a and 9–4b. However, boundary conditions need be stated explicitly only for differential equation (9–4a) since they are implicit in the integral equation (9–4b). In Eq. 9–4b, we have employed the influence function tensor Γ which describes deformations of the vehicle with respect to an x-y-z-system of axes with a point in the structure which we designate as P_0 located at the center of gravity of the undeformed airplane. In other words, we assume the vehicle clamped at this point P_0 and we evaluate the various influence functions which make up Γ by computing the deformations at (x, y, z) due to unit loads at (ξ, η, ζ). The center of gravity of the deformed structure, however, does not remain contiguous to the original point P_0. The latter moves linearly with respect to the center of gravity by an amount represented by the vector $\mathbf{q}_0$, and lines through the point rotate angularly by an amount represented by the vector $\tfrac{1}{2}(\nabla \times \mathbf{q})_0$. Therefore, since $\mathbf{q}$ represents a deformation with respect to an axis system with its origin at the center of gravity of the deformed vehicle and since Γ represents a tensor of influence coefficients measured with respect to an axis system clamped at P_0, we must introduce on the left-hand side of Eq. 9–4b the difference represented by

$$\mathbf{q} - \mathbf{q}_0 - \tfrac{1}{2}[(\nabla \times \mathbf{q})_0] \times \tilde{\mathbf{r}}$$

In the second term on the right-hand side of Eq. 9–4b, we have incorporated the effect of the surface tractions $\mathbf{F}^D + \mathbf{F}^M$ within a volume integral involving the same influence function tensor, Γ. This is made possible in a formal sense by introducing the Dirac delta function $\delta(\mathbf{r} - \mathbf{r}_s)$ with the property that

$$\iiint\limits_V g(x, y, z)\, \delta(\mathbf{r} - \mathbf{r}_s)\, dx\, dy\, dz = 0 \qquad (9\text{–}5a)$$

provided the surface, S, is not included in the volume of integration and

$$\iiint\limits_V g(x, y, z)\delta(\mathbf{r} - \mathbf{r}_s)\, dx\, dy\, dz = \iint\limits_S g(x, y, z)\, dx\, dy\, dz \qquad (9\text{–}5b)$$

providing S is included in the volume of integration.

Equations 9–2, 9–3, and 9–4 form the bases for mathematical analyses of the aeroelastic behavior of unrestrained three-dimensional vehicles. The latter portions of the present chapter are devoted to applications of these equations to aeroelastic phenomena of practical interest.

9–2 FREE VIBRATIONS IN A VACUUM

A fundamental question which arises in the dynamics of an unrestrained vehicle is that of defining its natural modes and frequencies of free vibrations. In addition to being important physical parameters of the system which can be measured experimentally, these quantities are useful in constructing solutions of practical aeroelastic problems.

In the absence of external forces the x-y-z-axis system remains inert, and the equations of free vibrations, derived from Eqs. 9–2, 9–3, and 9–4b, are the following:

$$\iiint\limits_V \frac{d^2\mathbf{q}}{dt^2}\, \rho\, dV = 0$$

$$\iiint\limits_V (\tilde{\mathbf{r}} + \mathbf{q}) \times \frac{d^2\mathbf{q}}{dt^2}\, \rho\, dV = 0 \qquad (9\text{–}6)$$

$$\mathbf{q} - \mathbf{q}_0 - \tfrac{1}{2}[(\nabla \times \mathbf{q})_0] \times \tilde{\mathbf{r}} = -\iiint\limits_V \Gamma \cdot \frac{d^2\mathbf{q}}{dt^2}\, \rho\, dV$$

We may take, as a solution of these equations, the product of a spatial function and a time function

$$q(x, y, z, t) = \mathbf{\phi}(x, y, z)T(t) \qquad (9\text{--}7)$$

where $\mathbf{\phi}(x, y, z) = \phi_x(x, y, z)\mathbf{i} + \phi_y(x, y, z)\mathbf{j} + \phi_z(x, y, z)\mathbf{k}$ is an eigenfunction in vector form which represents a natural mode shape and $T(t)$ is a function of time. Upon substitution of Eq. 9–7 into Eqs. 9–6, we obtain the following:

$$\iiint_V \mathbf{\phi}\rho \, dV = 0 \qquad (9\text{--}8)$$

$$\iiint_V \tilde{\mathbf{r}} \times \mathbf{\phi}\rho \, dV = 0 \qquad (9\text{--}9)$$

$$T[\mathbf{\phi} - \mathbf{\phi}_0 - \tfrac{1}{2}[(\nabla \times \mathbf{\phi})_0] \times \tilde{\mathbf{r}} = -\ddot{T} \iiint_V \Gamma \cdot \mathbf{\phi}\rho \, dV \qquad (9\text{--}10)$$

Equation 9–10 reduces to separate equations in space and time, as follows:

$$\mathbf{\phi} - \mathbf{\phi}_0 - \tfrac{1}{2}[(\nabla \times \mathbf{\phi})_0] \times \tilde{\mathbf{r}} = \omega^2 \iiint_V \Gamma \cdot \mathbf{\phi}\rho \, dV \qquad (9\text{--}11)$$

$$\ddot{T} + \omega^2 T = 0 \qquad (9\text{--}12)$$

where ω^2 is a separation constant which represents physically the frequencies of the natural vibration modes.

Equations 9–8, 9–9, and 9–11 can be combined into a single equation in which the amplitude $\mathbf{\phi}_0$ is eliminated. This is done in several steps. First, multiplying both sides of Eq. 9–11 by ρ and integrating over the volume, we obtain, after making use of Eq. 9–8 and the fact that the origin of co-ordinates remains at the center of gravity, the following result:

$$\mathbf{\phi}_0 = -\frac{\omega^2}{M} \iiint_V \rho \iiint_V \Gamma \cdot \mathbf{\phi}\rho \, dV \, dV \qquad (9\text{--}13)$$

Second, multiplying both sides of Eq. 9–11 by $\rho \tilde{\mathbf{r}} \times$ and integrating over the volume, we obtain by employing Eq. 9–9 and the fact that the origin remains at the center of gravity

$$-\tfrac{1}{2}\iiint_V \rho \tilde{\mathbf{r}} \times [(\nabla \times \mathbf{\phi})_0 \times \tilde{\mathbf{r}}] \, dV = \omega^2 \iiint_V \rho \tilde{\mathbf{r}} \times \iiint_V \Gamma \cdot \mathbf{\phi}\rho \, dV \, dV \qquad (9\text{--}14)$$

Making use of the vector triple product expansion law, the left-hand side of Eq. 9–14 may be altered so that the equation reads

$$-\tfrac{1}{2}\Psi \cdot (\nabla \times \mathbf{\phi})_0 = \omega^2 \iiint_V \rho \tilde{\mathbf{r}} \times \iiint_V \Gamma \cdot \mathbf{\phi}\rho \, dV \, dV \qquad (9\text{--}15)$$

where Ψ is a second-order inertial tensor defined by

$$\Psi = \iiint\limits_V \rho[(\tilde{\mathbf{r}} \cdot \tilde{\mathbf{r}})I - \tilde{\mathbf{r}}\tilde{\mathbf{r}}]\, dV = I_{xx}\mathbf{ii} + I_{yy}\mathbf{jj} + I_{zz}\mathbf{kk}$$

$$- I_{xy}(\mathbf{ij} + \mathbf{ji}) - I_{xz}(\mathbf{ik} + \mathbf{ki}) - I_{yz}(\mathbf{jk} + \mathbf{kj}) \qquad (9\text{–}16)$$

In Eq. 9–16, I is the idem factor ($I = \mathbf{ii} + \mathbf{jj} + \mathbf{kk}$) and the quantities I_{pq} are moments and products of inertia of the vehicle with respect to its body axis system.

Equation 9–15 can be transformed to read

$$-\tfrac{1}{2}[(\nabla \times \boldsymbol{\phi})_0] = \omega^2 \Psi^{-1} \cdot \iiint\limits_V \rho\tilde{\mathbf{r}} \times \iiint\limits_V \Gamma \cdot \boldsymbol{\phi} \rho\, dV\, dV \qquad (9\text{–}17)$$

By making use of Eqs. 9–13 and 9–17, we can eliminate $\boldsymbol{\phi}_0$ and $(\nabla \times \boldsymbol{\phi})_0$ in Eq. 9–11, and the latter becomes simply

$$\boldsymbol{\phi}(x, y, z) = \omega^2 \iiint\limits_V G(x, y, z; \xi, \eta, \zeta) \cdot \boldsymbol{\phi}(\xi, \eta, \zeta) \rho\, d\xi\, d\eta\, d\zeta \qquad (9\text{–}18)$$

where G is a second-order influence function tensor having the form

$$G(x, y, z; \xi, \eta, \zeta) = \Gamma(x, y, z; \xi, \eta, \zeta)$$

$$- \frac{1}{M} \iiint\limits_V \Gamma(r, s, t; \xi, \eta, \zeta)\rho(r, s, t)\, dr\, ds\, dt + \tilde{\mathbf{r}}(x, y, z)$$

$$\times \left\{ \Psi^{-1} \cdot \left[\iiint\limits_V \tilde{\mathbf{r}}(r, s, t) \times \Gamma(r, s, t; \xi, \eta, \zeta)\rho(r, s, t)\, dr\, ds\, dt \right] \right\}$$

$$(9\text{–}19)$$

It is easily shown by substitution into Eq. 9–18 that two different mode shapes, denoted by $\boldsymbol{\phi}_i$ and $\boldsymbol{\phi}_j$, are orthogonal.

$$\iiint\limits_V \boldsymbol{\phi}_i \cdot \boldsymbol{\phi}_j \rho\, dV = 0, \qquad (i \neq j) \qquad (9\text{–}20)$$

We may consider that a free vehicle has three translational modes of zero frequency which can be represented by the vector forms

$$\begin{aligned} \boldsymbol{\phi}_1 &= a_1\mathbf{i} \\ \boldsymbol{\phi}_2 &= b_2\mathbf{j} \\ \boldsymbol{\phi}_3 &= c_3\mathbf{k} \end{aligned} \qquad (9\text{–}21)$$

and three rotational modes, also of zero frequency, which are

$$\boldsymbol{\phi}_4 = -z\mathbf{j} + y\mathbf{k}$$
$$\boldsymbol{\phi}_5 = z\mathbf{i} - x\mathbf{k} \qquad (9\text{--}22)$$
$$\boldsymbol{\phi}_6 = -y\mathbf{i} + x\mathbf{j}$$

There are, in addition, an infinite number of modes of finite frequency defined by solutions to Eq. 9–18. Each pair of these modes satisfies the orthogonality relation (9–20). In addition to the orthogonality relation between deformation modes, there are orthogonality relations between the rigid-body modes and between the rigid-body and deformation modes. Thus, Eq. 9–20 may hold for the rigid-body modes of Eqs. 9–21 and 9–22 as well as combinations of these modes with the deformation modes. This assertion may be verified by substitution into Eq. 9–20 and making use of Eqs. 9–8, 9–9, and the assumption that the x-y-z-axis system is a centroidal principal axis system.

The reader will observe that, although the above development was carried out using Eq. 9–4b, it could have been carried out just as well by making use of Eq. 9–4a. In fact, it is easily shown that, when this approach is used, the three equations of free vibrations are

$$\iiint\limits_V \boldsymbol{\phi}\rho \, dV = 0; \qquad \iiint\limits_V \tilde{\mathbf{r}} \times \boldsymbol{\phi}\rho \, dV = 0; \qquad \tilde{\mathscr{S}}(\boldsymbol{\phi}) = \rho\boldsymbol{\phi}\omega^2 \qquad (9\text{--}23)$$

plus boundary conditions (5–89) with $F_x = F_y = F_z = 0$. The orthogonality conditions (9–20) can be derived also from this set of equations.

9–3 FORCED MOTION IN TERMS OF NATURAL VIBRATION MODES

A common approach to the calculation of the forced motion of an unrestrained vehicle is that of formulating the problem in terms of the natural vibration modes of the free, unrestrained vehicle. We include the possibility that the vehicle may undergo large rigid translations and rotations with small elastic deformations. The displacement vector of a particle in the structure with respect to the primed axis system (cf. Fig. 2–1) is written

$$\mathbf{r}' = \mathbf{r}_0' + \mathbf{r} \qquad (9\text{--}24)$$

where

$$\mathbf{r} = \tilde{\mathbf{r}} + \mathbf{q}$$

and where

$$\mathbf{q} = \sum_{i=1}^{\infty} \boldsymbol{\phi}_i(x, y, z)\xi_i(t)$$

where $\boldsymbol{\phi}_i(x, y, z)$ are the vector forms of the natural modes of the free unrestrained vehicle (solutions of Eq. 9–18). The normal coordinates, ξ_i, and the vectors $\mathbf{r}_0'$ and $\bar{\mathbf{r}}$ are regarded as the unknown quantities.*

The first of the defining vector equations of equilibrium is given by Eq. 9–2, as follows:

$$M \frac{d^2\mathbf{r}_0'}{dt^2} = \iint_S (\mathbf{F}^D + \mathbf{F}^M)\, dS \qquad (9\text{-}25)$$

The second vector equation is obtained by introducing Eq. 9–24 into Eq. 9–3. This results, after making use of Eq. 9–9 and neglecting higher-order terms, in the following:

$$\frac{d}{dt}(\Psi \cdot \boldsymbol{\omega}) = \iint_S \mathbf{r} \times (\mathbf{F}^D + \mathbf{F}^M)\, dS \qquad (9\text{-}26)$$

where Ψ is the inertial tensor defined by Eq. 9–16 and $\boldsymbol{\omega}$ is the angular velocity vector of the x-y-z-system. Equations 9–25 and 9–26 provide the basis for computing the space location and orientation of the x-y-z-axis system. The deformational displacement with respect to this axis system is obtained by substituting Eqs. 9–24 into Eq. 9–4b. When this substitution is made, and when Eq. 9–11 is employed, we obtain

$$\sum_{i=1}^{\infty} \{\boldsymbol{\phi}_i - \boldsymbol{\phi}_i(0) - \tfrac{1}{2}[\nabla \times \boldsymbol{\phi}_i(0)] \times \bar{\mathbf{r}}\}\left(\xi_i + \frac{\ddot{\xi}_i}{\omega_i^2}\right)$$

$$= -\iiint_V \Gamma \cdot (\ddot{\mathbf{r}}_0' + \ddot{\bar{\mathbf{r}}})\, \rho\, dV$$

$$+ \iiint_V \Gamma \cdot (\mathbf{F}^D + \mathbf{F}^M)\, \delta(\mathbf{r} - \mathbf{r}_s)\, dV \qquad (9\text{-}27)$$

where $\mathbf{q}_0 = \sum_{i=1}^{\infty} \boldsymbol{\phi}_i(0)\xi_i$ and where ω_i are the natural frequencies associated with the natural mode shapes, $\boldsymbol{\phi}_i$. Applying the dot product of $\rho\boldsymbol{\phi}_j$ to both sides of Eq. 9–27 and integrating over the volume, we obtain, after making use of the orthogonality conditions and Eq. 9–9, the result

$$M_j \ddot{\xi}_j + M_j \omega_j^2 \xi_j = \Xi_j^D + \Xi_j^M, \qquad (j = 1, 2, \cdots, \infty) \qquad (9\text{-}28)$$

---

* The vector $\bar{\mathbf{r}}$, although it is a position vector of a particle in the structure of the undeformed airplane with respect to the center of gravity, is nevertheless regarded as an unknown since its direction must be determined.

where M_j is the generalized mass

$$M_j = \iiint\limits_V \rho \, |\boldsymbol{\phi}_j|^2 \, dV$$

$\Xi_j{}^D$ is the explicitly defined component of the generalized force

$$\Xi_j{}^D = \iint\limits_S (\mathbf{F}^D \cdot \boldsymbol{\phi}_j) \, dS$$

and $\Xi_j{}^M$ is the component of the generalized force which depends upon the motion

$$\Xi_j{}^M = \iint\limits_S (\mathbf{F}^M \cdot \boldsymbol{\phi}_j) \, dS$$

Thus Eqs. 9–25, 9–26, and 9–28 provide in concise form the necessary equations for computing the aeroelastic behavior of an unrestrained three-dimensional elastic vehicle in terms of the eigenvalues and eigenfunctions of the fiee vehicle.

Although the preceding derivation was based upon the equations of equilibrium (9–2, 9–3, and 9–4a) as a starting point, it is evident that Hamilton's principle could also be employed as an alternative method of deriving the same results. We shall illustrate how this may be accomplished. Hamilton's principle is applied by computing the kinetic and strain energies in terms of the coordinates employed in the above derivation. The kinetic energy is computed from (cf. Eq. 2–28) the following expression:

$$\tau = \frac{1}{2} \iiint\limits_V \frac{d\mathbf{r}'}{dt} \cdot \frac{d\mathbf{r}'}{dt} \, \rho \, dV \tag{9–29}$$

Putting $\dfrac{d\mathbf{r}'}{dt} = \dfrac{d\mathbf{r}_0{}'}{dt} + \dfrac{d\bar{\mathbf{r}}}{dt} + \dfrac{d\mathbf{q}}{dt}$, where $\dfrac{d\bar{\mathbf{r}}}{dt} = \boldsymbol{\omega} \times \bar{\mathbf{r}}$ and $\mathbf{q} = \sum\limits_{i=1}^{\infty} \boldsymbol{\phi}_i \xi_i$, we obtain by invoking Eqs. 9–8, 9–9, 9–20, and the fact that the origin of the x-y-z-axis system is at the center of gravity, the following:

$$\tau = \frac{1}{2} M \frac{d\mathbf{r}_0{}'}{dt} \cdot \frac{d\mathbf{r}_0{}'}{dt} + \frac{1}{2} \boldsymbol{\omega} \cdot \boldsymbol{\Psi} \cdot \boldsymbol{\omega} + \frac{1}{2} \sum_{i=1}^{\infty} M_i \dot{\xi}_i^2 \tag{9–30}$$

where $\boldsymbol{\Psi}$ is the inertial tensor (cf. Eq. 9–16) and M_i are the generalized masses (cf. Eq. 9–28). The internal strain energy can be expressed in the form

$$U = \frac{1}{2} \iiint\limits_V \mathbf{q} \cdot \tilde{\mathscr{S}}(\mathbf{q}) \, dV \tag{9–31}$$

where $\mathscr{S}$ is a differential operator. Introducing $\mathbf{q} = \sum\limits_{i=1}^{\infty} \boldsymbol{\Phi}_i \xi_i$ into Eq. 9–31 yields

$$U = \frac{1}{2} \sum_{i=1}^{\infty} \sum_{j=1}^{\infty} \iiint\limits_{V} \boldsymbol{\Phi}_i \cdot \mathscr{S}(\boldsymbol{\Phi}_j)\, dV\, \xi_i \xi_j \qquad (9\text{–}32)$$

which reduces by making use of $\tilde{\mathscr{S}}(\boldsymbol{\Phi}_j) = \rho \omega_j{}^2 \boldsymbol{\Phi}_j$ (cf. Eq. 9–23) to

$$U = \frac{1}{2} \sum_{i=1}^{\infty} \sum_{j=1}^{\infty} \omega_j{}^2 \iiint\limits_{V} \boldsymbol{\Phi}_i \cdot \boldsymbol{\Phi}_j \rho\, dV\, \xi_i \xi_j \qquad (9\text{–}33)$$

The orthogonality condition, Eq. 9–20, reduces Eq. 9–33 to

$$U = \frac{1}{2} \sum_{i=1}^{\infty} M_i \omega_i{}^2 \xi_i{}^2 \qquad (9\text{–}34)$$

Hamilton's principle is stated by Eq. 2–46 in the form

$$\delta \int_{t_0}^{t_1} (\tau - U)\, dt = - \int_{t_0}^{t_1} \delta W\, dt$$

If we regard $\mathbf{r}_0{}'$, $\int_0^t \boldsymbol{\omega}\, dt$, and ξ_i $(i = 1, 2, \cdots, \infty)$ as the degrees of freedom, the virtual work quantity, δW, is

$$\delta W = \iint\limits_{S} (\mathbf{F}^D + \mathbf{F}^M) \cdot \delta \mathbf{r}_0{}'\, dS + \iint\limits_{S} \tilde{\mathbf{r}} \times (\mathbf{F}^D + \mathbf{F}^M) \cdot \delta \int_0^t \boldsymbol{\omega}\, dt\, dS$$

$$+ \sum_{i=1}^{\infty} \iint\limits_{S} (\mathbf{F}^D + \mathbf{F}^M) \cdot \boldsymbol{\Phi}_i\, \delta \xi_i\, dS, \qquad (i = 1, 2, \cdots, \infty) \qquad (9\text{–}35)$$

Introducing Eqs. 9–30, 9–34, and 9–35 into Eq. 2–46, carrying out the variations, and integrating by parts, we obtain

$$\int_{t_0}^{t_1} \left\{ \left[M \frac{d^2 \mathbf{r}_0{}'}{dt^2} - \iint\limits_{S} (\mathbf{F}^D + \mathbf{F}^M)\, dS \right] \cdot \delta \mathbf{r}_0{}' + \left[\frac{d}{dt} (\Psi \cdot \boldsymbol{\omega}) \right. \right.$$

$$- \iint\limits_{S} \tilde{\mathbf{r}} \times (\mathbf{F}^D + \mathbf{F}^M)\, dS \left. \right] \cdot \delta \int_0^t \boldsymbol{\omega}\, dt \qquad (9\text{–}36)$$

$$+ \left[M_j \ddot{\xi}_j + M_j \omega_j{}^2 \xi_j - \iint\limits_{S} (\mathbf{F}^D + \mathbf{F}^M) \cdot \boldsymbol{\Phi}_j\, dS \right] \delta \xi_j \left. \right\} = 0$$

Since $\delta \mathbf{r}_0{}'$, $\delta \int_0^t \boldsymbol{\omega}\, dt$, and $\delta \xi_j$ are perfectly arbitrary, Eq. 9–36 can be satisfied only if the square-bracketed terms within the integrand are put individually equal to zero. This procedure leads, then, to the three equations previously recorded as (9–25), (9–26), and (9–28).

9-4   EQUATIONS OF MOTION IN SCALAR FORM

In the previous sections, we have developed the equations of motion of the unrestrained vehicle in vector form. In order to carry out analyses and numerical computation it will be necessary to reduce the equations to their component scalar form.

We assume that the linear velocity vector of the center of gravity, that is the origin "O," is represented by

$$\frac{d\mathbf{r}_0'}{dt} = U_x\mathbf{i} + U_y\mathbf{j} + U_z\mathbf{k} \tag{9-37}$$

where U_x, U_y, and U_z are the magnitudes of the linear velocity vector components along the body axes, as illustrated by Fig. 9-1. In addition, the angular velocity vector of the body axis system is given by

$$\boldsymbol{\omega} = p\mathbf{i} + q\mathbf{j} + r\mathbf{k} \tag{9-38}$$

where p, q, and r are the magnitudes of the angular velocity vector components along the body axes, as illustrated also by Fig. 9-1.

Introducing Eqs. 9-37 and 9-38 into Eq. 9-2 and making use of

$$\frac{d\mathbf{i}}{dt} = r\mathbf{j} - q\mathbf{k}$$

$$\frac{d\mathbf{j}}{dt} = -r\mathbf{i} + p\mathbf{k} \tag{9-39}$$

$$\frac{d\mathbf{k}}{dt} = q\mathbf{i} - p\mathbf{j}$$

we obtain the scalar equations

$$M(\dot{U}_x - U_y r + U_z q) = P_x$$
$$M(\dot{U}_y + U_x r - U_z p) = P_y \tag{9-40}$$
$$M(\dot{U}_z - U_x q + U_y p) = P_z$$

where P_x, P_y, and P_z are components along the body axes of the force vector

$$\mathbf{P} = \iint_S (\mathbf{F}^D + \mathbf{F}^M)\, dS \tag{9-41}$$

Equation 9-3 can be expanded into component form by introducing $\mathbf{r} = \tilde{\mathbf{r}} + \mathbf{q}$. This yields, by making use of the facts that $\dfrac{d\tilde{\mathbf{r}}}{dt} = \boldsymbol{\omega} \times \tilde{\mathbf{r}}$

and $\dfrac{d\mathbf{q}}{dt} = \dfrac{\delta\mathbf{q}}{\delta t} + \boldsymbol{\omega} \times \mathbf{q}$, the following result:*

$$\frac{d}{dt}\iiint\limits_{V}\left[\bar{\mathbf{r}}\times(\boldsymbol{\omega}\times\bar{\mathbf{r}}) + \bar{\mathbf{r}}\times(\boldsymbol{\omega}\times\mathbf{q}) + \mathbf{q}\times(\boldsymbol{\omega}\times\bar{\mathbf{r}}) + \bar{\mathbf{r}}\times\frac{\delta\mathbf{q}}{\delta t}\right.$$
$$\left. + \mathbf{q}\times\frac{\delta\mathbf{q}}{\delta t} + \mathbf{q}\times\boldsymbol{\omega}\times\mathbf{q}\right]\rho\,dV = \iint\limits_{S}\bar{\mathbf{r}}\times(\mathbf{F}^{D} + \mathbf{F}^{M})\,dS \qquad (9\text{-}42)$$

For simplification of the present discussion, we shall assume that the elastic deformations are sufficiently small so that terms involving their products may be neglected and that the moment of momentum is unaffected by deformational changes in geometry. Thus, we neglect the second, third, fifth, and sixth terms on the left-hand side of Eq. 9–42 in comparison with the first and fourth terms. Expanding the vector triple product of the first term, Eq. 9–42 becomes

$$\frac{d}{dt}\left[\boldsymbol{\Psi}\cdot\boldsymbol{\omega} + \iiint\limits_{V}\bar{\mathbf{r}}\times\frac{\delta\mathbf{q}}{\delta t}\,\rho\,dV\right] = \iint\limits_{S}\bar{\mathbf{r}}\times(\mathbf{F}^{D} + \mathbf{F}^{M})\,dS \qquad (9\text{-}43)$$

where $\boldsymbol{\Psi}$ is the inertial tensor. Introducing into Eq. 9–43 the quantities

$$\bar{\mathbf{r}} = x\mathbf{i} + y\mathbf{j} + z\mathbf{k}$$
$$\mathbf{q} = u\mathbf{i} + v\mathbf{j} + w\mathbf{k} \qquad (9\text{-}44)$$
$$\boldsymbol{\omega} = p\mathbf{i} + q\mathbf{j} + r\mathbf{k}$$

and assuming for brevity that the product of inertia terms are zero, that is, the body axes are principal inertial axes, we obtain the following component scalar forms:

$$I_{xx}\dot{p} + (I_{zz} - I_{yy})qr$$
$$+ \iiint\limits_{V}[y\dot{w} - z\dot{v} + (x\dot{w} - z\dot{u})r + (x\dot{v} - y\dot{u})q]\,\rho\,dV = L_{x}$$

$$I_{yy}\dot{q} + (I_{xx} - I_{zz})pr$$
$$+ \iiint\limits_{V}[z\ddot{u} - x\dot{w} + (y\dot{w} - z\dot{v})r + (y\dot{u} - x\dot{v})p]\,\rho\,dV = L_{y}$$

$$I_{zz}\dot{r} + (I_{yy} - I_{xx})pq$$
$$+ \iiint\limits_{V}[x\ddot{v} - y\ddot{u} + (z\dot{v} - y\dot{w})q + (z\dot{u} - x\dot{w})p]\,\rho\,dV = L_{z}$$

$$(9\text{-}45)$$

* The symbol $\delta/\delta t$ is employed to denote a partial differentiation in which the unit vectors $\mathbf{i}$, $\mathbf{j}$, and $\mathbf{k}$ are held fixed.

where L_x, L_y, and L_z are components along the body axes of the moment vector

$$\mathbf{L} = \iiint_V \bar{\mathbf{r}} \times (\mathbf{F}^D + \mathbf{F}^M) \, dV \qquad (9\text{--}46)$$

In addition to the six equations (9–40 through 9–45), we require three more equations obtained by putting Eq. 9–4b in component form. These are the following:

$$u - u_0 - \frac{1}{2}\left[z\left(\frac{\partial u_0}{\partial z} - \frac{\partial w_0}{\partial x}\right) - y\left(\frac{\partial v_0}{\partial x} - \frac{\partial u_0}{\partial y}\right)\right]$$

$$= \iiint_V \{C^{xx}(x,y,z;\xi,\eta,\zeta)[(F_x^{\,D} + F_x^{\,M})\delta(x - x_s) - \rho a_x]$$

$$+ C^{xy}(x,y,z;\xi,\eta,\zeta)[(F_y^{\,D} + F_y^{\,M})\delta(y - y_s) - \rho a_y]$$

$$+ C^{xz}(x,y,z;\xi,\eta,\zeta)[(F_z^{\,D} + F_z^{\,M})\delta(z - z_s) - \rho a_z]\} d\xi\, d\eta\, d\zeta$$

$$v - v_0 - \frac{1}{2}\left[x\left(\frac{\partial v_0}{\partial x} - \frac{\partial u_0}{\partial y}\right) - z\left(\frac{\partial w_0}{\partial y} - \frac{\partial v_0}{\partial z}\right)\right]$$

$$= \iiint_V \{C^{yx}(x,y,z;\xi,\eta,\zeta)[(F_x^{\,D} + F_x^{\,M})\delta(x - x_s) - \rho a_x]$$

$$+ C^{yy}(x,y,z;\xi,\eta,\zeta)[(F_y^{\,D} + F_y^{\,M})\delta(y - y_s) - \rho a_y]$$

$$+ C^{yz}(x,y,z;\xi,\eta,\zeta)[(F_z^{\,D} + F_z^{\,M})\delta(z - z_s) - \rho a_z]\} d\xi\, d\eta\, d\zeta$$

$$w - w_0 - \frac{1}{2}\left[y\left(\frac{\partial w_0}{\partial y} - \frac{\partial v_0}{\partial z}\right) - x\left(\frac{\partial u_0}{\partial z} - \frac{\partial w_0}{\partial x}\right)\right]$$

$$= \iiint_V \{C^{zx}(x,y,z;\xi,\eta,\zeta)[(F_x^{\,D} + F_x^{\,M})\delta(x - x_s) - \rho a_x]$$

$$+ C^{zy}(x,y,z;\xi,\eta,\zeta)[(F_y^{\,D} + F_y^{\,M})\delta(y - y_s) - \rho a_y]$$

$$+ C^{zz}(x,y,z;\xi,\eta,\zeta)[(F_z^{\,D} + F_z^{\,M})\delta(z - z_s) - \rho a_z]\} d\xi\, d\eta\, d\zeta$$

$$(9\text{--}47)$$

where a_x, a_y, and a_z are acceleration components; and $(F_x^{\,D} + F_x^{\,M})\delta(x - x_s)$, $(F_y^{\,D} + F_y^{\,M})\delta(y - y_s)$, and $(F_z^{\,D} + F_z^{\,M})\delta(z - z_s)$ are surface force components along the body axes. The acceleration components are obtained by observing that

$$\mathbf{a} = \frac{d}{dt}\left(\frac{d\mathbf{r_0}'}{dt} + \frac{d\tilde{\mathbf{r}}}{dt} + \frac{d\mathbf{q}}{dt}\right) \tag{9-48}$$

where

$$\frac{d\mathbf{r_0}'}{dt} = U_x\mathbf{i} + U_y\mathbf{j} + U_z\mathbf{k}$$

$$\frac{d\tilde{\mathbf{r}}}{dt} = \boldsymbol{\omega} \times \tilde{\mathbf{r}}$$

$$\frac{d\mathbf{q}}{dt} = \frac{\delta\mathbf{q}}{\delta t} + \boldsymbol{\omega} \times \mathbf{q}$$

Equations 9–40, 9–45, and 9–47 represent the component form of the vector equations (9–2, 9–3, and 9–4b). If we express the elastic deformations in terms of the normal modes of the unrestrained vehicle, as we did in Sec. 9–3, we state the following:

$$u = \sum_{i=1}^{\infty} \phi_{x_i}(x, y, z)\xi_i(t)$$

$$v = \sum_{i=1}^{\infty} \phi_{y_i}(x, y, z)\xi_i(t) \tag{9-49}$$

$$w = \sum_{i=1}^{\infty} \phi_{z_i}(x, y, z)\xi_i(t)$$

where ϕ_{x_i}, ϕ_{y_i}, and ϕ_{z_i} are components of the vector function [cf. Sec. 3–5(d)]

$$\boldsymbol{\phi}_i(x, y, z) = \phi_{x_i}(x, y, z)\mathbf{i} + \phi_{y_i}(x, y, z)\mathbf{j} + \phi_{z_i}(x, y, z)\mathbf{k} \tag{9-50}$$

When the substitutions of Eqs. 9–49 and 9–50 are made, Eqs. 9–40 remain unchanged, and Eqs. 9–45 and 9–47 reduce to the following:

$$I_{xx}\dot{p} + (I_{zz} - I_{yy})\,qr = L_x$$

$$I_{yy}\dot{q} + (I_{xx} - I_{zz})\,pr = L_y \tag{9-51}$$

$$I_{zz}\dot{r} + (I_{yy} - I_{xx})\,pq = L_z$$

$$M_j\ddot{\xi}_j + M_j\omega_j^2\xi_j = \Xi_j^D + \Xi_j^M, \quad (j = 1, 2, \cdots, n) \tag{9-52}$$

The generalized mass is (cf. Eq. 3–117a)

$$M_j = \iiint\limits_V (\phi_{x_j}^2 + \phi_{y_j}^2 + \phi_{z_j}^2)\rho \, dV \tag{9-53}$$

and the generalized forces are (cf. Eq. 3–118)

$$\Xi_j{}^D = \iint_S (F_x{}^D \phi_{x_j} + F_y{}^D \phi_{y_j} + F_z{}^D \phi_{z_j}) \, dS$$

$$\Xi_j{}^M = \iint_S (F_x{}^M \phi_{x_j} + F_y{}^M \phi_{y_j} + F_z{}^M \phi_{z_j}) \, dS$$

(9–54)

We recognize Eqs. 9–40 and 9–51 as comprising the six Euler equations of motion of a rigid body that is free in space. The angular position of such a body is described by the Eulerian angles θ, ϕ, and ψ, as shown by Fig. 9–1. In fact, when the angular velocities p, q, and r are known functions of time, the angular position may be computed by solving the following equations for θ, ϕ, and ψ (Ref. 9–1):

$$\dot{\theta} \sin \psi - \dot{\phi} \sin \theta \cos \psi = p$$
$$\dot{\theta} \cos \psi + \dot{\phi} \sin \theta \cos \psi = q \qquad (9\text{–}55)$$
$$\dot{\phi} \cos \theta + \dot{\psi} \qquad\qquad = r$$

9–5 THE EQUATIONS OF SMALL DISTURBED MOTION FROM STEADY RECTILINEAR FLIGHT

The complex nature of the equations of motion of an elastic vehicle free to undergo large angles of rotation is evident from the preceding sections of this chapter. These equations may, of course, be solved numerically to obtain the responses of an elastic vehicle subjected to any flight condition. However, for purposes of discussion here, it will be desirable to introduce some simplifications. One possible simplification is that of linearization to represent the case of small disturbed motion from steady, rectilinear flight. Let us suppose that the vehicle is climbing with steady forward velocity U, making a path angle with the horizontal of γ_0, as shown by Fig. 9–2, before it is subjected to a disturbance. The x-z-plane of the vehicle is the plane of symmetry, and the velocity vector U is assumed collinear with the x-axis.

If we define Eulerian angles ϕ, θ, and ψ as rotation angles of the x-y-z-coordinate system from the direction of rectilinear flight, the assumption of small disturbances assures that ϕ, θ, and ψ are small; and we can put

$$p = \dot{\phi}, \qquad q = \dot{\theta}, \qquad r = \dot{\psi} \qquad (9\text{–}56)$$

The components of the velocity of the center of gravity with respect to the

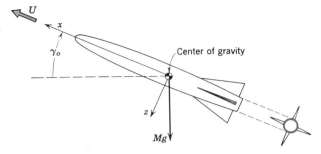

Fig. 9-2. Vehicle in steady climbing flight.

x-y-z-axis system are denoted by $U + U_x$, U_y, and U_z. The assumption of small disturbances requires that U_x, U_y, and U_z be small compared to U. The linearized form of Eqs. 9-40 provides

$$M(\dot{U}_x) = -(Mg \cos \gamma_0)\theta + P_{xa}$$

$$M(\dot{U}_y + U\dot{\psi}) = (Mg \sin \gamma_0)\psi + (Mg \cos \gamma_0)\phi + P_{ya} \qquad (9\text{-}57)$$

$$M(\dot{U}_z - U\dot{\theta}) = -(Mg \sin \gamma_0)\theta + P_{za}$$

where P_{xa}, P_{ya}, and P_{za} are aerodynamic forces.
The linearized form of Eqs. 9-51 becomes

$$I_{xx}\ddot{\phi} = L_{xa}$$

$$I_{yy}\ddot{\theta} = L_{ya} \qquad (9\text{-}58)$$

$$I_{zz}\ddot{\psi} = L_{za}$$

where L_{xa}, L_{ya}, and L_{za} are aerodynamic moments.
Equations 9-52 remain unaltered in their form for the case of small disturbances from rectilinear flight.
We may take as an even further simplification the case of a structure represented by an unrestrained elastic plate lying in the x-y-plane, as shown by Fig. 9-3. Such a mathematical model is often adopted for aeroelastic studies of aircraft. We assume that the vehicle is in steady rectilinear horizontal flight and that its rigid degrees of freedom are those of pitching, rolling, and vertical translation, in addition to its plate-like elastic deformations. Under the circumstances of small disturbances, it is more convenient to adopt coordinate axes that are fixed in space. Designating ξ_1 as the disturbed vertical displacement, $\xi_2 = \phi$ as the disturbed rolling displacement, and $\xi_3 = \theta$ as the disturbed pitching displacement, the

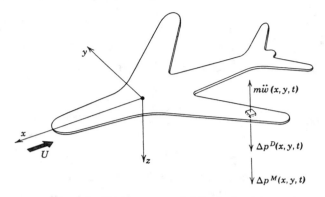

Fig. 9–3. Airplane represented by elastic plate.

equations of motion which represent the disturbed rigid and elastic displacements become

$$M_j \ddot{\xi}_j + M_j \omega_j^2 \xi_j = \Xi_j^D + \Xi_j^M, \quad (j = 1, 2, \cdots, \infty) \quad (9\text{–}59)$$

where

$$M_1 = M; \quad M_2 = I_{xx}; \quad M_3 = I_{yy}; \quad \omega_1 = \omega_2 = \omega_3 = 0$$

are the generalized masses and the natural frequencies of the rigid-body modes of motion. The quantities ω_j and M_j $(j = 4, 5\ 6, \cdots, \infty)$ are the natural frequencies and the generalized masses of the deformational modes. The generalized masses, in the present simplified example, are defined by

$$M_j = \iint\limits_S \phi_j^2(x, y) m(x, y)\, dx\, dy, \quad (j = 1, 2, \cdots, \infty) \quad (9\text{–}60)$$

where $m(x, y)$ is the mass density per unit of projected area of the vehicle on the x-y-plane and $\phi_1 = 1$, $\phi_2 = y$, and $\phi_3 = -x$.

The natural mode shapes and frequencies of a model represented by an elastic surface may be obtained by applying the homogeneous forms of Eqs. 9–2, 9–3, and 9–4b directly. If $w(x, y, t)$ represents the total displacement in the z-direction of the weaving elastic surface with respect to the fixed x-y-plane, we have from Eq. 9–2 a single scalar equation:

$$\iint\limits_S \ddot{w}(x, y, t) m(x, y)\, dx\, dy = 0 \quad (9\text{–}61a)$$

From Eq. 9–3, we have the two scalar equations

$$\iint_S \ddot{w}(x, y, t)xm(x, y)\, dx\, dy = 0$$

$$\iint_S \ddot{w}(x, y, t)ym(x, y)\, dx\, dy = 0 \tag{9–61b}$$

A fourth scalar equilibrium equation, derived from Eq. 9–4b, reads

$$w(x, y, t) - w(0, 0, t) - x\frac{\partial w(0, 0, t)}{\partial x} - y\frac{\partial w(0, 0, t)}{\partial y}$$

$$= -\iint_S C^{zz}(x, y; \xi, \eta)m\ddot{w}\, d\xi\, d\eta \tag{9–61c}$$

In deriving Eq. 9–61c, we have made use of the assumption that the normals to the midplane reference surface before deformation remain normal after deformation and that $\dfrac{\partial v}{\partial z} = -\dfrac{\partial w}{\partial y}$ and $\dfrac{\partial u}{\partial z} = -\dfrac{\partial w}{\partial x}$. By introducing as a free vibration solution

$$w(x, y, t) = \phi(x, y)\, T(t) \tag{9–61d}$$

and by combining Eqs. 9–61a, 9–61b, and 9–61c in a manner similar to that employed in obtaining Eq. 9–18 of Sec. 9–2, we can derive the homogeneous integral equation

$$\phi(x, y) = \omega^2\iint_S G(x, y; \xi, \eta)\phi(\xi, \eta)m(\xi, \eta)\, d\xi\, d\eta \tag{9–61e}$$

where

$$G(x, y; \xi, \eta) = C(x, y; \xi, \eta)$$

$$- \iint_S C(r, s; \xi, \eta)\left[\frac{1}{M} + \frac{ys}{I_x} + \frac{xr}{I_y}\right]m(r, s)\, dr\, ds \text{ *}$$

is the influence function of the unrestrained airplane. Equation 9–61e is satisfied by an infinite number of pairs of deformation mode shapes $\phi_j(x, y)$ and frequencies ω_j. The same remarks concerning orthogonality of mode shapes in Sec. 9–2 apply also to the natural mode shapes of the elastic plate derived from Eq. 9–61e.

The generalized forces of the unrestrained elastic plate are computed by referring to the forces which act during the disturbance. We assume that

---

* This form of the influence function of a free elastic plate may also be derived directly by breaking Eq. 9–19 down into its component form.

a pressure $\Delta p^D(x, y, t)$, with arbitrary spatial and time dependence, is applied by an atmospheric disturbance. As a result, there are disturbed displacements, velocities, and accelerations; and aerodynamic pressures denoted by $\Delta p^M(x, y, t)$ are brought into play.

The generalized force component producing the disturbance is an explicit function of time defined by

$$\Xi_j{}^D(t) = \iint\limits_S \Delta p^D(x, y, t)\phi_j(x, y) \, dx \, dy \qquad (9\text{-}62)$$

The component resulting from the disturbed motion, $\Xi_j{}^M$, serves not only to damp the motion, but also introduces coupling among the normal coordinates

$$\Xi_j{}^M(\xi_1, \cdots, \xi_n; \dot{\xi}_1, \cdots, \dot{\xi}_n; \ddot{\xi}_1, \cdots, \ddot{\xi}_n) = \iint\limits_S \Delta p^M \phi_j(x, y) \, dx \, dy \quad (9\text{-}63)$$

Thus, it is evident that although the normal coordinates uncouple the system elastically and dynamically, $\Xi_j{}^M$ terms may provide very strong aerodynamic coupling.

9–6 EXAMPLES OF THE DISTURBED MOTION OF UNRESTRAINED ELASTIC VEHICLES

In previous sections of the present chapter we have derived the equations of motion which are appropriate to a three-dimensional unrestrained aeroelastic system. Following the practice of Chaps. 6, 7, and 8, it would be possible to derive in detail various aeroelastic phenomena in connection with these equations. However, lack of space precludes the detailed treatment which is presented, for example, in Chap. 6; and we shall restrict our attention to three examples in which the rigid-body degrees of freedom are present.

(a) Dynamic response to a discrete atmospheric disturbance

Let us take up first the application of Eq. 9–59 to the problem of the small disturbed motion of an elastic vehicle which is in steady level rectilinear flight and subjected to a discrete atmospheric disturbance. We shall make use of the theory presented in Sec. 6–3 pertaining to the typical section. The disturbance generalized force is assumed to be expressible as

$$\Xi_i{}^D(t) = C_i f(t) \qquad (9\text{-}64)$$

where $f(t)$ is a nondimensional function representing the time variation of the disturbing force. The differential equations, based on Eqs. 9–59, are

$$M_j\ddot{\xi}_j + M_j\omega_j{}^2\xi_j = \Xi_j{}^M(\xi_1, \cdots, \xi_n; \dot{\xi}_1, \cdots, \dot{\xi}_n; \ddot{\xi}_1, \cdots, \ddot{\xi}_n)$$
$$+ C_j f(t), \qquad (j = 1, 2, \cdots, n) \quad (9\text{–}65)$$

with initial conditions $\xi_j(0) = \dot{\xi}_j(0) = 0$. In Eqs. 9–65, ξ_1, ξ_2, and ξ_3 represent the small disturbed plunging, rolling, and pitching motions from steady rectilinear flight; $\xi_4, \cdots, \xi_n$ represent the disturbed quantities which are to be superimposed upon the quantities corresponding to steady level flight prior to the onset of the disturbance. Of the various methods of solution of transient problems mentioned in Chap. 6, the Laplace transform (cf. Sec. 6–3b) is selected for the present illustration. The Laplacian operator $\mathscr{L}\{\quad\}$, defined by Eq. 6–52, applied to Eq. 9–65 yields

$$M_j p^2 \bar{\xi}_j(p) + M_j\omega_j{}^2 \bar{\xi}_j(p) = \mathscr{L}\{\Xi_j{}^M + C_j f(t)\}, \qquad (j = 1, 2, \cdots, n)$$
$$(9\text{–}66)$$

where

$$\bar{\xi}_j(p) = \mathscr{L}\{\xi_j(t)\}$$

Since Eq. 9–66 is linear, it can be put into the form

$$[(p^2 + \omega_j{}^2)M_j - \bar{\Xi}_{jj}{}^M(p)]\bar{\xi}_j(p) - \sum_{i \neq j}^{n} \bar{\Xi}_{ji}{}^M(p)\bar{\xi}_i(p) = C_j \bar{f}(p)$$
$$(j = 1, 2, \cdots, n) \quad (9\text{–}67)$$

where $\bar{\Xi}_{jj}{}^M \bar{\xi}_j(p)$ is the transformed component of the generalized force $\Xi_j{}^M$ due to the motion in the coordinate, ξ_j; $\bar{\Xi}_{ji}{}^M(p)\bar{\xi}_i(p)$, ($i \neq j$), are the transformed coordinates of the generalized force $\Xi_j{}^M$ due to the motion in coordinates other than ξ_j; and $\bar{f}(p)$ is the transform of $f(t)$.

Equation 9–67 represents a set of n simultaneous linear algebraic equations in the unknown functions $\bar{\xi}_1(p), \cdots, \bar{\xi}_n(p)$, as follows:

$$\sum_{i=1}^{n} C_{ji}\bar{\xi}_i = C_j\bar{f}(p), \qquad (j = 1, 2, \cdots, n) \qquad (9\text{–}68)$$

Solving Eqs. 9–68 for $\bar{\xi}_j(p)$, we obtain

$$\bar{\xi}_j(p) = \frac{M_j(p)}{N(p)}\bar{f}(p), \qquad (j = 1, 2, \cdots, n) \qquad (9\text{–}69)$$

where

$$M_j(p) = \sum_{i=1}^{n} a_{ji}(p)C_i$$

$N(p)$ is the determinant of the matrix of coefficients C_{ji}, and $a_{ji}(p)$ is the cofactor of the term C_{ji}.

The process of computing the response $\xi_j(t)$ requires that the inverse transformation of Eq. 9–69 be taken. This is accomplished by applying the convolution integral (cf. Eq. 6–65) to Eq. 9–69, which gives

$$\xi_j(t) = \mathscr{L}^{-1}\left\{\frac{M_j(p)}{N(p)}\bar{f}(p)\right\}$$

$$= \int_0^t \mathscr{L}^{-1}\left\{\frac{M_j(p)}{N(p)}\right\} f(t - \tau)\, d\tau, \qquad (j = 1, 2, \cdots, n) \quad (9\text{–}70)$$

Since the elements C_{ji} in the determinant $N(p)$ are algebraic functions of p, the determinant can be expanded into a polynomial which may be factored to

$$N(p) = \prod_j^m (p - p_j) \qquad (9\text{–}71)$$

where m is the number of roots of the polynomial.

The inverse transformation of $M_j(p)/N(p)$ is carried out by Heaviside's partial-fraction expansion (Ref. 9–2)

$$\mathscr{L}^{-1}\left\{\frac{M_j(p)}{N(p)}\right\} = \sum_{k=1}^m \frac{M(_jp_k)}{N'(p_k)} e^{p_k \tau}, \qquad (j = 1, 2, \cdots, n) \quad (9\text{–}72)$$

where

$$N'(p_k) = \left[\frac{d}{dp}N(p)\right]_{p=p_k}$$

Substituting Eq. 9–72 into Eq. 9–70, we obtain the final solution for the jth normal coordinate:

$$\xi_j(t) = \sum_{k=1}^m \frac{M_j(p_k)}{N'(p_k)} \int_0^t f(t - \tau)e^{p_k \tau}\, d\tau, \qquad (j = 1, 2, \cdots, n) \quad (9\text{–}73)$$

In obtaining this result, it has been necessary to factor the denominator polynomial, $N(p)$. This is often an exceptionally difficult task since the polynomial is usually of a high degree with a number of complex roots.

After the motion of the airplane has been computed in terms of the normal coordinates, $\xi_i(t)$, the analyst may have several objectives in mind. Perhaps the simplest is that of assigning an arbitrary form to the disturbance function $f(t)$, say, $1(t)$ (unit step function) or $\delta(t)$ (unit impulse function), and studying the stability or the damping characteristics of the response. Such results may be especially valuable in those cases where there is a feedback loop in the control system which involves a sensing element mounted on the structure. Another objective may be that of computing the stresses for structural design purposes when $f(t)$ is assigned explicit forms. Such forms may arise from a variety of sources such as discrete atmospheric gusts, shock waves, blast waves, landing loads, or abrupt control surface reactions. When stresses are the end objective, several methods are available for computation, the most prominent of which is the so-called mode-displacement method. In the mode-displacement method, the stresses are computed by summing the stresses due to the displacement of each normal mode. Thus the stresses at a point (x, y) in the airplane may be expressed as

$$\sigma(x, y, t) = \sum_{i=4}^{n} A_i(x, y)\xi_i(t) \tag{9-74}$$

The coefficient $A_i(x, y)$ is the stress at (x, y) due to a unit displacement of the ith mode and may be conveniently computed from the inertia loading required to maintain a free vibration of unit amplitude in that mode. Thus,

$$A_i(x, y) = \omega_i^2 \iint\limits_{S} W(x, y; \xi, \eta) m(\xi, \eta) \phi_i(\xi, \eta) \, d\xi \, d\eta, \qquad (i = 4, 5, \cdots, n) \tag{9-75}$$

where $W(x, y; \xi, \eta)$ is a weighting function which gives the stress at (x, y) due to a unit load at (ξ, η).

Another method of stress computation, the so-called mode-acceleration method (Ref. 9–3), is derived by rearranging the equations of motion as

$$\xi_i(t) = \frac{\Xi_i^D}{M_i\omega_i^2} + \frac{\Xi_i^M}{M_i\omega_i^2} - \frac{\ddot{\xi}_i}{\omega_i^2} \tag{9-76}$$

When this expression for $\xi_i(t)$ is substituted into Eq. 9–74, there results

$$\sigma(x, y, t) = \sigma_s(x, y, t) + \sum_{i=4}^{n} A_i(x, y)\left(\frac{\Xi_i^M}{M_i\omega_i^2} - \frac{\ddot{\xi}_i}{\omega_i^2}\right) \tag{9-77}$$

where

$$\sigma_s(x, y, t) = \sum_{i=4}^{n} A_i(x, y) \frac{\Xi_i^D}{M_i\omega_i^2}$$

is the so-called "static stress." The latter is equivalent to the stress which would be produced by the disturbance if the velocities and accelerations of the vibratory natural modes could be suppressed. It can be computed more accurately by the following, which in effect carries the summation on i to ∞:

$$\sigma_s(x, y, t) = \iint\limits_{S} W(x, y; \xi, \eta) \, \Delta p^D(\xi, \eta) \, d\xi \, d\eta$$

$$- f(t) \sum_{i=1}^{3} \frac{C_i}{M_i} \iint\limits_{S} W(x, y; \xi, \eta) m(\xi, \eta) \phi_i(\xi, \eta) \, d\xi \, d\eta \tag{9-78}$$

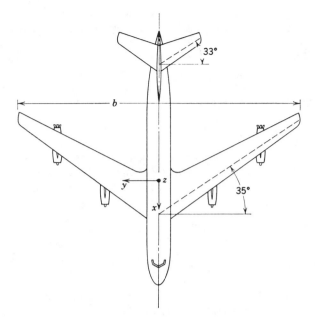

Fig. 9–4. Hypothetical jet transport.

$M = 4{,}500$ slugs	$S = 1430$ ft²
$I_{yy} = 1.6 \times 10^6$ slug ft²	$\mathcal{R} = 9.4$
$b = 120$ ft	$\bar{c} = 12.5$ ft

The "static stress" is in reality a pseudo-static stress since it includes the effects of the inertial forces associated with the rigid-body accelerations. In computing the stress by means of Eq. 9–77, since $\Xi_i{}^M$ is itself a sum of integrated motion forces, the summation is very laborious to form. It has been pointed out, however, (Ref. 9–4) that Eq. 9–77 can be replaced by an equivalent form

$$\sigma(x, y, t) = \sigma_s(x, y, t) + \sum_{i=4}^{n} A_i(x, y)\left[\xi_i - \frac{\Xi_i{}^D}{M_i \omega_i{}^2}\right] \qquad (9\text{–}79)$$

where the unwieldly motion terms have been eliminated. The advantage of the mode-acceleration over the mode-displacement method lies primarily in the fact that the former accounts for the static stresses directly; this provides more rapid convergence of the modal series.

As an example, let us consider the hypothetical jet transport airplane shown in planform by Fig. 9–4 and illustrate its response to a sharp-edged gust. In Fig. 9–5, we show the *acceleration ratio* of the center of gravity of the airplane, as a function of the dimensionless time parameter $s = 2Ut/\bar{c}$, after the airplane strikes a vertically rising sharp-edged gust (Ref. 9–5). U is the airplane speed, assumed to be 460 mph at 11,000 ft altitude, t is the time, and $\bar{c}$ is the mean geometric chord. The *acceleration ratio*, A.R., is defined by

$$\text{A.R.} = \frac{a_{cg}}{a_{ss}} \qquad (9\text{–}80)$$

a_{cg} is the acceleration of the center of gravity of the deformed airplane and a_{ss} is the acceleration given by the sharp-edged gust formula as

$$a_{ss} = \frac{wU}{\mu\bar{c}} \qquad (9\text{–}81)$$

where μ is the dimensionless mass parameter, $M/\rho S \dfrac{\bar{c}}{2}\dfrac{dC_L}{d\alpha}$, and w is the gust velocity. The influences of the rigid-body modes of heaving and pitching and of a single vibratory mode (predominantly wing bending) are evident from Fig. 9–5. Whereas the heaving degree of freedom alone produces an alleviation in acceleration (A.R. < 1), the inclusion of a pitching degree of freedom cancels out this relief for a swept-wing airplane.

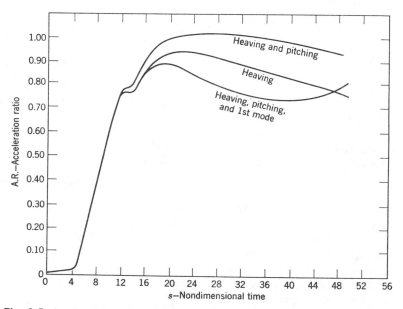

Fig. 9–5. Acceleration ratios of deformed swept-wing airplane for vertically rising sharp-edged gust.

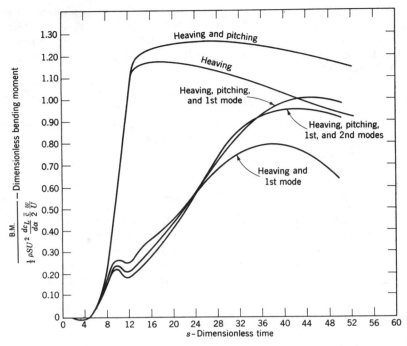

Fig. 9–6. Wing root bending moment for a vertically rising sharp-edged gust.

When the first vibratory mode is added, however, there is a marked relieving effect, the peak acceleration with heaving, pitching, and a single vibratory mode included being about 0.9.

In Fig. 9–6, the dimensionless root bending moment is shown as a function of dimensionless time, s, after the airplane strikes a sharp-edged gust (Ref. 9–5). The bending moment is normalized in terms of the parameter $\frac{1}{2}\rho U^2 S \frac{dC_L}{d\alpha} \frac{\bar{c}}{2} \frac{w}{U}$. In this figure, the large alleviating influence of the elastic degrees of freedom is evident. The figure also shows clearly that the alleviation in peak stresses commonly attributed to flexible sweptback wings is partially canceled by the increased pitching motion of the airplane. The inclusion of a second vibratory mode (predominantly wing twisting) is seen to provide further alleviation. For the particular airplane under consideration, the second mode frequency is 3.30 times the first mode frequency.

In Figs. 9–5 and 9–6, we have illustrated the response characteristics of an elastic swept-wing airplane subjected to a sharp-edged gust. To find the stress response due to an arbitrary gust profile, the time history of the stress response to a sharp-edged gust must be integrated in convolution with the desired gust profile (cf. Eq. 6–65). If $B(s)$ is the response or stress due to a sharp-edged gust per unit w/U, the corresponding response or stress due to an arbitrary gust profile per unit w_{max}/U is

$$C(s) = \int_0^s B(\sigma)F'(s - \sigma) \, d\sigma \qquad (9\text{–}82)$$

where the prime indicates a first derivative, and where $F'(s) = w'(s)/w_{max}$. A frequently used gust profile in gust-loading studies is the one-minus-cosine profile

$$F(s) = \tfrac{1}{2}\left(1 - \cos\frac{\pi s}{s_G}\right) \qquad (9\text{–}83)$$

where s_G is the value of s at the gust peak. The response to a one-minus-cosine gust profile is then

$$C(s) = \frac{\pi}{2s_G} \int_0^s B(\sigma) \sin\frac{\pi}{s_G}(s - \sigma) \, d\sigma \qquad (9\text{–}84)$$

It is apparent that the response to such a gust profile will depend upon s_G, a quantity which is difficult to define explicitly. This dependence is illustrated by Fig. 9–7, where the tendency for the airplane to be tuned to certain values of s_G is shown (Ref. 9–5). In this figure, the peak dimensionless bending moment at the root is plotted as a function of s_G. The figure again illustrates the alleviating influence of elasticity and the aggravating influence of the pitching degree of freedom. The preferred value of s_G, producing peak stresses, for the rigid airplane is of the order of 30; whereas, in the case of the elastic airplane, it is of the order of 50.

(b) Dynamic response to continuous atmospheric turbulence

We turn our attention next to a solution of the system of equations (9–59) when it is assumed that the disturbance pressure $\Delta p^D(x, y, t)$ results from continuous and random atmospheric turbulence. That is, the character of the turbulence is such that it is not susceptible to explicit Fourier analysis. In fact, the following remarks on the mathematical processes involved are limited to an airplane with a linear admittance function. With this approach and with these assumptions, we must state certain mean properties of the atmosphere for which the airplane is to be studied. We also recognize that the important properties of the atmosphere are indeed known in a statistical sense. However, at the same time we must be prepared as engineers to accept the fact that such a procedure can give us explicitly only certain mean values of the response and that any further information can be extracted only in a statistical sense.

In Sec. 6–3(c), we have outlined the statistical functions and parameters

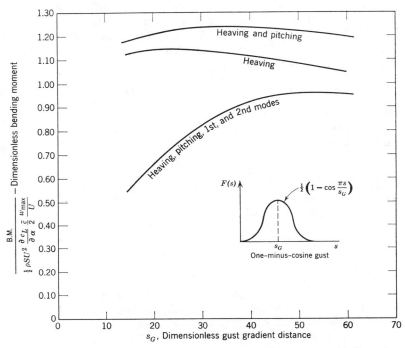

Fig. 9–7. Variation of peak dimensionless wing-root bending moment with dimensionless gust gradient distance.

commonly used to describe random processes and have seen how a typical section wing reacts to continuous atmospheric turbulence. In the present section, we shall merely extend these concepts to the more elaborate case of a complete unrestrained airplane. When the turbulence can be assumed one-dimensional, that is, when the vertical gust velocity, w_G, is either constant along the wing span or the scale of turbulence is very large compared to the wing span, the mean-squared value of the stress, $\overline{\sigma^2}$, at some point in the airplane is related to the mean-squared vertical gust velocity, $\overline{w_G{}^2}$, through the equation

$$\overline{\sigma^2} = \int_0^\infty \Phi_\sigma(\omega)\, d\omega \tag{9–85}$$

where

$$\Phi_\sigma(\omega) = |H_\sigma(\omega)|^2\, \Phi_w(\omega) \tag{9–86}$$

and

$$\overline{w_G{}^2} = \int_0^\infty \Phi_w(\omega)\, d\omega \tag{9–87}$$

The functions $\Phi_\sigma(\omega)$ and $\Phi_w(\omega)$ are the power spectra of $\sigma(t)$ and $w_G(t)$, respectively; and $H_\sigma(\omega)$ is the transfer or frequency response function of the system. The spectral elements $\Phi_\sigma(\omega)\,d\omega$ and $\Phi_w(\omega)\,d\omega$ give the contributions of the Fourier components of $\sigma(t)$ and $w_G(t)$ to $\overline{\sigma^2}$ and $\overline{w_G{}^2}$, respectively, for frequencies ranging from ω to $\omega + d\omega$; and $H_\sigma(\omega)$ is the complex amplitude of the steady-state response to a sinusoidal gust of unit amplitude and of frequency ω. The transfer function is obtained by computing the stress response at a point in the structure due to the vertical gust pattern

$$w_G(x, y, t) = e^{i\,\omega(t - x/U)} \tag{9–88}$$

where x is measured aft from some reference point and U is the forward speed.

The preceding remarks apply when spanwise variations of gust velocity are unimportant. The same methods are applicable when variations in vertical gust velocity over the x-y-plane become important, although their form is more complex. Liepmann (Ref. 9–6) has derived the following expression for the mean-squared lift $\overline{L^2}$ on a rigid wing flying through homogeneous turbulence, that is, for the case where the vertical gust velocity is a stationary random function of space in the three directions x, y, and z at a particular instant of time:

$$\overline{L^2} = \iiint\limits_{-\infty}^{+\infty} |\Gamma_L(k_1, k_2)|^2 \, \phi_{33}(k_1, k_2, k_3)\, dk_1\, dk_2\, dk_3 \tag{9–89}$$

The spectral function ϕ_{33} is a vertical gust component of the spectrum tensor for homogeneous turbulence, and Γ_L is a transfer function for the lift in a two-dimensional sinusoidal gust field. The variables k_1, k_2, and k_3 are wave numbers in the x, y, and z-directions, respectively. Employing Liepmann's expression, we may replace Eqs. 9–86 and 9–87, respectively, by*

$$\overline{\sigma^2} = \iiint\limits_{-\infty}^{+\infty} |\Gamma_\sigma(k_1, k_2)|^2 \phi_{33}(k_1, k_2, k_3)\, dk_1\, dk_2\, dk_3 \tag{9–90}$$

$$\overline{w_G{}^2} = \iiint\limits_{-\infty}^{+\infty} \phi_{33}(k_1, k_2, k_3)\, dk_1\, dk_2\, dk_3 \tag{9–91}$$

The transfer function Γ_σ is a two-dimensional generalization of the transfer

---

* Diederich and Richardson have formulated the same problem in terms of correlation functions corresponding to ϕ_{33}, in which case the aerodynamic and structural influence functions can be employed directly.

function H_σ in which $H_\sigma(\omega) = \Gamma_\sigma(k_1, 0)$. It is computed in a similar manner by finding the steady-state stress response at a point in the airplane due to the vertical gust pattern (Ref. 9–7)

$$w_G(x, y, t) = \exp\{i[k_1(Ut - x) + k_2 y]\} = \exp(ik_2 y) \cdot \exp\left[i\omega\left(t - \frac{x}{U}\right)\right]$$
(9–92)

Because the steady-state response of a linear system to a harmonic input must also be harmonic, we may express ξ_j, Δp^D, and Δp^M as

$$\xi_j(t) = \bar{\xi}_j(\omega, k_2)e^{i\omega t}$$
(9–93)

$$\Delta p^D = \bar{p}^D(x, y, \omega, k_2)e^{i\omega t}$$
(9–94)

and

$$\Delta p^M = \sum_{j=1}^{n} \bar{p}_j{}^M(x, y, \omega)\bar{\xi}_j(\omega, k_2)e^{i\omega t}$$
(9–95)

The quantities $\bar{\xi}_j$, $\bar{p}^D$ and $\bar{p}^M$ are complex quantities which represent the amplitudes and phases of the response and the aerodynamic pressures. In terms of these new variables, the equations of motion (Eqs. 9–59) become

$$M_i(\omega_i{}^2 - \omega^2)\bar{\xi}_i = \iint_S \bar{p}^D(x, y, \omega, k_2)\phi_i(x, y)\, dx\, dy$$

$$+ \sum_{j=1}^{n} \bar{\xi}_j \iint_S \bar{p}_j{}^M(x, y, \omega)\phi_i(x, y)\, dx\, dy, \quad (i = 1, 2, \cdots, n) \quad (9\text{–}96)$$

We define the generalized force integration in terms of the notation

$$\tfrac{1}{2}\rho U^2 S \frac{dC_L}{d\alpha} C_{ij}(\omega) = -\frac{\bar{c}}{2} \iint_S \bar{p}_j{}^M(x, y, \omega)\phi_i(x, y)\, dx\, dy \quad (9\text{–}97)$$

and

$$\tfrac{1}{2}\rho U^2 S \frac{dC_L}{d\alpha} K_i(\omega, k_2) = U \iint_S \bar{p}^D(x, y, \omega, k_2)\phi_i(x, y)\, dx\, dy \quad (9\text{–}98)$$

The equations of motion (9–96) can then be rewritten as

$$M_i(\omega_i{}^2 - \omega^2)\bar{\xi}_i + \tfrac{1}{2}\rho U^2 S \frac{dC_L}{d\alpha} \sum_{j=1}^{n} C_{ij}(\omega)\frac{\bar{\xi}_j}{\bar{c}/2} = \tfrac{1}{2}\rho U S \frac{dC_L}{d\alpha} K_i(\omega, k_2)$$
$$(i = 1, 2, \cdots, n) \quad (9\text{–}99)$$

Equation 9–99 is reduced to dimensionless form by making use of the mass parameter, μ, the reduced frequency parameter, $k = \omega \bar{c}/2U$, and the parameters $\Omega_i = \omega_i \bar{c}/2U$ and $\kappa = bk_2/2$.

$$\frac{M_i}{M} \mu(\Omega_i^2 - k^2) \frac{\bar{\xi}_i}{\bar{c}/2} + \frac{1}{2} \sum_{j=1}^{n} C_{ij}(k) \frac{\bar{\xi}_j}{\bar{c}/2} = \frac{1}{2U} K_i(k, \kappa)$$

$$(i = 1, 2, \cdots, n; \Omega_1 = \Omega_2 = \Omega_3 = 0) \qquad (9\text{--}100)$$

The dimensionless functions $C_{ij}(k)$ and $K_i(k, \kappa)$ may be regarded as generalized unsteady lift functions analogous to Theodorsen's function and Sears' gust function for the unsteady lift on a two-dimensional airfoil. These functions could, in general, include such effects as those due to wing sweep, spanwise induction, and interference among wing, fuselage, and tail.

Equations 9–100 are a set of simultaneous linear algebraic equations with complex coefficients, which can be written concisely in the matrix notation

$$[a_{ij}(k)]\{\bar{\xi}_j\} = \frac{\bar{c}}{4U} \{K_i(k, \kappa)\} \qquad (9\text{--}101)$$

where $a_{ij}(k) = \dfrac{M_i}{M} \mu(\Omega_i^2 - k^2)\delta_{ij} + \frac{1}{2}C_{ij}(k)$ and δ_{ij} is the Kronecker delta. The solution of Eq. 9–101 is obtained by matrix inversion as

$$\{\bar{\xi}_i\} = \frac{\bar{c}}{4U} [b_{ij}(k)]\{K_j(k, \kappa)\} \qquad (9\text{--}102)$$

where $[b_{ij}] = [a_{ij}]^{-1}$; and the solution for each normal coordinate is, therefore,

$$\bar{\xi}_i = \frac{\bar{c}}{4U} \sum_{j=1}^{n} b_{ij}(k)K_j(k, \kappa) \qquad (9\text{--}103)$$

The quantities $\bar{\xi}_i$ are transfer functions of the system with respect to displacements of the normal modes. It is evident that when the matrix $[a_{ij}]$ is singular, a condition of dynamic aeroelastic instability or flutter exists. Transfer functions may be constructed for any output quantity that is desired. For example, suppose that our principal interest is in stresses. The steady-state stress response, in terms of its transfer function $\Gamma_\sigma(x, y, k, \kappa)$, is

$$\sigma(x, y, t) = \Gamma_\sigma(x, y, k, \kappa)e^{i\omega t} \qquad (9\text{--}104)$$

The stress transfer function computed by means of the mode-displacement method has the form

$$\Gamma_\sigma(x, y, k, \kappa) = \sum_{i=4}^{n} A_i(x, y)\bar{\xi}_i(k, \kappa)$$

$$= \frac{\bar{c}}{4U} \sum_{i=4}^{n} A_i(x, y) \sum_{j=1}^{n} b_{ij}(k)K_j(k, \kappa) \qquad (9\text{--}105)$$

It is evident that in computing the transfer functions it is necessary to invert the matrix $[a_{ij}]$ as many times as necessary to define the variations of $b_{ij}(k)$ with k. The matrix $[a_{ij}]$ is of order $n \times n$ with complex coefficients, and its inversion is equivalent to the inversion of a $2n \times 2n$ matrix with real coefficients.

When the transfer function Γ_σ has been computed for the stress at a given location (x, y), the frequency spectrum of the time variation of the stress is found from

$$\Phi_\sigma(\omega) = \frac{2}{U} \iint_{-\infty}^{+\infty} |\Gamma_\sigma(k_1, k_2)|^2 \, \phi_{33}(k_1, k_2, k_3) \, dk_2 \, dk_3 \qquad (9\text{--}106)$$

or by defining a two-dimensional gust spectrum as

$$\phi_{ww}(k_1, k_2) = \int_{-\infty}^{+\infty} \phi_{33}(k_1, k_2, k_3) \, dk_3 \qquad (9\text{--}107)$$

the stress spectrum can be computed from*

$$\Phi_\sigma(\omega) = \frac{2}{U} \int_{-\infty}^{+\infty} |\Gamma k_\sigma(_1, k_2)|^2 \, \phi_{ww}(k_1, k_2) \, dk_2 \qquad (9\text{--}108)$$

The mean-squared value of the stress is obtained by substituting Eq. 9–108 into Eq. 9–86; and, in fact, other statistical parameters such as

$$\overline{\left(\frac{d\sigma}{dt}\right)^2} = \int_0^\infty \omega^2 \Phi_\sigma(\omega) \, d\omega \qquad (9\text{--}109)$$

can be found which are useful in fatigue studies. For example, if the random stress variation is a Gaussian process, the average number of stress peaks per second above a given stress level σ can be found from (Ref. 9–8)

$$N = \frac{1}{2\pi} \sqrt{\frac{1}{\overline{\sigma^2}} \overline{\left(\frac{d\sigma}{dt}\right)^2}} \, e^{-\sigma^2/2\overline{\sigma^2}} \qquad (9\text{--}110)$$

when $\sigma > 2\sqrt{\overline{\sigma^2}}$.

Let us now illustrate some of the factors involved when we apply the analysis outlined above to simple examples.

For theoretical analysis, the turbulence in the gust field usually is considered as isotropic, and it is shown by Liepmann (Ref. 9–6) that the

---

* Except for simple cases, this integration cannot be accomplished in closed form. Furthermore, the asymptotic behavior of the integrand usually resembles that of a damped sinusoid, making difficult any numerical integration technique. One method of circumventing these difficulties by a series expansion of the transfer function is described in Ref. 9–7.

power spectrum of vertical gust velocity may be approximated by

$$\phi_{33}(k_1, k_2, k_3) = \frac{2}{\pi^2} \overline{w_G^2} L^5 \frac{k_1^2 + k_2^2}{[1 + L^2(k_1^2 + k_2^2 + k_3^2)]^3} \qquad (9\text{--}111)$$

and the limiting case for one-dimensional turbulence by

$$\phi_w(\omega) = \frac{2}{U} \iint\limits_{-\infty}^{+\infty} \phi_{33}(k_1, k_2, k_3)\, dk_2\, dk_3$$

$$= \overline{w_G^2} \frac{L}{\pi U} \frac{1 + 3\left(\dfrac{\omega L}{U}\right)^2}{\left[1 + \left(\dfrac{\omega L}{U}\right)^2\right]^2} \qquad (9\text{--}112)$$

where L is the integral scale of turbulence, which can be considered as a measure of the average eddy size in the turbulence. It is generally agreed through experimental evidence that the scale of turbulence is somewhere between several hundred to over 1000 ft.

It will be instructive at first to call attention by means of a simple model to a comparison of the essential differences between the discrete-gust and power-spectrum approaches. Let us consider a simplified model of an airplane with two degrees of freedom: freedom to translate vertically and freedom to distort in the first symmetrical bending mode (Ref. 9–9). By neglecting the aerodynamic coupling between the rigid-body and the bending mode, and by using quasi-steady aerodynamics, it is a simple matter to calculate the root-mean-square value of the bending displacement due to one-dimensional continuous turbulence. The result is indicated by Fig. 9–8(a). The dynamic overstress is represented by a dynamic response factor which is the ratio of the root-mean-squared value of the dynamic displacement to that of the static displacement. The dynamic response factor is a function of the damping ratio ζ and the frequency ratio $\omega_1 L/U$, where ω_1 is the frequency of the fundamental bending mode. The damping ratio is a function of the mass density, the reduced frequency, and the mode shape of the airplane.

Figure 9–8(b) is a plot of the corresponding dynamic response factor under a discrete gust of $(1 - \text{cosine})$-shape. Here the time ratio t_1/T is defined as the ratio of the time required to reach peak gust intensity to the period of the fundamental bending mode of the wing.

Thus we have a side-by-side comparison of the more familiar discrete gust approach on the right with the power spectrum approach on the left. Several interesting deductions may be made. We should first inquire as to the portion of the abscissas of the two graphs over which we will be

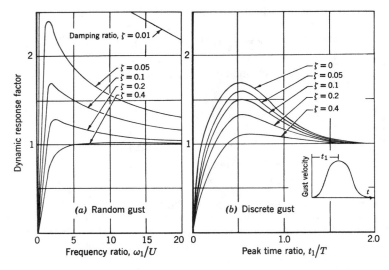

Fig. 9–8. Dynamic response factors.

concerned as aeronautical engineers. In order to decide this, we must consider the range of turbulence scale, L, to be encountered in the atmosphere. Although this question is not fully resolved, a range of from 300 ft to 2000 ft may be estimated at the present time. Using these numbers, it would appear that we can expect airplanes to encompass frequency ratios of roughly 1 to 10. Similarly, a range of peak time ratios of roughly 0.25 to 1.5 is not unreasonable. Thus, we may expect airplanes to be operating in the vicinity of the peaks of the curves in both graphs.

In comparing the values of the dynamic response factors in the two graphs, we see that the damping ratio plays a major role. It is of interest to observe that, for large transports and bombers (where gust design criteria are of most importance), the damping values are of the order of 0.1 to 0.3. For this range, the maximum dynamic response factors computed by the power spectrum and discrete gust approaches are of the same order of magnitude. For lower damping ratios, however, the maximum factor computed by the power spectrum approach is higher, and in the extreme case of no damping, the dynamic response factor obtained under a disturbance of $(1 - \text{cosine})$-shape is only 1.7, whereas it is infinite for continuous turbulence. This points up what is perhaps the most serious deficiency of the discrete gust approach, which would be remedied by considering the atmosphere in its proper perspective as a continuous phenomenon.

(c) Flutter of low-aspect-ratio delta wing free to heave and pitch

Finally, we consider an example of the flutter of a free-free delta wing configuration simulating a complete vehicle and employing the coordinate system shown by Fig. 7–12. This vehicle is of such low aspect ratio that only chordwise deformation modes are assumed to exist. In Sec. 7–4(b), we have already discussed the computation of the first three normal deformation mode shapes and frequencies of free vibrations of such a vehicle having linearly varying mass and stiffness distributions along its length. We shall make use of the rigid-body displacements of heaving and pitching and the first three normal deformation modes as degrees of freedom in a flutter analysis based upon piston theory.

Employing Eqs. 9–59 through 9–63 and the notation of Chap. 8, the equations of free motion are, in dimensionless form, (cf. also Eq. 8–14),

$$2\bar{\mu}M\bar{M}_i\frac{\partial^2\xi_i}{\partial s^2} + \sum_{j=1}^{5}\bar{a}_{ij}\frac{\partial\xi_j}{\partial s} + 2\bar{\mu}M\bar{M}_ik_i^2\xi_i$$

$$+ \sum_{j=1}^{5}\bar{b}_{ij}\xi_j = 0, \quad (i = 1, 2, \cdots, 5) \quad (9\text{–}113)$$

Since we have assumed linearly varying mass per unit length along the x-direction, it is reasonable to assume a constant mass distribution m per unit of area. Then, making use of the mode shapes of Sec. 7–4(b), we have

$$\bar{M}_1 = \iint_S d\xi\,d\eta = 0.5 \qquad \text{(Heaving mode)}$$

$$\bar{M}_2 = \iint_S (1 - \tfrac{3}{2}\xi)\,d\xi\,d\eta = 0.0625 \quad \text{(Pitching mode)} \quad (9\text{–}114)$$

$$\bar{M}_i = \iint_S \phi_i^2\,d\xi\,d\eta, \quad (i = 3, 4, 5) \quad \text{(Deformation modes)}$$

where the deformation mode shapes ϕ_i are those given by Eq. 7–177. Upon insertion of the latter, there are obtained

$$\bar{M}_3 = 0.02065, \qquad \bar{M}_4 = 0.01321, \qquad \bar{M}_5 = 0.00969$$

Similarly, the elements of the $\bar{a}_{ij}$ and $\bar{b}_{ij}$ matrices may be computed by employing the piston theory formulas (cf. Eq. 8–14)

$$[\bar{a}_{ij}] = \begin{bmatrix} 1.00000 & 0 & 0 & 0 & 0 \\ 0 & 0.12500 & 0 & 0 & 0 \\ 0 & 0 & 0.04130 & 0 & 0 \\ 0 & 0 & 0 & 0.02642 & 0 \\ 0 & 0 & 0 & 0 & 0.01938 \end{bmatrix} \quad (9\text{–}115)$$

$$[\bar{b}_{ij}] = \begin{bmatrix} 0 & -0.75000 & 0.17003 & -0.29887 & 0.14572 \\ 0 & 0 & -0.26103 & 0.04591 & -0.15008 \\ 0 & 0 & -0.02092 & -0.16030 & 0.00658 \\ 0 & 0 & 0.02071 & -0.01594 & -0.13181 \\ 0 & 0 & -0.00098 & 0.03156 & -0.01309 \end{bmatrix} \qquad (9\text{--}116)$$

where we have assumed also that the wing thickness is constant over the wing area, that is, $\Gamma = 0$. The integrations were performed by use of a fifteen-point Gauss quadrature formula, as given in Ref. 9–13. The use of piston theory in this example having only chordwise modes reaches some exactitude only at rather high Mach numbers where the leading edges are supersonic. At lower Mach numbers, where the vertex angle of the delta wing lies well within the leading edge Mach cone, a slender-body aerodynamic theory would be more appropriate (cf. Sec. 4–7).

Equation 9–113 is specialized to the case of flutter by putting $\xi_i = \bar{\xi}_i e^{iks}$, which yields the determinant condition

$$|2\bar{\mu}Mk^2\bar{M}_i(\Lambda_i^2\Omega - 1)\delta_{ij} + ik\bar{a}_{ij} + \bar{b}_{ij}| = 0, \qquad (i, j = 1, 2, \cdots, 5)$$

$$(9\text{--}117)$$

If we take ω_3 as the reference frequency, the frequency ratios, $\Lambda_i = \omega_i/\omega_3$, are

$$\Lambda_1 = \Lambda_2 = 0; \qquad \Lambda_3 = 1.0; \qquad \Lambda_4 = 2.5412; \qquad \Lambda_5 = 4.7700$$

$$(9\text{--}118)$$

Examination of the form of the coefficients $\bar{M}_i$, $\bar{a}_{ij}$, and $\bar{b}_{ij}$ shows that, since natural mode shapes are employed as degrees of freedom, the modal coupling of the delta wing arises entirely from the aerodynamic stiffness terms. In addition, the elastic modes are completely decoupled from the rigid-body modes. The latter fact makes it necessary to examine only the elastic modes for aeroelastic instability. It is important to observe that the independence of flutter from the rigid-body modes is due to the assumption in the present idealized case that the mass and aerodynamic lift distributions are similar. For a delta wing where the mass and thickness distributions are not uniform over the wing area, some coupling among elastic and rigid-body modes will exist. The latter coupling may, however, not be large in most cases.

Making use of the determinant condition of Eq. 9–117, and the numerical coefficients given by Eqs. 9–114, 9–115, and 9–116, Dugundji and Crisp (Ref. 9–12) have conducted three-degree-of-freedom flutter analyses, using elastic modes 3, 4, and 5 for values of $\bar{\mu}M = 0.5, 2, 5, 50,$ and 500. Their results are reproduced in Figs. 9–9(a) through (e) as plots of damping, g, versus $U/b\omega_3\sqrt{2\bar{\mu}M}$. At low values of $\bar{\mu}M$, the unstable mode is

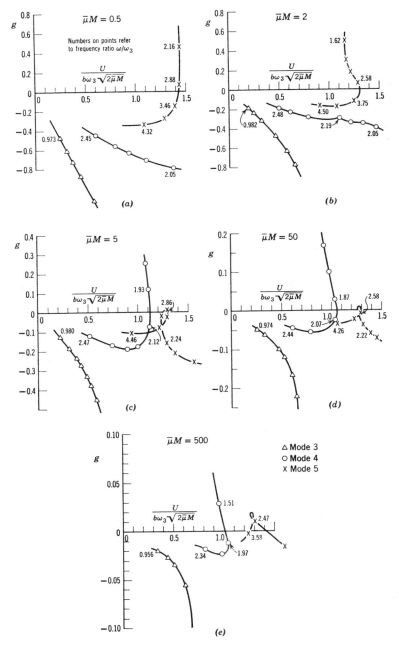

Fig. 9-9. U-g plot for delta wing (piston theory).

485

seen to be mode 5; whereas, when $\bar{\mu}M$ is increased, mode 4 becomes unstable. The nature of some of the U-g-curves is a factor worthy of consideration. At very high values of $\bar{\mu}M$, the unstable branch tends to double back on itself after crossing the zero damping line. Such behavior indicates that an increase in damping of the system will decrease its stability. This somewhat anomalous behavior has also been observed in flutter analyses of swept wings and in panel flutter.

REFERENCES

9–1. Whittaker, E. T., *Analytical Dynamics*, Cambridge University Press, 1937.

9–2. Hildebrand, F. G., *Advanced Calculus for Engineers*, Prentice-Hall, New York, 1949.

9–3. Williams, D., *Dynamic Loads in Aeroplanes Under Given Impulsive Load with Particular Reference to Landing and Gust Loads on a Large Flying Boat*, Great Britain Royal Aircraft Establishment Reports SME 3309 and 3316, 1945.

9–4. Mar, J. W., T. H. H. Pian, and J. M. Calligeros, "A Note on Methods for the Determination of Transient Stresses," *J. Aero. Sciences*, Readers' Forum, Vol. 23, No. 1, January 1956.

9–5. Kirsch,.A. A., J. M. Calligeros, and K. A. Foss, *Effects of Structural Flexibility on Gust Loading of Aircraft*, WADC Technical Report 54-592, Part 2, August 1955.

9–6. Liepmann, H. W., "Extension of the Statistical Approach to Buffeting and Gust Response of Wings of Finite Span," *J. Aero. Sciences*, Vol. 22, No. 3, March 1955.

9–7. Foss, K. A., and W. L. McCabe, *Gust Loading of Rigid and Flexible Aircraft in Continuous Atmospheric Turbulence*, WADC Technical Report 57-704, January 1958.

9–8. Liepmann, H. W., "An Approach to the Buffeting Problem from Turbulence Considerations," *J. Aero. Sciences*, Vol. 19, No. 12, December 1952.

9–9. Bisplinghoff, R. L., T. H. H. Pian, and K. A. Foss, *Response of Elastic Aircraft to Continuous Turbulence*, AGARD Report No. 117, April–May 1957.

9–10. Bisplinghoff, R. L., "Some Structural and Aeroelastic Considerations of High Speed Flight," *J. Aero. Sciences*, Vol. 23, No. 4, April 1956.

9–11. Diederich, F. W., "The Dynamic Response of a Large Airplane to Continuous Random Atmospheric Turbulence," *J. Aero. Sciences*, Vol. 23, No. 10, October 1956.

9–12. Dugundji, John, and J. D. C. Crisp, *On the Aeroelastic Characteristics of Low Aspect Ratio Wings with Chordwise Deformation*, AFOSR T.N. 59-787, July 1959.

9–13. Lowan, A. N., N. Davids, and A. Levenson, "Tables of the Zeros of Legendre Polynomials of Order 1–16 and the Weight Coefficients for Gauss Mechanical Quadrature Formula," *Bul. Amer. Math. Soc.*, Vol. 48, 1942.

10

SYSTEMS WITH TIME-VARYING COEFFICIENTS OR NONLINEARITIES

10–1 INTRODUCTION

The methods of aeroelastic analysis described in the preceding chapters have in common the requirement of linear systems with properties independent of time. Although these limitations appear quite restrictive, experience has shown that operational flight vehicles, with very few exceptions, can be represented satisfactorily by means of such elementary mathematical models.

The success of these simplified analyses is attributable to two things: (1) The effects of nonlinearities of aerodynamic or structural origin are virtually absent (from the elastic degrees of freedom) or, at worst, nonlinearity assumes importance only after an instability has started in a linear fashion. (2) Thrust-to-drag or thrust-to-weight ratios are generally small enough to preclude large accelerations. Coupled with the small to moderate rates of fuel consumption, this fact prevents rapid changes in inertial, stiffness, and ambient atmospheric parameters.

Deviations from the usual techniques are necessary for very high speeds and highly accelerated configurations such as anti-missile missiles and boost-glide vehicles. On these it is not difficult to foresee large deformations due to combinations of high dynamic pressure, thermal effects, and rapid changes in mass, stiffness, and ambient conditions. The consequences of these complexities will become apparent as we discuss typical situations in the remainder of this chapter.

As in many dynamical problems, we have at our disposal two avenues of "theoretical" attack: analog and analytical solutions. The former is a useful tool for treating rather complicated situations (e.g., Refs. 10–1, 10–2) but has the disadvantage that it must be reformulated for each individual example and requires an extensive parametric study to gain physical understanding of systems of any particular class. In contrast, the analytical approach, on which we shall concentrate here, has procedural limitations but gives a better insight into the system behavior.

Our treatments will have to be confined to the simplest of situations, for the state of the art has not progressed to the point where greater realism can be attained. Thus, the typical section, which has served so well in illustrating many of the useful ideas in previous chapters, will underlie some of the subsequent discussion.

10–2 RESPONSE AND STABILITY

In the analysis of a dynamical problem, we wish to compute, over a suitably long interval of time, the history of the motion (response) for arbitrary inputs and initial conditions. This affords the analyst all the necessary information to adjudge performance or structural integrity. Unfortunately, response calculations for systems with time-varying or nonlinear elements present formidable difficulties. There are situations in practice, however, where the principal concern is with the single important question of stability. In such cases we may seek simpler means to determine the stability or instability without direct recourse to complete solutions of the equations of motion. An example of this kind is the flutter problem, where we desire first to find some sort of stability boundary, although admittedly subcritical and slightly supercritical phenomena have a certain interest.

The appearance of time-varying coefficients or nonlinearities in the equation of motion portend various analytical complexities. There are no general and established ways of solving them, and an accepted criterion of stability for linear systems has little, if any, meaning without suitable qualification and amplification. To clarify this point further, let us look at two systems for which solutions are available. First consider the nonlinear van der Pol equation (e.g., Ref. 10–3),

$$\ddot{q} + \mu(q^2 - 1)\dot{q} + q = 0, \quad \mu = \text{positive constant} \quad (10\text{–}1)$$

It is well-known that if we impart to this system a "small" initial disturbance, say $q(0) = 0$, $\dot{q}(0) = \dot{q}_0$, the ensuing motion will exhibit increasing amplitudes at the start but will eventually stabilize to a "limit cycle." Adopting the usual stability criterion (for linear systems with constant coefficients), we would have to say that this system is unstable in its earlier phase and neutrally stable as it reaches its limit cycle. On the other hand, if we start out with a "sufficiently large" initial disturbance, the amplitudes of the motion will decrease with time until once more it reaches a limit cycle. For the latter situation, we would have to call the system stable in its initial phase. In this case the stability criterion has no useful meaning. A more meaningful question would be what is the maximum value of q or $\dot{q}$ over all time.

As a second example, consider the first-order linear system with a variable coefficient

$$\dot{q} + f(t)q = 0 \qquad (10\text{-}2)$$

the solution of which is

$$q = q_0 \exp\left(-\int_0^t f(t)\,dt\right) \qquad (10\text{-}3)$$

where q_0 is the initial displacement. Now, if $f(t) > 0$ for all times in the interval of interest $0 \le t \le T$, then q will decrease monotonically with time, and the system may be termed stable. If $f(t) < 0$ for all times in the interval, the system is unstable. Here the stability criterion has a definite meaning in that, following even a small disturbance, the motion could build up sufficiently to render the system unsatisfactory.

Let us next examine the situation when $f(t)$ can be both positive and negative. Then, during part of the interval, the motion will be unstable (when $f(t) < 0$), and it will be stable over the rest of the interval. Whether the system is satisfactory or not will now depend on $f(t)$ and the extent of the initial disturbance; to answer this question, we must evaluate a more meaningful quantity, such as q_{max} or $\dot{q}_{max}$.

Another anomaly associated with the stability of a system with time-varying coefficients arises from the dissimilar behaviors of the q, $\dot{q}$, and $\ddot{q}$ responses, as we shall see in Sec. 10-3.

We need not detail further these points here. Lucid expositions have been given by Collar (Ref. 10-4) and Tsien (Ref. 10-5), among others. Moreover, the reader will find it easy to recognize in the discussions and examples to follow many of the important features which distinguish these cases from those treated in the foregoing chapters.

10-3 THE SINGLE-DEGREE-OF-FREEDOM SYSTEM WITH TIME-VARYING COEFFICIENTS

With an occasional exception, exact analytical solutions even of problems of the sort considered in this section are not possible. Hence, we must turn for the most part to approximate treatments. Within the bounds of validity of such approximate solutions, a fair amount of information can be extracted about both response and stability.

The most general homogeneous form of the equation describing a linear, single-degree-of-freedom system is

$$a(t)\ddot{q} + b(t)\dot{q} + c(t)q = 0 \qquad (10\text{-}4)$$

Here q represents the displacement from the undisturbed position. q might, for example, be the elastic twist α about a pinned axis of rotation, as in Fig. 2–3. $a(t)$, $b(t)$, and $c(t)$ represent the inertial, damping, and elastic characteristics of the system, respectively.

The solution of Eq. 10–4 has received considerable attention in the literature. We mention only a few of the alternative approaches. The first is a classical one (cf. Refs. 10–6, 10–7), which assumes a solution of the form*

$$q = q_0 \exp\left\{\int^t \left[v(\tau) - \frac{b(\tau)}{2a(\tau)}\right] d\tau\right\} \qquad (10\text{-}5)$$

If Eq. 10–5 is substituted into Eq. 10–4, we find that $v(t)$ must satisfy the Riccati equation

$$v^2 + \dot{v} = R(t) \qquad (10\text{-}6)$$

where $R(t)$, the invariant of the differential equation, is given by

$$R(t) = \frac{1}{4}\left[\frac{b(t)}{a(t)}\right]^2 + \frac{1}{2}\frac{d}{dt}\left[\frac{b(t)}{a(t)}\right] - \left[\frac{c(t)}{a(t)}\right] \qquad (10\text{-}7)$$

Only for special forms of $R(t)$ is Eq. 10–6 integrable directly. One such case is when

$$a(t) = 1, \quad b(t) = b_0(1 + \epsilon t)^{-1}, \quad c(t) = c_0(1 + \epsilon t)^{-2} \qquad (10\text{-}8)$$

This situation arises, for instance, in the dynamic stability problem during a decelerating flight with speed given by

$$V(t) = V_0(1 + \epsilon t)^{-1}, \quad (\epsilon > 0)$$

where the aerodynamic damping and stiffness terms are assumed proportional to V and V^2, respectively (see Ref. 10–4). Here $R(t) \sim (1 + \epsilon t)^{-2}$,

---

* The lower limit of the integral in Eq. 10–5 may be assigned any constant value $\tau = t_0$. If t_0 is the time at which the initial conditions apply, q_0 will then represent the initial displacement. We prefer to leave the lower limit unassigned, so that the integral represents the indefinite integral evaluated at time $\tau = t$. q_0 will then be an arbitrary constant to be evaluated eventually from the initial conditions.

allowing Eq. 10–6 to be reduced to the form

$$v(t) = (1 + \epsilon t)^{-1}\left[\frac{\epsilon}{2} \pm \sqrt{\frac{b_0^2}{4} - c_0 + \frac{\epsilon}{2}\left(\frac{\epsilon}{2} - b_0\right)}\,\right] \qquad (10\text{–}9)$$

For the motion to be oscillatory, the quantity under the radical sign must be negative, i.e.,

$$\frac{b_0^2}{4} + \frac{\epsilon^2}{4} < c_0 + \frac{\epsilon b_0}{2}$$

The total solution is then

$$q = A(1 + \epsilon t)^{\frac{1}{2} - (b_0/2\epsilon)}$$

$$\times \cos\left[\phi + \frac{1}{\epsilon}\sqrt{\left(c_0 - \frac{b_0^2}{4}\right) - \epsilon\left(\frac{\epsilon}{4} - \frac{b_0}{2}\right)}\,\ln(1 + \epsilon t)\right] \qquad (10\text{–}10)$$

where A and φ are constants to be determined from the initial conditions. It is easily verified that, as $\epsilon \to 0$, this reduces to the classical result for constant coefficients.

The envelopes of the displacement, velocity, and acceleration responses are given, respectively, by

$$\bar{q} \sim (1 + \epsilon t)^{(1/2) - (b_0/2\epsilon)}, \qquad \bar{\dot{q}} \sim (1 + \epsilon t)^{-[(1/2) + (b_0/2\epsilon)]},$$

$$\bar{\ddot{q}} \sim (1 + \epsilon t)^{-[(3/2) + (b_0/2\epsilon)]}$$

When $0 < \dfrac{1}{2} - \dfrac{b_0}{2\epsilon} < 1$, the system is unstable in the sense that $\bar{q}$ increases with time; on the other hand, $\bar{\dot{q}}$ and $\bar{\ddot{q}}$ decrease with time and show "stable" responses. Here we see one dissimilarity between the behaviors with and without time-varying coefficients. The controlling parameter is $b_0/2\epsilon$. If there is sufficient damping and $\epsilon \ll 1$, so that $b_0/2\epsilon > 1$, then the displacement velocity, and acceleration all appear to be damped, although in different degrees. In the limiting case of $b_0/2\epsilon \gg 1$, the solution approaches that from quasi-steady considerations in the vicinity of some preassigned $t = t_1$, where $q = q_0 \cos \phi$:

$$q = A \exp\left(-\frac{b_0}{2}\frac{t - t_1}{1 + \epsilon t_1}\right)\cos\left[\varphi + \frac{b_0}{2(1 + \epsilon t_1)}\sqrt{\frac{4c_0}{b_0^2} - 1}\,(t - t_1)\right]$$

$$(10\text{–}11)$$

In many aeronautical applications, the rate parameter ϵ is sufficiently small to assure engineering accuracy from the quasi-steady approach.

We return once more to Eq. 10–6 and inquire about other forms of R which might make this nonlinear formula for $v(t)$ integrable. One case obviously occurs wheh $R = $ constant, for which we obtain

$$v = \pm\kappa, \quad -\kappa \left\{ \frac{1 - \dfrac{\kappa + v_0}{\kappa - v_0} \exp\left[2\kappa(t - t_0)\right]}{1 + \dfrac{\kappa + v_0}{\kappa - v_0} \exp\left[2\kappa(t - t_0)\right]} \right\},$$

$$\text{for } R = \kappa^2 > 0 \quad (10\text{--}12a)$$

$$v = 0, \quad \frac{1}{t + c}, \qquad \text{for } R = 0 \quad (10\text{--}12b)$$

$$v = \pm i\kappa, \quad \frac{-\kappa^2 \tan \kappa(t - t_0) + \kappa v_0}{\kappa + v_0 \tan \kappa(t - t_0)}, \qquad \text{for } R = -\kappa^2 < 0 \quad (10\text{--}12c)$$

Here c and v_0 are arbitrary constants. Equations 10–12a, b, c represent the overdamped, critically damped, and underdamped conditions, respectively. It is not difficult to show that, as far as the q-solution is concerned, the third and most complicated expressions for v in Eqs. 10–12a–c yield forms which are linearly dependent on those from $v = \pm\kappa$, and $v = \pm i\kappa$. We shall, therefore, consider the linearly independent solutions $(\pm\kappa)$, $[0, 1/(t + c)]$, and $(\pm i\kappa)$, choosing the one appropriate to the sign of R.

If R is a slowly varying function of time, it is reasonable to assume that the character of the solution will be as in $R = $ constant, i.e., $v = \pm\sqrt{R}$, and we may write as an approximation to Eq. 10–6

$$v(t) \cong \pm\sqrt{R(t)} \left(\sqrt{1 \mp \frac{1}{2R\sqrt{R}} \dot{R}} \right) \quad (10\text{--}13)$$

(The special case of $R(t) = 0$ need not be considered further, since its exact solution turns out to be

$$q = (At + B) \exp\left[-\int^t \frac{b(\tau)}{2a(\tau)} d\tau \right] \quad (10\text{--}14)$$

where A and B are again arbitrary constants.)

The quantity $\exp\left[\int^t v(\tau)\, d\tau \right]$ can be evaluated only for very special forms of $a(t)$, $b(t)$, and $c(t)$, so it is necessary to make further approximations in Eq. 10–13. If $|\dot{R}/2R\sqrt{R}|^2 \ll 1$, we expand that expression in a binomial series and retain the leading terms.

$$v(t) \cong \pm\sqrt{R} - \frac{1}{4}\frac{\dot{R}}{R} \mp \frac{1}{32}\frac{\dot{R}^2}{R^2\sqrt{R}} + \cdots \quad (10\text{--}15)$$

This in turn yields the solutions (with $R = R(\tau)$)

$$q \cong \exp\left\{-\frac{1}{2}\int^t\left[\frac{1}{2}\frac{\dot{R}}{R} + \frac{b(\tau)}{a(\tau)}\right]d\tau\right\}\left\{A\exp\left[\int^t\left(\sqrt{R} - \frac{1}{32}\frac{\dot{R}^2}{R^2\sqrt{R}}\right)d\tau\right]\right.$$

$$\left. + B\exp\left[-\int^t\left(\sqrt{R} - \frac{1}{32}\frac{\dot{R}^2}{R^2\sqrt{R}}\right)d\tau\right]\right\}, \quad R > 0 \qquad (10\text{-}16a)$$

and

$$q \cong A\left\{\cos\left[\varphi + \int^t\left(\sqrt{-R} + \frac{1}{32}\frac{\dot{R}^2}{R^2\sqrt{-R}}\right)d\tau\right]\right\}$$

$$\cdot\left\{\exp\left(\frac{1}{2}\int^t\left[\frac{1}{2}\frac{\dot{R}}{R} + \frac{b(\tau)}{a(\tau)}\right]d\tau\right)\right\}, \quad R < 0 \qquad (10\text{-}16b)$$

We must emphasize that, for the regions of R where the condition

$$\left|\frac{\dot{R}}{2R\sqrt{R}}\right|^2 \ll 1 \qquad (10\text{-}17)$$

is violated, Eqs. 10–16a, b are inapplicable. Subject to further limitations, these solutions correspond to the results of other investigators as follows:

(1) If the $\dot{R}^2$-term is neglected, the equations reduce identically to those of Brunelle (Ref. 10–7).

(2) With the assumption of Eq. 10–17, Squire's formulas (Ref. 10–8) correspond to the present ones.

(3) When $\dfrac{c(t)}{a(t)} \gg \left[\dfrac{b(t)}{2a(t)}\right]^2 + \dfrac{1}{2}\dfrac{d}{dt}\left[\dfrac{b(t)}{a(t)}\right]$, and the $\dot{R}^2$-term is neglected, these expressions are exactly those from the *WKB* approximation (cf. Collar, Ref. 10–4, Eq. 25).

The foregoing furnish some information as to the system's stability during a prescribed interval of time. To illustrate, let us take the underdamped condition, for which the velocity may be obtained directly from Eq. 10–16b. Defining the total phase as

$$\Phi = \varphi + \int^t \langle\sqrt{-R(\tau)}\rangle \, d\tau$$

and an auxiliary quantity

$$Q(t) = \frac{1}{2}\frac{\dot{R}(t)}{R(t)} + \frac{b(t)}{a(t)}$$

and neglecting the $\dot{R}^2$-term at the outset, we have

$$\dot{q} \cong A\{-\tfrac{1}{2}Q(t)\cos\Phi - \sqrt{-R(t)}\sin\Phi\}\left\{\exp\left[-\frac{1}{2}\int^t Q(\tau)\,d\tau\right]\right\}$$

$$(10\text{--}18)$$

$$\equiv A\left\{\sqrt{\frac{Q^2(t)}{4} - R(t)}\cos[\Phi + \psi(t)]\right\}\left\{\exp\left[-\frac{1}{2}\int^t Q(\tau)\,d\tau\right]\right\}$$

Here ψ is an additional phase angle dependent on time, which is related to the ratio $Q(t)/\sqrt{-R(t)}$. Consider any interval of time $t_1 \le t \le t_2$. If the quantity $Q(t)$ is always positive, the envelope of the q-response decays, while if $Q(t)$ is negative, the amplitude of q increases with time. If on the other hand, $Q(t)$ is positive and then negative during this interval (which is assumed to be sufficiently large to include several cycles of oscillations), the envelope decays first and then grows. Similar studies can be made on the behavior of $\dot{q}$ or $\ddot{q}$. Caution is needed, however, because the time dependence of the $\dot{q}$- and $\ddot{q}$-envelopes is not solely governed by the exponential form $\exp\left[-\dfrac{1}{2}\int^t Q(\tau)\,d\tau\right]$. However, Eq. 10–18 can be written in the alternative form $\left[\text{since }\dfrac{Q^2(t)}{4} - R(t) \equiv P(t) > 0\right]$

$$\dot{q} = A\cos(\Phi + \psi(t))\left\{\exp\left(-\frac{1}{2}\int^t\left[Q(\tau) - \frac{\dot{P}(\tau)}{P(\tau)}\right]d\tau\right)\right\} \quad (10\text{--}19)$$

where A is a new arbitrary constant. We thus find that the $\dot{q}$-envelope depends on the quantity $Q(t) - \dfrac{\dot{P}(t)}{P(t)}$ rather than $Q(t)$ alone.

Let us analyze by this approach the special vibrating system considered by Reed (Ref. 10–9) which, according to the present notation, is described by

$$\ddot{q} + (b_0 + b_1 t)\dot{q} + (c_0 + c_1 t)q = 0, \qquad b_1, c_1 \text{ "small"} \quad (10\text{--}20)$$

in the neighborhood of $t = 0$. From Eq. 10–16b, the oscillatory motion ($R < 0$) is given by

$$q \cong A(R)^{-1/4}\exp\left[-\left(\frac{b_0}{2} + \frac{b_1 t}{4}\right)t\right]$$

$$\times \cos\left[\varphi + \frac{2c_0^{3/2}}{3c_1}\left(1 - \frac{b_1}{2c_0} + \frac{c_1}{c_0}t\right)^{3/2}\right] \quad (10\text{--}21)$$

where A and φ are new constants. Here the same approximation as in

Ref. 10–9, the lightly damped condition of $b_0/2\sqrt{c_0} \ll 1$, has been utilized
with the consequence that

$$\sqrt{-R} \cong \sqrt{c_0}\sqrt{1 - \frac{b_1}{2c_0} + \frac{c_1}{c_0}t}$$

The result, Eq. 10–21, agrees with Eq. 15 of Ref. 10–9. According to our
procedure, the q-envelope behaves as

$$\bar{q} \sim (-R)^{-\frac{1}{4}} \exp\left[-\left(\frac{b_0}{2} + \frac{b_1 t}{4}\right)t\right] \tag{10–22}$$

Also, from Eq. 10–19, we get

$$\bar{\dot{q}} \sim (-R)^{-\frac{1}{4}}\sqrt{\frac{Q^2}{4} - R}\, \exp\left[-\left(\frac{b_0}{2} + \frac{b_1 t}{4}\right)t\right] \tag{10–23a}$$

which in turn yields, for $Q^2/4 \ll (-R)$,

$$\bar{\dot{q}} \sim (-R)^{+\frac{1}{4}} \exp\left[-\left(\frac{b_0}{2} + \frac{b_1 t}{4}\right)t\right] \tag{10–23b}$$

Again Eqs. 10–22 and 10–23b are in agreement with Reed's results and may
be employed to deduce the statements on stability given in his paper.

Another procedure for solving Eq. 10–4 is the iterative scheme proposed
by Garber (Ref. 10–10). We again start with an assumed form like Eq.
10–5. If the ensuing motion is oscillatory, v must read

$$v(t) = \sigma(t) + i\omega(t) \tag{10–24}$$

When this is substituted into Eq. 10–6, we have from the real and imaginary
parts

$$\omega^2(t) = -R(t) - \frac{\ddot{\omega}}{2\omega} + \frac{3\dot{\omega}^2}{4\omega^2} \tag{10–25}$$

and

$$\sigma(t) = -\frac{\dot{\omega}}{2\omega} \tag{10–26}$$

The resulting displacement is

$$q(t) = A[\omega(t)]^{-\frac{1}{2}}\left\{\cos\left[\varphi + \int^t \omega(\tau)\,d\tau\right]\right\}\left\{\exp\left[-\frac{1}{2}\int^t \frac{b(\tau)}{a(\tau)}\,d\tau\right]\right\} \tag{10–27}$$

Garber proceeds by improving the estimate of ω^2 with a sequence of values

$$\omega_{i+1}^2 = -R - \frac{\ddot{\omega}_i}{2\omega_i} + \frac{3\dot{\omega}_i^2}{4\omega_i^2}, \quad (i = 1, 2, \cdots) \tag{10–28}$$

The initial $\omega_1{}^2$ is given by

$$\omega_1{}^2 = -R = \frac{c}{a} - \left(\frac{b}{2a}\right)^2 - \frac{d}{dt}\left(\frac{b}{2a}\right), \quad \text{or} \quad \omega_1 = \sqrt{-R} * \quad (10\text{–}29)$$

It is of interest to note that, if we adopt for ω the initial value ω_1, Garber's result is exactly that of Eq. 10–16b with the $\dot{R}^2$-term omitted.

So far we have considered the homogeneous solution of Eq. 10–4. If the system is subjected to a specified applied force $f(t)$, we must add to the transient solution the particular integral, which is easily derivable once two linearly independent homogeneous solutions, say $q = q_1, q_2$, are known. The total solution is found by the method of variation of parameters (cf. Ref. 10–6, Sec. 1–9):

$$q(t) = \int^t \frac{f(\eta)}{a(\eta)} [q_1(\eta)q_2(t) - q_2(\eta)q_1(t)] W^{-1}[q_1(\eta), q_2(\eta)]\, d\eta$$
$$+ Aq_1(t) + Bq_2(t) \quad (10\text{–}30)$$

where W^{-1} is the reciprocal of the Wronskian

$$W(q_1, q_2) = (q_1\dot{q}_2 - q_2\dot{q}_1) \quad (10\text{–}31)$$

Of particular interest to us later is the impulsive response $\mathcal{T}(t, t_1)$ to a forcing function $f(t) = \delta(t - t_1)$ at time t_1, $\delta(t - t_1)$ being the Dirac delta function. Adopting the small-$\dot{R}^2$ approximation, we get from Eq. 10–16b

$$\begin{Bmatrix} q_1 \\ q_2 \end{Bmatrix} = \begin{Bmatrix} \cos\left[\int^t \sqrt{-R}\, d\tau\right] \\ \sin\left[\int^t \sqrt{-R}\, d\tau\right] \end{Bmatrix} \left(\exp\left\{-\frac{1}{2}\int^t \left[\frac{1}{2}\frac{\dot{R}}{R} + \frac{b(\tau)}{a(\tau)}\right] d\tau\right\}\right) \quad (10\text{–}32)$$

From the special property of the Dirac delta function, we compute the response $\mathcal{T}$ at $t > t_1$,

$$\mathcal{T}(t, t_1) = \frac{1}{a(t_1)} [q_1(t_1)q_2(t) - q_2(t_1)q_1(t)] W^{-1}[q_1(t_1), q_2(t_1)]$$
$$+ Aq_1(t) + Bq_2(t) \quad (10\text{–}33)$$

After some algebra, we find

$$W^{-1}[q_1(t_1), q_2(t_1)] = \exp\left[\int^{t_1} \frac{b(\tau)}{a(\tau)}\, d\tau\right] \quad (10\text{–}34)$$

* The root $\omega_1 = -\sqrt{-R}$ is included in Eq. 10–27 by inserting the phase angle φ.

which, in conjunction with Eq. 10–32, reduces Eq. 10–33 to

$$\mathscr{T}(t, t_1) = \begin{cases} \left[\dfrac{1}{a(t_1)} [R(t)R(t_1)]^{-\frac{1}{4}} \left\{ \sin \left[\displaystyle\int_{t_1}^t \sqrt{-R(\tau)} \, d\tau \right] \right\} \right. \\ \qquad\qquad\qquad\qquad\qquad \times \left\{ \exp \left[-\dfrac{1}{2} \displaystyle\int_{t_1}^t \dfrac{b(\tau)}{a(\tau)} \, d\tau \right] \right\} \\ \left. + Aq_1(t) + Bq_2(t), \qquad t \geq t_1 \right. \\ +0, \qquad\qquad\qquad\qquad t < t_1 \end{cases}$$

(10–35)

Applying the initial conditions,

$$\mathscr{T}(t_1, t_1) = 0 \qquad \left[\frac{\partial}{\partial t} \mathscr{T}(t, t_1) \right]_{t = t_1} = \frac{1}{a(t_1)} \qquad (10\text{–}36)$$

we obtain

$$\left. \begin{aligned} Aq_1(t_1) + Bq_2(t_1) = 0 \\ A\dot{q}_1(t_1) + B\dot{q}_2(t_1) = 0 \end{aligned} \right\} \qquad (10\text{–}37)$$

Since $q_1(t_1)\dot{q}_2(t_1) - q_2(t_1)\dot{q}_1(t_1) = W \neq 0$, A and B must be zero; and the complete solution is Eq. 10–35 with $Aq_1(t)$ and $Bq_2(t)$ dropped.

The impulsive response allows us to determine the effects of other inputs by Duhamel's superposition principle (cf. Ref. 10–11). Let $f(t)$ be the general forcing function; then

$$q(t) = \int_{t_0}^t f(t_1)\mathscr{T}(t, t_1) \, dt_1 + Aq_1(t) + Bq_2(t) \qquad (10\text{–}38)$$

Here A and B are to be evaluated from the initial conditions at time $t = t_0$, and they vanish if the motion starts from rest.

10–4 STABILITY OF SINGLE-DEGREE-OF-FREEDOM LINEAR SYSTEMS WITH SLOWLY TIME-VARYING COEFFICIENTS

In this section we shall look into the possibility of establishing a method to determine the stability of a single-degree-of-freedom linear system, of second or higher order, without detailed analysis of the response. We follow Grensted's treatment (Ref. 10–12) and assume the motion to be governed by

$$L(q) = \sum_{n=0}^N a_n(t) \frac{d^n}{dt^n} q = 0, \qquad (N = 2, 3, 4) \qquad (10\text{–}39)$$

Since the homogeneous equation is under study, we can set $a_0 = 1$ with no loss in generality. We seek a new variable $\tau = \tau(t)$ which satisfies the requirement that $d\tau/dt > 0$ and which transforms $L(q)$ into a new operator

$\mathscr{L}(q)$ with constant coefficients having τ as the independent variable. If $\mathscr{L}(q)$ shows stability in the sense that the amplitude of q decays as τ increases (in the neighborhood of a given time t_1), the same will be true of $L(q)$ as t increases in consequence of the specification $d\tau/dt > 0$. Following Ref. 10–12, we let primes denote differentiation with respect to τ; thus $\dot{q} = \dot{\tau}q'$, $\ddot{q} = \dot{\tau}^2 q'' + \ddot{\tau}q'$. Hence, Eq. 10–39, for $N = 2$, transforms to

$$(a_2\dot{\tau}^2)q'' + (a_1\dot{\tau} + a_2\ddot{\tau})q' + q$$
$$\equiv A_2 q'' + A_1 q' + q = 0 \qquad (10\text{–}40)$$

Let us confine our attention to the response in the neighborhood of time $t = t_1$. We can write, for small $|t - t_1|$,

$$\left.\begin{aligned}
\dot{\tau} &= 1 + v_1(t - t_1) + \tfrac{1}{2}v_2(t - t_1)^2 + \cdots \\
a_2(t) &= a_2(t_1) + \left(\frac{da_2}{dt}\right)_{t_1}(t - t_1) + \cdots \\
a_1(t) &= a_1(t_1) + \left(\frac{da_1}{dt}\right)_{t_1}(t - t_1) + \cdots
\end{aligned}\right\} \qquad (10\text{–}41a\text{–}c)$$

We obtain for A_2 and A_1

$$A_2 = a_2\dot{\tau}^2 = a_2(t_1) + \left[\left(\frac{da_2}{dt}\right)_{t_1} + 2v_1 a_2(t_1)\right](t - t_1) + 0[(t - t_1)^2]$$

$$A_1 = a_1\dot{\tau} + a_2\ddot{\tau} = [a_1(t_1) + v_1 a_1(t_1)]$$
$$+ \left[\left(\frac{da_1}{dt}\right)_{t_1} + v_1 a_1(t_1) + v_1\left(\frac{da_2}{dt}\right)_{t_1} + v_2 a_2(t_1)\right](t - t_1) + 0[(t - t_1)^2]$$
$$(10\text{–}42a,b)$$

We require that $\dot{A}_2 = \dot{A}_1 = 0$ (thus $A_2{'} = A_1{'} = 0$) at $t = t_1$; that is, in the neighborhood of $t = t_1$, A_2 and A_1 are constant. We must have

$$\left(\frac{da_2}{dt}\right)_{t_1} + 2v_1 a_2(t_1) = 0, \qquad v_1 = -\frac{(da_2/dt)_{t_1}}{2a_2(t_1)} \qquad (10\text{–}43a)$$

and

$$\left(\frac{da_1}{dt}\right)_{t_1} + v_1 a_1(t_1) + v_1\left(\frac{da_2}{dt}\right)_{t_1} + v_2 a_2(t_1) = 0 \qquad (10\text{–}43b)$$

These requirements yield

$$A_2 = a_2(t_1)$$
$$A_1 = a_1(t_1) - \frac{1}{2}\left(\frac{da_2}{dt}\right)_{t_1} \qquad (10\text{–}44\,a,\,b)$$

Since A_2 and A_1 are constants near $t = t_1$, we can say that we have stability if $(A_2 = a_2(t_1) > 0)$

$$A_1 > 0 \quad \text{or} \quad a_1(t_1) > \frac{1}{2}\left(\frac{da_2}{dt}\right)_{t_1}$$

Moreover, since t_1 is arbitrary, we can state that the system is stable in the neighborhood of time t as long as

$$a_1(t) > \frac{1}{2}\frac{da_2}{dt} \tag{10–45}$$

when $a_2(t_1) > 0$. A closer examination of this derivation reveals that the parameters $a_2(t)$ and $a_1(t)$ cannot vary too rapidly, lest the coefficients v_1 and v_2 get too large. Also, the assertion regarding stability is valid for the displacement q and not necessarily for the $\dot{q}$- or $\ddot{q}$-responses. To determine the stability of the $\dot{q}$-response, we can differentiate Eq. 10–39 and obtain (for $N = 2$)

$$\left(\frac{a_2(t)}{\dot{a}_1(t)+1}\right)\ddot{p} + \left(\frac{\dot{a}_2 + a_1}{\dot{a}_1(t)+1}\right)\dot{p} + p \equiv \bar{a}_2(t)\ddot{p} + \bar{a}_1(t)\dot{p} + p = 0$$

where $p = \ddot{q}$. This is of the same form as Eq. 10–34 with modified $\bar{a}_2(t)$ and $\bar{a}_1(t)$, and the test will then be

$$\bar{a}_1(t) > \frac{1}{2}\frac{d\bar{a}_2}{dt}$$

Applied to the system of Eq. 10–4, that is, to

$$a_2(t) = \frac{a(t)}{c(t)}, \quad a_1(t) = \frac{b(t)}{c(t)}, \quad \text{and} \quad a_0 = 1$$

the condition of stability from Eq. 10–45 is

$$\frac{b(t)}{c(t)} > \frac{1}{2}\frac{d}{dt}\left(\frac{a(t)}{c(t)}\right), \quad \text{or} \quad \frac{b(t)}{c(t)} > \frac{1}{2}\left(\frac{\dot{a}}{c} - \frac{a\dot{c}}{c^2}\right) \tag{10–46}$$

Since it is assumed that $a_2 = a/c > 0$, Eq. 10–46 is equivalent to

$$\frac{b}{a} > -\frac{1}{2}\left(-\frac{\dot{a}}{a} + \frac{\dot{c}}{c}\right) \tag{10–47}$$

which is in agreement with Sonine's theorem (cf. Ref. 10–12). For comparison, the discussion following Eq. 10–18 requires

$$Q(t) > 0, \quad \text{or} \quad \frac{1}{2}\frac{\dot{R}}{R} + \frac{b}{a} > 0, \quad \text{or} \quad \frac{b}{a} > -\frac{1}{2}\frac{\dot{R}}{R} \tag{10–48}$$

Conditions (10–47) and (10–48) are equivalent, if

$$-\frac{\dot{a}}{a} + \frac{\dot{c}}{c} = \frac{\dot{R}}{R} \qquad (10\text{–}49)$$

This equality is satisfied in the *WKB* approximation, where it is assumed that

$$\frac{c}{a} \gg \left(\frac{b}{2a}\right)^2 + \frac{1}{2}\frac{d}{dt}\left(\frac{b}{a}\right)$$

The foregoing can be extended to higher-order systems, and Routh's criterion (e.g., Ref. 10–13) can be used to test the stability. Without furnishing details, we summarize the results from Ref. 10–12. When $N > 2$ the expression corresponding to Eq. 10–40 is

$$\sum_{n=1}^{N} A_n q^{(n)} + q = 0 \qquad (10\text{–}50)$$

For a third-order system

$$A_3 = a_3, \quad A_2 = a_2 - \dot{a}_3, \quad A_1 = a_1 - \tfrac{1}{3}\dot{a}_2 - \tfrac{1}{9}a_2\frac{\dot{a}_3}{a_3} \qquad (10\text{–}51)$$

and for a fourth-order system we have

$$A_4 = a_4, \quad A_3 = a_3 - \tfrac{3}{2}\dot{a}_4, \quad A_2 = a_2 - \tfrac{2}{3}\dot{a}_3 - \tfrac{1}{4}\dot{a}_4\frac{a_3}{a_4},$$

$$A_1 = a_1 - \tfrac{1}{4}\dot{a}_2 - \tfrac{1}{24}\dot{a}_3\frac{a_3}{a_4} + \tfrac{1}{24}\dot{a}_4\frac{a_3^2}{a_4^2} - \tfrac{1}{8}\dot{a}_4\frac{a_2}{a_4} \qquad (10\text{–}52)$$

where terms containing products of derivatives have been neglected in Eqs. 10–51 and 10–52. In the vicinity of some $t = t_1$, Routh's criterion is applied to Eq. 10–50 with the "constant" coefficients indicated by Eqs. 10–51 or 10–52.

10–5 SOME AERONAUTICAL APPLICATIONS

Systems described by Eq. 10–4 are not uncommon and will soon be more frequently encountered in the aeronautical field, particularly in connection with the aeroelasticity and dynamic stability of rockets and other unmanned vehicles. We review two studies of this sort here.

As a first example, consider the free bending vibration of a slender beam, which might represent an elongated missile. If the mass per unit length m and the stiffness EI are functions of time as well as the spanwise location x, the pertinent equation of motion for the lateral deflection $q(x, t)$ is given by (cf. Refs. 10–7, 10–14)

$$m\ddot{q} + \dot{m}\dot{q} + \frac{\partial^2}{\partial x^2}\left(EI\frac{\partial^2 q}{\partial x^2}\right) = f(x, t) \qquad (10\text{--}53)$$

where $f(x, t)$ is a forcing function. Here rotary-inertia effects and shear deformations have been neglected. Consider first the case when $f(x, t) = 0$. If the m and EI are separable, in the sense that

$$m(x, t) = m_x(x)\, m_t(t) \equiv m_x m_t \qquad (10\text{--}54)$$

$$EI(x, t) = EI_x(x)\, EI_t(t) \equiv EI_x EI_t \qquad (10\text{--}55)$$

we may inquire as to the possibility of a separable solution

$$q = q_x(x)\, q_t(t) \equiv q_x q_t \qquad (10\text{--}56)$$

Such a solution is possible, if the boundary conditions are consistent with the form of Eq. 10-56. Upon substitution of Eqs. 10–54 through 10–56 into Eq. 10–53, we arrive at the ordinary differential equations

$$\frac{d^2}{dx^2}\left[EI_x\frac{d^2 q_x}{dx^2}\right] - \Omega^2 m_x q_x = 0 \qquad (10\text{--}57)$$

$$\ddot{q}_t + \frac{\dot{m}_t}{m_t}\dot{q}_t + \Omega^2\left(\frac{EI_t}{m_t}\right)q_t = 0 \qquad (10\text{--}58)$$

where Ω^2 is the separation constant. Equation 10–57 has the same form as the separated space equation for bending vibration of a slender beam with time-invariant inertia and stiffness properties. Supplied with a suitable set of boundary conditions, this equation yields the eigenfunction (mode shape) $(q_x)_n$ associated with each eigenvalue Ω_n, of which there is a denumerably infinite set. The concepts of eigenvalues and eigenfunctions being valid, we have also the orthogonality relation

$$\int_0^l m_x(q_x)_m(q_x)_n\, dx = \delta_{mn}\int_0^l m_x(q_x)_n^2\, dx \qquad (10\text{--}59)$$

provided the boundary conditions are of specific types among those listed in Eq. 3–19 of Ref. 10–15. Here δ_{mn} is the Kronecker delta and l the overall length.

For the forced motion, if we assume the solution to be

$$q(x, t) = \sum_{n=1}^{\infty}(q_x)_n\zeta_n(t) \qquad (10\text{--}60)$$

where $\zeta_n(t)$ are the generalized displacements, we obtain from Eq. 10–53

$$m_x m_t \sum_1^\infty (q_x)_n \ddot{\zeta}_n + m_x \dot{m}_t \sum_1^\infty (q_x)_n \dot{\zeta}_n$$
$$+ EI_t \sum_1^\infty \frac{d^2}{dx^2} \left(EI_x \frac{d^2 (q_x)_n}{dx^2} \right) \zeta_n(t) = f(x, t) \quad (10\text{–}61)$$

Multiplying by $(q_x)_j$, integrating between $x = 0$ and l, and utilizing the orthogonality relation (10–59), we arrive at the following relation for ζ_n:

$$\ddot{\zeta}_n + \frac{\dot{m}_t}{m_t} (\dot{\zeta}_n) + \frac{EI_t}{m_t} \Omega_n^2 \zeta_n = \frac{1}{m_t M_n} F_n(t), \quad (n = 1, 2, \cdots) \quad (10\text{–}62)$$

Here M_n and $F_n(t)$ are, respectively, the generalized mass and the generalized force

$$M_n = \int_0^l m_x (q_x)_n^2 dx \quad (10\text{–}63)$$

$$F_n(t) = \int_0^l f(x, t)(q_x)_n \, dx \quad (10\text{–}64)$$

If $f(x, t)$ is independent of the response, Eq. 10–62 constitutes a set of independent expressions for the ζ_n. Each is a second-order equation with time-varying coefficients, and the methods described in the previous sections apply. However, if $f(x, t)$ is motion-dependent, the sets of equations will in general be coupled.

A parallel investigation of the torsional vibration of a slender rod is given in Ref. 10–7.

A second example of practical interest is the closely related case of the bending vibration of a solid propellant rocket in powered flight (Ref. 10–14). To calculate the response of a rocket due to dynamic loading or control-system behavior as influenced by the structure, it is necessary to determine the natural frequencies of the rocket. Since the mass and stiffness properties vary with time, the case becomes one with time-varying coefficients. With certain simplifying assumptions, Birnbaum arrives at the following partial differential equation for the lateral oscillations of the rocket:

$$\frac{\partial^2}{\partial x^2} \left[EI \frac{\partial^2 q}{\partial x^2} \right] + \frac{\partial}{\partial t} \left[(m + \rho A) \frac{\partial q}{\partial t} \right] + (Q) = 0 \quad (10\text{–}65)$$

Here x is the distance along the rocket from the base towards the nose,
   EI the bending stiffness,
    q the bending (lateral) displacement,
   m the mass per unit length of unburned propellant plus structure,

A the total cross-sectional area minus the cross-sectional area of the propellant,

ρ the mass density of burning gases,

Q is an aggregate of terms which are assumed small and are neglected.

The effective reduction in bending stiffness because of axial compressive loading, not specifically mentioned by Birnbaum, is also assumed negligible.

Equation 10–56 is subject to the boundary conditions of vanishing bending moments at the two ends, zero shear at the nose, and a finite shear at the base of magnitude

$$\frac{\partial}{\partial x}\left(EI\,\frac{\partial^2 q}{\partial x^2}\right) = 0\,, \quad \text{at } x = 0 \tag{10–66}$$

Here T is the thrust. In Eq. 10–65 the quantities EI, q, m, A, and ρ are all functions of time as well as of x. This equation possesses a separable solution $q = q_x q_t$ provided

(1) EI and $(m + \rho A) = M$ are also separable, i.e., $EI = EI_x EI_t$, $M = M_x M_t$, and

(2) for compatability with the boundary condition (Eq. 10–66) $Tl^2/EI_t \equiv \bar{\gamma} = $ constant, where l is the length of the missile; i.e., the thrust and the bending stiffness must have the same time functionality. Both could, of course, remain constant.

We have thus reduced the problem to the same form as in the previous example, but with a special type of boundary condition. To illustrate, let us take the simple situation of EI_x and m_x independent of x; let $EI_x = EI_0$, $M_x = M_0$, $\gamma = \bar{\gamma}/EI_0$. After separation, the space equation associated with Eq. 10–65 becomes

$$\frac{d^4 q_x}{dx^4} - \frac{M_0 \Omega^2}{EI_0} q_x = 0 \tag{10–67}$$

with the boundary conditions

$$\frac{d^2 q_x}{dx^2} = 0, \quad \text{at } x = 0 \text{ and } l, \qquad \frac{d^3 q_x}{dx^3} = 0, \quad \text{at } x = l, \text{ and}$$

$$\frac{d^3 q_x}{dx^3} = 0\,, \quad \text{at } x = 0 \tag{10–68}$$

The eigenvalues $\lambda_n = l(\Omega^2 M_0/EI_0)^{1/4}$ must satisfy the transcendental relation

$$\gamma \sin \lambda \sinh \lambda - \lambda^2(1 - \cosh \lambda \cos \lambda) = 0 \tag{10–69}$$

When $\gamma = 0$ (thrust off, free-free condition), the λ_n's are well-known to be

$$(\lambda_n)_{\gamma=0} = 0,\, 0,\, 1.51\pi,\, \tfrac{5}{2}\pi,\, \tfrac{7}{2}\pi,\, \cdots, \qquad (n = 0, 1, 2, \cdots)$$

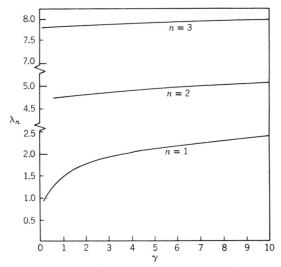

Fig. 10-1. Variations of the bending-frequency eigenvalues with the thrust parameter, γ, for the modes $n = 1, 2, 3$ of a variable mass rocket.

Here $\lambda_0 = 0$ corresponds to rigid-body translation, $\lambda_1 = 0$ to rigid-body rotation, and subsequent λ_n to elastic motions. For nonzero γ, we still have $(\lambda_0)_{\gamma \neq 0} = 0$. However, λ_1 is no longer zero and corresponds to the lowest nonvanishing value of λ satisfying the relation (10-69).* Referring to Fig, 10-1, which is essentially a reproduction of the results of Ref. 10-14 cast into the present notation, we see the influence of the thrust parameter γ on the natural frequencies of the system. We observe particularly the effect of γ on λ_1. It is important to note that for the $\gamma \neq 0$ case, the modes are no longer orthogonal since the boundary conditions are no longer of suitable type. Stated alternatively, the differential equation (10-67) and its boundary conditions (10-68) are no longer self-adjoint. However, if we denote $(q_x)_n{}^*$ as the solution to the adjoint equation, then we have

$$\int_0^l M_x(x)(q_x)_m(q_x)_n{}^* \, dx = 0, \qquad \text{if } m \neq n$$

This matter is discussed in more detail in Ref. 10-14.

* This particular root λ_1 is a direct result of the manner in which the shear condition is satisfied at $x = 0$. The approximations involved in this condition (Eq. 10-66) may not be realistic, thus casting some doubt as to the accuracy of this analysis. Nevertheless, it is included here as an illustration for lack of other suitable aeroelastic examples in the literature.

The time equation associated with Eq. 10–65 is as follows: (for $\gamma =$ constant, EI and M independent of x)

$$\ddot{q}_t + \frac{\dot{M}(t)}{M(t)} \dot{q}_t + \left(\frac{\lambda_n}{l}\right)^4 \frac{EI_t}{M(t)} q_t = 0 \qquad (10\text{–}70)$$

Again this is a second-order system with time-varying coefficients, and, subject to the stated restrictions, it can be treated by one of the methods of the previous section. If we specialize to the case of $EI_t =$ constant, and $M(t) = M_0(1 - \mu t)$, $\mu > 0$, as is done in Ref. 10–14, we can solve this equation directly, since it is then a form of Bessel's equation. The solution associated with the eigenvalue λ_n is

$$(q_t)_n = A J_0\left(\frac{\sqrt{1-\mu t}}{\mu} \frac{2\lambda_n^2}{l^2} \sqrt{\frac{EI_0}{M_0}}\right) + B Y_0\left(\frac{\sqrt{1-\mu t}}{\mu} \frac{2\lambda_n^2}{l^2} \sqrt{\frac{EI_0}{M_0}}\right) \quad (10\text{–}71)$$

It follows from the behavior of J_0 and Y_0 and from the condition $\mu > 0$ that the q_t response increases with time. (Note that the argument of J_0 and Y_0 is a decreasing quantity.)

10–6 TWO-DEGREE-OF-FREEDOM SYSTEM WITH TIME-VARYING COEFFICIENTS

We next consider the two-degree-of-freedom system, for which we may write the following general homogeneous form of the equations of motion:

$$L_1 x + L_2 y = 0$$
$$L_3 x + L_4 y = 0 \qquad (10\text{–}72a, b)$$

where the L's are the operators

$$L_m = \sum_{n=0}^{2} Q_{mn}(t) \frac{d^n}{dt^n} \qquad (10\text{–}73)$$

For convenience in subsequent derivations, we denote

$$Q_{1n} = A_n, Q_{2n} = B_n, Q_{3n} = C_n, Q_{4n} = D_n \qquad (10\text{–}74)$$

Since Eqs. 10–72a, b are homogeneous, we can divide by A_0 and D_0

(unless A_0 or $D_0 = 0$); thus, with no loss in generality, we can set $A_0 = D_0 = 1$. If the coefficients $Q_{mn}(t)$ were constants, the operators would be commutative, i.e.,

$$L_i L_j = L_j L_i \qquad (10\text{-}75)$$

There is then no difficulty in obtaining separate equations for x and y. The characteristic relations will be identical, and the x- and y-solutions will be of the same form but with different arbitrary constants (cf. Ref. 10–6).

With the Q_{mn}'s functions of time, however, the operators are not generally commutative; and we must resort to a direct and lengthy process of elimination. Brunelle (Ref. 10–7) suggests a systematic way of using conjugate operators $\tilde{L}$, such that

$$\begin{aligned} \tilde{L}_4 L_2 &= \tilde{L}_2 L_4 \\ \tilde{L}_3 L_1 &= \tilde{L}_1 L_3 \end{aligned} \qquad (10\text{-}76a, b)$$

These evidently permit the decoupling of Eqs. 10–72 for x and y. The same can be done, of course, by using proper combinations of Eqs. 10–72 and their first and second derivatives. We make no attempt to reproduce the details here; it suffices to outline the steps and give the final results. Let the conjugate operators be

$$\tilde{L}_m = \sum_0^2 q_{mn} \frac{d^n}{dt^n}; \quad q_{1n} = a_n, \quad q_{2n} = b_n, \quad q_{3n} = c_n, \quad q_{4n} = d_n \qquad (10\text{-}77)$$

where it is permissible once more to set $a_0 = d_0 = 1$. The q_{mn}'s are related to the Q_{mn}'s by the conditions of Eqs. 10–76a, b. The decoupled equations of motion take the form

$$\begin{aligned} (L_5)x &\equiv (\tilde{L}_4 L_1 - \tilde{L}_2 L_3)x = \sum_0^4 \nu_n \frac{d^n x}{dt^n} = 0 \\ (L_6)y &\equiv (-\tilde{L}_3 L_2 + \tilde{L}_1 L_4)y = \sum_0^4 \nu_n{}^* \frac{d^n y}{dt^n} = 0 \end{aligned} \qquad (10\text{-}78\,a, b)$$

After much algebra, using Eqs. 10–73, 10–76a, 10–77, and 10–78a, we find

$$\left.\begin{aligned} \nu_4 &= d_2 A_2 - b_2 C_2 \\ \nu_3 &= [d_2(2\dot{A}_2 + A_1) - b_2(2\dot{C}_2 + C_1) + d_1 A_2 - b_1 C_2] \\ \nu_2 &= [d_2(\ddot{A}_2 + 2\dot{A}_1 + 1) - b_2(\ddot{C}_2 + 2\dot{C}_1 + C_0) + d_1(\dot{A}_2 + A_1) \\ &\quad - b_1(\dot{C}_2 + C_1) + A_2 - b_0 C_2] \\ \nu_1 &= [d_2(\ddot{A}_1) - b_2(\ddot{C}_1 + 2\dot{C}_0) + d_1(\dot{A}_1 + 1) - b_1(\dot{C}_1 + C_0) \\ &\quad + A_1 - b_0 C_1] \\ \nu_0 &= [-b_2\ddot{C}_0 - b_1\dot{C}_0 - b_0 C_0 + 1] \end{aligned}\right\} \qquad (10\text{-}79a\text{-}e)$$

where $b_0 = B_0 + d_2 \ddot{B}_0 + d_1 \dot{B}_0$

$$b_1 = -\frac{d_2}{D_2}\left(F + \frac{EB_2}{D_2}\right) + \frac{d_1}{D_2} B_2$$

$$b_2 = \frac{d_2}{D_2} B_2$$

$$d_1 = \frac{1}{\beta}\left[B_1 - D_1 B_0 - \frac{\alpha}{\gamma}(B_2 - B_0 D_2)\right]$$

$$d_2 = \frac{1}{\gamma}(B_2 - B_0 D_2) - \frac{1}{\gamma}\left(K + \frac{GB_2}{D_2} + \dot{B}_0 D_2\right)$$

$$(10\text{--}80a\text{--}e)$$

and, in turn, $F = -2\dot{B}_2 - B_1$

$$E = 2\dot{D}_2 + D_1$$

$$G = \dot{D}_2 + D_1, \qquad L = -\ddot{B}_2 - 2\dot{B}_1 - B_0$$

$$H = \ddot{D}_2 + 2\dot{D}_1 + 1, \quad M = -\ddot{B}_1 - 2\dot{B}_0$$

$$J = 1 + \dot{D}_1, \qquad N = -\dot{B}_1 - B_0$$

$$K = -\dot{B}_2 - B_1$$

$$(10\text{--}81a\text{--}i)$$

$$\alpha = M + \frac{\ddot{D}_1 B_2}{D_2} - \frac{J}{D_2}\left(F + \frac{EB_2}{D_2}\right) + \ddot{B}_0 D_1$$

$$\gamma = L + \frac{HB_2}{D_2} - \frac{G}{D_2}\left(F + \frac{EB_2}{D_2}\right) + \ddot{B}_0 D_2$$

$$\beta = N + \frac{JB_2}{D_2} + \dot{B}_0 D_1 - \frac{\alpha}{\gamma}\left(K + \frac{GB_2}{D_2} + \dot{B}_0 D_2\right)$$

$$(10\text{--}82a\text{--}c)$$

We can obtain $\nu_i{}^*$ by interchanging A_n with D_n and B_n with C_n in all the expressions (10–79a) through (10–82c). As a check, if $A_i, \cdots, D_i$ are all independent of time, it can be easily shown from these expressions that

$$a_i = A_i, \quad b_i = B_i, \quad c_i = C_i, \quad d_i = D_i, \quad \text{and } \nu_i{}^* = \nu_i \quad (10\text{--}83)$$

Equations 10–78a, b, being linear and of the fourth order, have solutions of the form

$$x(t) = \sum_1^4 \xi_i x_i(t) \quad i = 1, 2, 3, 4$$
$$y(t) = \sum_1^4 \eta_i y_i(t) \quad \xi_i, \eta_i \text{ constants}$$

$$(10\text{--}84a, b)$$

Note that, when $A_i, \cdots, D_i$ are time-dependent, $x_i(t) \neq y_i(t)$. There are eight constants ξ_i, η_i, but only four initial values are available for Eqs. 10–84a, b. Let the given set be

$$x(0) = x_0, \quad \dot{x}(0) = \dot{x}_0, \quad y(0) = y_0, \quad \text{and } \dot{y}(0) = \dot{y}_0$$

As might be expected, the η_i's are dependent on the ξ_i's. As an illustration of how the full set of constants is evaluated, consider the ξ_i. Two determining equations are obviously

$$\left.\begin{aligned}
x_0 &= \sum_1^4 \xi_i x_i(0) \\
\dot{x}_0 &= \sum_1^4 \xi_i \dot{x}_i(0)
\end{aligned}\right\} \qquad (10\text{–}85\,a,b)$$

The other two conditions are furnished from Eqs. 10–72a, b and their derivatives.

$$\left.\begin{aligned}
A_2(0)\ddot{x}(0) + B_2(0)\ddot{y}(0) &= -A_1(0)\dot{x}_0 - x_0 - B_1(0)\dot{y}_0 - B_0(0)y_0 \\
C_2(0)\ddot{x}(0) + D_2(0)\ddot{y}(0) &= -C_1(0)\dot{x}_0 - C_0(0)x_0 - D_1(0)\dot{y}_0 - y_0
\end{aligned}\right\} \quad (10\text{–}86a, b)$$

We can solve Eqs. 10–86a, b simultaneously to obtain $\ddot{x}(0) = \ddot{x}_0$ and $\ddot{y}(0) = \ddot{y}_0$. Similarly, from

$$\left[\frac{d}{dt}\{L_1 x + L_2 y\}\right]_{t=0} = 0 \quad \text{and} \quad \left[\frac{d}{dt}\{L_3 x + L_4 y\}\right]_{t=0} = 0$$

the quantities $\dddot{x}(0) = \dddot{x}_0$ and $\dddot{y}(0) = \dddot{y}_0$ can be evaluated, since the lower derivatives are already available.

The main practical obstacle to carrying out the analysis just described arises when we seek the linearly independent solutions $x_i(t)$ and $y_i(t)$ of the decoupled equations (10–78a, b). Under certain special conditions, such as when the v's are constant or when Eqs. 10–78a, b are equidimensional, one can obtain results in closed form.* Moreover, we can assess the behavior of the solution in the neighborhood of a given time by Grensted's method (Eq. 10–47).

We mention here that, if $x_i(t)$ and $y_i(t)$ can be determined analytically, we can also solve in principle the nonhomogeneous problem by the method of variation of parameters (cf. Ref. 10–6, Sec. 1–9). The forcing functions of Eqs. 10–78a, b are, respectively,

$$f_1(t) = L_4\{F_1(t)\} - L_2\{F_2(t)\}$$
$$f_2(t) = -L_3\{F_1(t)\} + L_1\{F_2(t)\}$$

where $F_1(t)$ and $F_2(t)$ are the corresponding quantities for Eqs. 10–72 a, b.

---

* However, these are just the cases when the original operators $L_1, \cdots, L_4$ are commutative, and we need not resort to this tedious process of decoupling the equations.

To the procedure just described, there is an alternative approach which should receive thorough evaluation in practice. This is the more classical scheme (Ref. 10–16) of adopting $\dot{x}$ and $\dot{y}$ as distinct, additional dependent variables. Equations 10–72, supplemented by the two defining relations, constitute a set of four equations containing only first derivatives with respect to t. Suitable initial conditions are readily worked out, and such systems are the subject of extensive investigation in the mathematical literature. Incidentally, Ref. 10–2 presents an approximate solution along these lines for treating dynamical problems with time-varying coefficients.

10–7 AUTONOMOUS SECOND-ORDER NONLINEAR SYSTEMS

In the preceding sections, we have dealt with linear systems for which the superposition principle applies. There we have seen that the forced response can be determined readily in principle, once the homogeneous solutions are known. In addition, this important property has allowed us to effect solutions for continuous systems with constant or "separable" time-dependent coefficients by superposition of normal modes (cf. Sec. 10–5). In consequence, we are able to reduce, in a practical sense, a problem with an infinity of degrees of freedom to an excellent approximation having only a small finite number. When proceeding to the more difficult nonlinear analyses, these recourses are obviously no longer generally available. In the latter we must make clear distinctions as to "types" of nonlinearities, the forms of the forcing functions, etc. As stated earlier, the initial conditions may have a decided influence on the character of the response, so they must be considered as an integral part of the solution.

What follows is by no means a survey of the available techniques but, rather, an attempt to expose briefly a few possible approaches which have been applied to or appear useful for certain aeroelastic applications. Details of these and many other avenues of attack can be found in the profuse literature. Among the many treatises dealing with nonlinear dynamics, we mention particularly those of Minorsky (Ref. 10–17), Ku (Ref. 10–18), Stoker (Ref. 10–3), and Kryloff and Bogoliuboff (Ref. 10–19).

Apart from incidental remarks in the discussion, we shall restrict our attention to second-order autonomous systems, that is, to those where time does not enter explicitly into the equations of motion. This, of course, precludes time-dependent forcing functions. In addition, we shall assume that the coefficients of the highest-order derivative(s) is (are) constant. With these restrictions, we may represent the situations under consideration by the general forms:*

$$\ddot{x}_1 + c_1\ddot{x}_2 + F_1(\dot{x}_1, \dot{x}_2, x_1, x_2) = 0$$
$$\ddot{x}_2 + c_2\ddot{x}_1 + F_2(\dot{x}_1, \dot{x}_2, x_1, x_2) = 0$$

$$(10\text{-}87a, b)$$

Here F_1 and F_2 are linear or nonlinear functions of their arguments. Although these limitations appear quite restrictive, there are many practical problems which fall within the chosen class.

Consider, first, the simple one-degree-of-freedom case, for which we have

$$\ddot{x} + F(\dot{x}, x) = 0 \qquad (10\text{-}88)$$

This may be written as a set of two first-order equations

$$\frac{d\dot{x}}{dt} = -F(\dot{x}, x)$$
$$\frac{dx}{dt} = \dot{x}$$

$$(10\text{-}89a, b)$$

Dividing Eq. 10.88a by 10.88b, we get

$$\frac{d\dot{x}}{dx} = -\frac{F(\dot{x}, x)}{\dot{x}} \qquad (10\text{-}90)$$

which is a first-order equation with x as the independent variable. For a given set of initial conditions, this equation may be solved graphically in general [for instance, by the method of isoclines, see Ku (Ref. 10–18)] to yield $\dot{x}$ as a function of x. Once the relation $\dot{x} = \dot{x}(x)$ is established, an integration of Eq. 10–89b then yields x as a function of time. The process must be repeated for each new pair of initial conditions. If the results are plotted as $\dot{x}$ versus x in the phase plane ($\dot{x}$-x-plane), we obtain a series of curves, called solution curves or trajectories, which, when assigned directions of increasing time, give us a qualitative description of the system behavior. Associated with the differential equation (10–90) are singular points in the phase plane where the slopes of the curves are not unique. These situations arise when both the numerator and denominator of Eq. 10–89 vanish *simultaneously*. The character of these singularities, which play important roles in determining the nature of the responses, can be studied by the use of Poincaré's criteria (cf. Refs. 10–3, 10–18).

The phase-plane method of solution just described, apart from being somewhat tedious in numerical details, is indeed one of the most important tools for the treatment of nonlinear problems. It is capable of extension to higher-order systems or to systems of more than one degree of freedom

---

* For convenience, the problem is stated for a two-degree-of-freedom system. Its counterparts for one and several degrees of freedom are quite obvious.

at the expense of considerably more involved calculations and geometrical interpretations. For instance, for a third-order system, we must consider a three-dimensional phase space and three-dimensional trajectories. These matters are fully discussed by Ku in Ref. 10–18.

A sometimes more fruitful alternative approach is the approximate solution developed by Kryloff and Bogoliuboff (Ref. 10–19). It is strictly applicable only to systems with weak nonlinearities, small linear dampings, and couplings.* To make this restriction quantitative, we state that the aggregate of terms represented by F_1 and F_2 in Eqs. 10–87a, b are to be of the form

$$F_1 = \omega_1{}^2 x_1 + \mu f_1, \qquad F_2 = \omega_2{}^2 x_2 + \mu f_2$$

where μ is a small constant parameter indicative of the extent of the nonlinearity, and ω_1 and ω_2 are the "linear" uncoupled natural frequencies.

Consider once more the single-degree-of-freedom case, for which we have

$$\ddot{x} + \omega_1{}^2 x + \mu f(x, \dot{x}) = 0 \qquad (10\text{--}91)$$

As $\mu \to 0$, its solutions are given by

$$\bar{x} = a \sin (\omega_1 t + \varphi)$$

and

$$\dot{\bar{x}} = \omega_1 a \cos (\omega_1 t + \varphi) \qquad (10\text{--}92a, b)$$

For weak nonlinearities, it is logical to assume that the *form* of the solutions is the same as in Eqs. 10–92a, b, but with a and φ as slowly varying functions of t. We may briefly summarize the steps as follows: the differentiated form of Eq. 10–92a [with $a = a(t)$, $\varphi = \varphi(t)$], when compared with Eq. 10–92b yields the condition

$$\dot{a} \sin (\omega_1 t + \varphi) + a \dot{\varphi} \cos (\omega_1 t + \varphi) = 0 \qquad (10\text{--}93)$$

Furthermore, if the expressions for $\ddot{\bar{x}}$ (obtained by differentiation of Eq. 10–92b), $\dot{\bar{x}}$ (Eq. 10–92b), and $\bar{x}$ (Eq. 10–92a) are substituted into Eq. 10–91, we obtain a second condition

$$\dot{a}\omega_1 \cos (\omega_1 t + \varphi) - a\omega_1 \dot{\varphi} \sin (\omega_1 t + \varphi)$$
$$+ \mu f[a \sin (\omega_1 t + \varphi), a\omega_1 \cos (\omega_1 t + \varphi)] = 0 \quad (10\text{--}94)$$

From Eqs. 10–93 and 10–94, the expressions for $\dot{a}$ and $\dot{\varphi}$ are found to be

$$\dot{a} = -\frac{\mu}{\omega_1} f[a \sin (\omega_1 t + \varphi), a\omega_1 \cos (\omega_1 t + \varphi)] \cos (\omega_1 t + \varphi)$$

$$(10\text{--}95a, b)$$

$$\dot{\varphi} = \frac{\mu}{a\omega_1} f[a \sin (\omega_1 t + \varphi), a\omega_1 \cos (\omega_1 t + \varphi)] \sin (\omega_1 t + \varphi)$$

* See the comments in the concluding paragraph of this section.

As a and φ are assumed to be slowly varying functions of time, in the sense that they do not change appreciably during one cycle, it can be shown by an averaging process over the cycle that (to first order)

$$\frac{da}{dt} = -\frac{\mu}{\omega_1} g_1(a)$$

$$\frac{d\varphi}{dt} = \frac{\mu}{a\omega_1} g_2(a)$$

(10–96a, b)

Here

$$g_1(a) = \frac{1}{2\pi} \int_0^{2\pi} f(a \sin \psi, a\omega_1 \cos \psi) \cos \psi \, d\psi$$

(10–97)

and

$$g_2(a) = \frac{1}{2\pi} \int_0^{2\pi} f(a \sin \psi, a\omega_1 \cos \psi) \sin \psi \, d\psi$$

(10–98)

Once the functions $g_1(a)$ and $g_2(a)$ are determined, Eqs. 10–96a, b can be integrated to obtain a and φ as functions of time, and the solution will become

$$x(t) \cong \bar{x}(t) = a(t) \sin [\omega_1 t + \varphi(t)]$$

(10–99)

Let us apply this method to the vibration of a mass attached to a cubic spring (Duffing's problem), i.e.,

$$\ddot{x} + \omega_1^2 x + \mu x^3 = 0$$

(10–100)

Successively we have,

$$g_1(a) = \frac{1}{2\pi} \int_0^{2\pi} \{a^3 \sin^3 \psi\} \cos \psi \, d\psi = 0$$

$$g_2(a) = \frac{1}{2\pi} \int_0^{2\pi} \{a^3 \sin^3 \psi\} \sin \psi \, d\psi = \frac{3a^3}{8}$$

$$a = a_1 \quad \text{(constant)}$$

$$\varphi = \frac{\mu}{\omega_1} \frac{3a_1^2}{8} t + \varphi_0$$

So that

$$x \cong a_1 \sin \left(\left[\omega_1 + \frac{\mu}{\omega_1} \frac{3}{8} a_1^2 \right] t + \varphi_0 \right)$$

(10–101)

The effective frequency is to first order in μ

$$\omega^2 = \left[\omega_1 + \frac{\mu}{\omega_1} \frac{3a_1^2}{8} \right]^2 = \omega_1^2 + \frac{3}{4} a_1^2 \mu$$

(10–102)

which is in agreement with other methods. Note that the frequency ω is dependent on the motion amplitude a_1.

The above can be obtained by a slightly different, but closely related procedure (cf. Shen, Ref. 10–20). If the disturbed motion becomes *periodic* (not necessarily harmonic) eventually, we assume

$$x = \sum_{1}^{\infty} a_n \sin n\omega t \qquad (10\text{–}103)$$

Substitution of this expression into Eq. 10–100 yields

$$\sum_{n=1}^{\infty} (-n^2\omega^2 a_n + b_n) \sin n\omega t = 0 \qquad (10\text{–}104)$$

where b_n is the nth harmonic component of $\omega_1^2 x + \mu x^3$; i.e.,

$$\omega_1^2 x + \mu x^3 = \sum_{1}^{\infty} b_n \sin n\omega t \equiv F(x) \qquad (10\text{–}105)$$

If the first harmonic is balanced, we have

$$\omega^2 = b_1/a_1 \qquad (10\text{–}106)$$

Furthermore, if the first harmonic predominates, by Fourier analysis,

$$b_1 = \frac{2}{\pi} \int_0^{\pi} F(x) \sin \omega t \, d(\omega t) \approx \frac{2}{\pi} \int_0^{\pi} F(a_1 \sin \omega t) \sin \omega t \, d(\omega t)$$

$$= a_1[\omega_1^2 + \tfrac{3}{4}\mu a_1^2] \qquad (10\text{–}107)$$

whence, from Eq. 10–106,

$$\omega^2 = \omega_1^2 + \tfrac{3}{4}\mu a_1^2 \qquad (10\text{–}108)$$

which coincides with Eq. 10–102.

As stated earlier, the Kryloff and Bogoliuboff solution assumes weak nonlinearities and small damping and, when extended to a multi-degree-of-freedom system, it further requires weak coupling between the dependent variables. Although applied successfully to some "strong" nonlinear systems (see comments in Ref. 10–20), it has been shown to suffer considerably in accuracy when strong linear dampings are present. Brunelle (Ref. 10–7) suggests a modification which accounts more accurately for *linear* dampings and couplings. For instance, Eq. 10–91 is recast into the form

$$\ddot{x} + 2\beta\dot{x} + \omega_1^2 x + \mu f(\dot{x}, x) = 0$$

Hence, the original starting points would be

$$\bar{x} = ae^{-\beta t} \sin (\omega_2 t + \varphi)$$
$$\dot{\bar{x}} = ae^{-\beta t}\{\omega_2 \cos (\omega_2 t + \varphi) - \beta \sin (\omega_2 t + \varphi)\}$$

in place of Eqs. 10–92a, b. Parallel steps to those of Eqs. 10–93 through 10–99 are then followed to obtain the improved solutions.

10-8   AN APPLICATION TO FLUTTER AND RELATED PROBLEMS

For purposes of illustration of some of the ideas of the previous section, we turn once again to the a two-degree-of-freedom typical section wing, for which the equations of motion are (in the notation of Chap. 6)

$$m\ddot{h} + K_h h + S_\alpha \ddot{\alpha} = -F$$
$$I_\alpha \ddot{\alpha} + K_\alpha \alpha + S_\alpha \ddot{h} = Fd \qquad (10\text{--}109a,\ b)$$

In ordinary flutter or related analyses, we assume that all forces are linearly dependent on the amplitudes of motion. Stated alternatively, we assume the spring (or elastic) parameters K_h, K_α as well as the static unbalance S_α are constants; in addition, the aerodynamic force F and moment ($M = Fd$) are linear functions of the oscillation amplitudes h and α (and their derivatives $\dot{h}$, $\dot{\alpha}$, etc.). There are many potential situations, however, where these assumptions are no longer justified. Nonlinearities of aerodynamic origin can occur, for instance, when the wing is fluttering with moderate to large amplitudes near its stall range. Another example is large-amplitude flutter at very high Mach numbers (cf. Ref. 10–21). Of structural origin are such nonlinearities as backlash between elements, nonlinear elastic restoring forces, hysteresis, etc. The effects of these types of nonlinearities are discussed by Woolston et al. (Ref. 10–2) and by Shen (Ref. 10–20). In the former paper, the system is given initial disturbances in α (at various speeds), and the ensuing free motions are studied by analog means. In the latter, the eventual motions are assumed periodic, and by analytical means some sort of a flutter boundary is obtained in the form of critical speed versus "amplitude" of oscillation.

We now proceed to illustrate one of the cases analyzed by Shen, namely, that of bending-torsion flutter involving backlash in the torsional spring.* This situation is described by Fig. 10–2. In the absence of backlash, $\delta = 0$, the torsional spring will be linear ($K_\alpha = \bar{K}_\alpha$), and the flutter motion will be purely harmonic at some speed $\bar{U}_F$, regardless of the size of the amplitudes of oscillation. As δ is allowed to assume small positive values, it is reasonable to assume that the motion will remain periodic (though not necessarily harmonic) at a somewhat different speed than $\bar{U}_F$. Call the new eigenvalue U_F, a quantity which will shortly be shown to depend also on the torsional amplitude of oscillation α_0. Furthermore, the motion will be dominated by its first harmonic component, provided the amplitude of oscillation is somewhat larger than $\delta/2$.

---

* The presentation will be in a slightly different fashion from that given in Ref. 10–20.

Assuming the motions to be of the form

$$\alpha = \alpha_0 \sin \omega t \quad \text{and} \quad h = h_0 \sin (\omega t + \varphi)$$

we have, for the elastic force $K_\alpha \alpha$ (when $\alpha_0 > \delta/2$),

$$K_\alpha \alpha = 0, \quad \text{for} \quad 0 < \omega t < \psi, \quad \pi - \psi < \omega t < \pi + \psi,$$
$$2\pi - \psi < \omega t < 2\pi + \psi$$

$$K_\alpha \alpha = \bar{K}_\alpha \alpha_0 (\sin \omega t - \sin \psi), \quad \text{for } \psi < \omega t < \pi - \psi \qquad (10\text{-}110a\text{-}c)$$

$$K_\alpha \alpha = \bar{K}_\alpha \alpha_0 (\sin \omega t + \sin \psi), \quad \text{for } \pi + \psi < \omega t < 2\pi - \psi$$

Here $\psi = \sin^{-1} \delta/2\alpha_0 < \pi/2$. By Fourier analysis, as in Eq. 10–107, we obtain

$$K_\alpha \alpha = b_1 \sin \omega t + \sum_2^\infty b_n \sin n\omega t \qquad (10\text{-}111)$$

where

$$b_1 = \frac{1}{\pi} \int_0^{2\pi} K_\alpha \alpha \sin \omega t \, d(\omega t)$$

$$= \bar{K}_\alpha \alpha_0 \left[1 - \frac{2}{\pi} \psi - \frac{1}{\pi} \sin 2\psi \right] \qquad (10\text{-}112)$$

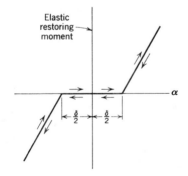

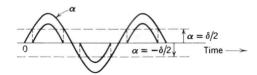

Fig. 10–2. Backlash-type nonlinearity in the torsional stiffness.

We need consider only the b_1-coefficient, as we are interested in balancing the first harmonic terms in the equations of motion. As the amplitude α_0 increases, $\psi \rightarrow 0$, $K_\alpha \alpha_0 \Rightarrow \bar{K}_\alpha \alpha_0$, and the effect of the backlash will decrease. On the other hand, if $\alpha_0 < \delta/2$, the wing will possess zero "effective" torsional frequency; in this case, the system is once more linear, and we can expect even a small backlash to yield a flutter speed considerably lower than $\bar{U}_F$. Above this minimum speed, however, the motion will build up so that the nonlinearity will come into play, and the backlash will have particularly strong effects in the neighborhood where α_0 is of the same order as $\delta/2$.

Substituting for $K_\alpha \alpha$ the quantity $b_1 \sin \omega t$ in the equations of motion and balancing the first harmonic, we obtain a set of characteristic equations, similar to the linear case, except that now the torsional stiffness will be dependent on α_0 (through ψ). If we fix the amplitude α_0, we can solve for the flutter speed and frequency.

For the sake of clarity, we have described first the simplest case treated by Shen. Unfortunately, in Ref. 10–20 he shows no numerical results pertaining to this example. He does, however, present other results for the system with preload and free play in the torsional spring. This case involves the situation depicted in Fig. 10–3. The procedure for obtaining the new solution is quite similar to the problem treated above, with the exception that the zeroth-order balance must also be effected (see Ref. 10–20).

Figure 10–4 is a reproduction of Fig. 7 in Ref. 10–20, which deals with a specific wing whose parameters are given in Table 1 of that reference. The abscissa is the amplitude of the fundamental harmonic of torsional

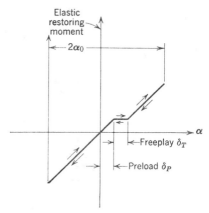

Fig. 10–3. Preload-backlash nonlinearity in the torsional stiffness of the typical section.

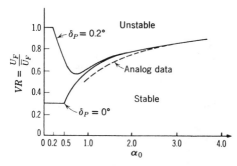

Fig. 10–4. Bending-torsion flutter of a typical wing section in incompressible flow. Dimensionless flutter speed vs. torsional oscillation amplitude.

oscillation α_0, while the ordinate represents dimensionless flutter speed $VR = U_F/\bar{U}_F$, with $\bar{U}_F$ corresponding to the true linear case of $\delta_T = 0$. The curves are for $\delta_T = 0.5°$ and for two values of δ_P. The analog data are taken from Fig. 2 of Ref. 10–22. As indicated there, δ_P for this curve varies from point to point in the approximate range $0.04° < \delta_P < 0.12°$.

The good agreement which may be noted between the analog and analytical results provides some confidence in the validity of the latter, at least for predicting the periodic behavior of the system. It remains to interpret these curves in terms of the stability of the system following an initial disturbance, say $\alpha(0) = \alpha_i$ and $\dot{\alpha}(0) = 0$. For this purpose, consider the curve for $\delta_P = 0.2$. With very small initial disturbances of $\alpha_0 < 0.2$, the system behaves linearly and will be stable if $VR < 1$, and unstable if $VR > 1$. The region above the curve is then an unstable one, and that below, a stable one. On the other hand, below a speed of about $VR \cong 0.55$–0.6, the motion is stable regardless of α_0. Slightly above $VR = 0.6$, flutter of small amplitude could conceivably occur if the initial disturbance α_i is sufficiently large. For higher velocities, and even with smaller initial disturbances, we would note flutter oscillations with much larger amplitudes. Thus we see that there is a direct relationship between the initial disturbance α_i and the flutter amplitude, which can be easily obtained from analog solutions. Unfortunately the analytical technique is not able to provide such transient-response data. But nevertheless it does give useful qualitative information. For a more complete description of the system behavior following initial inputs, we must return to a more general method (i.e., to a method capable of giving the complete response rather than the periodic behavior alone). The extended versions, for two degrees of freedom, of the phase-plane technique or the approximate Kryloff-Bogoliuboff procedure would be well-suited to the task.

REFERENCES

10–1. Woolston, D. S., H. L. Runyan, and R. E. Andrews, "An Investigation of Effects of Certain Types of Structural Nonlinearities on Wing and Control Surface Flutter," *J. Aero. Sciences*, Vol. 24, No. 1, January 1957, pp. 57–63.

10–2. MacNeal, R. H., J. H. Hill, and B. Mazelsky, *The Effects of Time Varying Aerodynamic Coefficents on Aeroelastic Response*, WADD Tech. Report 60–390, April 1960.

10–3. Stoker, J. J., *Nonlinear Vibrations in Mechanical and Electrical Systems*, Pure and Applied Mathematics Series, Vol. II, Interscience Publishers, New York, (2nd printing) 1954.

10–4. Collar, A. R., "On the Stability of Accelerated Motion: Some Thoughts on Linear Differential Equations with Variable Coefficients," *Aero. Quart.*, Vol. VIII, November 1957, pp . 309–330.

10–5. Tsien, H. S., *Engineering Cybernetics*, McGraw-Hill Book Company, New York, 1954.

10–6. Hildebrand, F. B., *Advanced Calculus for Engineers*, Prentice-Hall, Inc., New York, 1949.

10–7. Brunelle, E. J., *Transient and Nonlinear Effects on High Speed, Vibratory Thermoelastic Instability Phenomena, Part I—Theoretical Considerations*, WADD TR60-484, July 1960.

10–8. Squire, W., "Approximate Solution of Linear Second Order Differential Equations," *J. Royal Aero. Society*, Vol. 63, No. 582, June 1959, pp. 368–369.

10–9. Reed, W. H., "Effects of a Time-Varying Test Environment on the Evaluation of Dynamic Stability with Application to Flutter Testing," *J. Aero/Space Sciences*, Vol. 25, No. 7, July 1958, pp. 435–443.

10–10. Garber, T. B., "On the Rotational Motion of a Body Re-Entering the Atmosphere," *J. Aero/Space Sciences*, Vol. 26, No. 7, July 1959, pp. 443–449.

10–11. Duncan, W. J., "Indicial Admittances for Linear Systems with Variable Coefficients," *J. Royal Aero. Society*, Vol. 61, No. 553, January 1957, pp. 46–47.

10–12. Grensted, P. E. W., "Stability Criteria for Linear Equations with Time-Varying Coefficients," *J. Royal Aero. Society*, Vol. 60, No. 543, March 1956, pp. 205–208.

10–13. Gardner, M. F., and J. L. Barnes, *Transients in Linear Systems*, Vol. I., *Lumped-Constant Systems*, John Wiley and Sons, New York, 1950.

10–14. Birnbaum, S., *Bending Vibrations of a Perforated Grain Solid Propellant Rocket During Powered Flight*, Institute of Aerospace Sciences Preprint No . 61-30, January 1961.

10–15. Bisplinghoff, R. L., H. Ashley, and R. L. Halfman, *Aeroelasticity*, Addison-Wesley Publishing Company, Cambridge, Mass., 1955.

10–16. Ince, E. L., *Ordinary Differential Equations*, Paperback Publication, Dover Publications, New York, 1956.

10–17. Minorsky, N., *Introduction to Non-Linear Mechanics*, J. W. Edwards, Ann Arbor, Mich., 1947.

10–18. Ku, Y. H., *Analysis of Control of Nonlinear Systems, Nonlinear Vibrations and Oscillations of Physical Systems*, The Ronald Press, New York, 1958.

10–19. Kryloff, N., and N. Bogoliuboff, *Introduction to Non-Linear Mechanics*, Translated from Russian by S. Lefschetz, Princeton University Press, 1943.

10–20. Shen, S. F., "An Approximate Analysis of Nonlinear Flutter Problems," *J. Aero/Space Sciences*, Vol. 26, No. 1, January 1959, pp. 25–32, 45.

10–21. Zartarian, G., and P. T. Hsu, *Theoretical and Experimental Studies on Airloads Related to Hypersonic Aeroelastic Problems of General Slender Pointed Configurations*, U.S.A.F Aeronautical Systems Division Technical Report 61-7, April 1961.

10–22. Woolston, D. S., and R. E. Andrews, "Remarks on Analytical Results of Certain Nonlinear Flutter Problems," *J. Aero/Space Sciences*, Vol. 26, No. 1, January 1959, Readers' Forum, pp. 51–52.

INDEX